THE POSTMORTEM BRAIN IN PSYCHIATRIC RESEARCH

NEUROBIOLOGICAL FOUNDATION OF ABERRANT BEHAVIORS

THE POSTMORTEM BRAIN IN PSYCHIATRIC RESEARCH

edited by

Galila Agam
Ben-Gurion University of the Negev

Ian P Everall
Institute of Psychiatry, London

R.H. Belmaker
Ben-Gurion University of the Negev

KLUWER ACADEMIC PUBLISHERS
Boston / Dordrecht / London

B5

Distributors for North, Central and South America:
Kluwer Academic Publishers
101 Philip Drive
Assinippi Park
Norwell, Massachusetts 02061 USA
Telephone (781) 871-6600
Fax (781) 681-9045
E-Mail <kluwer@wkap.com>

Distributors for all other countries:
Kluwer Academic Publishers Group
Distribution Centre
Post Office Box 322
3300 AH Dordrecht, THE NETHERLANDS
Telephone 31 78 6392 392
Fax 31 78 6546 474
E-Mail <services@wkap.nl>

Electronic Services <http://www.wkap.nl>

Library of Congress Cataloging-in-Publication Data

A C.I.P. Catalogue record for this book is available
from the Library of Congress.

Printed on acid-free paper. Printed in the United States of America

TABLE OF CONTENTS

CONTRIBUTORS

Agam, Galila: Stanley Foundation International Research Center and Department of Clinical Biochemistry, Faculty of Health Sciences, Ben-Gurion University of the Negev, Beersheva, Israel, galila@bgumail.bgu.ac.il

Bachus, Susan E.: Section on Neuropathology, Clinical Brain Disorders Branch, Intramural Research Program, National Institute of Mental Health, National Institutes of Health, Bethesda, MD, 20892, USA, bachuss@intra.nimh.nih.gov

Barci, Beata M.: Stanley Foundation Research Programs, 5430 Grosvenor Lane, Suite 200, Bethesda, MD 20814, USA, fax: 301-571-0769

Belmaker, R.H.: Stanley Foundation International Research Center, Ministry of Health Mental Health Center, Faculty of Health Sciences, Ben-Gurion University of the Negev, Beersheva, Israel, belmaker@bgumail.bgu.ac.il

Benes Francine M.: The Laboratories for Structural Neuroscience, McLean Hospital, Belmont, MA; Program in Neuroscience and Department of Psychiatry, Harvard Medical School, Boston, MA, McLean Hospital, 115 Mill Street, Belmont, MA, 02478, USA , benesf@mclean.harvard.edu

Berretta, Sabina: The Laboratories for Structural Neuroscience, McLean Hospital, Belmont, MA; Program in Neuroscience and Department of Psychiatry, Harvard Medical School, Boston, MAMcLean Hospital, 115 Mill Street, Belmont, MA 02478, USA

Bunney, Blynn G.: Department of Psychiatry, College of Medicine, University of California, Irvine, CA 92697, USA

Bunney, William E.: Department of Psychiatry, College of Medicine, University of California, Irvine, CA 92697, USA, webunney@uci.edu

Chang, Annisa: Laboratory of Cellular and Molecular Pathophysiology, Centre for Addiction and Mental Health and Departments of Pharmacology and Psychiatry, Institute of Medical Science, University of Toronto, Toronto, Canada

Cotter', David R.: Section of Experimental Neuropathology and Psychiatry and Section of Clinical Neuropharmacology, Institute of Psychiatry, King's College, DeCrespigny Park, London SE5 8AF, United Kingdom, spkadrc@iop.kcl.ac.uk

Dean, Brian: The Rebecca L. Cooper Research Laboratories, The Mental Health Research Institute of Victoria, Parkville, Victoria 3052, Australia, B.Dean@papyrus.mhri.edu.au

Dwork, Andrew J.: Departments of Pathology and Psychiatry, Columbia University. Department of Neuroscience, New York State Psychiatric Institute, New York, NY, USA, ajd6@columbia.edu

Everall', Ian Paul: Departments of Neuropathology and Psychological Medicine, Institute of Psychiatry, DeCrespigny Park, London SE5 8AF, United Kingdom, i.everall@iop.kcl.ac.uk

Falkai, Peter: Department of Psychiatry, University of Bonn, Bonn, Germany, falkai@uni-bonn.de

Harrison, Paul J.: University of Oxford, Oxford. OX1 2JD, United Kingdom, paul.Harrison@psych.ox.ac.uk

Hashimoto, T.: Department of Psychiatry, University of Pittsburgh, Pittsburgh, PA 15213, USA

Honer, William G.: Molecular Psychiatry and Therapeutics Laboratory, University of British Columbia, Canada, honer@interchange.ubc.ca

Jope, Richard S.: Department of Psychiatry and Behavioral Neurobiology, University of Alabama at Birmingham, Birmingham, AL, USA, neuo033@uabdpo.dpo.uab.edu

Jurjus, George: Louis Stokes Cleveland Department of Veterans Affairs Medical Center, Department of Psychiatry, Case Western Reserve University, Cleveland, Ohio, 44106 USA

Kleinman, Joel E.: Section on Neuropathology, Clinical Brain Disorders Branch, Intramural Research Program, National Institute of Mental Health, National Institutes of Health, 9000 Rockville Pike, Bethesda, Maryland, 20892, USA, kleinmaj@intra.nimh.nih.gov

Knable, Michael B.: Stanley Foundation Research Programs, 5430 Grosvenor Lane, Suite 200, Bethesda, MD 20814 USA, knablem@stanleyresearch.org

Kozlovsky', Nitsan: Stanley Foundation International Research Center, Faculty of Health Sciences, Ben-Gurion University of the Negev, Beersheva, Israel

Lewis, D. A.: Department of Psychiatry, University of Pittsburgh, Pittsburgh, PA 15213, USA, lewisda@msx.upmc.edu

Li, Peter P.: Laboratory of Cellular and Molecular Pathophysiology, Centre for Addiction and Mental Health and Departments of Pharmacology and Psychiatry, Institute of Medical Science, University of Toronto, Toronto, Canada

Matsumoto, Izuru: Department of Neuropsychiatry, Fukushima Medical University, School of Medicine, 1 Hikarigaoka, Fukushima, Fukushima Prefecture, Japan, 960-1295, psyizuru@.fmu.ac.jp

Niwa, S.I.: Department of Neuropsychiatry, School of Medicine, Fukushima Medical University, Fukushima, 960-1295 Japan , fax: +81-24-548-6735

Pariante, Carmine M.: Section of Experimental Neuropathology and Psychiatry and Section of Clinical Neuropharmacology, Institute of Psychiatry, King's College, DeCrespigny Park, London SE5 8AF, United Kingdom.

Pearlson G.D.: The Johns Hopkins University School of Medicine, Baltimore, MD, USA, godfr@jhmi.edu

Pilowsky, Lyn: Psychiatry Medical Research Council, Institute of Psychiatry and Institute of Nuclear Medicine UCL, London, United Kingdom, spkakaw@iop.kcl.ac.uk

Potkin, Steven G.: University of California, Irvine, Psychiatry & Human Behavior, Irvine, CA 92697-3960, USA, spotkin@uci.edu

Rajkowska, Grazyna: Department of Psychiatry and Human Behavior, University of Mississippi Medical Center, Jackson, Mississippi 39216, USA, grajkowska@psychiatry.umsmed.edu

Riederer, Peter: Clinical Neurochemistry, University Clinic for Psychiatry, Wurzburg, Germany, peter.riederer@mail.uni-wuerzburg.de

Rujescu, Dan: Molecular Neurobiology, Department of Psychiatry, Ludwig Maximilians University, Nussbaumstr. 7, 80336 Munchen, Germany, Dan.Rujescu@psy.med.uni-muenchen.de

Stein, Richard: University of California, Irvine, Psychiatry & Human Behavior, Irvine, CA 92697-3960, USA, rstein@uci.edu

Torrey, E. Fuller: Stanley Foundation Research Programs, 5430 Grosvenor Lane, Suite 200, Bethesda, MD 20814, USA, fax 301-571-0769

Ravid, Rivka: Netherlands Brainbank, Meibergdreef 33, 1105 AZ, Amsterdam, The Netherlands, r.ravid@nih.knaw.nl

Stockmeier, Craig A.: Department of Psychiatry and Human Behavior, University of Mississippi Medical Center, Jackson, Mississippi, 39216 USA, cstockmeier@psychiatry.umsmed.edu

Warsh, Jerry J.: Laboratory of Cellular and Molecular Pathophysiology, Centre for Addiction and Mental Health and Departments of Pharmacology and Psychiatry, Institute of Medical Science, University of Toronto, Toronto, Canada, Jerry_Warsh@camh.net

Webster, Maree J.: Stanley Brain Research Laboratory, Uniformed Services University of the Health Sciences, 4301 Jones Bridge Road, Bethesda, MD, 20814, USA

Weinberger, Daniel R.: Clinical Brain Disorders Branch, Intramural Research Program, National Institute of Mental Health, National Institutes of Health, 10 Center Drive, Bethesda, Md. 20892-1379, USA, Weinberg@intra.nimh.nih.gov

ACKNOWLEDGEMENTS

This book was conceived during a visit of Ian Everall to Beersheba and Abraham's well, where the Biblical concept of emotion as resident in "heart and kidneys" was examined in the light of brain physiology and molecular biology. The foresight of the founders of the Stanley Foundation Brain Consortium as well as the generosity of its benefactors are deeply appreciated. Without the help of Yehudit Curiel, who edited and proofread, this manuscript could never have been completed.

INTRODUCTION

E. Fuller Torrey, and Michael B. Knable

This book represents the exploration of one of medicine's greatest frontiers – the causes of psychiatric disease. In 1848, when Phineas Gage was injured in a construction accident, it became clear that brain damage could profoundly affect a person's thinking and behavior. In 1871 Camillo Golgi described neurons, and in 1889 Santiago Ramon y Cajal showed how neurons are linked with each other through axons and dendrites. In 1897 Charles Sherrington described the synapses and chemical transmission the neurons use to communicate. It must have seemed at that time that science was on the threshold of understanding the brain dysfunction that underlies diseases such as schizophrenia, bipolar disorder, and severe depression.

Alas, most of a century was to pass with little more progress. The understanding of psychiatric disorders remained as a distant city, glimpsed only from afar. We devised tests for peripheral blood cells, but these are a far distance from the brain, as if in a neighboring country. We sampled the cerebrospinal fluid, which was closer to the ultimate target but still as a river, downstream from the desired site. In more recent years we took pictures of the brain using CT and MRI scans and measured its metabolism using PET and fMRI, but still this was as if we were on an airplane, flying high above the desired destination. The city remained on the horizon, mostly unknown and apparently unknowable.

The chapters of this book testify that this is no longer the case. We have now entered the brain and, armed with newly devised high-tech weapons, are exploring it block by block. As we do so, the causes of severe psychiatric disorders will be clarified and this will lead, inevitably, to better treatments and perhaps even to prevention of these diseases.

It will be interesting for the historians to speculate why a century passed with so little progress. The psychodynamic theories of Sigmund Freud and his followers certainly played a major role; why look at the brain if childhood events are the causes of these disorders? That psychologic cul-de-sac is ironic in that Freud himself was a respected researcher of the brain before he became seduced by the attractions of Oedipus and Id.

The lack of sophisticated technology also played a role. The microscope used by Golgi changed remarkably little for almost a century, until the electron microscope was introduced. The lack of well-characterized brain tissue available for researchers also played a role. Even as the technology for exploring the brain became available, postmortem tissue samples were difficult to obtain and often taken from elderly patients whose age and concurrent diseases complicated the analysis of any findings. Because of the lack of tissue, much of the research carried out on severe psychiatric disorders in the late twentieth century was done on rat brains, a necessary intermediate step but a poor substitute for the real human thing.

The excitement of entering a new frontier cannot be overstated. Surprises are already evident – the role of glial cells, the intricate structure of the second messenger system, the similarities in findings between schizophrenia and bipolar disorder. There will be many more surprises as we wander around each new corner and ultimately connect the brain structures with the genes and proteins associated with their functions. Schizophrenia, bipolar disorder, severe depression, and the other psychiatric disorders whose symptoms are caused by abnormal brain function will inevitably yield the secrets of their etiology to the exploration that is now underway.

We have entered the frontier now, and we will not turn back. What awaits is an intellectual feast, as Macdonald Critchley so articulately noted two decades ago:

> We must admit that the divine banquet of the brain was, and still is, a feast with dishes that remain elusive in their blending, and with sauces whose ingredients are even now a secret.

Reference

Critchley, M. The Divine Banquet of the Brain. Raven Press, New York, NY, 1979; p 267.

1 PSYCHIATRIC BRAIN BANKS: SITUATION IN EUROPE AND ASIA

I. Matsumoto, S.I. Niwa and R. Ravid

I THE FIRST SYSTEMATIC BRAIN BANK NETWORK FOR PSYCHIATRIC DISORDERS IN JAPAN

II BRAIN BANKING IN PSYCHIATRIC DISORDERS: THE AMSTERDAM AND EUROPEAN EXPERIENCE

General Introduction

Brain banking systems and networks in various parts of the world exhibit a large variety; they differ in many aspects which affect and influence the main fields of scientific research, as the flow of dessiminated material strongly depends on the legal, ethical, cultural and religious background and reflects the local situation in each country. The daily practice and routine procedures of brain banks in three continents (Europe, Asia and America) represent basic differences in the nations involved, scientific communities and the interactions between these two. European brain banks differ from the American and Asian groups collecting postmortem material for research. The various networks collaborate on the standardisation of protocols and diagnostic criteria and try to establish solid concensus criteria. Brain banking is not well advanced in Asian countries. There are several lines of reasons for this. Firstly, the biological approach to psychiatric disorders are not well accepted among psychiatrists compared with western countries. Secondly, spontaneous participation of users and families to understand and to support biological researches are not encouraged enough. However, as initiated by Japan, there is certainly a movement to form a nation-wide brain bank network with internationalized criteria in various Asian countries.

Brain banks have developed protocols for tissue collection and cryopreservation which vary considerably so as to meet the needs of the requesting scientist. A brain bank is essentially a prospective project for the collection of human CNS material with the underlying support of donor programs. Essential to a brain bank are high quality clinical and neuropathological support services; the activities of a brain bank are also directed towards research. This consideration allows us to discerne brain banks from neuropathological laboratories and from clinically based laboratories where material handling and diagnostic service are directed to specific tests with no reasons to point to prospective standardized protocols for multidisciplinary research. A brain bank also permits the collection of control tissue which could otherwise be difficult to obtain. The absence of homogeneous criteria to collect and preserve tissue is an aspect which has been one of the reasons why standardization of procedures and criteria have been proposed for brain banking.

European Brain Bank networks can deal with coordination of brain banks based on local activities and the national output and later on also at an international level by collaboration with Asian and American Brain Banks. This will hopefully lead to a scientific output which will exceed the summed results of the local brain banks and will make it possible to have a large amount of tissue for research, comprehensive clinical data, reliable statistical analysis of the obtained results and the possibility to apply sophisticated molecular biological techniques. It will also make it possible to efficiently use the specific skills of the local banks and in this way help to improve the state of the art and fill information gaps. A close collaboration between networks in Europe and Asia can match the objectives mentioned later in this chapter by allowing the necessary infrastructure which will connect and liaise the various Brain Banks in a single on-line network, simultaneously meeting the needs of many researches.

This collaboration is also necessary for the pooling of knowledge and expertise of modern neurobiological techniques as well as resources and material which are certainly not evenly distributed at the present time between the various European countries due to ethical problems, cultural and religious backgrounds or due to socio-economical differences which make brain banking and research involved with it incredibly difficult. This approach is the only feasible way to perform high quality reseach and cannot be achieved on national or regional level.

I THE FIRST SYSTEMATIC BRAIN BANK NETWORK FOR PSYCHIATRIC DISORDERS IN JAPAN

Abstract

It is our intention to establish the first systematic brain bank for various psychiatric disorders in Japan and in so doing, develop innovative collaborative research programs between neuroscience and psychiatric researchers, based on the shared premise that the biological basis of brain function is the key to understanding the cause of this mental illness. Specimens of psychiatric brain tissue for research have been collected and appropriately preserved to interact with diagnostic assessment to ensure appropriate identification and documentation of tissue held. We have established a strong network between our Department and 39 associated hospitals within Fukushima district in Japan, and precise protocols for brain collection following the internationalized procedures and the site to store the materials. In the last two years, we have collected 13 schizophrenic and bipolar disordered brains. With more spontaneous collaboration directly given by patients and their family groups and increased public awareness, activity of brain collection has been dramatically enhanced. Several lines of research have already started in collaboration with brain banks in other countries and research institutes with sophisticated molecular biological tools, producing variety of scientific data. This project represents a unique opportunity to explore the biologically and socially important question of the changes associated with psychiatric illness in the human brain. It brings together the resources contained in the brain tissue collection maintained by a team of enthusiastic psychiatrists and pathologists, and the expertise of experienced neurochemists to promote/increase current Japanese psychiatric research projects to internationally competitive levels, which lead to the development of new therapeutic agents. We also intend to expand this activity to other Asian countries to form Asian Brain Network as conducted in other western countries, such as Europe, USA and Australia.

Background

Although various animal models have been developed and studied to understand the pathogenesis of psychiatric disorders, delineating this human-specific brain disorder has been largely dependent upon the availability of postmortem human brain. Brain banks are important in psychiatric research due to the following reasons. Firstly, psychiatric disorders are chronic and common diseases. However, mechanisms of disease processes are still to be fully understood. Secondly, utilization of animal experiments is limited, because psychiatric diseases are mostly specific to humans. Thirdly, *in vivo* imaging studies have demonstrated various abnormal findings of the affected brains. Therefore, it is required to investigate human postmortem brains. Through such investigations we will be able to obtain scientific bases for therapeutic interventions at cellular and molecular levels.

Despite the emphasis placed on establishing brain banks, no systematic brain bank existed in Japan until recently. Considering the lack of availability of human psychiatric disordered brain samples, since 1997 we have been working to establish a brain bank in Fukushima prefecture. Fukushima prefecture is a relatively large state located 250km north from Tokyo with a population of 2,133,000. Compared with other prefectures in Japan, there are a large number of psychiatric hospitals established in Fukushima prefecture (total 39), holding more than eight thousand psychiatric beds (Table1). The total number of in-patients in these hospitals is 7,944 and

Table 1: Number of Psychiatric Hospitals and Beds in Northeast Japan

	Population (x1,000)	No of psychiatric hospitals	No of psychiatric beds	Beds/10,000	No of psychiatric inpatients
Aomori	1,482	26	4,929	33.3	4,468
Iwate	1,420	22	4,915	34.6	4,678
Miyagi	1,358	20	3,578	26.3	3,410
Akita	1,214	26	4,458	36.7	4,357
Yamagata	1,257	18	3,226	25.7	3,089
Fukushima	2,133	39	8,600	40.3	7,942
Tokyo	11,772	117	26,428	22.4	4,130

of that number 4,658 are schizophrenic patients (males 2,802, females 1,856) (Table2). This large catchment area creates an ideal situation for establishing a brain bank. Our Department of Neuropsychiatry plays a significant role in improving the collaboration between these psychiatric hospitals, pathology and other departments of Fukushima Medical University (FMU) involved in biologically oriented psychiatric research. We have established a strong network system and precise protocols for collecting brain samples.

Table 2: Disease Category of Psychiatric In-Patients in Fukushima

Diagnosis	Patients
Schizophrenia	4,658 (58.6%)
Organic Mental Disorders	1,228 (15.5%)
Mood Disorders	557 (7.0%)
Other Psychotic Disorders	467 (5.9%)
Mental Retardation	418 (5.3%)
Substance Related Disorders	376 (4.7%)
Anxiety Disorders	113 (1.4%)
Personality Disorders	26 (0.3%)
Others	99 (1.3%)
Total	**7,944**

In collecting brains for brain banks, several conditions are necessary to be met. First, satisfaction of strict ethical guidelines is necessary. Second, internationally standardized procedures are required for systematic brain banking, for example symptom assessment collection and storage method. Third, and most importantly, spontaneous collaboration between well-educated users, care givers and researchers are necessary. The brain bank in Fukushima has been carefully focusing onto these three issues, resulting in successful sample collection, which can be exchangeable between brain banks in other countries, without causing any ethical problems and any confliction with patients, scientific community and further general publics.

Organization and aims of the Fukushima Brain Bank for Psychiatric Disorders

To establish and maintain a human brain bank, the following issues need to be addressed.

Recruitment and Information of Donors and Controls

The basic concept for the brain bank is "working together with users and their families". The following several points are thought to be important: (1) is the voluntary participation of users which should be based upon informed consent of the illness. (2) is psychoeducation of the illness for users and their family members which will facilitate their voluntary participation to the brain bank. (3) is facilitation of cooperative participation of users, their families and the general public in advancing psychiatric research. (4) is organizing the brain bank foundation which is called Tsubame-kai. (5) is consent ethical control of the bank. In our case, the bank is controlled by the supervisory ethics board which consists of representatives of local family support groups of mentally ill and a critical reviewer from academic standpoints. Our research project is now reviewed and supported by the National Board of the Family Association of the Mentally Ill in Japan.

By educating and informing the family support groups and patient self-supporting groups that exist in the majority of the hospitals (total number of families and patients registered is approximately 1,500) and by placing a qualified and authorized brain bank co-coordinator in each hospital, we have established a strong network for brain collection. In addition to caring for these families and patients, brain bank coordinators are responsible for educating both patients and families to understand the necessity and importance of the bank for psychiatric research. As results from these efforts, we have now several family members and patients, who are supportive of our activity, working as brain bank staffs to co-ordinate bi-directional relationship between researchers and users. With their support, in the last two years, we have organized annual meetings to understand the nature of psychiatric illness and the importance of advanced research. Interest from not only users, but also from general public was extremely high, collecting more than 500 people to each meeting. It is certainly a difficult task to collect normal control samples, but we found the above approach a very effective method to draw attention from general public and to increase the number of ante-mortem registrations of normal controls. The concept and importance of our brain bank has been slowly, but surely penetrating into people's minds to encourage their spontaneous collaboration. The number of members joining the foundation association called "Tsubame-kai" continues to increase (exceeding 100 members) and this supports the bank activity psychologically and financially. The consent form to comply with the proceedings of the bank, which will also include storage of personal data, is given in written form and will be signed by the next of kin and if possible by patients and witnessed by caregivers and coordinators. Confidentiality for the registered donors must be strictly maintained with full responsibility onto the brain bank.

Clinical and Pathological Documentation of Disease Status

We have obtained a Diagnostic Instrument for Brain Studies, Psychosis and Mood Disorder "version" through our established relationship with the Mental Health Research Institute of Victoria, Australia, which is actively engaged in schizophrenia research using more than 100 human brain specimens. The instrument includes autopsy summary, patient's background (sociodemographics), substance use, psychotic symptoms, affective symptoms and diagnostic summary. Using a modified version of this instrument, it is possible to give precise diagnosis, following the diagnostic criteria of DSM-III-R, DSM-IV, ICD-10, RDC, Schneider and Feighner (multidiagnostic evaluation). Upon completion of the neuropathological report, all of the necessary information is entered into the database. Because most of our donors are registered ante-mortem and are monitored prospectively, it is possible to obtain precise and detail clinical information. Considering the heterogenity of psychiatric disorders, it is obviously important to store clinical information as much as possible and to correlate it with scientific data arising from postmortem studies.

Provisions in Case of Death of a Donor and Tissue Preparation

In case of death of a donor the responsible brain bank coordinator will be notified immediately and initiate the proper handling of the body. We have established a hot line directly to the hospital, and outside of working hours, switchboard operators will answer calls and page the brain bank staff member on call. The staff member then contacts the caller to arrange for the donation. The body will be refrigerated at 4°C within the first four hours after death. The brain will be removed within 24 hours in the autopsy room of the pathology department of the FMU. The brain will be cut mid-sagittally in half. Coronal, whole hemisphere slices cut at standardized landmarks are frozen with a cooling device at -80°C. The hemi brainstem and hemi cerebellum from half of the brain are detached from the cerebral hemisphere. The brainstem and cerebellar slices are then frozen with the cooling device. Each frozen slice is transferred to a pre-labeled plastic bag and stored at -80°C. For neuropathological examination, the other half is immersed in 10% formalin. After 3 weeks a macroscopical examination will be carried out. The brain is serially sectioned in the coronal plane using the same landmark for fresh brains. Sections are routinely taken from various regions. Blocks are processed through paraffin and sectioned at 10 microns. Routine staining with haematoxyline and eosin is undertaken on all sections. Several silver stains and cresyl violet for cellular evaluation are performed on the temporal lobe, frontal cortex and caudal hippocampal sections as a routine. For frozen materials, RNA is extracted using rapid extraction kit from occipital cortex to quantify the state of tissues for molecular biological experiments, such as differential display and DNA chip evaluation. We have not collected DNA samples at the moment because another group collects DNA samples from blood (DNA bank). Collaboration with that group is always possible.

Transport and Storage of Tissue to the Bank

The brain bank coordinator will supervise the process of brain removal, intermediate storage and transport. The brain tissue is then put into storage in the bank established in our department. The freezers have already been set in a large, well-ventilated room. Freezers have been connected to an automated alarm system and in case of power failure an emergency generator will be activated so that a security guard can alert the person in charge of the freezers. As important as proper tissue storage is for prospective studies, so too is proper data banking. Data will be stored both in written form on file and digitized in a computer databank. Because of the small number of brains collected so far, we have not started the tissue dissemination program. However, when collaborative research is necessary, we are ready to send the necessary amount of tissue after confirming that the institutes have the ability to conduct experiments and appropriate ethical clearance for the project. We have a specific format ready for tissue request. All of the requests will be assessed by our organizing committee for approval.

Ethics and Public Relations Program

To conduct this research project, we have been following the guideline (released in 1997) of the Japanese Society of Psychiatry and Neurology (JSPN), which is based upon the Declaration of Helsinki. Our research project described in this manuscript has been ethically approved by a research ethics committee organized within the FMU and an independent assessment committee made up of individuals representing the patients and families and various interest groups. During the recent years, our Department has greatly expanded its Public Relations program. We have established a close relationship with each hospital's family support group and patient self-supporting group through periodical meetings and lectures, which are coordinated by the bank coordinator and bank staff from users. Furthermore, our proposed research programs have been reported in lay language so that they are well understood by the wider community through newsletters.

Conclusion

To develop postmortem human brain studies of psychiatric illnesses following the recent movement in western countries, there is an urgent necessity for all users, caregivers and researchers in Asian countries to recognize the need for intensive collaboration to establish systematic brain banks. Our medical staff must give strong support and education to patients and families so that they can understand the significance of the research and spontaneously join the research projects without being afraid of social discrimination against the mentally ill. We should be sensitive to the opinion and needs of the donor and family at all times during the process. The Mitsubishi Foundation grant awarded to our department this year emphasizes all of the above and funding will be available to promote further collaboration between Asian countries for the purpose of establishing a Brain Bank Network. In 2000, we organized Asian Pacific Forum for Human Brain Banking in Fukushima, inviting the researchers, representative of families and patients from various Asian countries, such as China, Taiwan, Korea, Hong Kong, New Zealand and Australia to discuss possible stand-up of brain bank network in these regions. Although situation of each county is of course very much different from others, we intend to keep exchanging the information to find future direction.

References

Hill C, Keks N, Roberts, S et al. Problem of diagnosis in postmortem brain studies of schizophrenia. Am J Psychiatry 1996;153: 533-537.

Niwa, S, Matsumoto I, Tago H, Mashiko H. "Establishment of the first systematic brain bank network for psychiatric disorders in Japan", The 9th Scienctific Meeting of the Pacific Rim College of Psychiatrists, Oct. 1999 Seoul, Korea.

Matsumoto I, Niwa S, Tago H, Mashiko H, Ito M, Iwasaki T and Shibata I. "Establishment of the First Systematic Brain Bank Network for Psychiatric Disorders in Japan" in symposia of the 3rd international congress of neuropsychiatry. Kyoto, Japan, April 2000. Title of symposia: "The Use of Postmortem Brain Tissue for the Study of Psychiatric Disorders Organized by Agam G.

Abstract of Asian Pacific Forum for Brain Banking, printed by Department of Neuropsychiatry, Fukushima Medical University, 2000.

Gift of Hope Tissue donor consent kit, The NISAD tissue donor program, The University of Sydney, June 2000

Matsumoto I, Niwa, S, Ito M Iwasaki T and Shibata I. "Establishment of the First Systematic Brain Bank Network for Psychiatric Disorders in Japan" in Contemporary Neureopsychiatry, Koho Miyoshi, Colin Shapiro, Moises Gaviria, Yoshio Morita (Eds) Title of symposia: "The Use of Postmortem Brain Tissue for the Study of Psychiatric Disorders. 2001; pp.310-313.

Shibata I, Iwasaki T, Ito M, Matsumoto I. and Niwa S. Neuronal orientation variability in the hippocampal subfield CA4 and reduction of chromogranin A immunoreactivities in the dentate gyrus in schizophrenia, International congress on schizophrenia research, April 2001, British Columbia, Canada.

Matsumoto I. "Ethical Issues for Schizophrenia Research in Japan", in plenary session at International Congress on Schizophrenia Research, Whistler, Canada, 2001.

II

BRAIN BANKING IN PSYCHIATRIC DISORDERS: THE AMSTERDAM AND EUROPEAN EXPERIENCE

Abstract

European Brain Bank organisations are an essential repository for basic scientists interested in the field of psychiatric disorders by using postmortem human brain specimens obtained at autopsy.The growing number of sophisticated neurobiological techniques which can be applied on postmortem brain obtained from psychiatric patients increases the pressure on brain banks to supply autopsy material to the scientific community. The active European brain banks collecting postmortem tissue from patients with neurological and psychiatric disorders have been established in the past decade and form at present an important link between donors, their relatives, personnel in psychiatric clinics, clinicians, neuropathologists and scientists. These brain banks are confronted with a need for a consensus for clinical and neuropathological criteria which will make the tissues suitable for high quality scientific research.

The Netherlands Brain Bank (NBB) fosters research in neurology and psychiatry by collecting and supplying clinically and neuropathologically well documented specimens for scientific research. The rapid autopsy system we established in Amsterdam for collecting brain specimens is based on a donor program set up with the permission of the families. The specimens are collected from cases with a diagnosis of depression, bipolar disorder, schizophrenia and controls.The history of our understanding of neurological and psychiatric disorders is anchored in the numerous European brain banks and their systematic well documented collections of specimens. The European brain bank network coordinates various activities which clearly lead to a better internation scientific collaboration which has as a spin-off the elucidation of the mechanisms underlying neurological and psychiatric disorders.

Background

The possible causes and underlying pathologies of most psychiatric disorders are still unknown and current scientific research is focusing on the neurobiological entities and the possible changes occuring in the brains of affected patients. The European brain banks collecting specimens of psychiatric cases are fairly young; there is no standard protocol for collection and handling of tissues and the consensus for the diagnostic criteria is still incomplete (Cruz-Sanchez and Tolosa 1993; Cruz-Sanchez et al. 1995; Riederer et al. 1995). In order to be able to closely follow and continuously update techniques, protocols and diagnostic procedures, brain banks need to be in the immediate proximity of an academic center with a strong neuroscience tradition. The brain banks also need to establish a well functioning donor program and have an ongoing collaboration with a large number of patient's associations, clinicians, hospitals, mental institutions, nursing homes, pathologists, undertakers, autopsy assistants and scientists.

The Netherlands Brain Bank (NBB) was established in 1985 in the Netherlands Institute for Brain Research to support neurobiological and medical research by supplying clinically and neuropathologically well characterized postmortem specimens from patients with neurological and psychiatric disorders. The NBB has been since 1985 an active participant in the various of the European Brain Banking activities (Eurage, European Brain Bank Network-EBBN; Brain-Net Europe) and has a close collaboration with other brain banks worldwide to improve the compatibility of the diagnostic criteria, handling procedures and dissection protocols and make exchange of specimens between banks and researchers feasible. The NBB's unique procedures of rapid autopsy with a short postmortem delay and fresh dissection protocol provide high quality specimens for research (illustration in figure 1). We use the pH of the brain as a measure for agonal state and this guarantees the quality of the tissue and m-RNA for scientific research (Ravid et al. 1992; Kingsbury et al. 1995; Johnston et al. 1997; Schramm et al.1999).

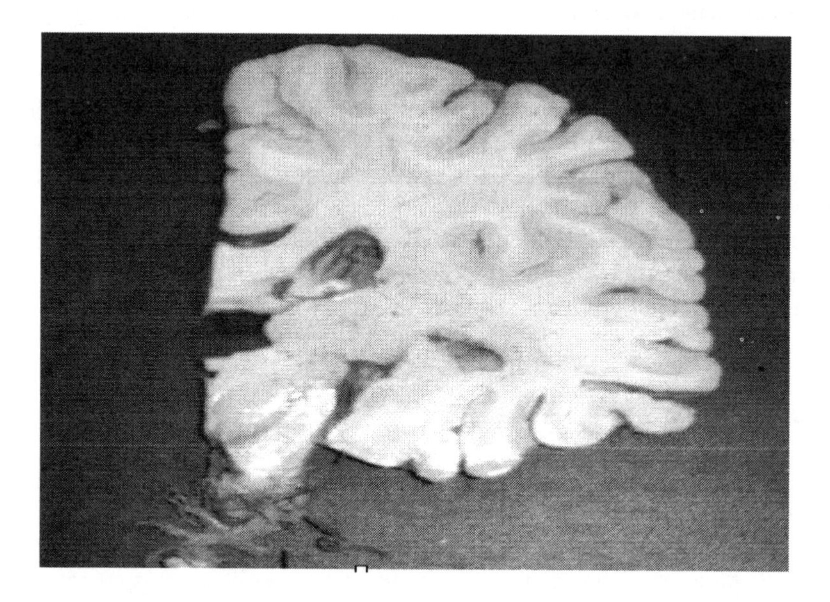

Figure 1. Coronal section of the cerebral left hemisphere of a 48 years old schizophrenic patient, obtained via the rapid autopsy system of the Netherlands Brain Bank (Case NBB99-037)

All cases in the disease groups and controls are well matched for various factors, both ante-mortem and postmortem. Ante-mortem factors include age, sex, agonal state, seasonal alterations, circadian variation, clock time of death and medication. Postmortem factors include postmortem delay, fixation and storage time and laterality (Swaab et al. 1989; Ravid et al. 1992; Ravid and Swaab 1983; Ravid and Winblad 1993; Ravid and Swaab 1995a; Ravid et al. 1995b,c) .

The availability of postmortem material from patients with psychiatric disorders is still limited while the interest from the international scientific community is ever-growing. In order to provide a solution to this problem, the NBB, in collaboration with several psychiatric departments and institutions, started to implement a postmortem research programme for depression, bipolar disorders and schizophrenia, including the possibility of genetic research through simultaneously approaching patients and their relatives. The psychiatrist approaches the potential candidates and conducts the ante-mortem and retrospective diagnostics. Our donor program is based on informed consent of the patients and their next of kin obtained during life. The NBB makes all possible effort to operate a well functioning autopsy system which works within a well defined ethical code of conduct in addition to the essential internationally accepted safety measures against hazardous agents for brain bank personnel and researchers using the tissues.

Postmortem research on autopsy material creates the possibility of investigation of a possible hereditary base through changes in gene expression in the differrent brain structures in patients, family members and controls. Several gene profiling techniques are used at present at NBB to search for differential gene expression on anatomically well defined subdivisions of the thalamus and cortex in specimens of controls and schizophrenic patients. In this manner insight may be gained into the underlying mechanisms of the abnormalities found with the in-vivo techniques, and possibly in the pathogeneisis of schizophrenia. As there is no generally accepted animal model for schizophrenia regarding anatomical changes, postmortem research will require human cerebral tissue.

In order to enable postmortem research in The Netherlands in the field of schizophrenia, the Psychiatry Department of the Academic hospital Utrecht and the Netherlands Brain Bank decided in 1998 to implement a postmortem research programme for schizophrenia, including the possibility of genetic research through simultaneuosly approaching patients and their parents. The Psychiatry department, with its large experience in schizophrenia research, approaches the young schizophrenia patients and their parents, and conducts the ante-mortem and retrospective diagnostics. The information gathered includes information on the donor's birth, development, education, work situations, family history, hospitalizations, alcohol and drug abuse and medication. The NBB, having gained large experience in similar procedures for neurological diseases, takes care of donor registration, autopsy procedures and neuropathological diagnosis. The donors/next of kin also sign informed consent for the release of all medical records. The records are summerized and stored in a computerized D-base (Visual Fox-Pro, 6.0) after anonimization by means of Brain Bank numbers. This D-base also includes all postmortem information and autopsy data.

We are at present putting continuous effort in recruiting more patients to our donor program and have a better collaboration with the various psychiatric institutions to convince psychiatrists that basic research is timely and essential for the development of appropriate therapies.

Organisation and Aims of the European Brain Banks

Brain banks have been set up in many European countries in order to support research on psychiatric disorders. Through close collaboration between banks and a large number of research groups in Europe, the individual efforts are intensified and the total effect is maximized.

A. The main task is to foster research in clinical and basic neuroscience and serve as a dynamic system with ongoing consultations on the many various daily issues of brain banking. The banks support the exchange of control cases on European level. Finding the suitable and sufficient numbers of controls is one of the most difficult objectives of every brain bank. The optimal strategy is to try to get the spouse of the donor to sign up as a control. The brain bank has to try to get exact clinical records of all patients, preferably soon after the patient signed up as a donor. The cause of death and time of death must be known and registered. Clinical data from patients who suffered from a psychiatric disorder should be obtained in a standardized way confom the internationally accepted clinical criteria. The agonal state of the patient must be assessed pre-mortem and the agonal state of the tissue must be assessed postmortem by measuring the pH of the ventricular CSF obtained at autopsy (Ravid et al.1992; Kingsbury et al. 1995).

B. Achieve standardization between banks of commonly accepted criteria for the neuropathological diagnosis and compatibility of protocols for tissue procurement, management and preparation. Over the past 15 years, several European meetings aiming at standardization of diagnostic and preparative protocols and to enhance collaborative postmortem brain research have been organized. A persisent issue has been whether there is a need for a single central European bank. The consensus view is that this would be neither efficient nor feasible, as the effectiveness of tissue banking is very much driven by local motivation (Swaab et al. 1989; Duyckaerts et al. 1990). The enormous variability in methods within the Europaean brain banking community has been demonstrated by various concordance studies and concerted actions supported by the European Community. Through a series of meetings organized by the NBB this collaboration has been widened to inlcude North American brain banks (Bidaut-Russel et al. 1995).In the future the banks will also make sure to have a compatible dissection protocols and develop widely applicable procedures and techniques of using postmortem specimens .

C. The European brain banks network has to facilitate multi-center concordance studies and studies of risk factors in psychiatric diseases which have genetic end environmental components. These factors play an important role in psychiatric diseases, but their exact role in schizophrenia and affective disorders is still quite vague. Studies performed on a European basis will enable screening a larger number of well documented cases. Furthermore, the genetic basis can be investigated in populations that are known to be susceptible in different dergrees. Most brain banks started collecting DNA samples and the isolation and storage of DNA is essential for future research; this can be done from both frozen and fixed tissues.

D. Safety measures - all European brain banks collect a broad variety of cases and store frozen and fixed specimens. Possible infectivity is a substantial problem, especially in respect to HIV and prion diseases. Secure storage is important and needs to take into account each specimen as possibly hazardous (Bell 1997; Budka et al. 1995; Hauw 1995). The safety procedures must abide by established international guidelines. The staff should be immunized against Hepatitis B and strictly follow the procedures for safe handling.

E. Ethical and legal issues - ethical codes of conduct of the Brain Banks, the laws regulating autopsy procedures, and many other important relevant issues are significantly different in the various brain banks of the member states of the European Union (Cruz-Sanchez et al. 1997). The use of human tissue in medical research is the focus of public and professional concern and has recently been the subject of several reports on bioethics. Brain banks are a source of adequately collected and preserved tissues and fluids of the central nervous system obtained in part via a donor programme, which are supplied for neurobiological research. To create and develop the adequate infrastructure underlying brain bank activities, one should have medico-legal and ethical support according to local legislation.

From an ethical point of view, European Brain Banks take the following aspects into consideration in their daily practice:

Tissue procurement: includes all factors related to the donor program . Written consent of the donor and/of next-of-kin at time of death or earlier. This is mandatory for use of tissues for medical and basic research. Respect for the dignity of human remains is considered at all times. All staff working with human remains are properly trained and qualified.

Tissue management: includes factors related to collection, handling and preservation of tissues. All human tissues should be regarded as potentially biohazardous and should be handled accordingly by appropriately trained personnel. Specific precautions should be taken against known hazards and general precautions against unknown hazards. All clinical documentation and information concerning the tissues should be made available to users as quickly as possible. All those potentially exposed to human brain tissues and fluids should be vaccinated against hepatitis-B virus and be regularly checked for level of immunity. All disposable equipment used in conjunction with human brain tissues and fluids should be sterilized according to a recommended procedure before transport for incineration.

Tissue dissemination: factors related to scientific research and supplying samples of high scientific quality and properly matched for specific projects. Users of the brain samples should undergo a process of accreditation, to ensure that a minimum set of criteria is fulfilled. These criteria include safe tissues, scientific reputation, ethical credibility, proper training, suitable facilities for safe handling and disposal of human tissues and confidentiality. Whenever possible, this accreditation should be performed by an appropriately qualified and independent expert group. As the legislation and documentation concerning shipment of human autopsy specimens varies between the European countries, the information regarding regulations of packing and national and international transport of human brain tissues and fluids should be made available to all parties involved in brain bank activities for research.

Confidentiality: the anonymity of tissues and patient records must be protected at all times. Samples supplied for research should be coded by the brain bank and a tissue tracking system restricted to local brain banks must be established to guarantee the anonymity of the donor. Research using human tissues can bring new information about the donor which is also relevant to his family. This option should play an important role in the ethical conduct. The acquired confidence with the donor may become irrelevant and the possibility of giving information about new findings should be taken into account.

"Financial gain": from a legal and ethical point of view it is highly important to establish brain banks as nonprofit sources of human tissues for scientific research. The brain banks act as a custodian of brain tissue and the tissues must not be commercially handled. On the other hand, it is extremely important to have a reasonable coverage for the costs involved in procurement, handling and transport. All technical and scientific activities and expenses related to handling and management of tissues as well as the costs involved in the acquisition and/or preservation of the tissues can be considered as a budget to be paid when tissues are requested for research. The various professionals who are involved in obtaining, handling and diagnosing the tissues have the right to be reimbursed for their services. It must be made clear in writing to all users that any possible payments are only made to cover the costs of brain banks that are forbidden to make any financial profit.

Genetic testing: the tremendous advances in genetic research in our decade raise serious ethical problems and complexities; there is sometimes conflicting interest in the need for knowledge and information on the one hand and the use and implications of this information for the people involved. The link found between certain genes and neurological diseases, creates a heavy burden on physicians, health care workers and brain bankers who test for these genetic factors as they are essential for the diagnostics and scientific interpretation of the results obtained on postmortem human brain tissues and fluids. This knowledge poses many difficult questions to concerned or afflicted individuals.

The rapid linkage between genes and diseases will have many future implications on the international legal and ethical systems. It would be advisable to have a consensus between European brain banks on that issue, as some of the genetic testing is done by collaboration between experts in different member countries of the European Union.

In applying for local ethical approval, care must be taken to include permission not only for diagnostic purposes but also for use of tissue in research projects and for shipment of tissues to other labs. Permission must also include the use of (anonymous) clinical data and DNA for research purposes.

When working in the framework of the European Union, each brain bank should be well aware of the current international discussions being considered with respect to ownership and use of postmortem tissues for scientific research. In practice, when necessary, changes to the legal frameworks operating in the various European countries are sought in conjunction with the relevant authorities, in order to facilitate the procurement of human tissues and its supply in a suitable form to end-users.

A European network of tissue banks is legally and ethically the optimal approach to the procurement and distribution of tissue for scientific research (Anderson et al. 2001). This in turn will ensure the progression towards understanding of the mechanisms of the various psychiatric disorders and will advance and enhance the development of appropriate therapeutic strategies.

All European brain banks abide by the national legal and ethical requirements and respect the principles expressed in the Helsinki Declaration, the conventions of the Council of Europe on human rights and biomedicine, and the UNESCO Declaration on the human genome.

Investigating the biological basis of psychiatric diseases requires a professional battery of techniques applied by brain banks. The dissection, handling, storage and dissimination of samples should be much better co-ordinated among the various European brain banks.

Since the various European brain banks do not use identical protocols for tissue acquisition, dissection and storage, they will have to look for a final common pathway to increase the efficiency of postmortem research. One such possible pathway would be development of compatibility of procedures and protocols for the various collections. Future research will apply among others genomics, proteomics and morphometry as possible methodologies; supplying clinically and neuropathologically well documented material by European brain banks can significantly contribute to the understanding of the causes and pathologies of the various disorders and may soon lead to new targets for therapeutic agents.

Acknowledgments

The authors gratefully acknowledge the work of the NBB technicians: Anne Holtrop, Michiel Kooreman and Jose Wouda and the editorial asssistance for this manuscript provided by Marina Kahlmann, secretariat of the NBB.
We would like to thank Solvay Duphar for supporting our research on schizophrenia.

References

Anderson R, Balls M, Burke MD, Cummins M, Fehily D, Gray N, de MG, Helin H, Hunt C, Jones D, Price D, Richert L, Ravid R, Shute D, Sladowski D, Stone H, Thasler W, Trafford J, van der Valk J, Weiss T, Womack C, Ylikomi T. The establishment of human research tissue banking in the UK and several western european countries: the report and recommendations of ecvam workshop 44. Altern Lab Anim 2001; 29: 125-134.

Andreasen NC, Arndt S, Swayze V II, Cizadlo T, Falum M, O'Leary D, Ehrhardt JC, Yuh, WTC. Thalamic abnormalities in schizophrenia visualized through magnetic resonance image averaging. Science 1994; 266: 294-298.

Bell J, Ironside J. Principles and practice of "high risk" brain banking. Neuropathol Appl Neurobiol 1997; 23: 281-288

Bidaut-Russell M, Ravid R, Cruz-Sánchez FF, Grossberg GT, McKeel DW. Survey of North American and European dementia brain banks: a 1994 directory. Alzheimer's disease and Related Disorders 1995; 9: 193-202.

Bogerts B. Recent advances in the neuropathology of schizophrenia. Schizophr Bull 1993; 19: 43-45.

Bogerts B. The neuropathology of schizophrenic disease: historical aspects and present knowledge. Eur Arch Psychiatry Clin Neurosci 1999; 249 Supll 4: 2-13.

Budka, H, Aguzzi A, Brown P, Brucher JM, Bugiani O, Collinge J, Diringer H, Gillotta F, Haltia M, Hauw JJ, Ironside JW, Kretzschmar HA, Lantos PL, Masullo C, Pocchiari M, Schlote W, Tateishi J, Will RG. Tissue handling in suspected Creutzfeldt-Jakob disease (CJD) and other human spongiform encephalopathies (Prion diseases). Brain Pathol 1995; 5: 319-322.

Cruz-Sanchez FF and E. Tolosa – The need for a consensus for brain banking. How to run a brain bank. J Neural Transm 1993;39Suppl: 1-4.

Cruz-Sánchez FF, Ravid R, Cuzner ML. The European Brain Bank Network (EBBN) and the need of standardized neuropathological criteria for brain tissue cataloguing. In: Neuropathological Diagnostic Criteria for Brain Banking, European Union Biomedical and Health Research, Vol. 10, F.F. Cruz-Sánchez, M.L. Cuzner and R. Ravid (Eds), IOS Press, 1995, Amsterdam, The Netherlands, pp. 1-3 .

Cruz-Sánchez FF, Mordini E, Ravid R. Ethical aspects to be considered in brain banking. Ann Ist Super Sanità 1997; 33: 477-482.

Duyckaerts, C, Delaère P, Hauw JJ, Abbamondi-Pinto AL, Sorbi S, Allen I, Brion JP, Flament-Durand J, Duchen L, Kauss J, Schlote W, Lowe J, Probst A, Ravid R, Swaab DF, Renkawek K, Tomlinson B. Rating of the lesions in senile dementia of the Alzheimer type: concordance between laboratories. A European multicenter study under the auspices of EURAGE. J Neurol Sci 1990; 97: 295-323.

Hardy JA, Wester P, Winblad B, Gezelius C, Bring G, Eriksson A. The patients dying after long terminal phase have acidotic brains; implications for biochemical measurements on autopsy tissue. J Neural Transm 1985; 61: 253-264.

Hauw, JJ, Sazdovitch V, Maokhtari K, Seilhaen D, Camilleri S, Kondo H, Duyckaerts C. Techniques de prevention des agents trasmissibles non conventionnels au laboratoire. L'exemple de l'anatomie pathologique. In: J. Grosset, M. Kitzis, N Lambert, M Sinegre eds: Prevention des infections nosocomiales. Protection contre les germes multiresistants,.Arnette Blackwell, Paris 1995; pp 171 – 177.

Johnston NL, Cervenak J, Shore AD, Torrey EF, Yolken RH, Cerevnak J. The StanelyNeuropathology Consortium. Multivariate analysis of RNA levels from postmortem human brains as measured by three different methods of RT-PCR. J Neurosci Meth. 1997; 77:83-92.

Kahn J, Bittner ML, Chen Y, Meltzer PS, Trent JM. DNA microarray technology; theanticipated impact on the study of human disease. Biochem Biophys Acta 1999; 1423(2): M17-M28.

Kingsbury AE , Foster OJ, Nisbet AP, Cairns NJ, Bray L, Eve DJ. Tissue pH as indicator of mRNA preservation in human postmortem brain. Brain Res Mol Brain Res 1995; 28:311-318.

Kromkamp M, Ravid R , Swaab DF. Protocol for the initiation of a postmortem research programme in schizophrenia in the Netherlands, Netherlands Brain Bank, Annual Report 1998-2000.

Pakkenberg, B. Pronounced reduction of total neuron number in mediodorsal thalamic nucleus and nucleus accumbens in schzophrenics. Arch Gen Psychiatry 1990; 47:1023-1028.

Perry EK, Perry RH, Tomlinson BE. The influence of agonal state on some neurochemical activities of postmortem human brain tissue. Neurosi Lett 1982; 29: 303-309.

Ravid R, van Zwieten EJ, Swaab DF. Brain Banking and the human hypothalammus –factors to match for, pitfalls and potentials.In: DF Swaab, M.A. Hofman, M. Mirmiran, et al. (eds) The human hypothalamus in health and disease. Elsevier, Amsterdam, 1992; pp 83-95.

Ravid, R, Winblad B. In: Alzheimer's Disease: Advances in Clinical and BasicResearch, B.Corain, K. Iqbal, M. Nicolini, B. Winblad, H. Wisniewski and P. Zatta (eds.) 1993, John Wiley & Sons, Sussex, U.K., pp. 213-218.

Ravid R, Swaab, DF. The Netherlands Brain Bank; a clinico-pathological link in aging and dementia research. J. Neural Transm. 1993, Suppl. 39: 143-153, Springer-Verlag, Vienna,

Ravid R, Swaab DF. Brain Banking in Alzheimer's disease: pitfalls and potentials. J Neuropath Appl Neurobiol 1995a; 21, Suppl 1: 18-19.

Ravid R, Swaab DF, Kamphorst W, Van Zwieten EJ. A golden standard protocol for the brain banking society? The Amsterdam Experience. J Neuropathol Exp Neurol 1995b;54, Suppl: 25- 26S.

Ravid, R., Swaab DF, Van Zwieten EJ, Salehi A. Controls are what makes a brain bank go round. In: Neuropathological Diagnostic Criteria for Brain Banking. Cruz-Sanchez FF, Ravid R, Cuzner ML (eds), IOS Press, Amsterdam, 1995c; pp 4-13.

Riederer P, Gsell W, Calza L, Franzek E, Junkunz G, Jellinger K, Reynods, GP, Crow T, Cruz-Sanchez FF, Beckmann H. Consensus on minimal criteria of clinical and neuropathological diagnosis of schizophrenia and affective disorders for postmortem research. J Neural Transm 1995; 102:255-264.

Schramm M, Falkai P, Tepest R, Schneider-Axmann T, Przkora R, Waha A, Pietsch T, Bonte W, Bayer TA. Stability of RNA transcripts in postmortem psychiatric brains, J Neural Transm 1999; 106: 329-335.

Swaab, DF, Uylings HBM. Potentialities and pitfalls in the use of human brain material in molecular neuroanatomy. In: Molecular Neuroanatomy, F.W. van Leeuwen, R.M. Buys, C.W. Pool and O. Pach (Eds.), Elsevier Science Publishers, Amsterdam, 1988; pp 403-416.

Swaab, DF, Hauw JJ, Reynolds GP, Sorbi S. Tissue Banking and EURAGE. J Neurol Sci 1989; 93: 341-343.

2 METHODOLOGICAL AND STEREOLOGICAL CONSIDERATIONS IN POSTMORTEM PSYCHIATRIC BRAIN RESEARCH

Ian Paul Everall and Paul J Harrison

Abstract

There has been a renewal of interest in understanding the brain changes that may underlie schizophrenia. However, microscopic assessment of the brain is not a trivial task and there are many potential factors, which can affect experimental results. In this chapter we will review the major potential confounding variables to be considered and principles of quantifying microscopic indices such as cell number and size. In this regard the basic principles of stereology will be outlined as well as the basic ideas of quantifying cellular arrangement by the newly emerging tool of spatial pattern analysis. During these description examples from the published literature will be provided.

Introduction

Over the last century the prevailing view of the causation of psychiatric disorders such as schizophrenia and major depressive disorder has varied. Postulated causes have included purely psychological and social factors, such as family psychodynamics, an excess of stressful life events, etc. However in the last twenty years, biological views of causation have re-emerged, a conceptual shift which has both resulted from, and has contributed to, a resurgence of neuropathological studies which had become almost extinct after the failure to make progress in the earlier part of the century (Corsellis, 1976). The contemporary neuropathology research is no longer driven by a simplistic view that there is likely to be a diagnostic lesion which, once discovered, would allow disorders such as schizophrenia to be reclassified on neuropathological criteria - although this remains possible (Harrison and Lewis, 2001)! Rather, it is hoped that histological or molecular correlates of the clinical syndromes may be identified which can provide much needed clues as to the underlying pathogenic processes, including the causative genes, as well as suggesting potential novel therapeutic targets.

Both the historical and current phases of neuropathological research have been bedevilled by a range of methodological problems. Though many of these are not unique to psychiatric disorders, they take on a particular importance in this field, and require careful attention if mistakes of the past are not to be repeated. Fortunately, the problems are now recognised more clearly than before, and new techniques are available to help solve them. In this chapter we review three areas which together account for most of the difficulties - clinical diagnosis, peri-mortem factors/tissue handling, and the quantitation method. The latter is considered in greatest detail, given the current controversies which surround it.

Clinical Diagnosis

Psychiatry has a diagnostic schedule that is based almost entirely on clinical features and presentation. Thus, schizophrenia and depression are clinical entities whose diagnosis is made on the presence of a constellation of clinical symptoms. Apart from a relatively small number of organic neuropsychiatric disorders, there is no pathological examination or confirmation available. This results in a number of potential problems, which can confound the search for putative characteristic pathological features of the various psychiatric disorders. For example, even operationalised criteria cannot overcome the possibility that schizophrenia is not a unitary disorder but a spectrum of disorders with similar clinical features. If this is the case then neuropathological research groups will continue to produce conflicting and variable results as they may be investigating heterogeneous disorders that merely appear similar clinically – in other words, it is akin to searching for the pathology of dementia or of epilepsy.

In theory, this problem can be addressed by subdivision of cases based on additional features implicated by other research approaches, such as family history (genetic liability), obstetric complications, presence of developmental problems, age of onset, response to treatment, etc. Similarly, within schizophrenia; the negative and cognitive symptoms seem more attractive correlates of pathology than do the positive symptoms, given their stability and persistence. In practice, however, this approach is severely limited, if not precluded, by sample size, which has generally been barely adequate for overall group comparisons, let alone subdivisions of the kind advocated.

The small size of most studies in the field – rarely exceeding 15 subjects per group, and often many fewer - is enforced on neuropathological researchers by the difficulty in obtaining brain tissue samples of sufficient quantity and quality. Neuropathological studies based on epidemiological sample sizes are rare, but where they have been feasible they have been very useful in delineating specific disease features (Davies et al, 1998; Neuropathology Group of the Medical Research Council Cognitive Function and Ageing Study, 2001). Thus, one issue which will assist neuropathological investigations is to improve awareness of the importance of autopsy examination in the general psychiatric community, including clinicians, service users and the family and carers, and pathologists. One practical way forward has been the establishment of brain banks to acquire tissue of particular disorders, which can then be distributed to various research groups. These banks have undoubtedly contributed to furthering our understanding of a number of disorders including Alzheimer's disease (Cairns and Lantos, 1996) and human immunodeficiency virus associated brain pathology (Davies et al, 1993; 1997). Recently the Stanley Foundation in the United States has been successful in acquiring brain tissue from relatively young individuals with clinical diagnoses of schizophrenia, major depressive disorder, and bipolar disorder (Torrey et al, 2000), through a significant, concerted investment of money and resources. This collection is proving to be the primary tissue resource for the international neuropathological research community investigating these disorders. Without a comparable level of commitment and funding, availability of tissue will increasingly become the rate limiting step for research, with studies limited either to diminishing *ad hoc* collections, or to series of elderly institutionalised patients in whom autopsies are easier to arrange (e.g. Arnold et al, 1995) but whose age, chronicity and severity are disadvantageous in many respects. The other noteworthy feature of the Stanley Foundation collection is that, by including three psychiatric disorders, it allows the relationship between clinical syndrome and neuropathological features to be addressed and in turn the diagnostic specificity of findings to be ascertained. To date, there has been very little information of this kind; hence it remains possible that, for example, the various abnormalities reported in schizophrenia might also be seen in other psychoses, mood disorder, and so on. Inclusion of patients with different disorders also allows the question of medication effects to be addressed, since prescribing crosses diagnostic boundaries – e.g. many bipolar disorder patients have received antipsychotics.

Occasionally, patients turn out at autopsy to have a neurological disease which is the likely cause of their psychiatric symptoms, necessitating a retrospective change in diagnosis (e.g. metachromatic leukodystrophy causing a schizophrenia-like psychosis). Common sense dictates that these individuals are excluded from research. However, a more difficult decision arises from the commoner finding that patients (especially elderly ones) have neurodegenerative pathology, such as Alzheimer's disease changes, a cerebral infarct, or a region of focal gliosis. The question is whether this is coincidental, and thus a confounder best dealt with by omission of the brain; this is usually the approach adopted in schizophrenia research (Roberts and

Harrison, 2000); however, it does make the assumption that there is not a fundamental neuropathological heterogeneity including a neurodegenerative 'subtype' (e.g. Stevens, 1997). At the very least, researchers should state clearly what they have done, and the extent to which the brains studied have been neuropathologically 'purified'.

Confounding Variables

There are many factors known, and no doubt many more unknown, which can affect the histology or neurochemistry of the brain. These variables are not unique to psychiatric disorders, but they do take on particular significance (relative to neurological ones) because the neuropathology being sought is not obvious, and hence there is a much lower signal to noise ratio. The factors may be divided into demographic, pre mortem and postmortem influences (Table 1).

Use of medication, especially antipsychotics (neuroleptics), is sometimes said to be a major cause of the reported neuropathological findings in psychiatric disorders. Virtually all patients have been treated at some stage with psychotropic drugs, usually for prolonged periods, and many are taking them at the time of death. Hence, this belief is sometimes taken to undermine the whole field. In fact, the evidence both from human and experimental animal data is that medication causes relatively limited demonstrable histological effects (Harrison, 1999), and that it is highly unlikely to explain the findings seen in studies of schizophrenia or mood disorder. Neurochemical and gene expression alterations are perhaps more susceptible to medication effects, certainly for studies of monoamine systems, but otherwise these are also relatively discrete and do not cast doubt on most disease-associated findings. Nevertheless, continuing investigation of possible medication-induced effects is essential, and must be extended to each new parameter being measured, and as new drugs come into use.

Mode of death can have a deceptively large effect. This particularly applies to measurements of mRNAs, although proteins and morphometric studies can also be affected. Although such a relationship can be shown by clinical descriptions (e.g. duration of terminal coma influences levels of muscarinic receptor and amyloid precursor protein mRNAs, as well as glutamic acid decarboxylase activity; Harrison et al, 1991, 1994), it is more clearly demonstrated (and better measured) by the pH of the brain or cerebrospinal fluid. The pH is a biochemical correlate of pre-mortem hypoxia/acidosis; fortunately, it is stable after death and unaffected by freezer storage (Ravid and Swaab, 1993; Harrison et al, 1995). Tissue pH has a strong association not only with preservation of many mRNAs, but also of some proteins (Harrison et al, 1995; Kingsbury et al., 1995) and even morphometric indices such as cell density (Cotter et al, 2001), presumably due to oedema.

Table 1 Potential Variables That May Confound Neuropathological Studies

Demographic and long-term
Age
Gender
Hemisphere
Handedness
Height
Intelligence and education
Comorbid medical disorders
Medication
Recent environment
Alcohol and substance use
Genetic variation
Duration of illness
Age at onset of illness
Pre-mortem (Agonal state,mode of death)
Hypoxia/acidosis (pH)
Pyrexia
Seizures
Coma
On ventilator
Cortisol and other hormone levels
Terminal medication (e.g. opioids)
Time of death (c.f. circadian or seasonal variation)
Postmortem
Delay from death to fixation/freezing
Temperature during this time
Processing
Fixative composition
Duration of fixation
Use of sucrose and other agents
Embedding medium
Mode and thickness of sectioning
Post-processing treatments (e.g. microwaving, autoclaving)
Frozen tissue: mode of freezing, temperature and duration of storage

Variables which may confound neuropathological studies of psychiatric disorders. The variables differ in their known strength of effect, and their influence depends on the kind of study being performed. For example, morphometric studies are particularly influenced by fixation and processing variables, whereas molecular studies on frozen tissue are affected more by pre and postmortem factors. Adapted from Harrison (1996) and Harrison and Kleinman (2000).

Postmortem interval, the time between death and fixation or freezing of the brain, is another factor which must always be reported and controlled for, but where the effect is variable and often less important than sometimes assumed. It is a significant problem for ultrastructural studies and measurement of neurotransmitters and some enzymes, for which even the most rapid feasible postmortem intervals (e.g. 4 hours) may be inadequate. However, other parameters, including many mRNAs, proteins, binding sites, and morphometric measures, are stable over quite prolonged postmortem intervals (48-72 hours; Barton et al, 1993). When in doubt, a parallel rat study in which tissues is processed after varying delays can be informative; human biopsy tissue is also valuable, though of limited availability.

In general when the brain is removed the tissue is either frozen or fixed in a fixative solution. Freezing to -70°C is storage method of choice for molecular and enzymatic studies. Ideally the brain tissue is rapidly frozen, either by placing in liquid nitrogen or putting the tissue on a brass plate previously cooled in a -70°C freezer, in order to enhance preservation. However, the freezing often causes tissue fractures or ice crystal damage making it less than ideal for microscopic studies. Fixation of the tissue is best when the investigation involves quantiation of cellular structures as tissue morphology is well preserved. The type of fixative used will depend upon the investigation: for most light microscopic studies tissue is fixed in 10% phosphate buffered formaldehyde or paraformaldehyde solutions. Glutaraldehyde is the fixative of choice for electron microscopic studies and other fixative solutions that can be used include alcohols. The range of fixative solutions is an important variable as they can cause the brain tissue to expand or shrink depending on the solution. These size changes in brain weight and volume can be erratic during fixation (Bonin, 1973) and are influenced by not only the particular fixative used but also the age of the individual at death (Aherne and Dunnill, 1982; Haug, 1986). The duration of fixation can also influence the size change but this stabilises within a few months (Haug et al, 1984). As the reaction to fixation by each individual is idiosyncratic and unpredictable, Haug et al (1984) suggested calculating a theoretical value mean volume change from a collection of brains, by plotting the change in brain weight (fresh to fixed) for brains of different ages. Length of fixation markedly influences the antigenicity (and mRNA accessibility) of the tissue and hence the ability of immunocytochemical and histochemical methods to visualise and quantify cellular proteins and transcripts. Although this often be reversed by retrieval steps (e.g. microwaving, autoclaving, protease treatments), it is necessary to confirm that the distribution of the target molecule is the same as that seen in unfixed tissue, and also that the retrieval is valid for quantitative purposes.

The greater the attention to confounding factors, the less 'noise' will affect a study, and the greater its power to give robust positive - or negative - results. There are two main aspects to achieving this goal: firstly, the documentation and matching of groups for as many of the factors as possible; second, the judicious use of statistics, including correlation, regression and covariance methods, to show and control for such effects. The fruits of this kind of approach can be considerable. For example, Eastwood and Harrison (2000) investigated the expression of complexin mRNAs (synaptic markers) in the Stanley Consortium tissue mentioned above. Between 64 and 89% of the variance in these mRNAs between the 60 individuals could be explained by a combination of diagnosis, age, pH and postmortem interval. This shows the potential of postmortem studies, when conducted on well collected material, especially with a reasonable sample size. In many neuropathological studies, however, a large and unexplained individual variability often remains. Presumably this is due to other peri-mortem factors, as well as intrinsic differences, due to diverse genetic and environmental factors, acting throughout a person's life and which may be very difficult in practice to identify let alone control for. For example, dendritic spines (the density of which is reportedly altered in several psychiatric disorders) can appear and disappear depending on recent environmental stimuli (Harris, 2000) and may be affected by occupation (Black et al, 2001).

Preparation of Brain Tissue for Quantitation and the Quantitative Method Utilised

The issue of tissue sampling for quantitative investigations, of for example neuronal number or size, has in the last 10 years become hotly disputed with proponents of different schools having contrasting views and approaches. Traditionally, quantitative morphometry (Aherne and Dunnill, 1982) identified a brain region of interest for investigation, a tissue block of that region was remove from the fixed brain, embedded in paraffin-wax, and cut at approximately 7μm thickness placed on a glass slide and the structure of interest visualised either by histocyochemistry or imunocytochemistry. The stained tissue section was then examined microscopically with a two-dimensional test grid of known size projected down upon the microscopic field and the number of structures in the test grid counted. The grid was applied a number of times until a sufficient number of structures, e.g. cells, had been counted. The test-grid could also contain probes to derive further information, such as a series of points as the number of points which 'hit' the structure of interest would give a measure of size. Or the grid could contain a series of lines of known length with the number of times the structure intersected a line being proportional its surface area (for review Aherne and Dunnill, 1982).

From the early 1980's stereologists, who approach issues of quantitation from a background of mathematical probability theory, began tackling the issue of how to 'correctly' count cells in the brain in a manner that resulted in an accurate and 'unbiased' estimate of either their number or size. A number of issues were challenged by stereologists, these included the method of sampling the brain in order to obtain tissue for quantitative studies, and the exact method of quantitation to be utilised. Stereological principles argued that a standard tissue block should not be obtain from the brain region of interest, for example the superior temporal gyrus, but that the entire gyrus should be available and random samples obtained. Furthermore, the orientation of these derived tissue samples should also be random so that when embedded in paraffin-wax and sectioned to obtain microsopic tissue sections all planes of the tissue have an equal opportunity of being sampled. The over-riding principle was to obtain and treat the tissue in such a manner that all cells of interest in the region of interest, eg superior temporal gyrus, have an equal opportunity of being sampled. Furthermore, the microscopic probes used to derive the estimates of cell number or size were redesigned by stereologists to obtain estimates in three-dimensional space. Finally stereologists stressed the need to obtain the volume of the region under investigation, eg the superior temporal gyrus so that the number of cells within the whole gyrus could be derived rather than density estimate of the number of cells per mm^3. Density estimates are viewed as inferior as they can be altered by factors such as tissue shrinkage, which can result from a disease process or from the tissue handling process.

In addressing the issue of estimating the number or size of cells stereologists proposed a variety of geometrical probes that can be superimposed on brain tissue, whether brain slices or microscopic sections to obtain three-dimensional data on a particular parameter. These probes include **points, lines, planes** and the **disector** (Gundersen et al 1988a,b), which can be used to derive information on volume, surface area and cell number respectively.

For example, to estimate the volume of the superior temporal gyrus the entire gyrus is sliced from end to end in a series of slices cut at a fixed thickness. If a grid of points is superimposed on each slice then the number 'hitting' the gyrus will be proportional to its volume (Delesse, 1847). The volume of the neocortex can be derived from the following formula: $V(object) = t \cdot a(p) \cdot \Sigma P$. In which t is the slice thickness, a(p) is the area associated with each point, and ΣP is the sum of the points 'hitting' the gyrus over all slices. This is the Cavalieri principle and it is a time efficient method of quantitation (Gundersen et al, 1981) compared to tracing around the region of interest on each brain slice (Cotter et al, 1999). Similarly, instead of using a grid of point, if a grid containing lines of known length are superimposed on the brain slices the number of intersections with the gyrus will provide an estimate of the surface area. This approach to quantiation is limited though in situations where the boundaries of the object of interest are not always clear, for example the posterior boundary of the superior temporal gyrus or the limits of the dorsolateral prefrontal cotex.

With regard to the microscopic estimation of cell number a three-dimensional probe, the disector, was developed (Cruz-Orive, 1980; Sterio,

1984). Essentially, tissue sections are prepared at a greater thickness, 30-50μm, than is routine for traditional quantitative morphometry (7-14μm), and a grid of a known size is projected onto the microscopic region of interest, visualised at high magnification (x100 objective lens) using a lens that has a very short depth of field (numerical aperture 1.4). This allows the observer to 'optically section' through the tissue section and observe cells come in and out of focus at various depths. The distance travelled is measured by a depth gauge, and so the number of cells falling in the grid over a particular depth (e.g. 10μm) are counted, which facilitates counting in three-dimensions. This grid is then reapplied many times to the same case on several representative sections and group mean cell densities obtained. Finally, having estimated the mean cell density as well as the macroscopic volume of the region of interest by the Cavalieri principle described above, the product of the volume and cell density will yield an estimate of the total number of cells in the region under study. Estimation of the total number of cells is considered important if the macroscopic reference volume changes, for example due to shrinkage due to a disease process or because of differences between animal strains of the same species (Abusaad et al, 1999) who investigated the hippocampus on different mice strains.

Several stereological methods have been devised for the microscopic estimation of cell volume (Howard and Reed, 1998). These include the 'nucleator' (Gundersen, 1988), which is especially applicable to determination of neuronal volume. In this method, neurons sampled in the disector probe, have a line thrown at a random 360° angle across them, but through the nucleolus. The calibrated length of the line as it transects the cell approximates to a measure of the diameter, as this transect length is obtained many times for all sampled neurons a mean diameter is acquired, which can be used to calculate mean neuronal volume from stated formulae. The approach for implementing methods to determine cell size are fully explained by Howard and Reed (1998).

Stereological studies have in recent times been applied to quantitative neuropathological studies of psychiatric disorders such as schizophrenia, bipolar and major depressive disorders. For example Pakkenberg et al (1990) estimated the total number of neurons in the dorsomedial thalamic nucleus. These were found to be reduced by 40%, astrocytes by 44% and oligodendroglia by 45% in schizophrenics when compared to controls. This difference was mainly due to the smaller volume of this nucleus in the schizophrenic brains and so would not have been detected if only the cell density had been estimated and the volume of the nucleus not obtained. Three other studies have also reported the same finding (Jones et al, 1998; Popken et al, 2000; Young et al, 2000), while a recent stereological study failed to demonstrate the loss (Cullen et al, 2000). In the hippocampus various traditional morphometric studies have conflictingly revealed either the presence or absence of neuronal loss, while a recent stereological study failed to observe evidence for a change in total neuronal number in any of the regions of the hippocampus (Heckers et al, 1991). In a study of anterior cingulate cortex from individuals with schizophrenia, bipolar and major depressive disorders compared to controls, Cotter et al (2000a) estimated laminar neuronal and glial cell number, as well as neuronal volume. They found no change in neuronal number, but a significant reduction in glial cell density in the deeper layers Vb and VI in schizophrenia and major depressive disorder. The findings were marked in the depressed group in which there was also diminution of neuronal size. Glial cell loss together with neuronal shrinkage in depression are consistent as glia, especially astrocytes, regulate neuronal activity, which in turn influences neuronal size. This study illustrates how stereological analysis can provide insights into the underlying pathological mechanisms of a disorder. Importantly these cellular changes in depression have been observed by two other groups (Ongur et al, 1998; Rajkowska et al, 1999) and so represent a consistent abnormality.

However, these stringent criteria demanded by stereological principles cause problems in applying stereological methods to studying psychiatric disorders. Firstly, it is expected that the whole of the region of interest be available for the study, whether this be the dorsomedial thalamus, cingulate gyrus or dorsolateral prefrontal cortex. However, brain tissue from individuals with schizophrenia or other psychiatric disorders is precious as it is difficult to obtain and is rarely available whole as tissue blocks are often removed for clinical neuropathological assessment, to exclude other pathological processes. Secondly, the principle of arranging the excised tissue block to have a random orientation in three-dimensional space, so that all cells have an equal opportunity of being sampled, while correct for satisfying the needs of mathematical probability, fails to take into account that cells, especially in the cortex are not randomly arranged and even neurons represent a heterogeneous population. Thus, the biology of the brain begins to suffer under the constraints of trying to obtain the most accurate estimates of cell number. Furthermore, vital information on whether cell alterations affect a particular cortical layer are lost when tissue is sectioned in a random orientation. Finally, in order to obtain cell counts in three-dimensions the cells have to be identified at high magnification. This results in the microscopic field of view and measuring grid being very small and so there are doubts about whether enough cells can be sampled to provide a representative view of those present across the cortex (Benes and Lange, 2001). Currently a debate is opening up between those who consider the theoretical principles of brain cell quantitation and those practitioners who consider the practical limitations and needs. As of today, stereology still offers an approach that considers potential pitfalls as well as possible approaches to estimation, however this has then to be adapted to the practicalities of the tissue and disease under study (Hyman et al, 1998). It must be stressed that one further advantage of stereology is that it has assisted in the developing the field of analysis of the spatial arrangement of cells and this form of emerging quantitative analysis is considered briefly below.

Analysis of the Spatial Arrangement of Cells

This newly developing statistical field is allowing the quantitation of the spatial arrangement of a cellular population in the brain, for example the position of neurons across the cortical ribbon in a particular Brodmann region. Furthermore, it can also facilitate studying the interaction of two cell populations, such as neurons and astrocytes, by comparing their respective spatial patterns. These two forms of analysis are referred to as Univariate (one cell population analysis) and Bivariate (two cell population comparison analysis) respectively. These patterns can then be evaluated as to whether changes in the spatial cellular arrangement are found in particular diseases.

To examine cellular spatial pattern a computer-assisted image analysis system can be utilised to collect a two-dimensional set of 'points' coinciding with the x- and y-coordinates of each sampled cell across for example the cortical ribbon of a region of interest. Patterns are analysed in the first instance to examine to what extent they are distributed according to, or deviate from, the pattern that would be expected if the points were randomly arranged. This random arrangement, in which the points are located within a region independently of each other and the density of points does not vary over an area, is called complete spatial randomness (CSR). To put it in statistical terms CSR is a Poisson distribution in which the mean and variance are equal (Diggle 1983).

The spatial point pattern, consisting of several hundred cells, has to be carried out for all cases in the groups under study. The first step is to test that the spatial point pattern for each case differs from CSR. This is performed by finding the distance for each point to its nearest point neighbour, and this is repeated for all points in each case. The distribution of nearest neighbour distances is then compared to the distribution of nearest neighbour distances that would be expected from a computed generated random points that are CSR. This distribution of actual cell distances to expected random cell distances is the **G-function**, and it determines whether a cell pattern deviates from CSR, by either being more clustered or regularly arranged. A variant of the G-function is the **F-function**. In this situation the computer superimposes a grid of generated points on the observed cellular point pattern and the distance from each cellular point to the nearest computer generated point is analysed in an identical way to the G-function. It is stated that the G function is more sensitive to detecting clustering whereas the F function has greater power in detecting regularly spaced cells (Upton and Fingleton, 1985). Having shown that the cellular point patterns in all the cases being studied are not randomly arranged the cases can be analysed by group and compared statistically, using the **K-function**. The K-function is essentially a combined count-distance measure (Ripley 1976), in which for each cellular point the number of other cellular points occurring around it is noted over a range of distances. All cases in a group can be examined together, these observations are compared to that expected if the points were arranged under CSR. It provides a quantitative description of the clustering or regularity of arrangement of a spatial pattern for a group of cases (Stewart 1992), and different groups can then be compared for statistical differences in their respective K functions. Statistical differences between group K-functions can be analysed by bootstrapping and other Monte Carlo randomisation procedures (Efron, 1982).

To date there have only been a few studies applying spatial pattern analysis to neuropathological disorders. In schizophrenia Diggle et al (1991) found a significant alteration in the spatial pattern of pyramidal neurons in the cingulate cortex in individuals with schizophrenia compared to controls. Similarly Cotter et al (2000b) in the same region of the brain found that the spatial arrangement of neurons containing the calcium binding protein parvalbumin were discretely changed in schizophrenia and bipolar disorder. Diggle (1983) concluded that 'knowledge of the pattern arrangement of neurons in control cases, and deviations from this pattern in pathological cases, may demonstrate how the disruptions are caused and determine what pattern alteration may be significant clinically'.

The bivariate K-function is used to compare potential spatial interactions between two cell populations, for example between neurons and astrocytes. In this situation, for each neuron the number of astrocytes that are found within a particular distance, over a range of distances is determined and compared to that expected under CSR. Thus, if more astrocytes than expected are found around neurons at a particular distance then that is evidence of spatial 'attraction' between the two cell types, if less are present than expected under CSR then that indicates that the two cell types are spatially 'repulsed' from one another, and finally if the number of astrocytes around neurons are similar to that expected under CSR then the arrangement of the two cell populations are independent.

Conclusions

This chapter has outlined the main principles and problems inherent in the undertaking of neuropathological investigations of psychiatric disorders. Whatever the focus of the study, considerable care must be paid to the well known but still sometimes neglected issues of clinical definitions and confounding variables. These considerations are now augmented by the requirements of stereology which, though always desirable because of its intrinsic advantages, should be employed pragmatically and with recognition that studies not meeting strict stereological criteria continue to have an important role to play. If due attention is paid to all these aspects, and assuming that brain collection is given the financial and ethical support it needs, the prospects for neuropathological studies of psychiatric disorders are bright, given the increasing technical armamentarium at the researcher's disposal.

References

Abusaad I, MacKay D, Zhao J, Stanford P, Collier DA, Everall IP. Stereological estimation of the total number of neurons in the murine hippocampus using the optical dissector. Journal of Comparative Neurology 1999; 408: 560-566.

Aherne WA, Dunnill MS. Morphometry. London: Edward Arnold. 1982

Arnold SE, Gur RE, Shapiro RM, Fisher KR, Moberg PJ, Gibney MR, Gur RC, Blackwell P, Trojanowski JQ. Prospective clinicopathologic studies of schizophrenia: accrual and assessment of patients. American Journal of Psychiatry 1995; 152: 731-737.

Barton AJL, Pearson RCA, Najlerahim A, Harrison PJ. Pre- and postmortem influences on brain RNA. Journal of Neurochemistry 1993; 61: 1-11.

Benes FM, Lange N. Two-dimensional versus three-dimensional cell counting a practical perspective. Trends in Neuroscience 2001; 24: 11-17.

Black JE, Kodish IM, Klintsova AY, Greenough WT, Uranova NA Quantitative study of human prefrontal pyramidal and interneurons: synaptic connectivity effects of schizophrenia and experience [abstract]. Schizophrenia Research 2001; 49 (suppl) 59.

Bonin G von. About quantitative studies in the cerebral cortex, Part I. Journal of Microscopy 1973; 99: 75-83.

Cairns NJ, Lantos PL. Brain tissue banks in psychiatric and neurological research. Journal of Clinical Pathology 1996; 49: 870-873.

Corsellis JAN Psychoses of obscure pathology. In: Greenfield's Neuropathology, 3rd edition (eds Blackwood H, Corsellis JAN). Edward Arnold, London, 1976: pp 903-915.

Cotter D, Miskuel K, Al-Sarraj S, Wilkinson I, Paley M, Harrison MJ. & Everall, I. The investigation of postmortem brain volume using MRI: a comparison of planimetric and stereological approaches. J Neuroradiol 1999; 41: 493-496.

Cotter D, MacKay D, Beasley C, Kerwin R, Everall I. Reduced glial density and neuronal volume in major depressive disorder and schizophrenia in the anterior cingulate cortex. 10th Biennial Workshop on Schizophrenia, Schizophrenia Research 2000a; 41: 106.

Cotter D, Beasley CL, Kerwin R, Everall IP. The spatial pattern of parvalbumin immuno-reactive neurons in the anterior cingulate cortex (BA 24b/c) in schizophrenia and bipolar disorder. 10th Biennial Workshop on Schizophrenia, Schizophrenia Research 2000b; 41: 107.

Cruz Orive LM. On the estimation of particle number. Journal of Microscopy 1980; 120: 15-27.

Cullen TJ, Walker MA, Roberts H, Crow TJ, Harrison P, Esiri M. The mediodorsal nucleus of the thalamus in schizophrenia: a postmortem study. 10th Biennial Workshop on Schizophrenia, Schizophrenia Research 2000; 41: 5.

Davies J, Everall IP and Lantos PL. The contemporary AIDS database and brain bank - lessons from the past. Journal of Neural Transmission 1993; 39 77-85.

Davies J, Everall IP, Weich S, McLaughlin J, Scaravilli F and Lantos PL. HIV-associated brain pathology in the United Kingdom: an epidemiological study. AIDS 1997; 11 1145-1150.

Davies J, Everall, IP, Weich S, Glass J, Sharer LR, Cho ES, Bell JE, Mattenyi C, Gray F, Scaravilli F, Lantos PL. HIV-associated brain pathology: a comparative international study. Neuropathology and Applied Neurobiology 1998; 24 118-124.

Delesse MA. Procede mechanique pour determines la composition des roches (extrait). CR Academy of Science (Paris) 1847; 25: 544.

Diggle PJ, Lange N, Benes FM. Analysis of variance for replicated spatial point patterns in clinical neuroanatomy. Journal of the American Statistical Association 1991; 86: 618-625.

Diggle PJ. Statistical analysis of spatial point patterns. Academic Press. New York. 1983

Eastwood SL, Harrison PJ Hippocampal synaptic pathology in schizophrenia, bipolar disorder and unipolar depression: a study of complexin mRNAs. Molecular Psychiatry 2000; 5 425-432

Efron B. The jackknife, the bootstrap and other resampling plans. Society for Industrial and Applied Mathematics. CBMS-NSF Monograph 38. Philadelphia. 1983

Gundersen HJ. The nucleator. Journal of Microscopy 1988; 151: 3-21.

Gundersen HJ, Osterby R. Optimizing sampling efficiency of stereological studies in biology: or 'do more less well!'. Journal of Microscopy 1981; 121: 65-73.

Gundersen HJG, Bendtsen TF, Korbo L, Marcussen N, Moller A, Nielsen K, Nyengaard JR, Pakkenberg B, Sorensen FB, Vesterby A, West MJ. Some new simple efficient stereological methods and their use in pathological research and diagnosis. Acta Pathologica Microbiologica et Immunologica Scandinavica 1988a; 96: 379-394.

Gundersen HJG, Bagger P, Bendtsen TF, Evans SM, Korbo L, Marcussen N, Moller A, Nielsen K, Nyengaard JR, Pakkenberg B, Sorensen FB, Vesterby A, West MJ. The new stereological tools: disector, fractionator, nucleator and point sampled intercepts and their use in pathological research and diagnosis. Acta Pathologica Microbiologica et Immunologica Scandinavica 1988b; 96: 857-881.

Harris KM Structure, development, and plasticity of dendritic spines. Current Opinion in Neurobiology 2000; 9: 343-348

Harrison PJ Advances in postmortem molecular neurochemistry and neuropathology: Examples from schizophrenia research. British Medical Bulletin 1996; 52: 527-538

Harrison PJ The neuropathological effects of antipsychotic drugs. Schizophrenia Research 1999; 40: 87-99

Harrison PJ, Kleinman JE Methdological issues. In: The Neuropathology of Schizophrenia: Progress and interpretation (eds PJ Harrison and GW Roberts), Oxford University Press, Oxford, 2000: pp339-350.

Harrison PJ, Lewis DA The neuropathology of schizophrenia. In Schizophrenia, 2nd edition (eds Hirsch S, Weinberger DR). Blackwells, Oxford (in press), 2001

Harrison PJ, Proctor AW, Barton AJL et al. Terminal coma affects messenger RNA detection in postmortem human temporal cortex. Molecular Brain Research 1991; 9: 161-164.

Harrison PJ, Barton AJL, Procter AW, Bowen DM, Pearson RCA The effects of Alzheimer's disease, other dementias and premorten course upon amyloid ß precursor protein messenger RNAs in frontal cortex. Journal of Neurochemistry 1994; 62 635-644

Harrison PJ, Heath PR, Eastwood SL, Burnet PWJ, McDonald B, Pearson RCA The relative importance of premortem acidosis and postmortem interval for human brain gene expression studies: selective mRNA vulnerability and comparison with their encoded proteins. Neuroscience Letters 1995; 200: 151-154.

Haug H. History of Morphometry. Journal of Neuroscience Methods 1986; 18: 1-17.

Haug H, kuhl S, Mecke E, Sass N-L, Wasner K. The significance of morphometric procedures in the investigation of age changes in cytoarchitectonic structures of the human brain. Journal fur Hirnforschung 1984; 25: 353-374.

Heckers S, Heinsen H, Geiger B, Beckmann H. Hippocampal neuron number in schizophrenia. A stereological study. Archives of General Psychiatry 1991; 48: 1002-1008.

Howard CV, Reed MG.Unbiased Stereology. Three Dimensional Measurement in Microscopy. Bios Scientific Publishers, Oxord, UK. 1998

Hyman BT, Gomez-Isla T, Irizarry MC. Stereology: a practical primer for neuropathology. Journal of Neuropathology and Experimental Neurology 1998; 57: 305-310.

Neuropathology Group of the Medical Research Council Cognitive Function and Ageing Study (MRC CFAS) Pathological correlates of late-onset dementia in a multicentre, community-based population in England and Wales. Lancet 2001; 357, 169-175

Johnston NL, Cerevnak J, Shore AD, Torrey EF, Yolken RH. Multivariate analysis of RNA levels from postmortem human brain as measured by three different methods of RT-PCR. Journal of Neuroscience Methods 1997; 77: 83-92.
Jones L, Mall N, Byne W. Localization of schizophrenia-associated thalamic volume loss. Society of Neuroscience Abstracts 1998; 24: 985
Kingsbury AE, Foster OJ, Nisbet AP Cairns N, Bray L, Eve DJ, Lees AJ, Marsden CD.. Tissue pH as an indicator of mRNA preservation in human postmortem brain. Molecular Brain Research 1995; 28: 311-318.
Ongur D, Drevets WC, Price JL. Glial reduction in the subgenual prefrontal cortex in mood disorders. Proceedings of the Natlional Academy of Sciences USA. 1998; 95: 13290-13295.
Pakkenberg B. Pronounced reduction of total neuron number in mediodorsal thalamic nucleus and nucleus accumbens in schizophrenics. Archives of General Psychiatry. 1990; 47: 1023-1028.
Popken GJ, Bunney Jr WE, Potkin SG, Jones EG Subnucleus-specific loss of neurons in medial thalamus of schizophrenics. Proc Natl Acad Sci USA 2000; 97: 9276-9280
Rajkowska G, Miguel-Hidalgo JJ, Wei J, et al. Morphometric evidence for neuronal and glial prefrontal pathology in major depression. Biological Psychiatry 1999; 45: 1085-1098.
Ravid R, Swaab D The Netherlands brain bank – a clinico-pathological link in aging and dementia research. Journal of Neural Transmission 1993; 39: 143-153 supplement
Ripley BD. The second-order analysis of stationary point processes. Journal of Applied Probability. 1976; 13: 255-266.
Roberts GW, Harrison PJ Gliosis and its implications for the disease process. In: The Neuropathology of Schizophrenia: Progress and interpretation (eds PJ Harrison and GW Roberts), Oxford University Press, Oxford, 2000: pp137-150.
Selemon LD, Lidow MS, Goldman-Rakic PS Increased volume and glial density in primate prefrontal cortex associated with chronic antipsychotic drug exposure. Biological Psychiatry 1999: 46: 161-172.
Sterio DC. The unbiased estimation of number and sizes of arbitrary particles using the disector. Journal of Microscopy 1984; 134: 127-136.
Stevens JR Enough of pooled averages: been there, done that. Biological Psychiatry 1997; 41: 633-635
Stewart MG. Quantitative methods in neuroanatomy. John Wiley and Sons, Chichester. 1992
Torrey EF, Webster M, Knable M, Johnston N, Yolken RH. The Stanley Foundation Brain Collection and neuropathology consortium. Schizophrenia Research 2000; 44: 151- 155.
Upton GJG, Fingleton B. The Use of Mapped Plant Locations. In Spatial Data Analysis by Example Volume 1 (Upton GJG, Fingleton B). John Wiley & Sons, Chichester, 1985: pp 70-93.
Young KA, Manaye KF, Liang CL, Hicks PB, German DC Reduced number of mediodorsal and anterior thalamic neurons in schizophrenia. Biological Psychiatry 2000; 47: 944-953

3 IMAGING VS POSTMORTEM RECEPTOR STUDIES: What You See is What You Get?

Lyn Pilowsky

Abstract

Postmortem research historically provided the technical and evidential basis for *in vivo* receptor imaging. *In vitro* mapping of receptor distribution and density is an essential first step for novel tracer development in order to image brain receptors in a living human subject. The chapter reviews the technical distinctions between receptor estimation postmortem and *in vivo*. The advantages and disadvantages of each approach are also discussed with respect to psychiatric disorder in particular. The neuropharmacology of schizophrenia has been most studied with neuroreceptor imaging and is used as an instructive example of the contribution of both methodologies to understanding the role of dopamine in the disorder. These data include simple studies estimating dopamine D2 receptors in schizophrenia, dynamic *in vivo* challenges of the dopamine system, and examining the links between dopamine D2 receptor occupancy by antipsychotic drugs and their toxic and therapeutic effects. These studies reveal the necessity of bridging the gap between ante- and postmortem research in schizophrenia and other disorders.

Introduction

In Vivo Receptor Imaging and Postmortem Research are Interactive

The field of *in vivo* receptor imaging was built on the platform of knowledge developed by postmortem neurochemistry. Prior to the first successful report imaging dopamine D2 receptors in living human subjects by ^{18}F N-methyl-spiperone positron emission tomography (PET) (Wong et al 1984), at least a decade of work had successfully mapped receptor changes in a variety of neurotransmitter systems and psychiatric disorders. The massive expansion of in vivo receptor imaging probes and techniques led some to question the applicability of postmortem receptor research to psychiatry. It seemed to many that imaging neurochemistry in vivo overcame many methodological problems associated with postmortem studies. These included artefacts incurred by delay to necropsy and tissue fixation methods, as well as difficulties linking findings with human behaviour in life, and disentangling fundamental pathology from the effects of drug treatment on various neurochemical systems, perhaps best illustrated by contradictory findings of increased dopamine D2 receptor density schizophrenia (Lee et al 1980, Mackay et al 1982). However, it is now very clear that combining both approaches will powerfully enhance neurochemical research for the future.

This chapter will review the role of postmortem research in the technical aspects of in vivo neurochemical imaging, and will provide examples of questions in which both techniques have a role to play. The way in which these tools interact will be particularly discussed with respect to findings in schizophrenia.

Postmortem Research and In Vivo imaging: Technical Aspects

Ligand Development

In vivo receptor imaging builds on postmortem research. Before a ligand is administered to living humans, its distribution and density must be clearly understood *in vitro*. Postmortem autoradiography and competitive binding assays are essentially the first step in the development of any new radioligand. These methods permit multiple studies of ligand behaviour under very many conditions which can be varied at will by the researcher. By these means it is possible to understand the affinity and selectivity of the potential ligand for the receptor of interest, and to identify areas of the brain where no or negligible numbers of receptors are present (very helpful for receptor estimation *in vivo*).

Modelling Receptor Indices: In Vitro Versus in Vivo

Postmortem studies evaluate receptors in a tissue system devoid of influences present *in vivo* including blood flow, radioligand clearance from the plasma, dynamic changes in receptor conformation or competition with endogenous neurotransmitter (for excellent reviews see Ichise et al 2001, Laruelle 2001, Delforge et al 2001). Temperature and pH are also tightly controlled and various conditions including presence or absence of agonists or antagonists, or post receptor signalling proteins may be manipulated. In a typical saturation radioligand binding assay, increasing amounts of radioligand [L] are added to a fixed concentration of receptors and the concentration of the receptor ligand complex [RL] is measured as a function of ligand concentration. By these methods receptor concentration [R] or density (Bmax) and affinity (Kd) may be measured. For the purposes of modelling ligand states (bound or unbound), a compartmental model is assumed. In tissue homogenate studies, the radioligand is incubated with tissue in a buffer solution. The receptor rich tissue and buffer solution represent two compartments (or one tissue compartment), between which the radioligand may pass. This simplest model follows Michaelis and Menten kinetics, and the above indices of receptor density and affinity may be calculated.

In PET and SPET studies *in vivo*, important differences may be highlighted. Temperature and pH conditions may not be so closely controlled as the *in vitro* situation. Innis et al (1992) showed the evaluation of striatal dopamine D2 receptors in primates by the ligand [^{123}I]-IBZM was affected by the temperature of the animal, lower temperatures resulted in lower indices approximating to Bmax. Secondly, the system is not 'closed' in that radioligand is delivered and cleared during the time of the experiment. Brain regions containing receptors have at least 3 compartments (or two tissue compartments). Radioligand administered intravenously circulates to the heart and is delivered to the brain via the arterial blood. From the blood compartment, the radioligand passes through the blood brain barrier into the second compartment (where it may exist as free radioligand or bound non-specifically to interstitial tissue or cytoplasmic membrane). Though the radioligand may exist in these two states, in practice it rapidly equilibrates between non-specific and free states and this is still regarded as one compartment. The third compartment includes tissue containing high affinity receptors that the radioligand binds to specifically. The second compartment (non-specific binding + free ligand) is often referred to as the nondisplaceable compartment, as administration of a competitor for the radioligand will have no effect on radioligand concentration in this region. It may be seen that delivery of radioligand to the first compartment is dependent on blood flow and tracer is also cleared (dependent on metabolism), so this compartment is not 'closed' in the controlled manner of the *in vitro* system, and make analysis of *in vivo* receptor binding more challenging than that *in vitro*. Furthermore, radioligands or tracers may be metabolised to active compounds which if too lipophilic, may re-enter the brain and interfere with radioligand binding. The assumptions behind tracer kinetic modelling have been discussed in Ichise et al (2001).

In practice, two scans are required to arrive at quantitative estimates of Bmax (density) and Kd (affinity) independently, but from a single scan protocol the binding potential (Bmax/Kd), or volume of distribution (Vd-proportional to Bmax/Kd)), or a variant of these measures linearly related to the binding potential is reported. In this sense, *in vivo* measures do not attain the accuracy of *in vitro* studies, but strenuous efforts are made to achieve agreement between these modalities. If this cannot be reached, recent studies have directly addressed the reasons for the mismatch, and clearly, this helps the field advance further (Kapur et al 2001). Nevertheless, certain physical issues associated with *in vivo* imaging apart from those discussed above may lead to differences between results obtained ante- or postmortem. The main problem is the partial volume effect, which describes the limit of resolution of the scanning instrument to resolve two points separately. If receptor density is inhomogeneous, distributed densely in a cortical layer, but not throughout the whole cortical field, emission tomographic methods may 'see' an area as having relatively few receptors, compared to another area with perhaps fewer receptors distributed more homogeneously over a wider area. Similarly, small structures, beneath the resolution of the instrument may not be visualised separately at all. An example of the partial volume effect as a possible cause of discrepancy *in vitro* vs *in vivo* occurs with the single photon emission tomography ligand [^{123}I]epidepride which binds to dopamine D2/D3 receptors. Uptake of [^{123}I] epidepride is higher in the thalamus than in the pituitary/hypothalamic region and the amygdala, whereas postmortem studies demonstrate comparable or higher binding of the ligand in these regions (Kessler et al 1992). Kessler et al (1992) suggest the reason for this probably lies in the fact that the thalamus is larger in size than pituitary or amygdala. A low resolution instrument would reduce the apparent concentration of radioactivity in the smaller structures.

Dopamine Receptor Findings and the Neuropharmacology of Schizophrenia

There are many instances where postmortem research findings have driven *in vivo* receptor imaging studies. An elegant example is the history of dopamine D2 receptor studies in schizophrenia. Initial postmortem binding studies found elevated levels of D2 receptors in the striatum of patients with schizophrenia compared to controls (Lee and Seeman 1980). This finding was replicated by several groups but was never fully accepted. The possibility the effect was an artefact, the result of neuroleptic treatment upregulation of dopamine D2 receptor density (Clow et al 1987) could not be excluded unless rare, neuroleptic naïve brain tissue was studied (Mackay 1982). When the first *in vivo* ligand to image dopamine D2 receptors became available in the mid-1980's, it seemed the perfect technology to address this issue. The first study to appear was in Science by Wong and colleagues (Wong et al 1986). These investigators used a spiperone derivative ligand [^{11}C] N-methylspiperone, and studied 11 cases of drug naïve schizophrenic patients compared to 11 healthy controls. A sixfold increase in D2 dopamine receptor binding was found in the patients relative to the controls. Parallel to this study, the Karolinska psychiatry group were performing a larger investigation of 18 neuroleptic naïve patients with a lower affinity ligand [^{11}C] raclopride and PET. This group did not find an increase in dopamine D2 receptors in schizophrenic patients compared to controls. Much debate ensued after these findings. The most obvious difference between the two studies lay in the US group's use of a ligand which bound almost irreversibly during the time of the PET experiment, and the Swedish use of a ligand which attained reversible 'transient equilibrium' conditions during a scan. The former approach required greater complexity of modelling and active displacement of the ligand in a two-scan protocol to arrive at an estimate of the binding potential. Two subsequent PET and SPET studies of striatal D2 receptor binding in large groups of neuroleptic naïve patients by independent groups in the UK and France (Martinot et al 1990, Pilowsky et al 1994) substantiated the Swedish data, and Nordstrom et al (1995) repeated the study at the Karolinska with [^{11}C]NMSP (the same ligand as the US group) and failed to show a significant increase in striatal D2 dopamine receptor binding in schizophrenic patients. The consensus in the mid 1990's was that dopamine D2 receptor elevation was not a feature of schizophrenia prior to antipsychotic drug treatment. Silvestri et al (2000) demonstrated increased striatal dopamine D2 receptor density in patients taken off chronic antipsychotic medication. This adds further weight to the notion that these changes may be the result of treatment and not the disease itself.

Dynamic Challenges of Dopamine in Vivo

The case appeared closed, suggesting there was no fundamental pathology of striatal dopamine D2 receptors in schizophrenic patients *in vivo*. This was disappointing, in that the dopamine hypothesis of schizophrenia was still compelling, and certainly there were, and still are, no effective drugs for schizophrenia that do not act at dopamine D2 receptors. In the late 1990's a series of elegant studies revisited the issue, perfectly exploiting the potential of *in vivo* receptor imaging to measure dynamic changes in a neurotransmitter system following pharmacological challenge. Laruelle et al (1996) and colleagues studied striatal D2 dopamine receptors in humans before and after challenge with amphetamine. Amphetamine released dopamine into the synaptic cleft, which competed with the probe [^{123}I]IBZM (a substituted benzamide) for binding to the D2 receptor. This was visualised as a reduction in [^{123}I]IBZM binding to the receptor after amphetamine administration. The degree of the reduction was taken to represent the amount of dopamine stored in the cell and released into the synaptic cleft. Schizophrenic patients and controls were studied with the technique. Patients had a higher mean dopamine response to amphetamine than controls. Interestingly, many patients overlapped with controls. The level of the response positively correlated with the acute nature of the patient's clinical state. The more acutely and severely ill the patient, the higher the dopamine responses to amphetamine challenge. These data do not encourage a simple reading of the dopamine hypothesis, in which intrasynaptic levels of dopamine, or dopamine receptor density is persistently elevated in schizophrenia. The findings mesh well with suggestions that dopamine flux is poorly controlled or dysfunctional in psychosis, and that this may be state related, emerging in response to stressors, rather than a trait, or basal effect. Indeed, one theory postulates that there is a relative dopamine deficit in frontal cortical regions, leading to negative symptoms including poverty of thought or action, and difficulty with willed tasks or initiative (for review see Moore et al 1999). Suggestions by some researchers, that the primary pathophysiology of the disorder lies in neurotransmitter systems (for example glutamate or GABA) that control dopamine flux (rather than the dopaminergic system itself), receive support from these data, which have tremendous explanatory power in schizophrenia and will certainly drive more studies(Moore et al 1999).

Limitations for in Vivo Imaging: The Role of in Vitro Neuroscience

Dynamic challenge studies are instructive in terms of what is imparted about the limitations of *in vivo* receptor imaging versus *in vitro* postmortem studies. The above findings could only have been made with *in vivo* neurochemical imaging, and yet at this point, carrying the strategy further to understand the data (in terms of, for example, post –receptor signalling mechanisms that drive the production of endogenous dopamine, or cell membrane receptor production) is untenable. This has been illustrated very well in the mixed findings of depletion studies with different *in vivo* ligands for the dopamine D2 receptor (Laruelle 2000). The depletion challenge is an opposite strategy from the amphetamine challenge. The brain is depleted of endogenous dopamine, and two [^{123}I] IBZM scans are performed, one prior to depletion, and one after at least 24 hours treatment with reserpine, a dopamine-depleting agent, or αMPT, a dopamine synthesis inhibitor. Following dopamine depletion, an increase in striatal D2 dopamine receptor binding is seen, and the degree of the increase is taken to reflect levels of basal dopamine at the synapse. The fractional baseline occupancy of dopamine D2 receptors appears far higher than that calculated from amphetamine challenge (0.46 vs 0.05). These discrepancies have been discussed by Strange (2001), who suggested that each technique might be measuring dopamine from different pools, intra- or extra-synaptic. It is persuasively argued that the pool measured by depletion strategies is likely to be largely intrasynaptic because as extrasynaptic levels are low at baseline, lowering these still further will have little effect on the eventual dopamine receptor measure. These are questions to be resolved with *in vitro* studies examining the impact of changes in dopamine flux in living cell cultures.

The Concept of 'Receptor Occupancy': A Role for in Vitro and in Vivo Studies

A further, and probably more salient role for *in vitro* techniques lies in understanding variable results derived from the use of different probes to measure the same receptor population. Laruelle (2000) has critically discussed the whole concept of 'occupancy'. The *in vivo* literature has propagated this term, which loosely means that receptor availability for binding to tracer concentrations of the radioligand has declined. The decline in receptor availability for radioligand binding has, in the past been taken to represent *occupancy* of the receptors either by endogenous ligand or exogenous drug. Kapur et al (2001) suggest this construct is central to understanding mechanisms of drug action in the brain. However, several anomalies have emerged which question what this decline really represents. Several studies have shown that treatment with certain antipsychotic drugs (for example clozapine, risperidone and olanzapine) results in massive blockade of 5HT2A receptor binding as measured with a variety of ligands (Nyberg et al 1993, Travis et al 1998, Kapur et al 1999). The degree of blockade is not clearly dose related, as even very low (and clinically relevant) doses of these drugs produce almost a hundred percent occupancy of the receptor *in vivo*. These data suggest that another process may be occurring to decrease receptor availability for ligand binding, including for example, receptor internalisation inside the cell membrane, or a change in the conformational state of the receptor altering its binding properties. Internalisation of the 5HT2a receptor in living cell culture preparations *in vitro* in response to presence of endogenous or exogenous agonist in the extracellular fluid space has been shown. It is therefore evident that the concept of receptor 'occupancy' should be seen as summarising potentially complex cellular events that may only be disentangled *in vitro*.

A further complication lies in the relative sensitivity or insensitivity of probes to the internalised state of the receptor. Laruelle (2000) showed that probes derived from different molecular classes were more or less sensitive to receptor internalisation. An inconsistency was observed in findings following depletion challenge with different probes all apparently capable of measuring dopamine D2 receptor binding *in vitro*. Some probes (nonbenzamide derivatives, for example spiperone and pimozide), were either unaffected by changes in dopamine release or gave findings in the opposite direction than would be predicted by the simple occupancy model (and were observed with benzamide ligands). Laruelle (2000) argues these differences may be the result of greater or lesser capacity of a probe to continue to bind to the receptor once it is internalised. A great deal of this theory will need to rest on experiments performed *in vitro*, where receptor behaviour and cellular responses to the local environment may be freely manipulated and explored.

Dopamine Receptor Occupancy by Antipsychotic Drugs: In Vivo and in Vitro Research

The history of the dopamine D2 receptor blockade hypothesis of antipsychotic drug action is largely dependent on *in vitro* research performed in the 1970's. A variety of different groups found a direct correlation between the clinical potency of antipsychotic drugs (in terms of their average daily dose to control symptoms in clinical trials), and affinity for dopamine D2 receptors (Creese et al 1976, Peroutka et al 1980). This relationship was not described for an array of other receptor populations including serotonergic, adrenergic, muscarinic and histaminergic (Peroutka et al 1980). These data drove attempts at drug discovery until the 1990's, until it was noted that the antipsychotic drug clozapine, strikingly effective even in the most chronic and severely ill schizophrenic patients had only modest affinity for dopamine D2 receptors. The re-emergence of clozapine, which was beneficial without extrapyramidal side effects lead to the development of a group of drugs loosely described as 'atypical' antipsychotics. *In vivo* receptor imaging studies revealed that clozapine also had modest occupancy of dopamine D2 receptors (approximately 20-60%) at clinically relevant doses (and indeed across its entire dose range), while patients taking classical, high potency antipsychotic drugs showed dose related striatal dopamine D2 receptor occupancy, saturating at nearly 100% occupancy in the higher dose range (Farde et al 1992, Pilowsky et al 1992, Kapur et al 1999). Importantly, the degree of striatal dopamine D2 receptor occupancy was not inevitably associated with good clinical response in living patients studied *in vivo* (Pilowsky et al 1992, 1993). However, the propensity for extrapyramidal or Parkinsonian side effect induction did appear quantitatively related to the degree of striatal D2 dopamine receptor blockade *in vivo*. These studies reveal the utility of the *in vivo* approach in trying to link behavioural and neurochemical effects of a drug at a particular receptor, and in performing dose-ranging studies in the developmental stages of a new CNS drug.

Recently, however, a controversy has arisen as to whether the novel atypical antipsychotic drugs act selectively at limbic cortical D2 receptor populations (with relative sparing of the nigrostriatal dopamine pathway). This could explain their capacity to relieve psychotic symptoms without leading to parkinsonian side effects. Three groups find that the atypical antipsychotic drugs (which have a reduced propensity to cause parkinsonian side effects at clinically useful doses) clozapine and olanzapine show a greater level of blockade at temporal cortical, than striatal dopamine D2 receptors (see figure 1)(Pilowsky et al 1997, Bigliani et al 2000, Xiberas et al 1999, Meltzer et al 2000). One group does not, showing no tendency for clozapine to act selectively at temporal cortical dopamine D2 receptors (Talvik et al 2001). *In vitro* studies are instructive, and also support a

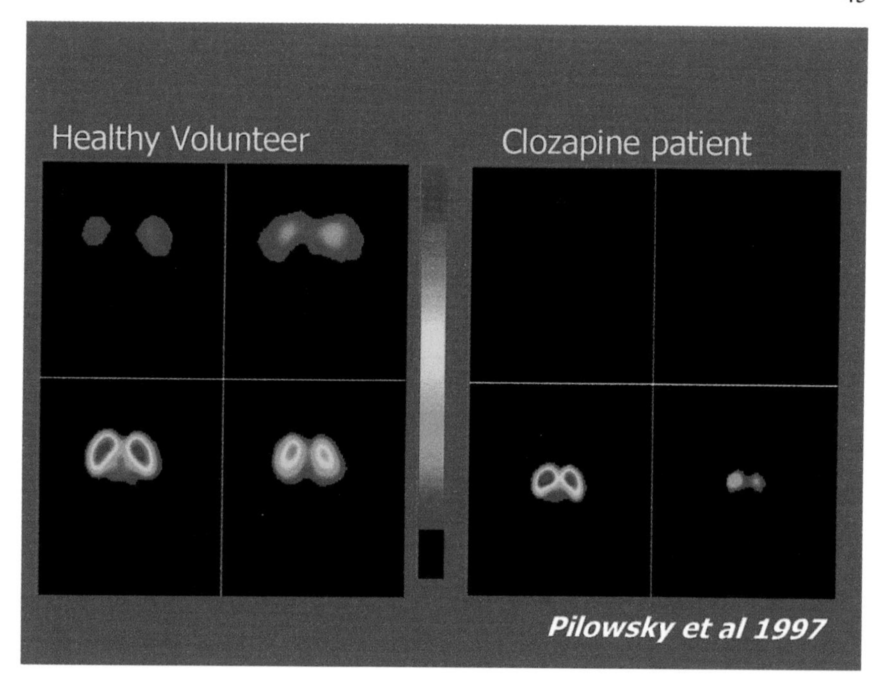

Figure 1: An *in vivo* neurochemical map of [123I] epidepride binding to dopamine D2/'D3 receptors in living humans as detected by single photon emission tomography. The images are in the transverse plane (slice thickness approximately 1cm, resolution 7-9mm). The upper image is at the level of the inferior temporal poles and cerebellum, the lower is at the level of striatum and thalamus. The healthy volunteer has measurable dopamine D2/D3 receptor levels in infero-medial temporal cortex, caudate-putamen and thalamus. The patient on clozapine shows marked blockade of temporal cortical dopamine D2/D3 receptors by the drug, with much less attenuation of the striatal and thalamic signal, indicating selective occupancy of limbic cortical receptors by clozapine.

selective action of the atypical antipsychotic drugs when compared to the standard typical antipsychotic drug, haloperidol (Lidow et al 1998, Janowsky et al 1992). The variable action of these drugs at dopamine sites in the brain has been discussed by Strange (2001), who points out that the cortical dopamine system is differentially specialised for widespread transmission over larger areas than striatal regions. Indeed, D2-like receptors in the cortex are differentially regulated by antipsychotic drugs, as are the neurons themselves. These mechanistic differences may give rise to an apparent selectivity of antipsychotic action and lend themselves to elucidation by *in vitro* methods.

Comparing Postmortem and in Vivo Approaches

The relevance of some of the issues discussed in this chapter (using schizophrenia as a model for the contribution of postmortem and imaging data) has been discussed at a recent US symposium. Contributors to the meeting summarised several important factors in each technique. Postmortem research has incredibly detailed spatial resolution in the micrometer range compared to 5 to 20mm in neuroimaging work. Quantification at the cellular level is only possible for postmortem research, and can alter interpretation of *in vivo* binding studies (Mann JJ 2001). However, *in vivo* studies permit repeated measures and careful sample control to unravel state-trait effects and determine causal relationships. Benes (2001) points out that while postmortem studies can inform understanding of disruptions in neurochemical circuitry, the ideal is a 'coordinated use of antemortem imaging and postmortem microscopy….to 'bridge the great divide' between clinic and bench'. This timely sentiment, if carried into practice will strengthen the explanatory power of both fields and lead to meaningful links between brain and behaviour in many disorders.

References

Benes FM. (2001) The study of neural circuitry in schizophrenia using postmortem and brain imaging approaches. Biol Psychiatry 2001; 49: 3S.

Bigliani V, Mulligan RS, Acton PD, Ohlsen RI, Pike VW, Ell PJ, Gacinovic S, Kerwin RW, Pilowsky LS. Striatal and temporal cortical D2/D3 receptor occupancy by olanzapine- a 123I epidepride single photon emission tomography (SPET) study. Psychopharmacology 2000;150:132-140.

Clow A, Theodoru A, Jenner P, et al. Changes in rat striatal dopamine turnover and receptor activity during one year's neuroleptic administration. Eur J Pharmacology 1980;63: 135-144.

Creese I, Burt DR, Snyder SH. Dopamine receptor binding predicts clinical and pharmacological potencies of antischizophrenic drugs. Science 1976;192:481-483.

Delforge J, Bottlaender M, Pappata S, Loc'h C, Syrota A. Absolute quantification by positron emission tomography of the endogenous ligand. J Cer Blood Flow Metab 2001;21:613-630.

Ichise M, Meyer J, Yonekura Y. An introduction to PET and SPECT neuroreceptor quantification models. J Nuc Med 2001;42:755-763.

Innis RB, Malison RT, Al-Tikriti M, Hoffer PB, Sybirska EH, Seibyl JP, Zoghbi SS. Amphetamine-Stimulated Dopamine Release competes In Vivo for (I}IBZM Binding to the D2 Receptor in Nonhuman Primates. Synapse 1992;10:177-184.

Janowsky A, Neve KA, Kinsie MJ, Taylor B, de Paulis T, Belknap J. Extrastriatal dopamine D2 receptors: distribution, pharmacological characterisation and region specific regulation by clozapine. J Pharm Exp Ther 1992;261:1282-1290.

Kapur S, Zipursky RB, Remington G. Clinical and theoretical implications of 5-HT2 and D2 receptor occupancy of clozapine, risperidone, and olanzapine in schizophrenia. Am J Psychiatry 1999;156:286-93.

Kapur S, Barlow K, Vanderspek SC, Javanmard M, Nobrega JN. Drug induced receptor occupancy: substantial differences in measurements made in vivo vs ex vivo. Psychopharmacology 2001; published online DOI 10.1007/s002130100790.

Kessler RM, Mason NS, Votaw JR, De Paulis TD, Clanton JA., Sib Ansari M, Schmidt DE, Manning RG, Bell RL. Visualisation of extrastriatal dopamine D2 receptors in the human brain. Eur J Pharmacol 1992;223:105-107.

Laruelle M, Abi-Dargham A, van Dyck CH, Gil R, D'Souza CD, Erdos J, McCance E, Rosenblatt W, Fingado C, Zoghbi SS, Baldwin RM, Seibyl JP, Krystal JH, Charney DS, Innis RB. Single photon emission computerized tomography imaging of amphetamine-induced dopamine release in drug-free schizophrenic subjects. Proc Natl Acad Sci U S A 1996;93:9235-40.

Laruelle M. Imaging synaptic neurotransmission with in vivo binding competition techniques- a criticial review, J Cer Blood Flow Metab 2000;20:423-451.

Lee T, Seeman P (1980) Elevation of Brain Neuroleptic/Dopamine receptors in Schizophrenia, Am J Psychiary 1980;137: 191-197.

Mackay AVP, Iversen LL, Rossor M, Spokes E, Bird E, Arregui A, Creese I, Snyder SH. Increased brain dopamine and dopamine receptors in schizophrenia. Arch Gen Psychiat 1982;39:991-997.

Manji HK, Lenox RH. Signaling: cellular insights into the pathophysiology of bipolar disorder. Biol Psychiatry 2000;48:518-30.

Mann JJ. Postmortem versus in vivo imaging: comparison of anatomical and temporal scale. Biol Psychiatry 2001;49: 2S.

Martinot J-L, Peron-Magnan P, Huret J-D et al. Striatal D2 dopaminergic receptors assessed with positron emission tomography and ^{76}Br Bromospiperone in untreated schizophrenic patients. Am J Psychiatry 1990;147:44-50.

Meltzer HY, Park S, Kessler R. Cognition, schizophrenia, and the atypical antipsychotic drugs. Proc Natl Acad Sci USA 1999;96:13591-3.

Moore H, West AR, Grace AA. The regulation of forebrain dopamine transmission: relevance to the pathophysiology and psychopathology of schizophrenia. Biol Psychiatry 1999;46:40-55.

Nordstrom AL, Farde L, Eriksson L, Halldin C. No elevated D2 dopamine receptors in neuroleptic-naive schizophrenic patients revealed by positron emission tomography and [11C]N-methylspiperone. Psych Res 1995;61:67-83.

Nyberg S, Farde L, Eriksson L, Halldin C, Eriksson B. 5HT2 and D2 dopamine receptor occupancy by risperidone in the living human brain. Psychopharmacology 1993;110:265-272.

Peroutka SJ, Snyder SH. Relationship of neuroleptic drug effects at brain dopamine, serotonin, alpha-adrenergic and histaminergic receptors to clinical potency. Am J Psychiatry 1980;137:1518-1522.

Pilowsky LS, Costa DC, Ell PJ, Murray R, Verhoeff N, Kerwin RW. Clozapine, single photon emission tomography and the D2 dopamine receptor blockade hypothesis of schizophrenia. Lancet 1992;340:199-202.

Pilowsky LS, Costa DC, Ell PJ, Verhoeff NPLG, Murray RM, Kerwin RW. D_2 dopamine receptor binding in the basal ganglia of antipsychotic free schizophrenic patients- a 123I IBZM single photon emission tomography (SPET) study. Br J Psychiatry 1994;164:16-26.

Pilowsky LS, Costa DC, Ell PJ, Murray R, Verhoeff N, Kerwin RW (1993) Antipsychotic medication, D2 dopamine receptor blockade and clinical response- a 123I IBZM SPET (single photon emission tomography) study. Psychol Med 1993;23:791-799.

Pilowsky LS, Mulligan R, Acton P, Costa D, Ell P, Kerwin RW. Limbic selectivity of clozapine. Lancet 1997;350:490-491.

Seeman P, Niznik HB, Guan HC. Elevation of dopamine D2 receptors in schizophrenia is underestimated by radioactive raclopride. Arch Gen Psychiatry 1990;47:1170-1172.

Silvestri S, Seeman MV, Negrete JC, Houle S, Shammi CM, Remington GJ, Kapur S, Zipursky RB, Wilson AA, Christensen BK, Seeman P. Increased dopamine D2 receptor binding after long-term treatment with antipsychotics in humans: a clinical PET study. Psychopharmacology (Berl) 2000;152:174-80.

Strange PG. Antipsychotic Drugs: Importance of dopamine receptors for mechanisms of therapeutic actions and side effects, Pharmacol Rev 2001;53:119-133.
Talvik M, Nordstrom AL, Nyberg S, Olsson H, Halldin C, Farde L. No support for regional selectivity in clozapine-treated patients: a PET study with [(11)C]raclopride and [(11)C]FLB 457. Am J Psychiatry 2001;158:926-30.
Travis MJ, Busatto GF, Pilowsky LS, Mulligan R, Acton PD, Gacinovic S, Mertens J, Terriere D, Costa DC, Ell PJ, Kerwin RW. 5-HT2A receptor blockade in patients with schizophrenia treated with risperidone or clozapine. A SPET study using the novel 5-HT2A ligand 123I-5-I-R-91150. Br J Psychiatry 1998;173:236-41.
Wong DF, Wagner HN, Dannals RF, et al. Effects of age on dopamine and serotonin receptors measured by positron tomography in the living human brain. Science 1984; 226:1393-1396.
Wong DF, Wagner Jr. HN, Tune LE et al. Positron emission tomography reveals elevated D2 dopamine receptors in drug-naive schizophrenics. Science 1986;234:1558-1563.
Xiberas X, Martinot JL, Mallet L, Artiges E, Canal M, Loc'h C, Maziere B, Paillere-Martinot ML. In vivo extrastriatal and striatal D2 dopamine receptor blockade by amisulpride in schizophrenia. J Clin Psychopharmacol 2001;21:207-14.

4 GLIAL PATHOLOGY IN MAJOR PSYCHIATRIC DISORDERS

David R. Cotter, Carmine M. Pariante and Grazyna Rajkowska

Abstract

Recent quantitative postmortem investigations of the cerebral cortex have demonstrated cortical glial cell loss and reduced density in subjects with major depression and bipolar disorder. There is also some evidence for a region-specific reductions in glial cell density in schizophrenia. These findings coincide with a re-evaluation of the importance of glial cells in normal cortical function; in addition to their traditional roles in neuronal migration and inflammatory processes, glia are now accepted to have roles in providing trophic support to neurons, neuronal metabolism, and the formation of synapses and neurotransmission. Consequently, reduced cortical glial cell numbers could be responsible for some of the pathological changes in schizophrenia and depression including reduced neuronal size, reduced levels of synaptic proteins, and abnormalities of cortical neurotransmission. Additionally, as astrocytes provide the energy requirements of neurons, deficient astrocyte function could account for aspects of the functional magnetic imaging abnormalities found in these disorders. We discuss the possible basis of glial cell pathology in these disorders and suggest that elevated levels of glucocorticoids, due to illness-related stress or to hyperactivity of the hypothalamic-pituitary-adrenal may down-regulate glial activity and so predispose to, or exacerbate psychiatric illness.

Introduction

'If the functional unit of the brain is not the neuron but rather the neuron-glial complex, then both neuronal and glial cells could be involved in the mental diseases' (Bogerts et al, 1983). This quote from Berhard Bogerts in 1983 was prescient, for during the two subsequent decades glia were viewed as 'passive handmaidens' to neurons and their central role in cortical and neuronal function was under-appreciated (Coyle and Schwarz, 2000). Interest in these cells was largely restricted to astrocytes and their use as markers of proposed inflammatory or degenerative processes in schizophrenia. The absence of gliosis in schizophrenia became a fundamental pillar of the developmental hypothesis (Weinberger, 1987) of schizophrenia and glia were typecast, their crucial role in other cortical functions overlooked.

However, recent quantitative postmortem investigations of the cerebral cortex have now demonstrated reduced cortical glial cell number and density in subjects with major depressive disorder (MDD) (Ongur et al, 1998; Rajkowska et al, 1999; Cotter et al, 2001), manic depressive (or bipolar, BPD illness (Ongur et al, 1998; Rajkowska et al., 2001) and possibly schizophrenia (Rajkowska-Markow et al, 1999c; Cotter et al, 2001). These findings coincide with a re-evaluation of the importance of glial cells in normal cortical function (Coyle and Schwarz, 2000). It is now recognised that glia have important roles in synaptic function (Peters, 1991; Araque et al, 1998; Araque et al, 1999; Ullian et al, 2001), the clearance of extracellular ions (Verkhratsky et al, 1998) and transmitters (Mennerick and Zorumski, 1994) and in neuronal metabolism (Connor and Dragunow, 1998; Magistretti et al, 1999) and migration (Rakic, 1988). Consequently, reduced cortical glial cell numbers could be responsible for some of the pathological changes in schizophrenia and MDD including reduced neuronal size, reduced levels of synaptic proteins, and abnormalities of cortical neurotransmission. Additionally, as astrocytes provide the energy requirements of neurons, deficient astrocyte function could account for aspects of the functional magnetic imaging abnormalities found in these disorders. Intriguingly, there have now been reports showing that cortical glial cell density is increased by neuroleptic medication in primates (Selemon, 1999) and that dopamine D2 receptors which are the primary targets of these medications are found on astroglia (Khan et al., 2001). Moreover, antidepressants and mood stabilizers increase neurogenesis in adult rodent brain (Chen et al., 2000; Malberg et al., 2000) and activate glial proliferation (Rocha et la., 1994, 1998; Levine et al., 2000). These findings suggest that glial cell dysfunction is involved in the pathophysiology of major psychiatric disorders. Furthermore, they indicate that in our search to identify the abnormal cellular cytoarchitecture of the diseases, we should assess not just neuronal or glial pathology in mental diseases, but the neuron-glial complex (Bogerts et al, 1983; Rajkowska, 2000).

In this review we will discuss the functions of the main glial cell populations with particular attention to astrocyte function. We then proceed to review the literature that has investigated glial cell populations in the major psychiatric disorders, i.e., schizophrenia, MDD, and bipolar disorder (BPD). Finally, we highlight the glial cell functions which may be relevant to the neuropathological and neuroimaging changes already described in the major psychiatric disorders, and discuss the potential causes of these changes.

Glial Cell Populations and Their Functions

Recent reviews of glial function (Pfrieger and Barres, 1997; Magistretti et al, 1999) highlight their critical roles in cortical function, and point to a 'gross neglect' of their role in neurobiological features of psychiatric disorders (Coyle and Schwarz, 2000). The glial populations of the central nervous system (CNS) are comprised of functionally distinct cells termed astrocytes, oligodendrocytes and microglia (Peters, 1991). Astrocytes are the predominant glial cell type and have many functions that could be relevant to abnormalities described in schizophrenia and mood disorders. For these reasons we will concentrate on this glial cell population in this overview, with particular attention to those astrocytic functions which may be relevant to psychiatric diseases. However, very recently oligodendroglial and microglial abnormalities have also been described in schizophrenia. For overviews of the functions of these populations readers are referred to previous reviews (Pfrieger and Barres, 1996).

Astrocytes

Virchow originally described astrocytes in 1859 as 'nerve glue'. They have numerous critical functions in the CNS, which include ensheathing areas of neurons that are not covered by oligodendrocytes, such as the nodes of Ranvier, neuronal cell bodies and synapses (Peters, 1991). A number of the most important functions which may be relevant to psychiatric disorder are outlined below.

Energy Requirements of Neurons and Astrocytes

Astrocytes have end-feet that are enriched in glucose transporters. Through these the energy for glial and for neuronal function is supplied, coupled with both glutamate and Na^+ uptake (Araque et al, 1999). Glycolytic processing of this glucose provides the energy for the conversion of glutamate to glutamine within astrocytes, and for the maintenance of the Na^+ gradient through Na^+/K^+ ATPase activity. Lactate is then released by glia and enters neurons to join the tricarboxylic acid cycle to yield ATP (Tsacopoulos and Magistretti, 1996). These activities together provide the observed signal in functional magnetic resonance imaging studies and positron emission tomography investigations (Magistretti et al, 1999).

Synaptic Function

The close association of astrocytes and synapses reflects their important role in synaptic function (Araque et al, 1999; Pfrieger and Barres, 1996). Astrocytes promote synapse formation *in vitro*, and are involved in the development and remodelling of synaptic connections (Pfrieger and Barres, 1997; Ullian et al, 2001). Through specific transporters systems they regulate the availability and uptake of neurotransmitters at synapses (Verkhratasky et al, 1998). Also, through glutamate uptake, they modulate the duration of the synaptic current and can protect from potential excitotoxic damage (Araque et al 1998). Furthermore, through control of astrocytic Ca^{++} levels astrocytes, which are connected through gap junctions, can regulate the spread of modulatory signals to neighbouring synapses and so modify synaptic transmission and plasticity.

Glutamatergic Functions

Astrocytes express transporters critical for glial synaptic uptake of glutamate and neuroprotection (Tanaka et al, 1997). This glutamate uptake by astrocytes terminates the postsynaptic action of glutamate, reduces the levels of extracellular glutamate, and protects neurons from cell death in mixed cultures (Rothstein et al, 1996). Additionally, mice with deletion of the glutamate transporter gene, GLT-I, show evidence of excitotoxic damage, demonstrating further the protective role of this transport process (Tanaka et al, 1997). A further glutamate-related function of astrocytes is the conversion of glutamate to glutamine through the activity of the enzyme glutamine synthetase (GS). Glutamine is then released and taken up by the neuronal terminals where it is reconverted to glutamate, and to GABA to replenish the neurotransmitter pool (Pfrieger and Barres, 1997).

NMDA Receptor Co-Activation

Astrocytes in the cortical grey matter contain the enzyme serine racemase which is responsible for the conversion of L-serine to D-serine (Wolosker et al, 1999). As coactivation by D-serine at the 'glycine site' is required for NMDA receptors activity (Johnson and Ascher, 1987), astrocytes may regulate NMDA receptor activity through regulating the amount of available D-serine (Wolosker et al, 1999). The importance of D-serine is underlined by its localisation to astrocyte foot processes in close contact with neurons in the synaptic cleft. Additionally, D-serine is more than three times more potent an NMDA receptor agonist than glycine (Miyazaki et al, 1999), suggesting that it is an important endogenous ligand for the glycine site of NMDA receptors. As only astrocytes contain serine racemase and as D-serine is present in similar concentrations to glycine in the prefrontal cortex, a crucial role for astrocytes in NMDA receptor activation is likely (Wolosker et al, 1999). Furthermore, the glial transporter GLT-1 may regulate concentrations of glycine at the synapse and so further influence NMDA activation (Bergeron et al, 1998). Other potential glial influences on NMDA receptor activation include agonism through quinolate (Foster et al, 1984, Speciale and Schwarz, 1993) and antagonism through N-acetyl-aspartyl glutamate (Carter et al, 1998)and kynurenate (Fletcher et al, 1997).

Neurotrophic Functions

Astrocytes synthesise and release many neurotrophic factors crucial to neuronal health (Connor and Dragunow, 1998). They also specifically release a neurotrophic factor, glial derived neurotrophic factor, which has a local effect in increasing synaptic plasticity and synaptic efficiency (Lindsay et al, 1995; Erickson, J.T., 2001). Interestingly BDNF, has recently been found to mediate at least some of the efficacy of antidepressant treatments. BDNF levels are raised by antidepressant medications (Duman, 1997) and intracerebral injection of BDNF in mice has antidepressant effects as evaluated using 'learned helplessness' paradigms (Siuiciak et al, 1996). Stressors such as immobilisation (Duman, 1997), or systemic injections of glucocorticoids (Schaff, 1997), which induce depression cause reduced BDNF mRNA, and this effect can be prevented by antidepressants drugs (Nibuya et al, 1995). BDNF and neurotrophin-3 have dramatic effects on serotonergic (5HT) neuron function, growth and regeneration, and therefore support the proposed link between these neurotrophins and psychiatric disorders (Mamounas et al, 1995).

Astrocytes also produce estrogen following injury by the activity of the enzyme aromatase which regulates the production of estrogen from precursors and increases local estrogen levels (Garcia-Segura et al, 1994). As estrogens contribute to the maintenance and recovery of normal brain function, and promote synaptic plasticity in the adult brain (Garcia-Segura et al, 1999), glial cell deficits could result in impaired estrogen mediated neuroprotection.

Inflammation

Astrocytes are implicated in neurodegenerative diseases and inflammatory processes and regeneration. In response to injury they become activated and increase their number, size and change their pattern of gene expression (Peters et al, 1991), such that the glial protein GFAP is greatly upregulated (Lapping et al, 1994). Activated glia may have roles in both inhibiting and promoting damage (Jordon, 1999). Activated astrocytes also release neurotrophic factors in response to injury and these promote neuronal survival and repair (Evans and Golden, 1987).

The Evidence for Glial Cell Loss in Major Psychiatric Disorder

Neuroanatomically, macroscopic investigations of schizophrenia, bipolar disorder (BPD) and major depressive disorder (MDD) show that there are similarities in their brain pathology, with some of the main macroscopic differences being ones of degree (Steffens and Krishnan, 1998). For example, ventricular dilatation and reduced hippocampal and neocortical volumes are seen in schizophrenia (Harrison, 1999), but they are also present to a lesser degree in MDD and BPD (McCarley et al, 1999, Steffens and Krishnan, 1998, Coffey, 1993). Similarly, microscopic investigations of schizophrenia point to abnormalities of neuronal cytoarchitecture (Harrison, 1999), with evidence for reductions in neuronal size (Rajkowska, 1998), dendritic spine density (Glantz and Lewis, 2000; Rosoklija et al, 2000), dendritic length (Glantz and Lewis, 2000) and synaptic proteins (Glantz and Lewis, 1997; Harrison and Eastwood, 1998; Honer et al, 1999; Perrone-Bizzozero et al, 1996). Such changes are mirrored to some degree in mood disorders where reduced dendritic spine density (Rosoklija et al, 2000) and reduced neuronal size (Cotter et al, 2001;Rajkowska et al, 1999) have been described. Additionally, among psychiatric control subjects, the majority of whom suffered from major depression, there is evidence for reduced dendritic length and some evidence for reduced spine density (Glantz and Lewis, 2000). More recently it has become apparent that glial cell loss and reduced density may also be feature of both disorders (Benes et al, 1986; 1991; Rajkowska-Markow et al, 1999c; Rajkowska et al, 1999a; Cotter et al, 2001). This similar pattern of changes in cortical cellular architecture in schizophrenia and MDD suggests that a common pathophysiology may underlie aspects of these psychiatric diseases. On the other hand, differences in the pattern of glial and neuronal cell pathology are also found between mood disorders and schizophrenic subjects in specific brain regions. For example, dorsolateral prefrontal cortex is reported to have unchanged or trend increases in glial densities and increased neuronal density in schizophrenia (Selemon et al., 1995, 1998), whereas the same region in major depression and bipolar disorder are characterized by prominent reductions in glial and neuronal cell density (Rajkowska et al., 1999b; 2001a). Thus, the regional differences in the pattern of cell pathology may account for unique clinical symptoms and treatments characterizing depression and schizophrenia.

Two tables are presented in which studies assessing glial cell populations in psychiatric disorder are summarised. Table 1, summarises data on investigations using stains, such as cresyl violet, that labels the Nissl substance of neurons and glia. Table 2, summarises data on investigations using other methods, largely immunohistochemistry, to identify and quantify glial populations in psychiatric disorders. While most of the data relates to schizophrenia, many of the more recent investigations have also assessed other psychiatric subgroups such as those of major depressive disorder and bipolar disorder, and have also utilised stereological methods. Where data is available on accompanying neuronal populations, this data was also presented. In the following paragraphs the data is summarised separately for mood disorder and for schizophrenia.

Mood Disorders (MDD and BPD)

The lack of quantitative investigations characterising the cortical glial and neuronal populations in mood disorder is surprising considering its importance and prevalence. In MDD and BPD there have been few stereological investigations of glial populations, yet despite this, the findings have been consistent, for each has found significant glial cell loss or reduced densities in the frontal cortex in MDD and BPD (for further details see also chapter by Rajkowska in this book).

Rajkowska et al (1999b) found decreased glial cell density in layers 3 and 5 of the dorsolateral prefrontal cortex and the deeper layers of the caudal orbitofrontal cortex in MDD. Similar reductions in glial cell density were recently detected in layer 3 of the dorsolateral prefrontal cortex in BPD subjects (Rajkowska et al, 2001a). These reductions were coupled with enlargement and changes in shape of glial nuclei which spanned multiple layers. These glial changes were accompanied by reduced mean neuronal size, and reduced density of large neurons, with no changes in mean neuronal density found in MDD (Rajkowska et al., 1999) and reductions in the density of large pyramidal neurons observed in BPD (Rajkowska et al., 2001a). Similarly, Ongur et al (1998) found reduced glial cell density in MDD in the subgenual part of the anterior cingulate cortex (ACC), and unchanged neuronal density and neuronal size. BPD subjects who had a family history of the disorder showed the same changes. Our own study of the supracallosal part of the anterior cingulate also found glial cell loss, unchanged mean neuronal density, but reduced neuronal size (Cotter et al, 2001). Another recent study found the density of oligodendroglial cells to be reduced in layer 6 of the prefrontal cortex of subjects with MDD (Orlovskaya et al, 2000). This reduction was not specific to MDD as it was also found in BPD and schizophrenia. Whether oligodendroglia in other cortical layers may also have been affected is not know for only layer 6 was assessed.

There are some important differences between these studies which helps to explain some of the divergent results. The study of Ongur et al (1998) although based on the same sample as our own, differed from ours, and that of Rajkowska et al (1999), in finding no reduction in neuronal size in MDD. There are several possible reasons for these differences. Firstly, Ongur and colleagues, (1998) assessed subgenual ACC while we assessed supracallosal ACC. It is possible that these proximate regions exhibit different vulnerabilities to cytoarchitectural changes in MDD. Additionally, Ongur and colleagues, (1998) sampled throughout the thickness of cortical width, rather than the sampling within individual cortical layers as we undertook. In contrast, our sampling, like that of Rajkowska et al (1999) allowed us to assess individual cortical layers, and to detect laminar specific changes. Another difference lies in the division, by Ongur and colleagues (1998) of the mood disordered groups into familial and non-familial cases, and the finding that glial cell loss was most prominent in familial MDD and BPD groups. In our investigation we found no evidence for any changes in microscopic neuroanatomy in BPD and we did not divide our patient group according to family history. However, Rajkowska and colleagues found glial and neuronal reductions in both MDD and BPD and in these studies the patients groups were also not divided according to family history. Nevertheless, all four studies utilising stereological methods and well-characterised and well-matched samples have found firm evidence for glial loss and reduced density in the frontal cortex in mood disorders (for further discussion on cell loss vs. cell atrophy see chapter by Rajkowska in this book).

Table 1: Summary of studies of major psychiatric disorders identifying glia by stain of Nissl substance

Authors	Disease	Cortical region	Stain	Glial density	Glial 'size'	Neuron density	Neuron size	Comments
Orlovskaya et al (2000)	Scz MDD BPD	PreF A9	Nissl	↓L6 ↓L6 ↓L6				Stereological methods Only layer 6 assessed and only oligodendroglia counted
Cotter et al 2001	Scz MDD BPD	ACC A24 (supra-callosal)	Nissl	↓ L6 ↓ L6 nc		nc nc nc	nc ↓L6 nc	Stereological methods
Rajkowska et al (2001)	BPD	DLPFC A9	Nissl	↓L3c, 5	↑L1, 3,5	↓L3, 5a		Stereological methods. Reduced density of glial cells and pyramidal neurons in specific cortical layers but not when all layers are combined
Rajkowska et al (1999a/c)	Scz	OFC(r)A10	Nissl	↓L2-5	↓	↓L1	↓L2-4	Stereological methods. Glial deficit in OFC suggested to relate to depressive symptoms
Rajkowska et al (1999b)	MDD	OFC(r)A10 OFC(c)A47 dlPFC A9	Nissl	nc ↓L3c,4-6 ↓L3b,5	nc nc ↑L3a	nc nc nc	↓L2 ↓L5,6 ↓L3,6	Stereological methods. Reduced density of large neurons L5-6 OFC(c), and L2,3,6 dlPFC
Ongur et al (1998)	Scz MDD BPD	ACC A24 (sub-genual)	Nissl	nc ↓ nc	nc nc nc	nc nc nc	nc nc nc	Stereological methods. No overall neuronal loss. Fewer large neurons in schizophrenia
Rajkowska et al (1998)	Scz	PreF A9 Occ A17	Nissl		nc nc		↓ L3c nc	Stereological methods. Reduced density of large neurons L3c paralleled by increased density of small neurons L2-6
Selemon et al (1998)	Scz	BA46	Nissl	nc		↑L2-4,6		Stereological methods

Table 1: continued

Authors	Disease	Cortical region	Stain	Glial density	Glial 'size'	Neuron density	Neuron size	Comments
Selemon et al (1995)	Scz	PreF A9	Nissl	nc		↑L3-6		Stereological methods.
		Occ A17		nc		↑		Neuronal density increases most marked L5 of Scz PreF
	Scz-Aff	PreF A9		nc		nc		
		Occ A17		nc		nc		
Benes et al (1991)	Scz	ACC A 24	Nissl	nc		nc		In Scz there is ↓density of small, non-pyramidal neurons in PreF. In Scz and Scz-Aff there is ↑ density of pyramidal neurons in L5 PreF
		PreF A10		nc		↓L2,↑L5		
	Scz-Aff	ACC A 24		↓L6 (25%)		↓L2-6		
		PreF A10		nc		↓L2		
Pakkenberg et al (1990)		MDT	GC/Nissl	↓		↓		Stereological methods. Glial loss from both astrocytes and oligodendroglia~40% of each population lost in MDT & NA
		NA		↓		↓		
		VP		nc		nc		
		BLA		nc		nc		
Benes et al (1986)	Scz	ACC A 24	Nissl	nc ↓14%		↓L5	nc	Presence of ↓ glial density in all regions in Scz noted, though to trend levels, except in motor cortex.
		PreF A10		nc ↓14%		↓L6	nc	
		PM A4		↓L3 ↓23%		↓L3	nc	
Falkai et al (1988)	Scz	EC	Nissl	nc		↓number		↓neuronal number
Falkai et al (1986)	Scz	Hipp	Nissl	↓ number		↓number		↓ glial number ↓ neuronal number (pyramidal cells)

The abbreviations listed in the table are as follows; cortical regions: ACC, anterior cingulate; EC, entorhinal cortex; PreF, prefrontal; OFC(r/c), orbitofrontal (rostral or caudal);dlPFC, dorsolateral prefrontal; Occ, occipital; MDT, mediodorsal thalamus; PM, primary motor; NA, nucleus accumbens; VP, ventral pallidum; EC, entorhinal; BLA, basolateral nucleus amygdala; MDD, major depressive disorder; Scz, schizophrenia; SczAff, schizoaffective disorder; nc, no change; L, layer; GC - Gallocyanin-chromalum.

Table 2: Summary of studies of major psychiatric disorders identifying glia by immunohistochemistry or by other techniques

Authors	Disease	Cortical region	Glial stain / marker	Count/ Level	Comment
Miguel-Hidalgo et al (2001)	MDD	PreF-A9	GFAP	↓	No overall differences in the density, or areal fraction of GFAP labelled cells, but reduced astroglial numbers in young subjects with MDD vs age matched controls
Radewicz et al (2000)	Scz	PreF-A9, STG, ACC	GFAP	nc	Activated microglia (for which HLA-DR is a marker) are increased in schizophrenia
		PreF-A9, STG,	HLA-DR	↑	
		ACC	HLA-DR	nc	
Webster et al (2000)	Scz	PreF A10	GFAP	↓	4 different phosphorylated forms of GFAP assessed.
	MDD			↓	
	BPD			↓	
Falkai et al(1999)	Scz	EC/Subiculum	GFAP	nc	
		White Matter	GFAP	nc	
Honer et al (1999)	Scz	Anterior Frontal	MBP	↓level	MBP a marker of oligodendroglia
	MDD		MBP	↓level	Reductions most marked in suicide
Orlovskaya et al (1999)	Scz & BPD	PreF A10 & Striatum	E.M.	↓	Dystrophic changes to oligodendroglial cells in Scz, in BPD oligodendroglial size reduced
Uranova et al (1999)	Scz & BPD	PreF A10 & Striatum	E.M.	↓	Myelin sheath damage in Scz only, not BPD
Bayer et al (1999)		Frontal & hippocampus	HLA-DR	↑	Activated microglia increased
Arnold et al (1996)	Scz	Hippocampus	GFAP	nc	In general, demented Scz subjects show ↑GFAP, non-demented show ↓GFAP
		EC, PreF, Occ	"	nc	

Table 2: continued

Authors	Disease	Cortical region	Glial stain / marker	Count/ Level	Comment
Johnston-Wilson et al (2000)	Scz MDD BPD	PreF A10	GFAP GFAP GFAP	↓ ↓ ↓	Proteomics used to identify 4 different forms of GFAP; 3↓ in MDD, 2↓ in Scz, 1↓ in BPD
Bruton et al (1990)	Scz	Grey, whitematter & periventricular	Holzer	nc	No gliosis in neuropathologically purified group, semi-quantitative
Crow et al (1989)	Scz	Grey, whitematter & periventricular	Holzer DBII	nc nc	No gliosis, semiquantitative DBII(diazepam binding inhibitor immunoreactivity) marker of gliosis
Stevens et al (1988)	Scz	Periventricular grey & ACC white matter	GFAP	nc	No evidence for gliosis
Owen et al (1987)	Scz	Frontal Temp	MAO-B	↓ ↓	MAO-B a marker of astrocytes - deficit suggested to reflect a disturbance of glial function
Roberts et al (1987)	Scz Aff	Multiple regions "	GFAP "	nc nc	Densitometric assessment only - no data on numbers, size of astrocytes
Roberts et al (1986)	Scz	Multiple regions	GFAP	nc	No gliosis or densitometric change in any brain region
Stevens (1982)	Scz	Multiple regions & Periventricular grey	Holzer stain	↑	Gliosis most marked in periventricular grey

The abbreviations listed in the table are as follows; cortical regions: ACC, anterior cingulate; PreF, prefrontal; STG, superior temporal gyrus; EC, entorhinal; Occ, occipital; HLA-DR, human leucocyte antigen; MBP, myelin basic protein; MDD, major depressive disorder; Scz, schizophrenia; BPD, bipolar disorder; Aff, affective disorder; nc, no change; L, layer; GFAP, glial fibrillary acidic protein; E.M., electron microscopy.

In summary, the above findings indicate that glial cell loss and reduced densities clearly characterize the histopathology of mood disorders. The glial cells analyzed in the above mentioned studies do not represent a homogenous population. They are comprised of distinct populations of oligodendrocytes, microglia and astrocytes, whose crucial role in brain function is being recently re-evaluated. These three distinct glial cell types cannot be identified in any of the above mentioned studies since they were conducted on tissue stained for Nissl substance. This staining does not distinguish between specific glial cell types as Nissl staining only visualizes morphological features of glial cell bodies and not glial cells processes. On the other hand, recent immunohistochemical examination of astroglial marker glial fibrillary acidic protein (GFAP) in the dorsolateral prefrontal cortex suggests the involvement of astroglia in overall glial pathology in MDD. Although no significant group differences in GFAP-reactive astroglial cell density were present overall between MDD and controls, marked reductions in the population of astroglial cells in the subgroup of young MDD subjects as compared to young controls and older MDD subjects were detected (Miguel-Hidalgo et la, 2000). Moreover, a significantpositive correlation between the number of immunopositive-astroglial cells and the density of glia with large cell body size (revealed by Nissl staining) was found in this study suggesting that the reductions in the population of astroglial cells in the subgroup of young MDD subjects account, at least in part, for the global glial deficit identified in this disorder. Alterations in GFAP levels in both BPD and MDD are also suggested by a recent proteomic study where four out of eight proteins displaying disease-specific alterations are forms of glial fibrillary acidic protein (GFAP, Johnstone-Wilson et al., 2000). Johnstone-Wilson and colleagues (2000), also demonstrated that reduced level of different forms of GFAP was not specific to mood disorders as it was also present in schizophrenia. It is not possible from this investigation to tell whether these reductions in GFAP protein reflects reduced cell numbers or diminished cell activation. However, since glial loss and reduced density have already been found in tissue samples from the same brain collection (Stanley Foundation Brain Consortium) highlighted above (Ongur et al, 1998; Orlovskaya et al, 2000; Cotter et al, 2001), the former is the likely explanation.

Schizophrenia

The literature on glial cell investigations of schizophrenia (see Tables 1&2) is one that has focused historically on ruling out a glial cell excess in the disorder, and not on identifying glial loss in the disorder. It was reasoned that cortical gliosis was evidence of cortical inflammation or degeneration and that, if present, excess gliosis in the brains of schizophrenic subjects could be taken as evidence of inflammation, or of a degenerative process, subsequent to or during the perinatal period. A caveat was that inflammation occurring prior to the third trimester of pregnancy would not produce cortical gliosis (for review, see Harrison, 1999). Thus, studies showing an absence of excess gliosis in the cortex in schizophrenia were taken as supportive of the developmental hypothesis on the basis that a degenerative process was believed to be out-ruled. Furthermore, they left open the possibility that second trimester processes had gone awry leading possibly to developmental brain changes. There are several good reasons to question this paradigm and these have been discussed in detail previously (Arnold and Trojanowski, 1996; Harrison, 1999). The consensus opinion is now that there is no increased gliosis within the cortex in schizophrenia, other perhaps than elderly demented subjects. However, the important point from the viewpoint of this review is that reduced glial cell density was not sought and the potential significance of the glial cell deficit was largely, though not completely (Bogerts, 1983, Owen et al, 1987) overlooked when it was noted.

Thus, although reported initially (Stevens et al, 1982; Bruton et al, 1990), gliosis is not now accepted as a characteristic feature of the histopathology of schizophrenia. In our own investigations in the anterior cingulate cortex in schizophrenia we found an estimated reduction in glial cell density of 15-20% and an estimated neuronal size reduction of 10%-15% in schizophrenia in layers 5 and 6, but found neuronal density unchanged (Cotter et al., 2001. Glial density is reported to be reduced in a number of other cortical regions; the orbitofrontal (Rajkowska-Markow, 1999) and the primary motor cortices (Benes et al, 1986), with trend reductions in the prefrontal (Benes et al, 1986)and anterior cingulate cortices (Benes et al, 1986; Benes et al, 1991). There are also reports of dystrophic alterations and reduced density of oligodendroglia (Orlovskaya et al, 2000) and layer-specific reductions in the fraction of GFAP-positive astroglia (Rajkowska et al., 2001b) in the dorsolateral prefrontal cortex in Brodmann area (BA) 9. However, in schizophrenia findings of reduced glial cell density are not consistent, for others have not found changes in glial density in either prefrontal BA 9 (Selemon et al, 1995; Rajkowska-Markow, 1999c), the adjacent prefrontal BA 10 (Benes et al, 1991), prefrontal BA 46 (Selemon et al 1998), the occipital cortex (Selemon et al, 1995; Rajkowska et al., 1998), or the ACC (Ongur et al, 1998). Similarly, astrocytes, labelled using GFAP are present in normal numbers in most studies (Arnold et al, 1996; Falkai et al, 1999, Radewicz et al, 2000;Roberts et al, 1986; 1987; Stevens et al, 1988), elevated in some (Stevens et al, 1982) and reduced in others (Webster et al, 2000). As discussed earlier, a recent study by Johnston-Wilson and colleagues (2000) found levels of GFAP to be reduced in the prefrontal cortex in schizophrenia, suggesting decreased astrocytic activation or cell numbers. Owen and colleagues (1987), found reduced levels of the enzyme monoamine oxidase (MAO-B) in astrocytes, in the frontal and temporal cortices of subjects with schizophrenia. It was suggested at the time that this deficit may reflect a disturbance of glial function in schizophrenia. Finally, two recent investigations have shown reductions in class II human leucocyte antigen (HLA-DR), a marker for activated microglia, in the frontal and superior temporal cortices in schizophrenia (Bayer et al, 1999; Radewicz et al, 2000). The basis of these changes is not yet known.

The overall picture is not a clear one. One proposal is that the overall lack of glial cell loss in the dorsolateral prefrontal cortex, despite its presence in the orbitofrontal cortex, indicates a region-specific distribution of glial cell abnormalities in schizophrenia and MDD (Rajkowska-Markow et al, 1999). It is also possible that specific, as yet unidentified, subgroups of subjects with schizophrenia may demonstrate more marked glial loss. Indeed, there is some evidence that glial loss in schizophrenia may confer a particular vulnerability to depressive mood. Schizophrenic subjects with affective symptoms may be more likely to show glial loss in the anterior cingulate cortex than those who do not (Benes et al, 1991), although this does not hold for the prefrontal or the occipital cortices (Selemon et al, 1995). Furthermore, myelin basic protein, a marker of the oligodendrocytic glial population, is reduced exclusively in the prefrontal cortex of MDD and schizophrenic subjects who died by suicide, but not in those who died from other causes (Honer et al, 1999). This lends further support to the suggestion that affective symptoms are associated with glial loss in schizophrenia, and may explain the inconsistent results of studies investigating glial cell pathology in schizophrenia. This is because depressive symptoms are common in schizophrenia and can be difficult to diagnose with certainty in many subjects. Consequently inconsistencies in the results of studies could reflect differences in the clinical characteristics of the sample, such as prevalence of depressive disorder, of which the investigators were unaware.

Reduced Glial Cell Density and Numbers: The Functional Consequences

As summarised above, the neuropathological data presented above provide evidence for glial loss and reduced glial cell density in both mood disorders and to some extent in schizophrenia. From our knowledge of glial functions these losses should have important functional consequences, which in turn may contribute to the clinical picture of these disorders.

One obvious astrocytic function which could be relevant to psychiatric disorders is the regulation of the glutamatergic system. Astrocytes are responsible for synaptic glutamate uptake, the glutamate-glutamine shift, maintenance of the neurotransmitter pool, and NMDA receptor agonism (through D-serine) (Verkhratasky et al, 1998; Pfrieger and Barres, 1997; Wolosker et al, 1999). Deficits in glutamatergic neurotransmission in schizophrenia, involving initial glutamatergic excititoxicity and subsequent NMDA receptor hypofunction, such as proposed by Olney and Farber (1995) and Benes (2000), are in keeping with the functional consequences of reduced astrocyte activity or numbers. Another relevant astrocytic function relates to their role in providing energy to neurons. This function can be monitored by PET and fMRI investigations which are based on the summed activities of glucose uptake, glycolysis and oxidative phosphorylation (Magistretti et al, 1999). Thus, the hypometabolism demonstrated in the prefrontal and anterior cingulate cortex in schizophrenia (Weinberger et al, 1986; Andreasen et al, 1992, Tamminga et al, 1992) and MDD (Baxter et al, 1989; Ebert and Ebmeier, 1996; Mayberg, 2000) may reflect glial cell loss or reduced astrocytic activation. Reduced metabolic support for neurons due to glial cell loss would also result in neurons being more vulnerable to excitotoxic damage, and less capable of supporting the metabolic demands of an extensive dendritic arboristion and a normal cell size. Similar changes could also result from the withdrawal of the normal trophic support provided by astrocytes and microglia through release of neurotrophic factors which are important in maintaining normal synaptic function and neuronal health generally (Connor and Dragunow, 1998).

Thus, disruption of one or several astrocytic functions could account for the functional and structural neuronal pathology observed in mood disorders and schizophrenia. Likewise, it is possible that at least some of the features of major psychiatric illness are due to glial cell deficits, although exactly which features or constellations of symptoms is unknown. Depressive symptomatology is an obvious candidate as it is a feature of most psychiatric disorders, and as mentioned previously, there is evidence that schizophrenia subjects with affective symptoms have the most marked glial cell changes.

Reduced Glial Cell Density and Numbers: The Causes

A Medication Effect?

Could the glial cell loss described in MDD, BPD and schizophrenia merely be a consequence of pharmacological treatment of the disorder? Unfortunately, few studies have investigated the effect of psychotropic agents on neuronal and glial cell density. The literature that is available indicates that pharmacological treatments may modulate glial cell densities. For example, a recent investigation in the monkey by Selemon and colleagues (1999), although small in sample size, suggests that chronic exposure to neuroleptics increases glial but not neuronal density in the primate prefrontal cortex. Interestingly, another recent study has discovered that dopamine D2 receptors are localized in astrocytes with a selective anatomical relationship to interneurons suggesting a new neuron/glia substrate for action of antipsychotic drugs (Khan et al., 2001). From a theoretical standpoint this could have a beneficial influence on neurons: increasing glial cell density would ensure better provision of neuronal energy needs, trophic support, augment glutamatergic neurotransmission and enhance synaptic function. The evidence for antidepressants and mood stabilisers are less clear. Lithium treatment is associated with increased gliosis in the, hippocampus (Rocha and Rodnight, 1994; Rocha et al, 1998) and the neural lobe of the pituitary (Levine et al., 2000) and antidepressants have been shown to reduce survival of hippocampal cells, an effect probably mediated by oxidative stress (Post et al, 2000). Moreover, Pariante et al. (1997) have demonstrated *in vitro* that the tricyclic antidepressant, desipramine, activates the glucocorticoid receptor translocation in mouse fibroblasts, a finding recently reproduced by Okugawa et al. (1999) in rat neurones. As discussed in the next paragraph, increased levels of glucocorticoids, whether due to stress or exogenous administration, induce neuronal atrophy (Sapolsky, 1994; Sapolsky, 2000), decrease dendritic spine density (Brown et al, 1999, McEwen, 1999) and reduced glial activation in animal brain (Crossin e al, 1997), and that this effect is very similar to the neuropathological abnormalities described in patients with depression and schizophrenia. Therefore, it is possible that the prolonged increase in glucocorticoid receptor function induced by antidepressants in both neurons and glial cells within glucocorticoid-sensitive brain areas, will result in enhanced intracellular effects of glucocorticoids that will affect the volume and/or number of these cells.

A Role for Glucocorticoids in Neurodegeneration and Glial Cell Inactivation?

Common features described in MDD and schizophrenia are hippocampal and frontal brain volume reductions (Steffans and Krishnan, 1998), and microscopically, the reduction of neuronal size (Cotter et al, 2001; Harrison, 1999; Rajkowska et al, 1999a; 1999b), decreased dendritic arborisation (Glantz and Lewis, 2000, Rosaklija et al, 2000) and some evidence for reduced glial cell density (Benes et al, 1986; Cotter et al, 2001; Orlovskaya et al 2000; Rajkowska-Markow et al, 1999c). Could these shared abnormalities be the consequence of a shared degenerative process? Glucocorticoid induced neurotoxicity is associated with reduced neuronal volume and reduced dendritic arborisation, and it is a mechanism that has been implicated in the neuropathology of MDD (Duman et al, 1997, Rajkowska, 2000). Elevated glucocorticoid levels are also associated with hippocampal volume reductions in MDD, post traumatic stress disorder, Cushing's Disease and normal aging. It is a challenging hypothesis that some of the neuropathological and macroscopic brain changes seen in subjects with MDD, but also in those with BPD and schizophrenia could be a consequence of glucocorticoid mediated neurotoxicity. This could be due either to a continuous hypersecretion of cortisol as part of the pathogenesis of the illness, or to stress-induced elevation of glucocorticoids during the acute phase of the illness, or to both. We do not suggest that schizophrenia, MDD and potentially BPD share the same primary pathophysiology. Rather, we are questioning whether there may be a secondary pathophysiological mechanism, involving stress-induced glucocorticoid mediate effects, which result in the some of the common features of these illnesses. Depressive symptoms may be the clinical manifestation of these changes.

Hyperactivity of the hypothalamic pituitary adrenal (HPA) axis in patients with MDD is a consistent finding. Specifically, patients with major depression have been shown to exhibit increased concentrations of cortisol in plasma, urine and cerebrospinal fluid, increased cerebrospinal fluid corticotropin releasing factor (CRF) levels, and an enlargement of both the pituitary and adrenal glands (Nemeroff, 1996). Although the mechanism by which the HPA axis is activated in depression has not been resolved, one possible pathway is through altered feedback inhibition by endogenous glucocorticoids. Data supporting the notion that glucocorticoid-mediated feedback inhibition is impaired in MDD comes from a multitude of studies demonstrating nonsuppression of cortisol secretion following administration of the synthetic glucocorticoid, dexamethasone (Evans and Golden, 1987; Nemeroff, 1996). Hyperactivity of the HPA axis is likely to be part of the pathogenesis of depressive symptoms. There is consistent evidence supporting the notion that CRF levels are elevated in the brain of depressed patients (Nemeroff et al, 1996). Moreover, animal studies have found that CRF induces behavioral responses which model depressive symptoms (Nemeroff et al, 1996). Depression is also a recognized feature of hypercortisolisemic states such as adrenal carcinoma or Cushing's syndrome, and lowering of plasma cortisol levels reduces the severity of depression in these patients (Checkley, 1996). Moreover, both in animals and in humans, elevated glucocorticoids levels are associated with neuropsychological changes, such as impairments in memory, that are similar to those described in depressed patients (McAllister-Williams et al, 1998).

There is however, no firm evidence that HPA hyperactivity may be part of the pathogenesis of schizophrenia. The dexamethasone nonsuppression rate in schizophrenia varies from 0 to 70 %, with a mean rate of approximately 20%. The range appears to reflect the type of patient, the activity of the patient, the presence of associated symptoms of depression and the effect of hospitalization (Evans and Golden, 1987). Moreover, dexamethasone non-suppression has been associated with the presence of negative symptoms, and persistent non-suppression has been associated with a poor outcome (Tandon et al, 1991). Cortisol responses to CRF administration in these patients appear to be normal, as are CRF concentrations in the cerebrospinal fluid. Therefore, stress-induced elevation of glucocorticoids during the acute phase of the illness may have a more relevant role in the brain changes occurring in these patients. Such changes are suggested by the acute reductions in hippocampal volume recently described during early psychosis (Giedd et al, 1999, Lawrie et al, 2000, Pantellis et al, 2000, Velakoulis et al, 2000) and by the reversibility of superior temporal gyrus volume reductions (Keshevan et al, 1998) and the progressive ventricular enlargement over time (DeLisi et al, 1995). While these findings have not been entirely consistent (Lawrie and Abukmeil, 1998) a picture is now emerging which supports the view that there are plastic changes in macroscopic (and therefore microscopic) cytoarchitecture, which occur throughout the illness. Interestingly, a recent study showing reversibility of brain volume reductions occurring following abstinence from alcohol (Liu et al, 2000), suggested it might be due to regeneration of glial cells in the hippocampus (Korbo, 1999). This supports the view that changes in glial cell populations may contribute in a reversible way in psychiatric disorder to macroscopic brain changes.

Corticosteroids mediate their actions on target tissues through an intracellular receptor referred to as the glucocorticoid receptor. The glucocorticoid receptor is present in glia cells and mediates the effects of glucocorticoids on these cells (Crossin e al, 1997). The mechanisms by which glucocorticoids may contribute to neuronal degeneration are not yet clarified but several options have been proposed (Porter and Landfield, 1998). Two major mechanisms seem to be glucocorticoid- induced elevation in serotonin and glutamate levels in the brain. In fact, glucocorticoids cause brain damage, which is prevented by agents that increase serotonin reuptake (such as the antidepressant, tianeptine) or inhibit glutamate release and block glutamate action (such as the antiepileptic, phenytoin)

(Brown et al, 1999). A further mechanism by which glucocorticoids may act to alter cortical function is through the regulation of astrocyte activation (Maurel et al, 2000). In the hippocampus glucocorticoids reduce astrocytic activation as marked by GFAP protein and mRNA expression (Lapping et al, 1994), and, because GFAP regulates astrocytic motility and shape and neuronal-glial interactions, this glucocorticoid-mediated activity may play a fundamental role in the plasticity of neuronal-glial complex . Glucocorticoids also inhibit glucose uptake by astrocytes, thus leading to impaired glial function, and this in turn may augment glucocorticoid-induced neuronal damage (49, Virgin et al, 1991). Thus, the potential effects of glucocorticoids may involve not only reduced activation of astrocytes but also cell atrophy or reduced cell survival (Nichols, 1998). Thus, regardless of whether glial cell loss or reduced glial cell density in MDD precedes illness, or is a consequence of an excess of glucocorticoids as we suggest it may be in schizophrenia, these changes could reduce astrocyte function and so alter cortical activity.

Overview and Future Directions

The data indicate that glial cell loss and reductions in glial density is a component of mood disorders and probably of schizophrenia and so provides a new insight into the neuroanatomy of these diseases. However, at present there are several important issues that need to be clarified. We do not know which cortical or indeed subcortical regions are particularly affected. Nor do we know which glial populations are affected, or whether there is a specific pattern of changes within each disorder.

However, the most important issue that needs to be clarified is the cause of the glial cell loss. While it is possible that glial cell loss could predate or predispose to psychiatric illness generally, the lack of diagnostic specificity of the change also suggests that glial cell loss could be an epiphenomenon and thus not of primary aetiological importance to the disease processes. For example, medication or some as yet unidentified aspects of clinical or institutional care could lead to these changes. We have discussed the potential role played by neuroleptic treatments and concluded that these are unlikely to be responsible as the data currently available indicates that they increase, rather decrease glial cell numbers. Similarly, we have discussed the role of antidepressants and mood stabilisers, and find no clear evidence indicating that these treatments cause glial cell loss, although it is possible that antidepressants, through upregulating glucocorticoid receptors, increase the vulnerability of cells to glucocorticoid induced damage.

Important clues regarding the basis of the finding may be provided by the consistency of the finding of glial cell loss in major depression, and the data suggesting that, among schizophrenic subjects, those with affective symptoms are particularly likely to show these changes. Furthermore, elevated glucocorticoid levels are a feature of depressive illness, and the cellular pathology of glucocorticoid mediated toxicity is in keeping with many of the brain changes, both microscopic and macroscopic, that are found in mood disorders and schizophrenia. Therefore, considering the important trophic influences and neurotransmitter functions of glial cells, we have therefore hypothesised that a combination of glial cell loss and direct glucocorticoid mediated neurotoxicity may be responsible for the some of the changes in microscopic and macroscopic neuroanatomy in schizophrenia and depression. We do known whether this glial cell loss in the human cortex is a consequence of glucocorticoid induced cell toxicity, or a primary vulnerability factor. Certainly, the effect of glucocorticoids on reducing astrocyte activation, would suggest the former. The neuropathological assessment of primate models of depression would help to clarify this issue. What is clear, however, and is of potentially great clinical relevance, is the likelihood that elevated levels of glucocorticoids can act to diminish astrocyte function and that regardless of whether glial cell loss is primary or secondary, the consequence could be abnormal neuronal function.

It is thus hypothesised that many of the brain changes seen in MDD, BPD and schizophrenia may indeed be epiphenomenon of the diseases, for they may be secondary to the depressive symptoms and elevated glucocorticoid levels which accompany the illnesses. However, although they may be epiphenomenon, these changes are hypothesised to have crucial clinical effects through diminishing neuronal and cortical function and so complicating recovery from the primary illness. Conversely, it is possible that these changes may be reversed by normalising glial cell numbers in the cortex. Consequently, it will be important in future to determine whether manipulation and upregulation of different glial populations by pharmacological treatments can normalise cortical function and so treat or prevent psychiatric disorder.

Acknowledgements

Funded by a UK MRC Clinician Scientist Fellowship to David Cotter, by an MRC Clinical Training Fellowship awarded to Carmine M. Pariante, by The Theodore and Vada Stanley Foundation and by National Institute of Mental Health Grant No 55872 and a National Alliance for Research on Schizophrenia and Depression grants awarded to Grazyna Rajkowska.

References

Andreasen NC, Rezai K, Alliger R, Swayze VW, Fllaum M, Kirchner P, Cohen G, O'Leary D.S. Hypofrontality in neuroleptic-naive patients and in patients with chronic schizophrenia: assessment with xenon 133 single-photon emmision computed tomography and the Tower of London. Arch Gen Psychiatry 1992; 49: 943-958.

Araque A, Parpura V, Sanzgiri RP, Haydon PG. Tripartite synapses: glia, the unacknowledged partner. Trends in Neuroscience 1999; 22: 208-215.

Araque A Parpura V, Sanzgiri RP, Haydon PG. Glutamate-dependent astrocyte modulation of synaptic transmission between cultured hippocampal neurons. Eur J Neuroscience 1998; 10: 2129-42.

Arnold SE, Trojanowski JQ. Recent advances in defining the neuropathology of schizophrenia, Acta Neuropathology 1996; 92: 217-231.

Arnold SE, Franz BR, Trojanowski JQ, Moberg PJ, Gur RE. Glial fibrillary acidic protein-immunoreactive astrocytosis in elderly patients with schizophrenia and dementia. Acta Neuropathology 1996; 91: 269-277.

Baxter LR, Schwartz JM, Phelps ME. Reduction of prefrontal cortex glucose metabolism common to three types of depression. Arch Gen Psychiatry 1989; 46: 243-250.

Bayer TA, Buslei R, Havas L, Falkai P. Evidence for activation of microglia in patients with psychiatric illness. Neurosci Letters 1999; 271: 126-128.

Bench CJ, Frackowiak RSJ, Dolan RJ. Changes in regional cerebral blood flow on recovery from depression. Psych Medicine 1995; 25: 247-251.

Benes FM, Davidson J, Bird ED. Quantitative cytoarchitectural studies of the cerebral cortex of schizophrenics. Arch Gen Psychiatry 1986; 43: 31-35.

Benes FM, McSparren J, Bird ED, SanGiovanni JP, Vincent SL. Deficits in small interneurons in prefrontal and cingulate cortices in schizophrenic and schizoaffective patients. Arch Gen Psychiatry 48; 996-1001: 1991.

Benes FM. Emerging principles of altered neural circuitry in schizophrenia. Brain Res Reviews 2000; 31: 251-269.

Bergeron R, Meyer TM, Coyle JT, Greene RW. Modulation of N-methyl-D-aspartate receptor function by glycine transport. Proc Natl Acad Sci USA 1998; 95:15730-4.

Bogerts B, Hantsch J, Herzer M. A morphometric study of the dopamine-containing cell groups in the mesencephalon of normals, Parkinson patients, and schizophrenics. Biol Psychiatry 1983; 18: 951-969.

Brown ES, Rush AJ, McEwen BS. Hippocampal remodelling and damage by corticosteroids: implications for mood disorders. Neuropsychopharmacology 1999; 21: 474-484.

Bruton CJ, Crow TJ, Frith CD, Johnstone EC, Owens DGC, Roberts GW. Schizophrenia and the brain: a prospective cliniconeuropathological study. Psych Medicine 1990; 20: 285-304.

Carter RL, Berger UV, Barczak AK, Enna M, Coyle JT. Isolation and expression of a rat brain cDNA encoding glutamate carboxypeptidase II. Proc Natl Acad Sci USA 1998; 52: 829-836.

Checkley S. The neuroendocrinology of depression and chronic stress. B Med Bulletin 1996; 52: 597-617.

Coffey CE, Wilkinson WE, Weiner RD, Parashos IA, Djang WT, Webb MC, Figiel GS, Spritzer CE. Quantitative cerebral anatomy in depression. A controlled magnetic resonance imaging study. Arch Gen Psychiatry 1993; 50: 7-16.

Coles JA, Abbot NJ. Signalling from neurones to glial cells in invertebrates. Trends in Neuroscience 1996; 19: 358-362.

Connor B, Dragunow M. The role of neuronal growth factors in neurodegenerative disorders of the human brain. Brain Res Reviews 1998; 27: 1-39.

Cotter D, Mackay D, Landau S. Kerwin R. Everall, I. Reduced glial cell density and neuronal volume in major depression in the anterior cingulate cortex. Arch Gen Psychiatry 2001 (in press).

Coyle JT, Schwarz R. Mind glue: Implications of glial cell biology for psychiatry. Arch. Gen. Psychiatry 2000; 57: 90-93.

Crossin KL, Tai M-H, Krushel LA, Mauro VP, Edelman GM. Glucocorticoid receptor pathways are involved in the inhibition of astrocyte proliferation. . Proc Natl Acad Science USA 1997; 94: 6, 2687-2692.

Crow T, Ball J, Bloom SR, Brown R, Bruton CJ, Colter N, Frith CD, Johnstone EC, Owens DG, Roberts GW. Schizophrenia as an anomaly of development of cerebral asymetry. A postmortem study and proposal concerning the genetic basis of the disease. Arch Gen Psychiatry 1989; 46: 1145-11150.

DeLisi LE, Tew W, Xie S, Hoff AL, Sakuma M, Kushner M, Lee G, Shedlack K, Smith AM. A prospective follow-up study of brain morphology and cognition in first-episode schizophrenic patients: preliminary findings. Biol. Psychiatry 1995; 38: 349-360.

Duman RS, Heninger GR, Nestler EJ. A molecular and cellular theory of depression. Arch Gen Psychiatry 1997 54: 597-606.

Ebert D, Ebmeier KP, The role of the cingulate gyrus in depression: from functional anatomy to neurochemistry. Biol Psychiatry 1996 39: 1044-1050.

Eddelston M, Mucke L. Molecular profile of recative astrocytes - implications for their role in neurologic disease. Neuroscience 1993 54: 15-36.

Erickson JT, Brosenitsch TA, Katz DM. Brain-Derived Neurotrophic Factor and Glial Cell Line-Derived Neurotrophic Factor Are Required Simultaneously for Survival of Dopaminergic Primary Sensory Neurons. In Vivo J Neuroscience 2001; 21: 581-589.

Evans DL, Golden RJ, The dexamethasone suppression test: A review. In Loosen PT Nemeroff CB (eds): Handbook of clinical psychoneuroendocrinology New York: Guilford Press 1987: 313- 350.

Falkai P, Bogerts B. Cell loss in the hippocampus. Eur Arch Psychiatr Neurol Science 1986; 236: 154-161

Falkai P, Bogerts B, Rozumek M. Limbic pathology in schizophrenia: the entorhinal region-a morphometric study. Biol Psychiatry 1998; 24: 515-521.

Falkai P, Honer WG, David S, Bogerts B, Majtenyis C, Bayer TA. No evidence for astrogliosis in brains of schizophrenic patients; A postmortem study. Neuropath Appl Neurobiology 1999; 25: 48-53.

Fletcher EJ, Millar JD, Zeman S, Lodge D. Non-competitive antagonism of N-methyl-D-aspartate by displacement of an endogenous glycine-like substance. Mol Chem Neuropathology 1997; 31: 97-118.

Foster AC, Miller LP, Oldendorf WH, Schwarcz R. Studies on the disposition of quinlinic acid after intracerebral or systemic administration in the rat. Exp Neurology 1984; 84: 428-440.

Garcia-Segura L, Chowen JA, Parducz A, Naftolin F. Gonadal hormones as promotypes of structural synaptic plasticity: cellular mechanisms. Prog Neurobiology 1994; 44: 279-307.

Garcia-Segura L, Naftolin F, Hutchinson JB, Azcoitia I, Chowen JA. Role of astroglia in estrogen regulation of synaptic plasticity and brain repair. J Neurobiology 1999; 40: 574-584.

Giedd JN, Jeffries NO, Blumenthal J, Castellanos FX, Vaituzis AC, Fernandez T, Hamburger SD, Lui H, Nelson J, Bedwell J, Tran L, Lenane M, Nicolson R, Rapaport JL. Childhood-onset schizophrenia: progressive brain changes during adolescence Biol Psychiatry 1999; 46: 7: 892-898.

Glantz LA, Lewis DA. Decreased dendritic spine density on prefrontal cortical pyramidal neurons in schizophrenia Arch Gen Psychiatry 2000; 57: 65-73.

Glantz LA, and Lewis DA. Reductions of synaptophysin immunoreactivity in the prefrontal cortex of subjects with schizophrenia. Arch Gen Psychiatry 1997; 54: 660-669.

Harrison PJ, Eastwood SL. Preferential involvement of excitatory neurons in the medial temporal lobe in schizophrenia. Lancet 1998; 352: 1669-73.

Harrison PJ. The neuropathology of schizophrenia: a critical review of the data and their interpretation Brain 1999; 122: 593-624.

Honer WG, Falkai P, Chen C, Arango V, Mann JJ, Dwork AJ. Synaptic and plasticity-associated proteins in anterior frontal cortex in severe mental illness. Neuroscience 1999; 91: 1247-55.

Johnson JW, Ascher P, Glycine potentiates the NMDA response in cultured mouse brain neurons. Nature 1987; 325: 529-31.

Johnston-Wilson N, Sims CD, Hofmann JP, Anderson L, Shore AD, Torrey EF, Yolken R. Disease-specific alterations in frontal cortex brain proteins in schizophrenia bipolar disorder and major depressive disorder. Mol Psychiatry 2000; 5: 142-149.

Jordon CL. Glia as mediators of steroid hormone action on the nervous system: an overview. J Neurobiology 1999; 40: 434- 445.

Kadekaro M, Ito M. Gross PM. Local cerebral glucose utilisation is increased in acutely adrenalectomised rats. Neuroendocrinology 1988; 47: 329-334.

Keshevan MS, Haas GL, Kahn CE, Aguilar E, Dick E, Schooler NR, Sweeney JA, Pettegrew JW. Superior temporal gyrus and the course of early schizophrenia: progressive static or reversible? J Psych Research 1998; 32: 161-167.

Kettenmann H Ransom BR (Eds) Neuroglia Oxford University Press New York 1995.

Khan ZU, Koulen P, Rubinstein M, Grandy DK, Goldman-Rakic PS. An astroglia-linked dopamine D2-receptor action in prefrontal cortex. Proc Natl Acad Sci U S A 2001; 98: 1964-1969.

Korbo L. Glial cell loss in the hippocampus of alcoholics. Alc Clin Exp Research 1999; 23: 164-168

Lapping NJ, Teter B ,Nichols NR, Rozovbsky I, Finch CE. Glial fibrillary acidic protein: regulation by hormones cytokines and growth factors. Brain Pathology 1994; 1: 259-275.

Lawrie SM, Abukmeil SS. Brain abnormality in schizophrenia: a systematic and quantitative review of volumetric magnetic resonance imaging studies. B J Psychiatry 1998; 172: 110-120.

Lawrie SM, Whalley H, Byrne M, Miller P, Best JJK, Johnstone E. Brain structure change and psychopathology in subjects at high risk of schizophrenia. Schiz Research 2000; 41 (1): 11.

Levine S, Saltzman A, Klein AW. Proliferation of glial cells in vivo induced in the neural lobe of the rat pituitary by lithiuml Cell Proliferation 2000; 33: 203-207.

Lindsay RM, Weigand SJ, Altar CA, DiStephano PS. Neurotrophic factors: from molocule to man TINS 1994; 17: 182-189.

Liu RSN, Lemieux L, Shorvon SD, Sisodiya SM, Duncan JS. Association between brain size and abstinence from alcohol. Lancet 2000; 355: 1969-1970.

Magistretti PJ, Pellerin L, Rothman DL, Shulman RG. Energy on Demand Science 1999; 283: 496-497.

Mayberg HS. Depression In Mazziotta JC Toga AW Frackowiak RSJ (eds) Brain Mapping The Disorders: Academic Press 2000 pp 485-505.

Mamounas LA, Blue ME, Siuciak JA, Altar CA. Brain-derived neurotrophic factor promotes the survival and sprouting of serotonergic axons in rat brain. J Neuroscience 1995; 15:7929-39.

Maurel D, Sage D, Mekaouche M, Bosler O. Glucocorticoids up-regulate the expression of glial fibrillary acidic protein in the rat suprachiasmatic nucleus. Glia 2000; 29: 121-221.

McAllister-Williams RH, Ferrier IN, Young AH. Mood and neuropsychological function in depression: the role of corticosteroids and serotonin. Psychol Medicine 1998; 28:573-584

McCarley RW, Wible CG, Frumin M, Hirayasu Y, Levitt JL, Fischer A, Shenton ME. MRI anatomy of schizophrenia. Biol Psychiatry 1999; 45: 1099-1119.

McEwen BS. Possible mechanisms for atrophy of the human hippocampus. Mol Psychiatry 1997; 2: 255-262.

Mennerick S, Zorumski CF. Glial contributions to excitatory neurotransmission in cultured hippocampal cells Nature 1994; 368: 59-62.

Miyazaki J , Nakanishi S, Jingami H. Expression and characterization of a glycine-binding fragment of the N-methyl-D-aspartate receptor subunit NR1. Biochemistry 1999, 340: 687-92.

Miguel-Hidalgo JJ, Baucom C, Dilley G, Overholser J, Meltzer HY, Stockmeier CA, Rajkowska G. Glial fibrillary acidic protein immunoreactivity in the prefrontal cortex distinguishes younger from older adults in major depressive disorder. Biol Psychiatry 48: 861-873 2000.

Schwab ME. Myelin-associated inhibitors of neurite growth. Exp Neurology 1990; 109: 2-5

Nakajima K, Kikuchi Y, Ikoma E, Honda S, Ishikawa M, Liu Y, Kohsaka S. Neurotrophins regulate the function of cultured microglia. Glia 1998; 24: 272-289.

Nemeroff CB. The corticotropin-releasing factor (CRF) hypothesis of depression: new findings and new directions. Mol Psychiatry 1996; 1:336-342.

Nibuya M, Morinobu S, Duman RS. Regulation of BDNF nd trkB mRNA in rat brain by chronic electroconvulsive and antidepressant drug treatments. J Neuroscience1995; 15: 7539-7547.

Nichols N. Glial responses to steroids as markers of brain aging. J Neurobiology 1998; 40: 585-610.

Okugawa G, Omori K, Suzukawa J, Fujiseki Y, Kinoshita T, Inagaki C. Long-term treatment with antidepressants increases glucocorticoid receptor binding and gene expression in cultured rat hippocampal neurones. J Neuroendocrinology 1999; 11: 887-895.

Olney JW, Farber NB. Glutamate receptor dysfunction and schizophrenia. Arch Gen Psychiatry 1995; 52: 998-1007.

Ongur D, Drevets WC, Price JL. Glial reduction in the subgenual prefrontal cortex in mood disorders. Proc Natl Acad Sci USA 1998; 95: 13290-13295.

Orlovskaya DD, Vikhreva OV, Zimina IS, Denisov DV, Uranova NA. Ultrastructural dystrophic changes of oligodendroglial cells in autopsied prefrontal cortex and striatum in schizophrenia: a morphometric study. Schiz Research 1999; 36: 82-83

Orlovskaya DD, Vostrikov VM, Rachmanova VI, Uranova NA. Decreased numberical density of oligodendroglial density cells in the prefrontal cortex area 9 in schizophrenia and mood disorders: a study of brain collection from the Stanley Foundation Neuropathology Consortium. Schiz Research 2000; 41: 105-106.

Owen F, Crow TJ, Frith CD, Johnson JA, Johnstone EC, Lofthouse R, Owens DG, Poulter M. Selective decreases in MAO-B activity in postmortem brains from schizophrenic patients with type II syndrome. B JPsychiatry 1987; 151: 514-519.

Pakkenberg B. Pronounced reduction of total neuron number in mediodorsal thalamic nucleus and nucleus accumbens in schizophrenics. Arch GenPsychiatry 1990; 47: 1023-1028.

Pantellis C, Velakoulis D, Suckling J, McGorry P, Philips L, Yung A, Wood S, Bullmore E, Brewer W ,Soulsby B, McGuire P. Left medial temporal lobe volume reduction occurs during the transition from high risk to first episode psychosis. Schiz Research 2000; 41 35.

Pariante CM, Pearce BD, Pisell TL, Owens MJ, Miller AH. Steroid-independent translocation of the glucocorticoid receptor by the antidepressant desipramine. Mol Pharmacology 1997; 52: 571-581.

Perrone- Bizzozero N, Sower AC, and Bird ED. Levels of the growth-associated protein GAP-43 are selectively increased in association cortices in schizophrenia. Proc Natl Acad Sci USA 1996; 93: 14128-14187.

Peters A, Palay SL, Webster HD. In: The fine structure of the nervous system. New York: Oxford University Press 1991.

Pfieger FW, Barres BA. New views on synapse-glia interactions. Curr Opin Neurobiology 1996; 6:615-621.

Pfrieger FW, Barres BA. Synaptic efficacy enhanced by glial cells in vitro. Science 1997; 277: 1684-1687.

Porter NM, Landfield PW. Stress hormones and brain aging: adding injury to insult. Nature Neuroscience 1998; 1: 3-4.

Post A, Crochemore C, Uhr M, Holsboer F, Behl C. Differential effects of antidepressants on the viability of clonal hippocampal cells. Biol Psychiatry 2000; 47: 138.

Radewicz K, Garey L, Gentleman S M, Reynolds R. Increased HLA-DR immunoreactive glia in frontal and temporal cortex of chronic schizophrenia. J Neuropathol Exp Neurology 2000; 59: 137-150.

Rajkowska G, Selemon LD, Goldman-Rakic PS. Neuronal and glial somal size in the prefrontal cortex: a postmortem morphometric study of schizophrenia and huntington disease. Arch Gen Psychiatry 1998; 55: 215-224.

Rajkowska G, Wei JJ, Miguel-Hidalgo JJ, Stockmeier R. Glial and neuronal pathology in rostral orbitofrontal cortex in schizophrenic postmortem brain. Schiz Research 1999a; 36: 84.

Rajkowska G, Miguel-Hidalgo JJ, Wei J. Morphometric evidence for neuronal and glial prefrontal cell pathology in major depression. Biol Psychiatry 1999b; 45: 1085-1098.

Rajkowska-Markow G, Miguel-Hidalgo JJ, Wei J, Stockmeir CA. Reductions in glia distinguish orbitofrontal region from dorsolateral prefrontal cortex in schizophrenia. Abs SocNeuroscience 1999c; 25: 818.

Rajkowska G. Postmortem studies in mood disorders indicate altered numbers of neurons and glial cells. Biol Psychiatry 2000; 48: 766-777.

Rajkowska G, Halaris A, Selemon LD. Reductions in neuronal and glial density characterize the dorsolateral prefrontal cortex in bipolar disorder. Biol Psychiatry 2001a; 49:741-752.

Rajkowska G, Miguel-Hidalgo JJ, Makkos Z, Dilley G, Meltzer HY, Overholser JC, Stockmeier CA. Layer-specific astroglia pathology in the dorsolateral prefrontal cortex in schizophrenia. Schizophrenia Research 2001b (in press).

Rakic P. Specification of the cerebral cortical areas. Science 1988; 241: 170-176.

Roberts GW, Colter N, Lofthouse R, Bogerts B, Zech M, Crow TJ . Gliosis in schizophrenia: a survey. Biol Psychiatry 1986; 21: 1043-1050.

Roberts GW, Colter N. Lofthouse R. Johnstone EC. Crow TJ. Is there gliosis in schizophrenia? Investigation of the temporal lobe Biol Psychiatry 1987; 22: 1459-1468.

Rocha E, Achaval M, Santos P, Rodnight P. Lithium treatment causes gliosis and modifies the morphology of hippocampal astrocytes in rats. Neuroreport 1998; 9: 3971-3974.

Rocha E, Rodnight R. Chronic administration of lithium chloride increases immunodetectable glial fibrillary acidic protein in the rat hippocampus. J Neurochemistry 1994; 63: 1582-1584.

Rosoklija G, Toomayan G, Ellis SP, Keilp J, Mann JJ, Latov N, Hays AP, Dwork AJ. Structural abnormalities of subicular dendrites in subjects with schizophrenia and mood disorders. Arch Gen Psychiatry 2000; 57: 349-356.

Rothstein JD, Dykes-Hoberg M, Pardo CA, Bristol LA, Jin L, Kuncl RW, Kanai Y, Hediger MA, Wang Y, Schielke JP,Welty DF. Knockout of glutamate transporters reveals a major role for astroglial transport in excitotoxicity and clearance of glutamate. Neuron 1996; 16:675-86.

Sapolsky R. Glucocorticoids stress and exacerbation of excitotoxic neuron death. Semin Neuroscience 1994; 6: 323-331.

Sapolsky R. The possibility of neurotoxicity in the hippocampus in major depression: a primer on neuron death. Biol Psychiatry 2000; 48 (8): 755-65.

Schaaf MJ, Hoetelmans RW, de Kloet ER, Vreugdenhil E. Corticosterone regulates expression of BDNF and trkB but not NT-3 and trkC mRNA in the rat hippocampus. J Neurosci Research 1997; 48: 334-341.

Selemon LD, Lidow MS, Goldman-Rakic PS. Increased volume and glial density in primate prefrontal cortex associated with chronic antipsychotic drug exposure. Biol Psychiatry 1999; 46: 161-172.

Selemon LD, Rajkowska G, Goldman-Rakic PS. Abnormally high neuronal density in the schizophrenic cortex: a morphometric analysis of prefrontal area 9 and occipital area 17. Arch Gen Psychiatry 1995; 52: 805-818.

Selemon LD, Rajkowska G, Goldman-Rakic PS. Elevated neuronal density in prefrontal area 46 in brains from schizophrenic patients - application of a three dimensional stereologic counting method. J Comp Neurology 1998; 392:402-12.

Siuiciak JA, Wiegand SJ, Lindsay RM. Antidepressant-like effect of brain-derived neurotrophic factor. Pharmacol Biochem Behaviour 1996; 56: 131-137.

Speciale C, Schwarcz R. On the production and disposition of quinolinic acid in rat brain and liver slices. J Neurochemistry 1993; 60: 212-218.

Steffens DC, Krishnan RR. Structural neuroimaging and mood disorders: recent findings implications for classification and future directions. Biol Psychiatry 1998; 43: 705-712.

Stevens CD, Altshuler LL, Bogerts B, Falkai P. Quantitative study of gliosis in schizophrenia and Huntington's Chorea. Biol Psychiatry 1988; 24: 697-700.

Stevens JR. Neuropathology of schizophrenia. Arch Gen Psychiatry 1982; 39: 1131-9.

Tamminga CA, Thaker GK, Buchanan R, Kirkpatrick B, Alphs LD, Chase TN, Carpenter WT. Limbic system abnormalities identified in schizophrenia using positron emission tomography with fluorodeoxyglucose and neocortical alterations with deficit syndrome. Arch Gen Psychiatry 1992; 49: 522-530.

Tanaka K, Watase K, Manabe T, Yamada K, Watanabe M, Takahashi K, Iwama H, Nishikawa T, Ichihara N, Kikuchi T, Okuyama S, Kawashima N, Hori S, Takimoto M, Wada K. Epilepsy and exacerbation of brain injury in mice lacking the glutamate transporter GLT-1. Science 1997; 276:1699-702.

Tandon R, Mazzara C, DeQuardo J. Dexamethasone suppression test in schizophrenia: relationship to symptomatology ventricular enlargement and outcome. Biol Psychiatry1991; 29: 953.

Tsacopoulos M, Magistretti PJ. Metabolic coupling between glia and neurons J Neuroscience 1996; 516: 877-88.

Ullian EM, Sapperstein S, KChristopherson KS, Barres B A. Control of Synapse Number by Glia. Science: 2001; 291: 5504: 657-661.

Uranova NA, Zimina IS, Vikhreva OV, Denisov DV, Orlovskaya DD. Morphometric study of ultrastructural alterations of myelinated fibers in postmortem schizophrenic brains. Schiz Research 1999; 36: 85.

Velakoulis D, Pantelis C, McGorry PD, Dudgeon P, Brewer W, Cook M, Desmond P, Bridle N, Tierney P, Murrie V, Singh B, Copolov D. Hippocampal volume in first episode psychoses and chronic schizophrenia: a high resolution magnetic resonance imaging study. Arch Gen Psychiatry 2000; 56: 133-141.

Verkhratsky A, Orkand RK, Kettenmann H. Glial calcium: homeostasis and signaling function. Physiol Reviews 1998; 78: 99-141.

Virgin CE, Ha TP, Packan DR, Tombaugh GC, Yang SH, Horner HC, Sapolsky RM. Glucocorticoids inhibit glucose transport and glutamate uptake in hippocampal astrocytes: Implications for glucocorticoid neurotoxicity J Neurochemistry 1991; 57: 1422-1428.

Wayne C, Drevets MD. Functional Neuroimaging studies of depression: the anatomy of melancholia. Ann Rev Medicine 1998; 49: 341-361.

Webster MJ, Johnston-Wislon N, Nagata K, Yolken RH .Alterations in the expression of phosphorylated glial fibrillary acidic proteins in the frontal cortex of individuals with schizophrenia bipolar disorder and depression. Schiz Reearch 2000; 41: 106

Weinberger DR. Implications of normal brain development for the pathogenesis of schizophrenia. Arch Gen Psychiatry 1987; 44: 660-9.

Weinberger DR, Berman KF, and Zec RF. Physiological dysfunction of dorsolateral prefrontal cortex in schizophrenia I: regional cerebral blood flow (rCBF) evidence. Arch Gen Psychiatry 1986; 43: 114-124.

Wolosker H, Blackshaw S, Snyder SH. Serine racemase: a glial enzyme synthesising D-serine to regulate glutamate-N-methyl-D-aspartae neurotransmission. Proc Natl Acad Sciences USA 1999 96: 13409-13414.

5 INDICATIONS OF ABNORMAL CONNECTIVITY IN NEUROPSYCHIATRIC DISORDERS IN POSTMORTEM STUDIES

William G. Honer

Abstract

Proteins enriched in presynaptic terminals are frequently used as postmortem markers for neural connectivity in neuropsychiatric disorders. This chapter describes the animal studies which form the foundation for interpreting results in humans, followed by comments on studies in dementia and other disorders. Studies of presynaptic proteins and their mRNAs in schizophrenia and affective disorders indicate that multiple proteins are abnormally expressed or regulated. In the future, an approach which considers interactions between presynaptic proteins involved in neurotransmission may be fruitful.

Introduction

The molecular and cellular biology of neural connectivity is of great importance for understanding neuropsychiatric disorders with minimal or no neuronal loss. Dysfunction at synaptic terminals, and disturbances in the organization of neuronal connections are likely to be the substrates for clinical symptomatology, and in some cases may limit the utility of current therapeutics. This chapter will focus on synaptic abnormalities in major mental disorders.

Postmortem Studies of Neural Connectivity: Usefullness

A limited number of features of synaptic terminals can be assessed in postmortem studies. If appropriate conditions of tissue preservation can be maintained, synaptic ultrastructure can be visualized with electron microscopy. This technique provides a wealth of detail on the pre- and postsynaptic structure of synapses, but is limited by the demands of tissue handling, and provides an extremely limited sampling frame. Histological studies using silver impregnation techniques (Golgi staining) allow dendrites and their spines to be assessed. A degree of artistry may be necessary in preparing specimens for these studies. The limited number of studies of connectivity in neuropsychiatric disorders using electron microscopy or Golgi impregnation were reviewed recently (Honer et al 2000).

Molecular approaches to studying connectivity use assessments of proteins enriched in synaptic terminals, or study their mRNA expression. These include molecules involved in neurosecretion: the synaptic vesicle proteins (synaptophysin, VAMP, synaptotagmin), presynaptic membrane proteins (SNAP-25, syntaxin) and cytoplasmic proteins (complexins, synapsin). As well, proteins which play a role in neural plasticity such as GAP-43 and NCAM have been investigated. The interpretation of protein and mRNA changes may not always be the same. In some hippocampal circuits, high synapsin mRNA in the soma was related to high synapsin immunoreactivity in the terminal fields, while in other circuits mRNA levels in the soma and protein levels in the terminals were poorly coupled (Melloni et al 1993). There is a suggestion that at least during development, the levels of several synaptic vesicle proteins are controlled by cytoplasmic mechanisms such as changes in half-life, rather than through signalling to the nucleus and altering mRNA expression (Daly and Ziff 1997). A brief review of selected animal and human studies follows to provide a context for the interpretation of results in neuropsychiatric disorders.

Simple Lesion Studies in Animals

Lesions of the entorhinal cortex result in predictable reductions in the immunoreactivity of presynaptic proteins involved with neurotransmission in the terminal fields of the projecting neurons. The affected proteins include synaptophysin (Cabalka et al 1992) and SNAP-25 (Geddes et al 1990; Patanow et al 1997). The reductions in synaptophysin were associated with fewer synaptophysin-immunoreactive terminals, demonstrated with confocal microscopy (Masliah et al 1991). In contrast to the protein results, the mRNA for SNAP-25 was observed to be increased in several populations of neurons which were the targets of lesioned projection systems (Geddes et al 1990). It is also difficult to generalize from effects on neurotransmission-related presynaptic proteins to other synapse-enriched molecules such as GAP-43 and NCAM. Entorhinal cortex lesions resulted in smaller effects on GAP-43 compared with SNAP-25 (Patanow et al 1997). Further, during the recovery process from an entorhinal lesion, while synaptophysin immunoreactivity remained reduced relative to control, GAP-43 immunoreactivity was massively increased (Masliah et al 1991).

Similar results occur in the striatum. Lesions of cortical or ascending fibre systems resulted in reduced immunoreactivity of synaptophysin or synapsin I (Poltorak et al 1993; Walaas et al 1988). In contrast, NCAM immunoreactivity was increased (Poltorak et al 1993).

Complex Lesions

Chemical injuries of the hippocampus (Brock and O'Callaghan 1987) or striatum (Walaas et al 1988) resulted in decreased immunoreactivity for presynaptic proteins in the terminal fields of the lesioned neurons, including effects on synaptophysin, synapsin I and SNAP-25. Again, increased SNAP-25 mRNA expression in the denervated neurons was observed (Geddes et al 1990), and some increases in immunoreactivity were reported in the terminal fields of these cells (Patanow et al 1997). Divergent results on syntaxin mRNA and protein were observed in response to kindling stimuli in the hippocampus: mRNA expression was increased while immunoreactivity was unchanged (Kamphuis et al 1995).

Effects of Electrophysiological or Behavioural Manipulations

Long-term potentiation in the hippocampus results in several changes in presynaptic protein mRNA expression. Syntaxin 1B and synapsin I mRNAs were increased, while syntaxin 1A, nSEC1, VAMP, synaptophysin, Rab3A and synapsin II mRNAs were unchanged (Hicks et al 1997; Richter-Levin et al 1998; Rodger et al 1998). Conflicting reports indicate either no change (Hicks et al 1997) or an increase in SNAP-25 mRNA (Roberts et al 1998). Syntaxin 1B immunoreactivity was increased in mossy fibre terminals following LTP (Helme-Guizon et al 1998). In the dentate gyrus, synapsin, synaptotagmin and synaptophysin immunoreactivities were increased (Lynch et al 1994). Increased tyrosine phosphorylation of synaptophysin was noted with LTP (Mullany and Lynch 1998). Syntaxin 3A mRNA and immunoreactivity increased in parallel following LTP, and while syntaxin 3B mRNA decreased, immunoreactivity appeared unchanged (Rodger et al 1998). Syntaxin 1B mRNA was increased in hippocampus by training on a working memory task, and increased in frontal cortex and the shell of the nucleus accumbens by a spatial reference memory task (Davis et al 1996).

Finally, an elegant series of experiments investigating hippocampal development demonstrated significant alterations in hippocampal synaptophysin in rat pups related to maternal behaviour (Liu et al 2000).

Effects of Drugs

Antipsychotic and other drug treatments may alter synaptic morphology and protein composition, as part of therapeutic actions or related to side effects of treatments. A review of ultrastructural studies in rodents indicated antipsychotic drugs resulted in pre- and postsynaptic reorganization in caudate and in layer VI of frontal cortex (Harrison 1999). Both the ratio of synaptic types, as well as changes in the properties of synapses were observed. A summary of animal studies of the effects of antipsychotic drugs on presynaptic proteins and mRNA appears in Table 1. Most studies investigated effects on mRNA rather than protein. Overall, where there are effects of antipsychotics on mRNA the trend appears to be to increase mRNA expression. However, there may be heterogeneity between brain regions, between effects on different presynaptic markers, and all antipsychotics may not share the same profile of effects. The extent to which effects may persist following termination of treatment is uncertain.

Presynaptic Proteins in Brain Disease

Studies of presynaptic proteins in Alzheimer's disease (AD) were the first application of antibody-based techniques to investigate syapses in human postmortem specimens (Hamos et al 1989; Masliah et al 1989). Significant progress has been made in understanding the nature of synaptic pathology in AD and other neurological diseases, with several implications for studies of other neuropsychiatric disorders. First, in AD studies were carried out with confocal microscopy and multiple presynaptic markers to demonstrate that reduced synaptophysin immunoreactivity was related to losses of terminal number (Masliah et al 1992). The fundamental issue of whether changes in synaptic protein immunoreactivity are due to changes in number of terminals, or altered immunoreactivity within terminals remains incompletely studied in most other neuropsychiatric disorders. Second, initially there was considerable emphasis placed on the putative relationship between synaptic terminal loss and cognitive change in AD (Terry et al 1991). However, as more studies were carried out with cases having different severity of pathology and of cognitive impairment, it became clear that until the later stages of AD, considerable plasticity is present in at least neocortical synapse populations (Lue et al 1996; Minger et al 2001; Mukaetova-Ladinska et al 2000). Synaptic proteins appeared to be elevated or unchanged in early to mid-AD relative to controls. These findings reinforce the concept that neural connections are dynamic, and capable of responding to pathological changes or stressors. On the other hand, compromise in some aspect of the capacity to respond to threats to synaptic integrity could have far-reaching consequences. Finally, one study established a relationship between postmortem assessment of presynaptic proteins and antemortem structural brain imaging (Heindel et al 1994). Loss of synaptophsyin was related to the extent of cortical gray matter loss on MRI in cases with HIV infection.

Table 1. Presynaptic proteins in rats treated with antipsychotics

Frontal cortex	Drug	Increased protein	Unchanged protein	Increased mRNA	Unchanged mRNA	Decreased mRNA
Synaptophysin	Haloperidol		(Eastwood et al 1994; Lidow et al 2001)	(Eastwood et al 1994; Eastwood et al 1997)	(Eastwood et al 2000a; Nakahara et al 1998)	
	CPZ Risperidone Olanzapine Clozapine				(Eastwood et al 2000a)	
SNAP-25, syntaxin, Rab3	Haloperidol				(Nakahara et al 1998)	
Complexin I	Olanzapine			(Eastwood et al 2000a)		
	CPZ, Risperidone Clozapine				(Eastwood et al 2000a)	
	Haloperidol				(Eastwood et al 2000a)	(Nakahara et al 2000)
Complexin II	CPZ			(Eastwood et al 2000a)		
	Risperidone Olanzapine Clozapine				(Eastwood et al 2000a)	
	Haloperidol				(Eastwood et al 2000a; Nakahara et al 2000)	
Synaptotagmin I, IV	Haloperidol				(Nakahara et al 1998)	
VAMP2	Haloperidol				(Nakahara et al 1998)	
GAP-43	Haloperidol				(Eastwood et al 1997)	

Table 1. continued

Striatum	Drug	Increased protein	Unchanged protein	Increased mRNA	Unchanged mRNA	Decreased mRNA
Synaptophysin	Haloperidol	(Marin and Tolosa 1997)	(Eastwood et al 1994; Eastwood et al 1997; Loessner et al 1988)	(Eastwood et al 1994; Eastwood et al 1997)	(Nakahara et al 1998)	
	Perphenazine		(Fog et al 1976)			
SNAP-25	Perphenazine		(Fog et al 1976)			
SNAP-25, Rab3a, syntaxin	Haloperidol				(Nakahara et al 1998)	
Complexin I	Olanzapine			(Eastwood et al 2000a)		
	CPZ Risperidone Clozapine				(Eastwood et al 2000a)	
	Haloperidol				(Eastwood et al 2000a; Nakahara et al 2000)	(Nakahara et al 2000) (accumbens)
Complexin II	CPZ			(Eastwood et al 2000a) (ventral)		
	Haloperidol				(Eastwood et al 2000a; Nakahara et al 2000)	
	Risperidone Olanzapine Clozapine				(Eastwood et al 2000a)	
Synaptotagmin I, IV	Haloperidol				(Nakahara et al 1998)	(Nakahara et al 1998) (accumbens)
VAMP2	Haloperidol				(Nakahara et al 1998)	
Secretogranin II, chromogranin	Haloperidol Clozapine			(Kroesen et al 1995)		
GAP-43	Haloperidol				(Eastwood et al 1997)	

Table 1. continued

Cingulate cortex	Drug	Increased protein	Unchanged protein	Increased mRNA	Unchanged mRNA	Decreased mRNA
Synaptophysin	Haloperidol		(Lidow et al 2001)			
Hippocampus						
Synaptophysin	Haloperidol		(Eastwood et al 1997; Lidow et al 2001; Loessner et al 1988) (entorhinal)		(Eastwood et al 1995; Eastwood et al 1997)	
GAP-43	Haloperidol				(Eastwood et al 1997)	
Cerebellum						
Synaptophysin	CPZ Haloperidol Risperidone Olanzapine Clozapine				(Eastwood et al 2001)	
Complexin I	CPZ Haloperidol Risperidone Olanzapine Clozapine				(Eastwood et al 2001)	
Complexin II	CPZ Haloperidol Risperidone Olanzapine Clozapine				(Eastwood et al 2001)	

A summary of studies of presynaptic proteins in schizophrenia appears in Table 2. About half of studies of both protein and mRNA expression show reduced synaptic proteins. Most of the rest of the reports indicate unchanged synaptic proteins, however there are several exceptions where increased immunoreactivity was observed. One of these studies represents an interesting series of elderly patients with antemortem assessments of cognition indicating severe impairment in many of the patients (Gabriel et al 1997). An Alzheimer's disease positive control group in this study did show selective reductions in presynaptic proteins, in contrast to the increased levels observed in the samples from patients with schizophrenia. Further studies of this group of patients demonstrated altered relationships between age and mRNAs coding for presynaptic proteins in samples from patients with schizophrenia (Sokolov et al 2000; Tcherepanov and Sokolov 1997). These may represent a distinct subgroup of schizophrenia, or a phase of the illness which is not widely represented in other samples.

There are few studies of multiple markers in the same region, or of one marker in multiple brain regions. Studies of multiple markers in the same region frequently report variability in the extent of changes observed between markers. Likewise, studies of multiple brain regions tend to indicate variability between regions in the extent of synaptic changes. No single synaptic protein is likely to be the key to mechanism of illness in schizophrenia, and generalizations concerning schizophrenia as a developmental deficit of synaptic connectivity may be premature.

Two recent studies suggest the importance of considering changes in networks of interacting presynaptic proteins. Using a cDNA screening strategy, reduced mRNAs for the synaptic proteins N-ethylmaleimide-sensitive factor (NSF) and synapsin II were observed consistently in schizophrenia, while mRNAs coding for other presynaptic were also affected, but less consistently (Mirnics et al 2000). Different patterns of altered presynaptic protein gene expression were observed in different cases. A second study investigated the in vitro interactions between the three SNARE proteins (SNAP-25, syntaxin and VAMP) in brain homogenates from cases of schizophrenia, depression with suicide and controls (Honer et al 2001). The tendency of the trimeric SNARE complex to form spontaneously was increased in cases with suicide as a cause of death. Future studies need to consider the functional implications of changes in presynaptic protein levels.

Table 2. Presynaptic proteins in selected brain regions in schizophrenia

Frontal cortex	Increased protein	Unchanged protein	Decreased protein	Unchanged mRNA	Decreased mRNA
Synaptophysin		(Davidsson et al 1999; Eastwood et al 2000b; Eastwood and Harrison 1998; Honer et al 1999: suicides)	(Glantz and Lewis 1997; Honer et al 1999; Karson et al 1999; Perrone-Bizzozero et al 1996)	(Eastwood et al 2000b; Glantz et al 2000; Karson et al 1999)	
SNAP-25	(Thompson et al 1998) (BA9)	(Gabriel et al 1997; Honer et al 2001: suicides)	(Honer et al 2001; Karson et al 1999; Stefan et al 1995; Thompson et al 1998) (BA10)	(Karson et al 1999)	
Syntaxin		(Gabriel et al 1997; Honer et al 2001)			
Rab3a		(Glantz and Lewis 1993; Glantz and Lewis 1997)	(Davidsson et al 1999)		
Complexin I			(Sawada et al 2001)		
Complexin II		(Sawada et al 2001)			
NCAM	(Vawter et al 1998)	(Barbeau et al 1995)			
GAP-43	(Perrone-Bizzozero et al 1996)	(Honer et al 1999)		(Eastwood and Harrison 1998)	(Weickert et al 2001)
Temporal neocortex					
Synaptophysin		(Davidsson et al 1999; Eastwood et al 2000b; Gabriel et al 1997)	(Perrone-Bizzozero et al 1996; Thompson et al 1998)		(Eastwood et al 2000b)
SNAP-25		(Gabriel et al 1997)	(Thompson et al 1998)		
Syntaxin		(Gabriel et al 1997)			
Rab3a		(Davidsson et al 1999)			
GAP-43	(Perrone-Bizzozero et al 1996)			(Eastwood and Harrison 1998)	

Table 2. continued

Cingulate cortex	Increased protein	Unchanged protein	Decreased protein	Unchanged mRNA	Decreased mRNA
Synaptophysin	(Gabriel et al 1997)	(Eastwood et al 2000b; Eastwood and Harrison 2001; Honer et al 1997)	(Davidsson et al 1999)	(Eastwood et al 2000b)	
SNAP-25	(Gabriel et al 1997)	(Honer et al 1997)			
Syntaxin	(Gabriel et al 1997; Honer et al 1997)				
Rab3a			(Davidsson et al 1999)		
Complexin I, II		(Eastwood and Harrison 2001)			
NCAM	(Honer et al 1997)				
GAP-43	(Blennow et al 1999)	(Eastwood and Harrison 2001)			(Eastwood and Harrison 1998)
Occipital cortex					
Synaptophysin		(Eastwood et al 2000b; Glantz and Lewis 1993; Glantz and Lewis 1997; Perrone-Bizzozero et al 1996)		(Eastwood et al 2000b)	
Rab3a		(Glantz and Lewis 1993; Glantz and Lewis 1997)			
GAP-43		(Perrone-Bizzozero et al 1996)		(Weickert et al 2001)	(Eastwood and Harrison 1998)

Table 2. continued

Hippocampus	Increased protein	Unchanged protein	Decreased protein	Unchanged mRNA	Decreased mRNA
Synaptophysin		(Browning et al 1993; Eastwood et al 1995; Vawter et al 1999; Young et al 1998)	(Davidsson et al 1999; Eastwood and Harrison 1995)		(Eastwood et al 1995; Eastwood and Harrison 1999; Webster et al 2001)
SNAP-25			(Young et al 1998)		
Rab3a			(Davidsson et al 1999)		
Complexin I		(Harrison and Eastwood 1998)			(Eastwood et al 2000a; Harrison and Eastwood 1998)
Complexin II			(Harrison and Eastwood 1998)		(Eastwood et al 2000a; Harrison and Eastwood 1998)
NCAM	(Vawter et al 1998)	(Breese et al 1995; Vawter et al 1999)	(Barbeau et al 1995)		
GAP-43	(Blennow et al 1999)			(Webster et al 2001)	(Eastwood and Harrison 1998)

Postmortem Studies of Connectivity in Affective Disorders

The results of recent morphometric and functional brain imaging studies set the stage for investigation of molecular aspects of connectivity in affective disorders. The early findings presented in Table 3 indicate some abnormalities may be present, particularly in molecules such as NCAM and GAP-43 which may contribute more to plasticity than synaptic neurotransmission directly.

Conclusions

Several models concerning syanptic connectivity as a mechanism of illness in schizophrenia are described. The most prominent of these are based on extrapolations from clinical features of schizophrenia (Feinberg 1982-83; McGlashan and Hoffman 2000), or on neuronal morphometric studies (Selemon and Goldman-Rakic 1999). The extent to which direct studies of presynaptic proteins in schizophrenia support these models is not entirely clear. Abnormalities of synaptic connectivity in schizophrenia near the time of death appear to be demonstrated in substantial numbers of studies, and the few studies in affective disorders indicate there may be altered connectivity in these illnesses also. However, widespread reductions in presynaptic markers do not appear to be consistently observed. No single marker appears to hold the key to understanding connectivity in mental disorders. Disruption can occur at many points in a neural circuit at a macroscopic level, and it may be that within synaptic terminals, circuits or pathways of interacting molecules need to be considered in a similar fashion.

Table 3. Presynaptic proteins in selected brain regions in affective disorders

Frontal cortex	Increased protein	Unchanged protein	Decreased protein	Unchanged mRNA	Decreased mRNA
Synaptophysin		(Honer et al 1999)			
SNAP-25		(Honer et al 2001; Jørgensen and Riederer 1985)			
Syntaxin		(Honer et al 2001)			
Complexin I		(Sawada et al 2001)			
Complexin II		(Sawada et al 2001)			
NCAM		(Jørgensen and Riederer 1985)			
GAP-43				(Weickert et al 2001)	
Cingulate cortex					
Synaptophysin		(Eastwood and Harrison 2001)	(Eastwood and Harrison 2001)		
Complexin I		(Eastwood and Harrison 2001)			
Complexin II			(Eastwood and Harrison 2001)		
GAP-43		(Eastwood and Harrison 2001)	(Eastwood and Harrison 2001)		
Hippocampus					
Synaptophysin		(Vawter et al 1999)			(Webster et al 2001)
SNAP-25	(Jørgensen and Riederer 1985)				
Complexin I					(Eastwood et al 2000a)
Complexin II					(Eastwood et al 2000a)
NCAM	(Jørgensen and Riederer 1985; Vawter et al 1999)	(Vawter et al 1998)			
GAP-43				(Webster et al 2001)	

Acknowledgements

Supported by a Scientist Award from the Canadian Institutes of Health Research, and CIHR grant MT14037.

References

Barbeau D, Liang JJ, Robitaille Y, Quirion R, Srivastava LK. Decreased expression of the embyronic form of the nerve cell adhesion molecule in schizophrenic brains. Proc Nat Acad Sciences USA 1995;92: 2785-2789.

Blennow K, Bogdanovic N, Gottfries CG, Davidsson P. The growth-associated protein GAP-43 is increased in the hippocampus and in the gyrus cinguli in schizophrenia. Journal of Molec Neurosci 1999;13: 101-109.

Breese CR, Freedman R, Leonard SS. Glutamate receptor subtype expression in human postmortem brain tissue from schizophrenics and alcohol abusers. Brain Res 1995;674: 82-90.

Brock T, O'Callaghan J. Quantitative changes in the synaptic vesicle proteins synapsin I and p38 and the astrocyte-specific protein glial fibrillary acidic protein are associated with chemical-induced injury to the rat central nervous system. J Neurosci 1987;7: 931-942.

Browning MD, Dudek EM, Rapier JL, Leonard S, Freedman R. Significant reductions in synapsin but not synaptophysin specific activity in the brains of some schizophrenics. Biol Psychiatry 1993;34: 529-535.

Cabalka L, Hyman B, Goodlett C, Ritchie T, Van Hoesen G. Alteration in the pattern of nerve terminal protein immunoreactivity in the perforant pathway in Alzheimer's disease and in rats after entorhinal lesions. Neurobiol Aging 1992;13: 283-291.

Daly C, Ziff EB. Post-transcriptional regulation of synaptic vesicle protein expression and the developmental control of synaptic vesicle formation. J Neurosci 1997;17: 2365-2375.

Davidsson P, Gottfries J, Bogdanovic N, et al. The synaptic-vesicle-specific proteins rab3a and synaptophysin are reduced in thalamus and related cortical brain regions in schizophrenic brains. Schizophr Res 1999;40: 23-29.

Davis S, Rodger J, Hicks A, Mallet J, Laroche S. Brain structure and task-specific increase in expression of the gene encoding syntaxin 1B during learning in the rat: a potential molecular marker for learning-induced synaptic plasticity in neural networks. Eur J Neurosci 1996;8: 2068-2074.

Eastwood SL, Burnet PWJ, Harrison PJ. Striatal synaptophysin and haloperidol-induced synaptic plasticity. NeuroReport 1994;5: 677-680.

Eastwood SL, Burnet PWJ, Harrison PJ. Altered synaptophysin expression as a marker of synaptic pathology in schizophrenia. Neuroscience 1995;66: 309-319.

Eastwood SL, Burnet PWJ, Harrison PJ. Expression of complexin I and II mRNAs and their regulation by antipsychotic drugs in the rat forebrain. Synapse 2000a;36: 167-177.

Eastwood SL, Cairns NJ, Harrison PJ. Synaptophysin gene expression in schizophrenia. Brit J Psychiatry 2000b;176: 236-242.

Eastwood SL, Cotter D, Harrison PJ. Cerebellar synaptic protein expression in schizophrenia. Neuroscience 2001;(in press).

Eastwood SL, Harrison PJ. Decreased synaptophysin in the medial temporal lobe in schizophrenia demonstrated using immunoautoradiography. Neuroscience 1995;69: 339-343.

Eastwood SL, Harrison PJ. Hippocampal and cortical growth-associated protein-43 messenger RNA in schizophrenia. Neuroscience 1998;86: 437-448.

Eastwood SL, Harrison PJ. Detection and quantification of hippocampal synaptophysin messenger RNA in schizophrenia using autoclaved, formalin-fixed, paraffin wax-embedded sections. Neuroscience 1999;93: 99-106.

Eastwood SL, Harrison PJ. Synaptic pathology in the anterior cingulate cortex in schizophrenia and mood disorders. Brain Res Bull 2001;(in press).

Eastwood SL, Heffernan J, Harrison PJ. Chronic haloperidol treatment differentially affects the expression of synaptic and neuronal plasticity-associated genes. Molec Psychiatry 1997;2: 322-329.

Feinberg I. Schizophrenia: caused by a fault in programmed synaptic elimination during adolescence? J Psychiatric Res 1982-83;17: 319-334.

Fog R, Pakkenberg H, Juul P, Bock E, Jørgensen OS, Andersen J. High-dose treatment of rats with perphenazine. Psychopharmacol 1976;50: 305-307.

Gabriel SM, Haroutunian V, Powchik P, et al. Increased concentrations of presynaptic proteins in the cingulate cortex of schizophrenics. Arch Gen Psychiatry 1997;54: 559-566.

Geddes JW, Hess EJ, Hart RA, Kesslak JP, Cotman CW, Wilson MC. Lesions of hippocampal circuitry define synaptosomal-associated protein-25 (SNAP-25) as a novel presynaptic marker. Neuroscience 1990;38: 515-525.

Glantz LA, Austin MC, Lewis DA. Normal cellular levels of synaptophysin mRNA expression in the prefrontal cortex of subjects with schizophrenia. Biol Psychiatry 2000;48: 389-397.

Glantz LA, Lewis DA. Synaptophysin and not rab3A is specifically reduced in the prefrontal cortex of schizophrenic subjects. Soc Neurosci Abstr 1993;20: 622.

Glantz LA, Lewis DA. Reduction of synaptophysin immunoreactivity in the prefrontal cortex of subjects with schizophrenia: regional and diagnstic specificity. Arch Gen Psychiatry 1997;54: 943-952.

Hamos JE, DeGennaro LJ, Drachman DA. Synaptic loss in Alzheimer's disease and other dementias. Neurology 1989;39: 355-361.

Harrison PJ. The neuropathological effects of antipsychotic drugs. Schizophr Res 1999;40: 87-99.

Harrison PJ, Eastwood SL. Preferential involvement of excitatory neurons in medial temporal lobe in schizophrenia. Lancet 1998;352: 1669-1673.

Heindel WC, Jernigan TL, Archibald SL, Achim CL, Masliah E, Wiley CA. The relationship of quantitative brain magnetic resonance imaging measures to neuropathologic indices of human immunodeficiency virus infection. Arch Neurology 1994;51: 1129-1135.

Helme-Guizon A, Davis S, Israel M, et al. Increase in syntaxin 1B and glutamate release in mossy fibre terminals following induction of LTP in the dentate gyrus: a candidate molecular mechanism underlying transsynaptic plasticity. Eur J Neurosci 1998;10: 2231-2237.

Hicks A, Davis S, Rodger J, Helme-Guizon A, Laroche S, Mallet J. Synapsin I and syntaxin 1B: key elements in the control of neurotransmitter release are regulated by neuronal activation and long-term potentiation in vivo. Neuroscience 1997;79: 329-340.

Honer WG, Falkai P, Bayer TA, et al. Abnormalities of SNARE mechanism proteins in anterior frontal cortex in severe mental illness. Cerebral Cortex 2001;(submitted).

Honer WG, Falkai P, Chen C, Arango V, Mann JJ, Dwork AJ. Synaptic and plasticity associated proteins in anterior frontal cortex in severe mental illness. Neuroscience 1999;91: 1247-1255.

Honer WG, Falkai P, Young C, et al. Cingulate cortex synaptic terminal proteins and neural cell adhesion molecule in schizophrenia. Neuroscience 1997;78: 99-110.

Honer WG, Young C, Falkai P. Synaptic pathology. In: Harrison PJ, Roberts GW (eds). The Neuropathology of Schizophrenia. Oxford University Press, Oxford, 2000; pp 105-136.

Jørgensen OS, Riederer P. Increased synaptic markers in hippocampus of depressed patients. J Neural Transmission 1985;64: 55-66.

Kamphuis W, Smirnova T, Hicks A, Hendriksen H, Mallet J, Lopes da Silva FH. The expression of syntaxin 1B/GR33 mRNA is enhanced in the hippocampal kindling model of epileptogenesis. J Neurochem 1995;65: 1974-1980.

Karson CN, Mrak RE, Schluterman KO, Stumer WQ, Sheng JG, Griffin WST. Alterations in synaptic proteins and their encoding mRNAs in prefrontal cortex in schizophrenia: a possible neurochemical basis for 'hypofrontality'. Molec Psychiatry 1999;4: 39-45.

Kroesen S, Marksteiner J, Mahata SK, et al. Effects of haloperidol, clozapine and citalopram on messenger RNA levels of chromogranins A and B and secretogranin II in various regions of rat brain. Neuroscience 1995;69: 881-891.

Lidow MS, Song Z-M, Castner SA, Allen PB, Greengard P, Goldman-Rakic PS. Antipsychotic treatment induces alterations in dendrite- and spine-associated proteins in dopamine-rich areas of the primate cerebral cortex. Biol Psychiatry 2001;49: 1-12.

Liu D, Diorio J, Day JC, Francis DD, Meaney MJ. Maternal care, hippocampal synaptogenesis and cognitive development in rats. Nature Neurosci 2000;3: 799-806.

Loessner B, Bullock S, Rose SPR. 411B: a monoclonal postsynaptic marker for modulations of synaptic connectivity in the rat brain. J Neurochem 1988;51: 385-390.

Lue L-F, Brachova L, Civin WH, Rogers J. Inflammation, Ab deposition, and neurofibrillary tangle formation as correlates of Alzheimer's disease neurodegeneration. Journal of Neuropathology and Experimental Neurology 1996;55: 1083-1088.

Lynch MA, Voss KL, Rodriguez J, Bliss TVP. Increase in synaptic vesicle proteins accompanies long-term potentiation in the dentate gyrus. Neuroscience 1994;60: 1-5.

Marin C, Tolosa E. Striatal synaptophysin levels are not indicative of dopaminergic supersensitivity. Neuropharmacol 1997;36: 1115-1117.

Masliah E, Ellisman M, Carragher B, et al. Three-dimensional analysis of the relationship between synaptic pathology and neuropil threads in Alzheimer disease. J Neuropathol Exp Neurol 1992;51: 404-414.

Masliah E, Fagan AM, Terry RD, DeTeresa R, Mallory M, Gage FH. Reactive synaptogenesis assessed by synaptophysin immunoreactivity is associated with GAP-43 in the dentate gyrus of the adult rat. Exp Neurol 1991;113: 131-142.

Masliah E, Terry RD, DeTeresa RM, Hansen LA. Immunohistochemical quantification of the synapse-related protein synaptophysin in Alzheimer disease. Neurosci Lett 1989;103: 234-239.

McGlashan TH, Hoffman RE. Schizophrenia as a disorder of developmentally reduced synaptic connectivity. Arch Gen Psychiatry 2000;57: 637-648.

Melloni RH, Hemmendinger LM, Hamos JE, DeGennnaro LJ. Synapsin I gene expression in the adult rat brain with comparative analysis of mRNA and protein in the hippocampus. J Comp Neurol 1993;327: 507-520.

Minger SL, Honer WG, Esiri MM, et al. Synaptic pathology in prefrontal cortex is present only with severe dementia in Alzheimer's disease. J Neuropath Exp Neurol 2001 (in press).

Mirnics K, Middleton FA, Marquez A, Lewis DA, Levitt P. Molecular characterization of schizophrenia viewed by microarray analysis of gene expression in prefrontal cortex. Neuron 2000;28: 53-67.

Mukaetova-Ladinska EB, Garcia-Siera F, Hurt J, et al. Staging of cytoskeletal and b-amyloid changes in human isocortex reveals biphasic synaptic protein response during progression of Alzheimer's disease. Am J Pathol 2000;157: 623-636.

Mullany PM, Lynch MA. Evidence for a role for synaptophysin in expression of long-term potentiation in rat dentate gyrus. NeuroReport 1998;9: 2489-2494.

Nakahara T, Motomura K, Hashimoto K, et al. Long-term treatment with haloperidol decreases the mRNA levels of complexin I, but not complexin II, in rat prefrontal cortex, nucleus acumbens and ventral tegmental area. Neurosci Lett 2000;290: 29-32.

Nakahara T, Nakamura K, Tsutsumi T, et al. Effect of chronic haloperidol treatment on synaptic protein mRNAs in the rat brain. Molec Brain Res 1998;61: 238-242.

Patanow CM, Day JR, Billingsley ML. Alterations in hippocampal expression of SNAP-25, GAP-43, stannin and glial fibrillary acidic protein following mechanical and trimethyltin-induced injury in the rat. Neuroscience 1997;76: 187-202.

Perrone-Bizzozero NI, Sower AC, Bird ED, Benowitz LI, Ivins KJ, Neve RL. Levels of the growth-associated protein GAP-43 are selectively increased in association cortices in schizophrenia. Proc Nat Acad Sci U.S.A. 1996;93: 14182-14187.

Poltorak M, Herranz AS, Williams J, Lauretti L, Freed WJ. Effects of frontal cortical lesions on mouse striatum: reorganizationof cell recognition molecule, glial fiber, and synaptic protein expression in the dorsomedial striatum. J Neurosci 1993;13: 2217-2223.

Richter-Levin G, Thomas KL, Hunt SP, Bliss TVP. Dissociation between genes activated in long-term potentiation and in spatial learning in the rat. Neurosci Lett 1998;251: 41-44.

Roberts LA, Morris BJ, O'Shaughnessey CT. Involvement of two isoforms of SNAP-25 in the expression of long-term potentiation in the rat hippocampus. NeuroReport 1998;9: 33-36.

Rodger J, Davis S, Laroche S, Mallet J, Hicks A. Induction of long-term potentiation in vivo regulates alternate splicing to alter syntaxin 3 isoform expression in rat dentate gyrus. J Neurochem 1998;71: 666-675.

Sawada K, Takahashi S, Dwork AJ, Li H-Y, Hu L, Falkai P. Complexins I and II in anterior frontal cortex in schizophrenia. Schizophr Res 2001;(in press).

Selemon LD, Goldman-Rakic PS. The reduced neuropil hypothesis: a circuit based model of schizophrenia. Biol Psychiatry 1999;45: 17-25.

Sokolov BP, Tcherepanov AA, Haroutunian V, Davis KL. Levels of mRNAs encoding synaptic vesicle and synaptic plasma membrane proteins in the temporal cortex of elderly schizophrenic patients. Biol Psychiatry 2000;48: 184-196.

Stefan MD, Horton K, Johnston P, Bruton CJ, Roberts GW, Royston MC. Synaptic pathology in schizophrenia: abnormalities of the prefrontal cortex. Schizophr Res 1995;15: 32.

Tcherepanov AA, Sokolov BP. Age-related abnormalities in expression of mRNAs encoding synapsin 1A, synapsin 1B, and synaptophysin in temporal cortex of schizophrenics. J Neurosci Res 1997;49: 639-644.

Terry RD, Masliah E, Salmon DP, et al. Physical basis of cognitive alterations in Alzheimer's disease: synapse loss is the major correlate of cognitive impairment. Ann Neurol 1991;30: 572-580.

Thompson PM, Sower AC, Perrone-Bizzozero NI. Altered levels of the synaptosomal associated protein SNAP-25 in schizophrenia. Biol Psychiatry 1998;43: 239-243.

Vawter MP, Cannon-Spoor HE, Hemperly JJ, et al. Abnormal expression of cell recognition molecules in schizophrenia. Exp Neurol 1998;149: 424-432.

Vawter MP, Howard AL, Hyde TM, Kleinman JE, Freed WJ. Alterations of hippocampal secreted N-CAM in bipolar disorder and synaptophysin in schizophrenia. Molec Psychiatry 1999;4: 467-475.

Walaas SI, Jahn R, Greengard P. Quantitation of nerve terminal populations: synaptic vesicle-asociated proteins as markers for synaptic density in the rat neostriatum. Synapse 1988;2: 516-520.

Webster MJ, Weickert CS, Herman MM, Hyde TM, Kleinman JE. Synaptophysin and GAP-43 mRNA levels in the hippocampus of subjects with schizophrenia. Schizophr Res 2001;49: 61-70.

Weickert CS, Webster MJ, Hyde TM, et al. Reduced GAP-43 mRNA in dorsolateral prefrontal cortex of patients with schizophrenia. Cereb Cortex 2001;11: 136-147.

Young CE, Arima K, Xie J, et al. SNAP-25 deficit and hippocampal connectivity in schizophrenia. Cereb Cortex 1998;8: 261-268.

6 STUDIES IN THE HUMAN FRONTAL CORTEX: EVIDENCE FOR CHANGES IN NEUROCHEMICAL MARKERS IN SCHIZOPHRENIA AND BIPOLAR DISORDER

Brian Dean

Abstract

Commonality of treatments in schizophrenia and bipolar disorder suggest that there may be a single pathology driving some of the symptoms of these illnesses. To test this hypothesis, the literature on the changes in the molecular neuroanatomy of postmortem brain tissue from subjects with schizophrenia and bipolar disorder has been reviewed to attempt to identify a common neurobiology. Such studies provide evidence to suggest that changes in the serotonergic system of the frontal cortex may be such a common factor. However, currently available data does not support the argument that the same changes in the serotonergic systems of the frontal cortex are present in schizophrenia and bipolar disorder. By contrast, such data would suggest that targeting receptor-G-protein interactions might be therapeutically beneficial in both illnesses.

Introduction

The underlying pathology of schizophrenia and bipolar disorder remain to be discovered. Given that the combined incidence of schizophrenia and bipolar disorder is approximately 2% of the population worldwide (Andrews et al., 1985;Keller and Baker, 1991), there is a clear need to identify the pathologies of these illnesses.

The symptomatology of schizophrenia and bipolar disorder appear distinct. In schizophrenia there are psychotic episodes and, in some individuals, prominent negative symptoms (affective flattening, alogia and avolition) (American Psychiatric Association, 1994). In bipolar disorder there are manic episodes, periods of major depression and hypomanic episodes (American Psychiatric Association, 1994). However, from the neuropharmacological standpoint, the psychoses of schizophrenia and the mania of bipolar disorder can both be treated with antipsychotic drugs (Calabrese et al., 1995;Wirshing et al., 1995). This would suggest that there might be some commonality in the biochemistry of psychoses and mania.

Both lithium and anticonvulsive drugs have also been shown to be effective in controlling mania (Calabrese et al., 1995). Significantly, both these drugs are used in combination with antipsychotic drug treatment to give improved outcomes in the treatment of some individuals with schizophrenia (Wirshing et al., 1995). These data would again support the hypothesis that there may be some common neurobiological abnormalities involved in the genesis of psychoses and mania.

Macro- or microscopic studies of the brain have failed to produce convincing data as to the pathology for schizophrenia (Stevens, 1997) or bipolar disorder (Vawter et al., 2000). This has led many to suggest that the pathology of both these illnesses are due to changes in the molecular cytoarchitecture of the brain. Moreover, in the absence of a known underlying pathology, the pharmacology of drugs used to treat the illness has been used to direct the search for molecular causes of both illnesses. Thus, in the case of psychotic disorders the improved outcomes from the use of atypical antipsychotic drugs has led to consideration of their pharmacology. As these compounds antagonise dopamine and serotonin receptors, it has been proposed that changes in the dopaminergic and serotonergic systems of the brain could be involved in the genesis of psychoses (Huttunen, 1995). This argument is supported by the observation that dopaminergic agonists such as amphetamine, and the serotonergic agonist lysergic acid, can both induce or exacerbate psychotic episodes (Meltzer, 1987).

In the case of bipolar disorder both lithium and anticonvulsant drugs have proven useful in treating mania. It is still not clear how these drugs achieve their therapeutic outcomes but both types of drugs affect key second messenger systems in the brain, particularly the inositol system (O'Donnell et al., 2000). These data suggest that changes in second messenger systems may be central to the pathology of bipolar disorder.

Whilst neuropharmacological observations can provide direction to systems in the brain that may be involved in psychiatric disorders, they cannot provide direction as to the regions of the brain affected by such disorders. However, it would seem logical that the regions of the brain that control key behaviours that seem to be affected in schizophrenia and bipolar disorder would be involved in the pathology of the illness. Notably, perturbation of cognitive function is apparent in both schizophrenia and bipolar disorder, and these functions are increasingly recognised as being controlled by the frontal cortex (Scott et al., 2000; Stevens et al., 1998). Hence, it would seem likely that changes in the cytoarchitecture of the frontal cortex would be important in the pathology of both schizophrenia and bipolar disorder. The frontal cortex of the human brain is large and multifunctional

(Fuster, 1989) meaning that the localisation of molecular changes within the frontal cortex may add to an understanding of their possible behavioural consequences. Confirming that abnormalities are localised to specific regions of the frontal cortex that control the functions that are abnormal in schizophrenia or bipolar disorder would support a role for such changes in the pathology of the illness.

Given the above, it would seem opportune to review data from the study of the frontal cortex obtained postmortem from subjects with schizophrenia and bipolar disorder. In particular, a review focusing on findings relating to the systems of the frontal cortex most likely affected by the currently available therapeutic agents used to treat the illnesses would be most appropriate. At the beginning of a new millennium it is important to consider how such data may be providing information on the pathology of schizophrenia and bipolar disorder and how it may influence the design of new, more effective therapies for both illnesses.

Studies in Schizophrenia

All antipsychotic drugs currently used to treat psychoses, be it associated with schizophrenia or bipolar disorder, are antagonists of the dopaminergic systems of the brain. Hence the long-standing dopamine hypothesis of schizophrenia, which suggests that hyperactive dopaminergic neurons are involved in the genesis of psychoses (Meltzer, 1987), is still worth considering in trying to understand the underlying pathology of psychoses. It is therefore of significance that it has been reported that there is a reduction in the size of dopaminergic neurons in lamina VI of the frontal cortex of subjects with schizophrenia that appears to be independent of antipsychotic drug treatment (Akil et al., 1999). This would suggest that there should be significant changes in cortical dopaminergic function in schizophrenia.

In contrast to the study of dopaminegic neurons using tyrosine hydroxylase as a marker (Akil et al., 1999) studies on another pre-synaptic dopaminergic marker, the dopamine transporter, have failed to show any differences in the frontal cortex of subjects with schizophrenia (Dean et al., 1999a;Hitri et al., 1995). These data could indicate that although there are morphological abnormalities in dopaminegic neurons in the frontal cortex of subjects with schizophrenia these abnormalities are not preventing the neurons functioning within normal neurochemical parameters. This hypothesis would be supported by the observations that other markers on dopaminergic neurons, the dopamine D_1-like and dopamine D_2-like receptors, are not altered in the frontal cortex of subjects with schizophrenia (Dean et al., 1999a;Knable et al., 1996).

The use of radioligand binding to identify changes in dopamine receptors is limited because radioligands specific for each of the five forms of dopamine receptor are not yet available. This has led to the study of levels of mRNA for specific dopamine receptors in the frontal cortex of subjects with schizophrenia. One such study reported a decrease in levels of mRNA for the D3 and D4 receptor that was restricted to Brodmann's area 11 of the cortex from subjects with schizophrenia (Meador-Woodruff et al., 1997). By contrast a second study has reported that mRNA for the D4 receptor is increased in the frontal cortex from subjects with schizophrenia (Stefanis et al., 1998) whilst a third study reported that levels of mRNA for that receptor were not changed in the frontal cortex (Mulcrone and Kerwin, 1996). These data could be indicating that there may be specific regions of the frontal cortex from subjects with schizophrenia that are expressing dopamine receptors at different levels to those seen in the same region from control subjects.

It is still not clear from studies of mRNA, receptor proteins and dopaminergic cell structure whether or not there are changes in the dopaminergic systems of the frontal cortex of subjects with schizophrenia. Given the prominent role of the dopamine hypothesis in our thinking on schizophrenia it will be important to generate more data to help clarify if a change in the dopaminergic system does have a role in the pathology of the illness.

The most consistent finding in the frontal cortex obtained postmortem from subjects with schizophrenia is a decrease in the density of the serotonin$_{2A}$ ($5HT_{2A}$) receptor (Arora and Meltzer, 1991;Bennett et al., 1979; Burnet et al., 1996; Dean et al., 1998; Dean and Hayes, 1996; Gurevich and Joyce, 1997; Mita et al., 1986). Significantly, the $5HT_{2A}$ receptor appears to be predominately localised in cortical laminae III and IV (Dean and Hayes, 1996) (Figure 1). This would suggest that changes in that receptor in subjects with schizophrenia may be confined to specific laminae within the frontal cortex. To explore this hypothesis further it will be important to determine which synapses in the laminae containing the $5HT_{2A}R$ have decreased levels of that receptor.

Some of the regions of the frontal cortex that have been shown to have decreases in $5HT_{2A}$ receptors have also been shown to have decreases in the levels of mRNA encoding for that receptor (Burnet et al., 1996). In these regions, it would appear that regulation of gene expression may have an important role in the changes in $5HT_{2A}$ receptor. However, it is less clear what mechanisms are at play in regions where there is altered receptor density with no change in levels of mRNA encoding for the receptor. In these regions, it would seem that the decrease in receptor density would be primarily due to an increase in the removal of the receptor from neuronal membranes. Thus, the changes in $5HT_{2A}$ receptors in different brain regions from subjects with schizophrenia may be complex and due to different underlying mechanisms.

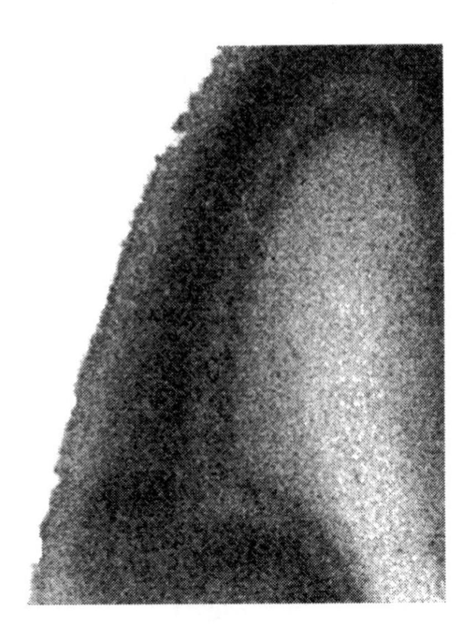

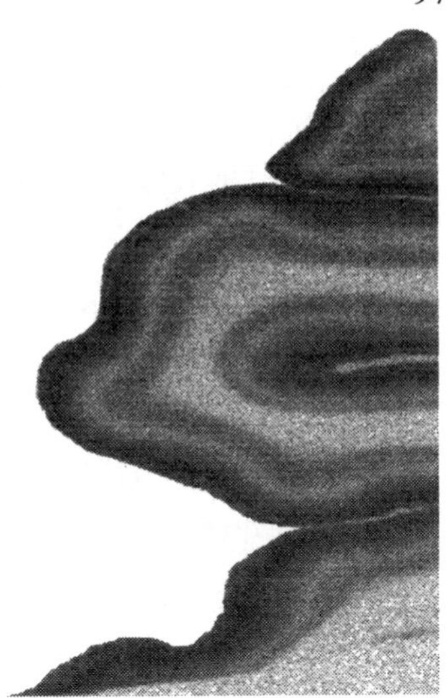

Figure 1: The distribution of $5HT_{2A}$ and $5HT_{1A}$ receptors in Brodmann's area 9 of the human frontal cortex.

There have also been a number of studies suggesting there is an increase in the density of $5HT_{1A}$ receptors in the frontal cortex form subjects with schizophrenia (Burnet et al., 1996; Hashimoto et al., 1991; Laruelle et al., 1993; Simpson et al., 1996; Sumiyoshi et al., 1996). Notably, the $5HT_{1A}$ receptor is localised to laminae I and II as well as V and VI of the human frontal cortex (Dean et al., 1999b) (Figure 1). Hence, in the main, there is a discrete separation of the two predominant 5HT receptors ($5HT_{2A}$ and $5HT_{1A}$) in the human frontal cortex. The cortical separation of the $5HT_{2A}$ and $5HT_{1A}$ receptors has led to the suggestion that the opposing changes in these receptors in schizophrenia is an attempt to compensate for an abnormal pathological process in schizophrenia (Hashimoto et al., 1993). It could therefore be significant that in Brodmann's Area 9 (BA 9) of the frontal cortex from subjects with schizophrenia, where there is a change in the $5HT_{2A}$ receptor (Dean et al., 1998), there does not appear to be a change in the density of the $5HT_{1A}$ receptor (Dean et al., 1999b; Joyce et al., 1993). These data could suggest that BA 9 is particularly important in the pathology of schizophrenia because opposing changes in $5HT_{2A}$ and $5HT_{1A}$ receptors have not successfully compensated for a pathological drive into that region of the cortex.

One source of abnormal serotonergic drive onto serotonergic receptors would be alterations in the uptake and release of serotonin. The uptake of serotonin is predominately controlled by the serotonin transporter on pre-synaptic neurons. There are two reports of a decrease in the density of the serotonin transporter in the frontal cortex of subjects with schizophrenia (Joyce et al., 1993; Laruelle et al., 1993). These data would support the hypothesis that the changes in serotonin receptors in the cortex from subjects with schizophrenia are reflecting attempts to compensate for an abnormal serotonergic drive because of a change in the serotonin transporter. However, against this is the fact that the reported changes in the serotonin transporter in the frontal cortex of subjects with schizophrenia are small and that these changes have not been found in all studies (Dean et al., 1995; Dean et al., 1999b).

In conclusion, there would appear to be strong evidence that there are changes in the serotonergic system of the frontal cortex from subjects with schizophrenia. At present neither the cause nor the effect of these changes on the functioning of the human cortex is known. Therefore, further work is required to determine the full extent and implications of changes in this important system within the frontal cortex. Hence the demonstration that the $5HT_4$ receptor is not changed in the frontal cortex of subjects with schizophrenia (Dean et al., 1999b) is an important finding as it shows there is not generalised changes in serotonin receptors in the cortex from subjects with the illness. There are at least 13 serotonin receptors thought to be present in the human brain (Hoyer et al., 1994), until the status of these receptors are examined in the brains of subjects with schizophrenia, the full extend of changes of the serotonergic system will not be fully understood.

Studies in Bipolar Disorder

The effectiveness of antipsychotic drug treatment in bipolar disorder could also suggest an involvement of the dopaminergic and serotonergic systems in the frontal cortex in the pathology of bipolar disorder. It has been proposed that changes in the dopaminergic system is important in mania (Gerner et al., 1976). However, the absence of changes in important markers on the dopaminergic system in the frontal cortex of subjects with bipolar disorder (Dean et al., 2000) could argue that changes in the dopaminergic system of the frontal cortex are not central to the pathology of bipolar disorder. However, more studies are necessary prior to making such a conclusion.

In the case of the serotonergic system, the suggestion that there was a decrease in the turnover of serotonin in frontal cortex of subjects with bipolar disorder (Young et al., 1994) appeared to support a role for changes in the serotonergic system in the pathology of the disorder. However, if there is a change in the serotonergic system of the frontal cortex of subjects with bipolar disorder it has not manifested itself in changes in important serotonergic markers within that brain region. Thus, changes in neither the density of the serotonin transporter (Dean et al., 2000; Leake et al., 1991) nor the $5HT_{2A}$ or $5HT_{1A}$ receptors (Dean et al., 2000) have been detected in frontal cortex obtained postmortem from subjects with bipolar disorder.

However, the exploration of serotonergic markers in the frontal cortex of subjects with bipolar disorder needs to be expanded. Until such research is completed, it will be difficult to draw any firm conclusions as to the role of changes in the serotonergic system in the frontal cortex of subjects with bipolar disorder.

The usefulness of lithium and anticonvulsant drugs in mood stabilisation in subjects with bipolar disorder has been an important signpost to direct research using tissue obtained postmortem. Both lithium and the anticonvulsants have been shown to act on the inositol signalling system (O'Donnell et al., 2000). This, plus the demonstration that there were decreased levels of inositol in the frontal cortex from subjects with bipolar disorder (Shimon et al., 1997) seem to support the hypothesis that changes in second messenger systems in the frontal cortex were important in the pathology of the illness.

It has been reported that phosphatidylinositol turnover, a marker of the activity of inositol-dependent second messenger systems, was not altered in the frontal cortex of subjects with bipolar disorder despite showing differences in the temporal cortex (Jope et al., 1996). These data could indicate that, despite the presence of increased levels of inositol, the overall activity of its related second messenger cascades are unaltered in the frontal cortex of subjects with bipolar disorder. If that conclusion is correct, then it is difficult to hypothesis how the changes in inositol may be of pathophysiological significance. More widespread changes in second messenger cascades were suggested by the finding that there were decreased levels of [^{3}H]cyclic AMP binding in the frontal cortex and other regions of the brain of subjects with bipolar disorder (Rahman et al., 1997). Clearly, more experimental data will be required to understand the extent and the mechanisms by which changes in second messenger cascades may be involved in bipolar disorder.

One interesting finding suggested that there was an increase in receptor-G protein coupled responses to various receptor agonists, including serotonin, in the frontal cortex from subjects with bipolar disorder (Friedman and Wang, 1996). This would suggest there could be an exaggerated response to serotonin despite normal levels of the neurotransmitter and its receptors in the frontal cortex of subjects with bipolar disorder. Support for this hypothesis comes from the report that there are increased levels of $G_{S\alpha}$ in the frontal cortex of subjects with bipolar disorder (Young et al., 1991). Significantly, this increase in G-protein is not accompanied by changes in levels of mRNA (Young et al., 1996). This finding could indicate that an increased level of $G_{s\alpha}$ protein would led to an increase in G-protein/coupled receptors complexes in response to a specific serotonin stimuli in the frontal cortex of subjects with bipolar disorder.

Summary

The posit to be tested in this review of the current literature was that the commonality of treatments for psychoses and mania suggested some underlying common neurobiological abnormality. Data from the study of frontal cortex obtained postmortem from subjects with schizophrenia and bipolar disorder would suggest that this common abnormality may be a change in the serotonergic system of the frontal cortex. However, these same data would suggest the change in the serotonergic system involve two separate mechanisms.

In schizophrenia, current data suggest that psychoses may be due to a hyperserotonergic drive in some regions of the frontal cortex (Dean, 2000), the cause of which remains to be discovered. In an attempt to compensate for overactivity, there is a decrease in the levels of $5HT_{2A}$ receptor with an associated increase in levels of $5HT_{1A}$ receptor (Hashimoto et al., 1993). It would seem possible that the physiological constraints on changes in serotonin receptor density fail to fully compensate for the increased drive. Therefore antagonising a critical receptor in this system, the $5HT_{2A}$ receptor, allows for the final compensation in the overactivity of the system to be attained.

In the mania associated with bipolar disorder, there may be an abnormality in serotonergic drive (Young et al., 1994) but this would not seem to be of sufficient magnitude to cause a change in other serotonergic markers in the frontal cortex (Dean et al., 2000). However, changes in the amplification of the signal from the serotonin receptors due to increases in $G_{S\alpha}$ protein, possibly due to an increase in G-protein/receptor coupling, could cause the equivalent of a hyperserotonergic state in the frontal cortex of subjects with bipolar disorder. In that situation, the use of atypical antipsychotic drugs would be appropriate, as antagonising the $5HT_{2A}$ receptor would partially dampen the amplified signal.

Whilst the above hypothesis would account for the effectiveness of antipsychotic drugs in schizophrenia and mania, can a similar hypothesis account for the effects of lithium in mania and psychoses? It has been shown that treatment with lithium increases levels of inhibitory G-proteins (Spleiss et al., 1998). This could explain why lithium is effective in bipolar disorder where, increasing levels of G_l proteins may assist in reducing the impact of a signal over-amplification due to excess G_s proteins. Similarly, in some individuals with schizophrenia, a dampening of an excess signal through the serotonin receptors could be achieved by a generalised increase in levels of G_l proteins following treatment with lithium. The same argument could be raised for anticonvulsant drugs which are also thought to act at the level of the G-proteins (Avissar et al., 1990).

It must be recognised that, at least in the case of schizophrenia, we are dealing with a syndrome and therefore any hypothesis will only likely apply to a sub-set of individuals. However, the argument that different mechanisms in the serotonergic pathways are involved in the genesis of both psychoses and mania provides an intriguing and testable hypothesis. Further studies using postmortem tissue and suitable peripheral tissue from subjects with schizophrenia and bipolar disorder as well as studies in animals and using cell culture will be required to allow the validity of the hypothesis to be tested. However, if the hypothesis is proven, new drugs designed to target receptor-G-protein interaction could prove to be the next generation of therapeutics in schizophrenia and bipolar disorder.

Acknowledgements

Associate Professor Brian Dean is the holder of a NARSAD Young Investigator Award.

References

Akil M, Pierri JN, Whitehead RE, Edgar CL, Mohila C, Sampson AR, and Lewis DA. Lamina-specific alterations in the dopamine innervation of the prefrontal cortex in schizophrenic subjects. A J Psychiatry 1999;156:1580-1589.

American Psychiatric Association. Diagnostic and statistical manual of mental disorders (Forth edition). American Psychiatric Association , Washington, D.C., 1994.

Andrews G, Hall W, Goldstein G, Lapsley H, Bartels R, and Silove D. The economic costs of schizophrenia. Implications for public policy. Arch Gen Psychiatry 1985;42:537-543.

Arora RC and Meltzer HY. Serotonin2 (5-HT2) receptor binding in the frontal cortex of schizophrenic patients. J Neural Transm Gen Sect 1991;85:19-29.

Avissar S, Schreiber G, Aulakh CS, Wozniak KM, and Murphy DL. Carbamazepine and electroconvulsive shock attenuate beta-adrenoceptor and muscarinic cholinoceptor coupling to G proteins in rat cortex. Eur J Pharmacol 1990;189:99-103.

Bennett JP, Enna SJ, Bylund DB, Gillin JC, Wyatt RJ, and Snyder SH. Neurotransmitter receptors in frontal cortex of schizophrenics. Arch Gen Psychiatry 1979;36:927-934.

Burnet PW, Eastwood SL, and Harrison PJ. 5-HT1A and 5-HT2A receptor mRNA and binding site densities are differentially altered in schizophrenia. Neuropsychopharmacology 1996;15:442-455.

Calabrese JR, Bowden V, and Woyshville MJ. Lithium and the anticonvilsants in bipolar disorder. In: Bloom FE and Kupfer DJ, (eds). Psychopharmacology: The Fourth Generation of Progress. Raven Press ,New York, 1995; pp 1099-1111.

Dean B. Signal transmission, rather than reception, is the underlying neurochemical abnormality in schizophrenia. Aust N Z J Psychiatry 2000;34:560-569.

Dean B and Hayes W. Decreased frontal cortical serotonin2A receptors in schizophrenia. Schizophr Res 1996;21:133-139.

Dean B, Hayes W, Hill C, and Copolov D. Decreased serotonin2A receptors in Brodmann's area 9 from schizophrenic subjects. A pathological or pharmacological phenomenon? Mol Chem Neuropathol 1998;34:133-145.

Dean B, Hussain T, Hayes W, Scarr E, Kitsoulis S, Hill C, Opeskin K, and Copolov DL. Changes in serotonin2A and GABA(A) receptors in schizophrenia: studies on the human dorsolateral prefrontal cortex. J Neurochem 1999a;72:1593-1599.

Dean B, Opeskin K, Pavey G, Naylor L, Hill C, Keks N, and Copolov DL. [3H]paroxetine binding is altered in the hippocampus but not the frontal cortex or caudate nucleus from subjects with schizophrenia. J Neurochem 1995;64:1197-1202.
Dean B, Pavey G, McLeod M, Opeskin K, Keks N, and Copolov D. Evidence for an abnormality in the assembly of the $GABA_A$ receptor in the prefrontal cortex of subjects with bipolar disorder. J Affect Dis 2000; (In press).
Dean B, Tomaskovic-Crook E, Opeskin K, Keks N, and Copolov D. No change in the density of the serotonin1A receptor, the serotonin4 receptor or the serotonin transporter in the dorsolateral prefrontal cortex from subjects with schizophrenia. Neurochem Int 1999b;34:109-115.
Friedman E and Wang HY. Receptor-mediated activation of G proteins is increased in postmortem brains of bipolar affective disorder subjects. J Neurochem 1996;67:1145-1152.
Fuster JM. The Prefrontal Cortex: Anatomy, Physiology and Neuropsychology of the Frontal Lobe. Raven Press, New York, 1989.
Gerner RH, Post RM, and Bunney WE, Jr. A dopaminergic mechanism in mania. Am J Psychiatry 1976;133:1177-1180.
Gurevich EV and Joyce JN. Alterations in the cortical serotonergic system in schizophrenia: A postmortem study. Biol Psychiatry 1997;42:529-545.
Hashimoto T, Kitamura N, Kajimoto Y, Shirai Y, Shirakawa O, Mita T, Nishino N, and Tanaka C. Differential changes in serotonin 5-HT1A and 5-HT2 receptor binding in patients with chronic schizophrenia. Psychopharmacology (Berl) 1993;112:S35-S39.
Hashimoto T, Nishino N, Nakai H, and Tanaka C. Increase in serotonin 5-HT1A receptors in prefrontal and temporal cortices of brains from patients with chronic schizophrenia. Life Sci 1991;48:355-363.
Hitri A, Casanova MF, Kleinman JE, Weinberger DR, and Wyatt RJ. Age-related changes in [3H]GBR 12935 binding site density in the prefrontal cortex of controls and schizophrenics. Biol Psychiatry 1995;37:175-182.
Hoyer D, Clarke DE, Fozard JR, Hartig PR, Martin GR, Mylecharane EJ, Saxena PR, and Humphrey PP. International Union of Pharmacology classification of receptors for 5-hydroxytryptamine (Serotonin). Pharmacol Rev 1994;46:157-203.
Huttunen M. The evolution of the serotonin-dopamine antagonist concept. J Clin Psychopharmacol 1995;1, Suppl 15:4S-10S.
Jope RS, Song L, Li PP, Young LT, Kish SJ, Pacheco MA, and Warsh JJ. The phosphoinositide signal transduction system is impaired in bipolar affective disorder brain. J Neurochem 1996;66:2402-2409.
Joyce JN, Shane A, Lexow N, Winokur A, Casanova MF, and Kleinman JE. Serotonin uptake sites and serotonin receptors are altered in the limbic system of schizophrenics. Neuropsychopharmacology 1993;8:315-336.
Keller MB and Baker LA. Bipolar disorder: epidemiology, course, diagnosis, and treatment. Bull Menninger Clin 1991;55:172-181.
Knable MB, Hyde TM, Murray AM, Herman MM, and Kleinman JE. A postmortem study of frontal cortical dopamine D1 receptors in schizophrenics, psychiatric controls, and normal controls. Biol Psychiatry 1996;40:1191-1199.
Laruelle M, Abi-Dargham A, Casanova MF, Toti R, Weinberger DR, and Kleinman JE. Selective abnormalities of prefrontal serotonergic receptors in schizophrenia: A postmortem study. Arch Gen Psychiatry 1993;50:810-818.
Leake A, Fairbairn AF, McKeith IG, and Ferrier IN. Studies on the serotonin uptake binding site in major depressive disorder and control postmortem brain: neurochemical and clinical correlates. Psychiatry Res 1991;39:155-165.
Meador-Woodruff JH, Haroutunian V, Powchik P, Davidson M, Davis KL, and Watson SJ. Dopamine receptor transcript expression in striatum and prefrontal and occipital cortex. Focal abnormalities in orbitofrontal cortex in schizophrenia. Arch Gen Psychiatry 1997;54:1089-1095.
Meltzer HY. Biological studies in schizophrenia. Schizophr Bull 1987;13:77-107.

Mita T, Hanada S, Nishino N, Kuno T, Nakai H, Yamadori Y, and Tanaka C. Decreased serotonin S2 and increased dopamine D2 receptors in chronic schizophrenics. Biol Psychiatry 1986;21:1407-1414.

Mulcrone J and Kerwin RW. No difference in the expression of the D4 gene in postmortem frontal cortex from controls and schizophrenics. Neurosci Lett 1996;219:163-166.

O'Donnell T, Rotzinger S, Nakashima TT, Hanstock CC, Ulrich M, and Silverstone PH. Chronic lithium and sodium valproate both decrease the concentration of myo-inositol and increase the concentration of inositol monophosphates in rat brain. Brain Res 2000;880:84-91.

Rahman S, Li PP, Young LT, Kofman O, Kish SJ, and Warsh JJ. Reduced [3H]cyclic AMP binding in postmortem brain from subjects with bipolar affective disorder. J Neurochem 1997;68:297-304.

Scott J, Stanton B, Garland A, and Ferrier IN. Cognitive vulnerability in patients with bipolar disorder. Psychol Med 2000;30:467-472.

Shimon H, Agam G, Belmaker RH, Hyde TM, and Kleinman JE. Reduced frontal cortex inositol levels in postmortem brain of suicide victims and patients with bipolar disorder. Am J Psychiatry 1997;154:1148-1150.

Simpson MDC, Lubman DI, Slater P, and Deakin JFW. Autoradiography with [3H]8-OH-DPAT reveals increases in 5-HT1A receptors in ventral prefrontal cortex in schizophrenia. Biol Psychiatry 1996;39:919-928.

Spleiss O, van Calker D, Scharer L, Adamovic K, Berger M, and Gebicke-Haerter PJ. Abnormal G protein alpha(s) - and alpha(i2)-subunit mRNA expression in bipolar affective disorder. Mol Psychiatry 1998;3:512-520.

Stefanis NC, Bresnick JN, Kerwin RW, Schofield WN, and McAllister G. Elevation of D4 dopamine receptor mRNA in postmortem schizophrenic brain. Brain Res Mol Brain Res 1998;53:112-119.

Stevens AA, Goldman-Rakic PS, Gore JC, Fulbright RK, and Wexler BE. Cortical dysfunction in schizophrenia during auditory word and tone working memory demonstrated by functional magnetic resonance imaging. Arch Gen Psychiatry 1998;55:1097-1103.

Stevens JR. Anatomy of schizophrenia revisited. Schizophr Bull 1997;23:373-383.

Sumiyoshi T, Stockmeier CA, Overholser JC, Dilley GE, and Meltzer HY. Serotonin1A receptors are increased in postmortem prefrontal cortex in schizophrenia. Brain Res 1996;708:209-214.

Vawter MP, Freed WJ, and Kleinman JE. Neuropathology of bipolar disorder. Biol Psychiatry 2000;48:486-504.

Wirshing WC, Marder SR, Van Putten T, and Ames D. Acute treatment of schizophrenia. In: Bloom FE and Kupfer DJ, (eds). Psychopharmacology: The Fourth Generation of Progress. Raven Press, New York, 1995; pp 1259-1275.

Young LT, Asghari V, Li PP, Kish SJ, Fahnestock M, and Warsh JJ. Stimulatory G-protein alpha-subunit mRNA levels are not increased in autopsied cerebral cortex from patients with bipolar disorder. Brain Res Mol Brain Res 1996,42:45-50.

Young LT, Li PP, Kish SJ, Siu KP, and Warsh JJ. Postmortem cerebral cortex Gs alpha-subunit levels are elevated in bipolar affective disorder. Brain Res 1991;553:323-326.

Young LT, Warsh JJ, Kish SJ, Shannak K, and Hornykeiwicz O. Reduced brain 5-HT and elevated NE turnover and metabolites in bipolar affective disorder. Biol Psychiatry 1994;35:121-127.

7 SUMMARY OF PREFRONTAL MOLECULAR ABNORMALITIES IN THE STANLEY FOUNDATION NEUROPATHOLOGY CONSORTIUM

Michael B. Knable, Beata M. Barci, Maree J. Webster, E. Fuller Torrey

Abstract

Postmortem specimens from the Stanley Foundation Neuropathology Consortium, which contains matched samples from patients with schizophrenia, bipolar disorder, non-psychotic depression, and normal controls (n=15 per group), have been distributed to many research groups around the world. This chapter provides a summary of abnormal markers found in prefrontal cortical areas from this collection between 1997 and 2000. From 69 separate data sets, a total of 17 abnormal markers were identified that pertained to a variety of neural systems and processes including neuronal plasticity, neurotransmission, signal transduction, inhibitory interneuron function, and glial cells. Schizophrenia was associated with the largest number of abnormalities, many of which were also present in bipolar disorder. Major depression was associated with relatively few abnormalities. Most abnormal findings represented a decrease in protein or mRNA levels that could not be fully explained by exposure to psychotropic or illicit drugs or by other confounding variables. It is argued that the abnormal findings are not due to stochastic processes but represent viable markers for independent replication and further study as candidate genes or targets for new treatments.

Introduction

The Stanley Foundation Brain Collection was established in 1994 in order to develop an international postmortem tissue resource for studies of the major psychiatric illnesses (schizophrenia, bipolar disorder, and major depression). The collection is maintained within the Department of

Psychiatry of the Uniformed Services University of the Health Sciences (USUHS) in Bethesda, Maryland, and presently consists of over 400 specimens. Brains specimens were obtained at autopsy by medical examiners in several U.S. cities. Details regarding the selection of cases, collection of clinical data, diagnostic procedure, and processing of postmortem tissue have been published previously (Torrey et al., 2000).

The Stanley Foundation Neuropathology Consortium is a selection of matched specimens from within the larger Brain Collection. It contains 15 cases each from individuals with schizophrenia, bipolar disorder, major depressive disorder without psychotic features, and normal controls. These groups are matched as shown in Table 1 and were all collected between September 1994 and February 1997. Table 2 contains data regarding cause of death, presence of psychosis, family history of psychosis, lifetime exposure to antipsychotic drugs, and co-morbid substance abuse.

Table 1. Matched Variables for the Stanley Foundation Neuropathology Consortium

	Schizophrenia	Bipolar disorder	Major depression	Normal controls
Age	44.2 (25–62)	42.3 (25–61)	46.4 (30–65)	48.1 (29–68)
Sex	9M, 6 F	9M, 6 F	9M, 6 F	9M, 6 F
Race	13 C, 2 A	14 C, 1 AA	15 C	14 C, 1 AA
PMI (hours)	33.7 (12–61)	32.5 (13–62)	27.5 (7–47)	23.7 (8–42)
mRNA quality	10 A 2 B 3 C	13 A 2 B	11 A 2 B 2 C	12 A 2 B 1 C
pH	6.1 (5.8–6.6)	6.2 (5.8–6.5)	6.2 (5.6–6.5)	6.3 (5.8–6.6)
Side of brain frozen	6 R, 9 L	8 R, 7 L	6 R, 9 L	7 R, 8 L

A = Asian, AA = African-American, C = Caucasian

Table 2. Clinical Characteristics of the Stanley Foundation Neuropathology Consortium

	Schizophrenia	Bipolar disorder	Major depression	Normal control
Cause of death				
Suicide	4	9	7	0
Cardiopulmonary	8	4	7	13
Accident	2	1	0	2
Other	1	1	1	0
Family history of psychosis				
First-degree	3	4	1	0
Second-degree	3	3	0	1
Family history not available	0	1	0	0
History of psychosis	15	11 with 4 without	0	0
Antipsychotic exposure (mg)	52,267±62,062 1 Never	20,827±24,016 3 Never 1>20 years 1 several mos.	0	0
Current alcohol/drug abuse or dependence	3	4	3	0
Past alcohol/drug abuse or dependence	3	3	1	2

The 60 matched specimens that constitute the Neuropathology Consortium have been made available without charge to research groups around the world. All investigators receive coded tissue samples and perform their assays in a blinded fashion. Investigators receive the code after submitting their data to the Stanley Foundation, and they are then free to individually interpret and publish their findings. Data obtained from researchers using this tissue is also stored in a master database within the Stanley Foundation. The Stanley Consortium is therefore the single most extensively characterized collection of pathological specimens from patients with major mental illnesses, and integrative analyses of available data will be performed in an ongoing manner.

In this report, we present a summary of all data produced in studies of this tissue carried out on prefrontal cortical areas between 1997 and 2000. The prefrontal cortex was chosen as the first brain region to summarize, since it has been heavily requested by investigators and has been repeatedly implicated in the pathophysiology of schizophrenia (Goldman-Rakic and Selemon, 1997; Lewis and Gonzalez-Burgos, 2000).

Global Assessment of Molecular Abnormalities

For the purposes of this summary, all data collected from Brodmann's areas 8, 9, 10 or 46 were examined. As of March 2000, 14 different laboratories had performed assays on these brain areas and had produced data sets for 69 different markers. This is not the entire set of data produced on the prefrontal cortex in this collection, as work in a number of laboratories is still ongoing. The markers included in this review are listed in Table 3. Data on these markers were produced with a variety of techniques (radioligand binding, immunoblot, in situ hybridization, for example) and were performed using several possible tissue conditions (frozen or fixed, sections or tissue blocks). Details regarding tissue preparation and specific assay methodology are beyond the scope of this chapter and have been or will be described by individual investigators.

Table 3. Univariate Analyses of Frontal Cortical Markers from the Stanley Foundation Neuropathology Consortium

Category	Marker	Molecular species	Method	Levels*	F
Apoptosis	Bcl2-membrane bound	protein	WB	1	0.96
	Bcl-X1	protein	WB	1	0.86
Calcium-binding proteins	Calbindin	protein	IH	2 x 6	
	Calretinin	protein	IH	2 x 6	
	Parvalbumin	protein	IH	2 x 6	
Enzymes	cytochrome oxidase-4	protein	WB	1	1.36
	cytochrome oxidase-7	protein	WB	1	1.27
	e-Nitrogen oxide synthetase	protein	WB	1	0.79
	GADPH	RNA	PCR	1	0.08
	GAD65 [Guidotti et al., 2000]	protein	WB	1	0.48
	GAD67 [Guidotti et al., 2000]	protein	WB	1	**6.43**
	glycogen synthetase kinase- 3b [Kozlovsky et al., 2000]	protein	WB	1	**3.08**
	i-Nitrogen oxide synthetase	protein	WB	1	**3.22**
	n-Nitrogen oxide synthetase	protein	WB	1	0.5
	protein kinase Ca - cytosolic	protein	WB	1	1.97

Category	Marker	Molecular species	Method	Levels*	F
	protein kinase Ca - cytosolic, phosphorylated	protein	WB	1	2.25
	protein kinase Ca - membrane bound	protein	WB	1	1.08
	protein kinase Ca - membrane bound, phosphorylated	protein	WB	1	1.01
	protein kinase Ca - cytosolic	protein	WB	1	**5.14**
	protein kinase Ca - cytosolic, phosphorylated	protein	WB	1	**3.5**†
	protein kinase Ca - membrane bound, phosphorylated	protein	WB	1	0.38
	protein kinase Ca -membrane boun	protein	WB	1	0.41
Glial markers	glial fibrillary acidic protein	protein	WB	1	**4.58**
	oligodendrocytes [Uranova et al., 2000]	whole cells	Nissl stain	1	**3.88**
Ion channels	Calcium channel - alpha subunit	RNA	ISH	3	1.35
	Calcium channel - beta subunit	RNA	ISH	3	0.21
Peptide transmitters	Dynorphin	RNA	ISH	1	**3.17**
	Neuropeptide Y [Caberlotto and Hurd, 1999]	RNA	ISH	1	1.86
Receptors	AMPA-GluR1	RNA	ISH	6	0.71
	AMPA-GluR2	RNA	ISH	6	1.09
	AMPA-GluR3	RNA	ISH	6	1.55
	AMPA-GluR4	RNA	ISH	6	1.41
	Corticotrophin Releasing Factor 1	protein	RB	1	0.93
	Corticotrophin Releasing Factor 2	protein	RB	1	2.49
	Dopamine D1	RNA	ISH	6	0.38
	Dopamine D2	RNA	ISH	6	1.42
	Dopamine D4	RNA	ISH	6	0.66
	Dopamine D5	RNA	ISH	6	0.44
	Glucocorticoid	RNA	ISH	6	**2.89**
	Kainate (^{3}H-Kainate)	protein	RB	5	0.63
	Kainate-GluR5	RNA	ISH	6	1
	Kainate-GluR6	RNA	ISH	6	0.78
	Kainate-GluR7	RNA	ISH	6	0.87
	Kainate-KA1	RNA	ISH	6	0.94
	Kainate-KA2	RNA	ISH	6	**2.32**
	Neuropeptide Y	RNA	ISH	1	0.09
	Neurotensin	protein	RB	1	1.47
	NMDA (^{3}H-CGP39653)	protein	RB	3	0.58
	NMDA (^{3}H-ifenprodil)	protein	RB	2	0.78
	NMDA (^{3}H-MDL105519)	protein	RB	2	0.09
	NMDA (^{3}H-MK-801)	protein	RB	2	1.4

Category	Marker	Molecular species	Method	Levels*	F
	NMDA-NR1	RNA	ISH	6	1.38
	NMDA-NR2	RNA	ISH	6	0.64
	NMDA-NR2B	RNA	ISH	6	0.75
	NMDA-NR2C	RNA	ISH	6	1
	NMDA-NR2D	RNA	ISH	6	0.83
	Nurr 77 [Xing et al., 2001]	protein	WB	1	**2.86**
	tyrosine receptor kinase B [Bayer et al., 2000]	RNA	PCR	1	0.73
	tyrosine receptor kinase C [Schramm et al., 1998]	RNA	PCR	1	0.73
Signal transduction	Phosphoinositide turnover	alcohol	biochemical detection	1	**3.29**
	Inositol [Shapiro et al., 2000]	alcohol	gas chromatography	1	0.47
Synaptic and developmental proteins	Synaptophysin	RNA	ISH	1	0.79
	Synaptophysin	RNA	PCR	1	0.68
	VAMP	RNA	PCR	1	0.47
	NCAP	RNA	PCR	1	0.45
	SNAP 25	RNA	PCR	1	0.39
	Synapsin	RNA	PCR	1	0.72
	Reelin [Guidotti et al., 2000]	RNA	PCR	1	**7.09**
	Reelin [Guidotti et al., 2000]	protein	IH	1	**23.8†**

df=(3,56) generally; † p<.05; ‡ df=(3,40)

Abbreviations: IH = immunohistochemistry; ISH = in situ hybridization; PCR = polymerase chain reaction; RB = radioligand binding; WB = Western blot.

** Number of separate measurements made for each marker (e.g., 1 = one data point for entire cortical thickness; 6 = data points for each of the cortical laminae; 2x6 = data points for each of the cortical laminae in two separate Brodman's areas).*

The calcium binding proteins (calbindin, calretinin, and parvalbumin) were measured in Brodmann's areas 9 and 46 and in all 6 cortical laminae; thus, they are represented in Table 3 as having been measured in 2 x 6 "levels." The remaining markers, for which multiple levels are indicated in Table 3, were measured in separate cortical laminae. Phosphoinositide turnover was measured with a variety of pharmacological probes (Song et al., 2001); raw and normalized data were entered into the master database. When "levels" of the 69 markers were considered separately, 231 variables were available for analysis.

Because standard statistical techniques are difficult to apply to this kind of data set in which there are more measurements than subjects, we first examined the data for global patterns of abnormality between the diagnostic groups. In order to summarize collectively data obtained using multiple techniques, we first converted all available data to z-scores. Each individual data point was expressed in units of the standard deviation of the normal control group for that particular variable. For example,

$$\frac{\text{Value for Calbindin for Case 1}}{\text{Standard Deviation for Calbindin from Controls}} = \textit{z-score} \text{ for Case 1}$$

As mentioned above, there were 231 possible z-scores for each individual case. This means that for each diagnostic group there are approximately 15 x 231= 3465 z-scores.

Figure 1 is a histogram showing the number of cases in each diagnostic group with z-scores > 1.0 or < -1.0 and z-scores > 2.0 or < -2.0. Clearly, if one considers all the available data together, the overall deviance from normal for the three diagnostic groups is similar. For schizophrenia, bipolar disorder, and major depression, roughly 2.5% of the 3465 z-scores are greater than one standard deviation from the control mean. For normal controls, only 1.7% of markers are greater than one standard deviation from the control mean. About 0.8% of all markers for schizophrenia, bipolar disorder, and depression are greater than two standard deviations from the control mean. This proportion is four times the number of normal control measurements that are more than two standard deviations from the control mean (0.2%).

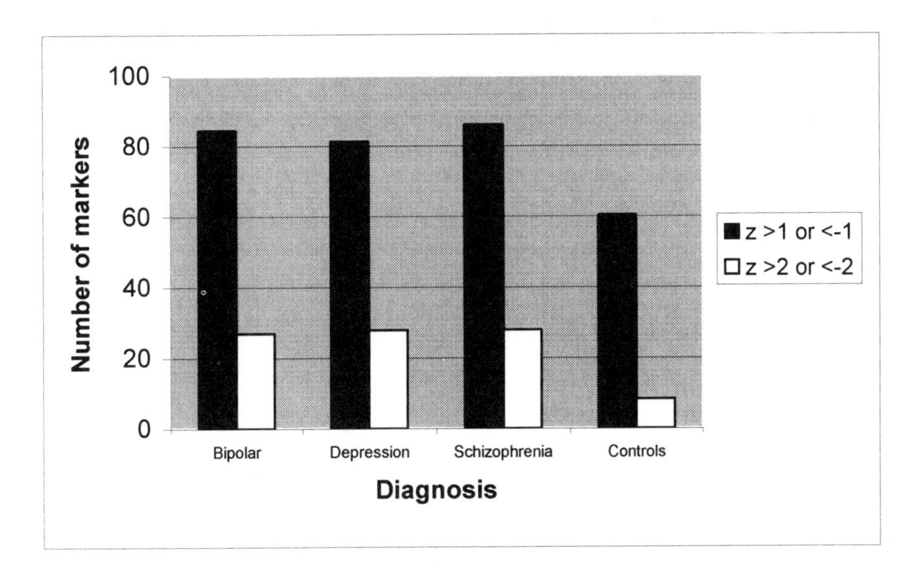

Figure 1. Frequency of Abnormal z-scores

Figure 1 suggests that each of the psychiatric groups contains similar numbers of markers that deviate markedly from controls. It also suggests that the frequency distribution of z-scores for the psychiatric groups is shifted only slightly from that of the normal controls and that there is broad overlap between groups for most of the markers. This is in keeping with most other biological studies of these disorders and suggests that rather severe psychopathology may be produced by subtle alterations in varying combinations of biological systems. In order to determine which particular markers are most useful in differentiating cases from control, we have also analyzed the variables individually with analysis of variance and collectively with a linear discriminant function.

Multiple Univariate Analyses

All markers listed in Table 3 were compared with individual analyses of variance (ANOVAs). When values for multiple cortical layers, or multiple prefrontal regions, were obtained with the same marker, a MANOVA was performed with diagnosis as a between-group factor and cortical level as a within-group factor. Post-hoc Newman-Keuls tests were performed to evaluate statistically significant ANOVAs. ANOVAs were performed using Statistica (StatSoft, Tulsa, Oklahoma, 1995).

The results of the univariate ANOVAs are displayed in Table 3. It should be stressed that these results may differ from those reported by individual investigators, who may have had *a priori* reasons to exclude some cases or to employ different statistical methods. From the 69 univariate ANOVAs there were 14 significantly abnormal results (21% of the total number). Table 4 displays the post-hoc statistical significance of these 14 variables. Statistical significance and effect sizes (according to Cohen, 1988) are given relative to the normal control group. The effect size is the difference between the diagnostic group mean and the control group mean in terms of the pooled standard deviation of the two groups. Effect sizes greater than 0.5 or less than -0.5 are generally considered to be robust in meta-analyses of biological data.

Linear Discriminant Function

In order to reduce the likelihood of spurious results due to multiple comparisons, all data (including data on individual cortical laminae) were examined using a stepwise linear discriminant function (LDF) model. The LDF seeks to find subsets of predictor variables that maximally classify subjects into predefined diagnostic groups. In the LDF model, the variable with the best F statistic from the univariate ANOVAs is selected as a first step. After this predictor, subsequent variables are selected conditionally, in a manner that maximally increases the accurate classification of subjects into diagnostic groups. Thus, variables that enter after the first step may not necessarily have the next best univariate ANOVA results. Since the results

Table 4. Summary of Abnormal Results and Effect Sizes

Marker	Abnormal group compared to control	Technique	p	Effect size
Calcium channel, beta subuni RNA, superficial layers	Schizophrenia	LDF	0.01	-1.4
Calcium stimulated phosphoinositide turnover	Schizophrenia	ANOVA	0.04	-0.9
Cytochrome oxidase 4	Bipolar	LDF	NS	0.6
	Depression	LDF	NS	0.1
	Schizophrenia	LDF	NS	0.2
Dopamine D2 RNA – Layer 3	Bipolar	LDF	0.04	-0.6
	Depression	LDF	0.007	-1.2
	Schizophrenia	LDF	0.005	-1.2
Dynorphin RNA	Bipolar	ANOVA	0.03	-1.1
Glial fibrillary acidic protein	Bipolar	ANOVA	0.05	-0.8
	Depression	ANOVA	0.02	-1.2
	Schizophrenia	ANOVA	0.005	-1.4
Glucocorticoid receptor RNA	Depression (mean values for layers 3-6)	ANOVA	0.02	-1.3
	Schizophrenia (mean values for layers 3-6)	ANOVA	0.01	-1.7
Glutamic acid decarboxylase 67	Bipolar	ANOVA	0.01	-1.2
	Schizophrenia	ANOVA	0.004	-1.9
Glycogen synthetase kinase -3b	Schizophrenia	ANOVA	0.04	-1.0
i-Nitrogen oxide synthetase	Depression	ANOVA	0.02	1.0
Kainate receptor - KA2 subunit RNA	Depression (layer 1)	ANOVA	0.01	-1.0
	Bipolar (layer 1)	ANOVA	0.01	-1.2
	Bipolar (layer 2)	ANOVA	0.04	-0.2
	Schizophrenia (layer 1)	ANOVA	0.02	-0.5
Nurr 77	Schizophrenia	ANOVA	0.04	-1.0
Oligodendrocytes	Bipolar	ANOVA	0.01	-1.1
	Depression	ANOVA	0.05	-0.7
	Schizophrenia	ANOVA	0.02	-0.9
Protein kinase Ce cytosolic	Bipolar	ANOVA	0.007	-0.5
	Schizophrenia	ANOVA	0.004	0.6
Protein kinase Ce cytosolic, phosphorylated	Bipolar	ANOVA	0.04	-0.8
	Schizophrenia	ANOVA	0.02	-1.0
Reelin immunohistochemistr	Bipolar	ANOVA	0.000	-3.0
	Schizophrenia	ANOVA	0.008	-1.1
Reelin RNA	Bipolar	ANOVA, LDF	0.008	-1.2
	Schizophrenia	ANOVA, LDF	0.001	-1.6

of the LDF can be changed dramatically by deleting individual data points, the LDF was performed only on data sets of n ≥ 58 using the NCSS program (Kaysville, Utah, 2000). Only four data sets in this collection had fewer than 58 observations (corticotrophin releasing factor receptor 1 and 2, neurotensin, and reelin immunohistochemistry).

The LDF model identified four significant variables that, when added together, correctly classified a maximum number of subjects as having schizophrenia, bipolar disorder, or depression, or being a normal control. These variables were reelin mRNA, dopamine D2 receptor mRNA in cortical layer 3, calcium channel β-subunit in superficial cortical layers, and cytochrome oxidase 4. The p-values and percentage of cases correctly classified are shown in Table 5, and the actual versus predicted group membership with the combination of these four variables is displayed in Table 6. The percentage of cases correctly classified by chance is assumed to be 25%. The correct classification of 62% of cases by the LDF method was statistically significant from chance by the binomial test ($p<0.00001$). The markers that emerged as significant on the LDF were tested post-hoc with univariate ANOVAS. These results are displayed in Table 4.

Table 5. Significant Variables with the 4 Diagnostic Group LDF Method

Step	Variable	P value at entry	% classified
1	Reelin mRNA	0.0004	38.0
2	Dopamine D2 receptor RNA (layer 3)	0.003	43.0
3	Calcium channel, β subunit (superficial layers)	0.002	57.0
4	Cytochrome oxidase 4	0.0007	62.0

Table 6. Actual and Predicted Group Membership with the LDF Method

Actual Group	Predicted Group Membership				Total
	Bipolar	Depression	Normal	Schizophrenia	
Bipolar	8	2	1	4	15
Depression	1	8	1	4	14
Normal	2	1	12	0	15
Schizophrenia	4	2	0	8	14
Total	15	13	14	16	58

The LDF analysis, like the multiple univariate analyses, is also not an ideal system for analyzing a large-scale data set. Although it reduces the number of comparisons and extracts highly correlated and redundant data, its results are extremely sensitive to the presence or absence of missing data cells. Table 5 displays the set of variables that maximally classifies the subjects, but other combinations of variables that classify the subjects "almost as well" also exist. Are these sets of second- or third-best variables intrinsically less valuable than those that produce the maximal classification? Given our current lack of knowledge of the molecular neuropathology of these diseases, it would seem premature to exclude variables that may emerge as important by independent replication or substantiation in other models of research. Clearly, as in the burgeoning field of micro-array genomic and proteomic studies, new statistical methods are needed to fully exploit such large-scale data sets. In subsequent studies, we plan to present analyses of these data with non-parametric techniques, such as classification and regression tree analyses, or with neural network modeling.

Effects of Descriptive and Demographic Variables

The effects of continuous descriptive variables, such as postmortem interval, age, illness duration, neuroleptic exposure, and brain pH, were examined using Spearman's rank order correlations for each molecular assay within each diagnostic group. This correlational analysis required 120 separate Spearman's rank-order correlation calculations. For the significant variables identified in the univariate analyses, there were a number of significant correlations, as displayed in Table 7. This analysis suggests that some descriptive variables affect the outcome of these assays differently amongst the diagnostic groups, even though the mean values for the descriptive variables did not differ between the diagnostic groups. There was only one significant correlation amongst the variables identified with the LDF. Cytochrome oxidase 4 levels were positively correlated with pH in bipolar samples (rho=0.58, p=0.03).

The presence of co-morbid substance abuse in the Stanley Consortium is a potential confounding factor for any significant result. ANOVA and Student's t tests were used to compare subjects with a present or past history of substance abuse to subjects who had not used drugs. To examine whether a history of substance abuse was a confounding factor for the significant variables, cases were divided into groups with no history of substance abuse (n=37), substance abuse at the time of death (n=15), and past substance abuse (n=8). Effects of specific substances could not be examined because of inadequate sample sizes or because subjects abused a variety of drugs. For current users, 9 abused multiple drugs, 2 abused only marijuana, 2 abused only alcohol, and 2 abused only cocaine. For past users, 7 abused alcohol and 1 abused multiple drugs. The three groups were compared for all significant variables shown in Table 4 with ANOVAs, and with multiple Student's t tests. No significant difference was found for any of variables.

Table 7. Spearman's Rank Order Correlations for Descriptive Variables

Variable	Group	Neuropathological variable	Rho	p
Age	Depression	Glucocorticoid receptor RNA (layer 3)	-0.53	0.04
	Normal	i-Nitrogen oxide synthetase	0.59	0.02
	Normal	Dynorphin RNA	-0.68	0.005
	Schizophrenia	i-Nitrogen oxide synthetase	0.59	0.02
Alcohol use severity	All	Nurr 77	0.27	0.04
		Calcium channel, beta subunit RNA	-0.3	0.03
		Reelin RNA	-0.36	0.01
		Dynorphin RNA	-0.31	0.02
Drug abuse severity	All	GFAP	-0.29	0.02
Fluphenazine equivalents	Schizophrenia	Phosphoinositide turnover	0.67	0.005
Illness onset	Depression	Nurr 77	-0.69	0.004
	Schizophrenia	Phosphoinositide turnover	0.61	0.02
pH	Bipolar	Glucocorticoid Receptor RNA (layers 3-6)	0.55-0.62	0.02-0.04
	Bipolar	Protein kinase Ce - cytosolic	0.59	0.02
	Bipolar	Protein kinase Ce - cytosolic, phosphorylated	0.51	0.05
	Normal	Kainate-KA2 RNA	0.6	0.02
	Schizophrenia	Oligodendrocytes	0.6	0.02
	Schizophrenia	Protein kinase Ce - cytosolic	0.53	0.04
PMI	Bipolar	Nurr 77	-0.57	0.04
	Schizophrenia	Oligodendrocytes	0.55	0.03
	Schizophrenia	Dynorphin RNA	-0.51	0.05

As described previously (Torrey et al., 2000), alcohol and substance use severity were rated qualitatively (0=none, 1=social or occasional, 2=moderate past use, 3=moderate present use, 4=heavy past use, 5=heavy present use). All significant variables found in Table 4 were analyzed using the Spearman's rank-order method for correlation with substance abuse severity. A number of weak, but statistically significant, correlations emerged and are listed in Table 7. Thus, substance abuse does not appear to be a major confounding factor for most of the significant variables that emerged in this analysis. However, significant effects of alcohol and drug abuse cannot be completely excluded.

Other potential confounding binary variables (gender and brain hemisphere analyzed) were evenly balanced in all four diagnostic groups. Differential effects of binary variables between the diagnostic groups were sought by testing each of the significant variables in a 3-way ANOVA (diagnosis x side x gender). The three-way ANOVA revealed significant interactions for 4 of the 17 variables in Table 4 (cytochrome oxidase 4, dopamine D2 receptors in layer 3, glial fibrillary acidic protein, and phosphorylated cytosolic protein kinase Cε). For cytochome oxidase 4 ($F(3,33)=3.13$, $p<0.04$), post-hoc testing revealed that females with depression had significantly lower levels than normal females for the right hemisphere ($p=0.05$), and males with schizophrenia had significantly lower levels than normals in the left hemisphere ($p=0.01$). For dopamine D2 receptors in layer 3 ($F(3,44)=3.69$, $p<0.02$), post-hoc analysis revealed that females with schizophrenia had statistically lower levels than normal females for the right hemisphere ($p=0.01$). For glial fibrillary acidic protein ($F(3,44)=4.08$, $p<0.01$), post-hoc testing revealed that males with bipolar disorder ($p=0.05$) and males with schizophrenia ($p=0.01$) had significantly lower levels than normal males in the left hemisphere. For phosphorylated cytosolic protein kinase Cε ($F(3,44)=3.77$, $p<0.03$) there were no statistically significant post-hoc differences between any of the diagnostic groups and normal controls.

In summary, although the degree of variance in molecular markers explained by descriptive and demographic variables is rather low, the effects of some of these variables, such as pH, PMI, and alcohol use severity, are not entirely negligible. Of note, drug abuse severity and exposure to neuroleptic drugs seemed to affect a very small minority of the molecular markers. It is hoped that future postmortem samples can be more precisely matched for some of these confounding variables.

Summary of Abnormal Markers by Disease

In Table 8, abnormal results from the Stanley Consortium are summarized for each of the major disease categories. It is clear from this table that schizophrenia and bipolar disorder are associated with the greatest number of abnormalities (14 and 11, respectively) and that depression is associated with fewer (6). Cases with schizophrenia also showed the highest number of abnormalities not seen in the other disorders (4), whereas those with bipolar disorder or depression had fewer specific abnormalities (2 and 1, respectively). The preponderance of findings in schizophrenia and bipolar disorder is consistent with clinical notions regarding the relative chronicity and severity of these illnesses.

Table 8. Summary of Abnormal Markers by Disease

Schizophrenia	Bipolar	Depression
↓ GFAP	↓ GFAP	↓ GFAP
↓ oligodendrocytes	↓ oligodendrocytes	↓ oligodendrocytes
↓ KA_2 subunit RNA (layer 1)	↓ KA_2 subunit RNA (layers 1 and 2)	↓ KA_2 subunit RNA (layer 1)
↓ D2 receptor RNA (layer 3)*	↓ D2 receptor RNA (layer 3)*	↓ D2 receptor RNA (layer 3)*
↓ GAD 67	↓ GAD 67	
↑ cytosolic PKCε	↓ cytosolic PKCε	
↓ phosphorylated cytosolic PKCε	↓ phosphorylated cytosolic PKCε	
↓ reelin + cells	↓ reelin + cells	
↓ reelin RNA*	↓ reelin RNA*	
↓ glucocorticoid receptor RNA (layers 3-6)		↓ glucocorticoid receptor RNA (layers 3-6)
		↑ iNOS
	↓ Dynorphin RNA	
	↑ Cytochrome oxidase 4*	
↓ GSK-3β		
↓ calcium channel, beta subunit, RNA		
↓ Nurr 77		
↓ Ca-stimulated phosphoinositide turnover		

**positive by LDF*

In Table 8, it can also be seen that bipolar disorder is neuropathologically more similar to schizophrenia than it is to major depression. Bipolar disorder and schizophrenia shared overlapping abnormalities in the same direction for 8 markers and shared 1 abnormal marker that was altered in opposite directions. Of the 6 abnormal markers seen in major depression, 5 overlapped with schizophrenia, and 4 overlapped with bipolar disorder.

Comparison with Existing Literature

In order to compare the results of the present summary with prior studies, we performed a MEDLINE search on all markers listed in Table 3. In Table 9 we have listed all available studies on these markers performed with quantitative methodology. Subanalyses (Student's t tests between subjects and controls) of consortium data were performed to determine the degree to which prior studies were replicated. These comparisons were then classed in the column labeled "Replication Status" as follows. If two or more studies demonstrated significant differences between group means ($p<0.05$) in the same direction, these markers were rated as confirmed (C). If at least one study demonstrated statistical significance ($p<0.05$) and at least one other study demonstrated a trend toward significance in the same direction ($p<0.10$), these markers were rated as potentially confirmed (PC). Studies were also considered potentially confirmed if they found congruent findings but with different methods of molecular detection (for RNA and for protein, for example). If all studies of a given marker failed to achieve statistical significance (at $p<0.10$ level), or showed significant differences between means but in opposite directions, these studies were rated as non-confirmed (NC). If at least two studies were negative, these were rated as confirmed negative (CN).

Table 9 displays results for 29 different markers for which at least one other published study exists.

Table 9. Frontal Cortical Marker Studies

Marker/Method/Author	Sample/ Control	p	Effect size	Consortium method	Consortium p	Effect Size	Replication status
CALCIUM-BINDING PROTEINS:							
Calbindin (BA 9)	5 S/5	NS (Layer 1)		IH	NS (Layer 1)		CN
IH		NS (Layer 2)			0.009 (Layer 2)	-1.030	NC
Daviss, 1995		≤0.004 (Layer 3)	1.60		0.053 (Layer 3)	-0.737	NC
		NS (Layer 4)			NS (Layer 4)		CN
		≤0.051 (Layer 5/6)	0.36		0.060 (Layer 5/6)	-0.716	NC
Calbindin (BA 46)	5 S/5	NS (Layer 1)		IH	NS (Layer 1)		CN
IH		NS (Layer 2)			NS (Layer 2)		CN
Daviss, 1995		≤ 0.004(Layer 3)	1.79		0.068 (Layer 3)	-0.693	NC
		NS (Layer 4)			NS (Layer 4)		CN
		≤0.051 (Layer 5/6)	1.28		NS (Layer 5/6)	-0.061	NC
Calretinin (BA 9)	5 S/5	NS (all layers)		IH	NS (all layers)		CN
IH							
Daviss, 1995							
Calretinin (BA 46)	5 S/5	NS		IH	NS (Layer 1)		CN
IH					NS (Layer 2)		CN
Daviss, 1995					NS (Layer 3)		CN
					NS (Layer 4)		CN
					0.031 (Layer 5/6)	-0.832	NC
Parvalbumin (BA 9)	15 S/15	NS (Layer 1-3)	0.13	IH	0.03 (Layer 1)	-0.842	NC
IH					NS (Layer 2)		NC
Woo, 1997					0.053 (Layer 3)	-0.739	NC
		NS (Layer 4)	0.38		0.076 (Layer 4)	-0.674	NC
		NS (Layer 5/6)			NS (Layer 5/6)		CN

Marker/Method/Author	Sample/ Control	p	Effect size	Consortium method	Consortium p	Effect Size	Replication status
Parvalbumin (BA 46)	15 S/15	NS (Layer 1-3)		IH	0.03 (Layer 1)	-0.842	NC
IH					NS (Layer 2)		CN
Woo, 1997					0.025 (Layer 3)	-0.865	NC
		NS (Layer 4)			NS (Layer 4)		CN
		NS (Layer 5/6)	0.13		0.031 (Layer 5/6)	-0.831	NC
ENZYMES:							
GAD67 (BA 9)	10 S/10	0.015 (full thickness)	NA	WB	0.004	-1.930	PC
ISH		0.05 (Layer 1)	NA				different species
Akbarian, 1995		0.001(Layer 2)	NA				
		0.05 (Layer 3)	NA				
		0.01 (Layer 4)	NA				
		0.05 (Layer 5)	NA				
		0.11 (Layer 6)	NA				
GAD67 (BA 9)	10 S/10	0.099 (Layer 1)	NA				PC (with Akbarian, 1995)
ISH		0.056 (Layer 2)	NA				PC (with Akbarian, 1995)
Volk, 2000		0.046 (Layer 3)	NA				C (with Akbarian, 1995)
		0.046 (L3&4 border)	NA				C (with Akbarian, 1995)
		0.044 (Layer 5)	NA				C (with Akbarian, 1995)
		NS (Layer 6)	NA				CN (with Akbarian, 1995
PKC soluble (BA 10) RB (Bmax)	12 D (drug-free)/12	< 0.05	0.913	WB	NS		NC
PKC particulate (BA 10) RB (Bmax)	12 D (drug-free)/12	NS					
PKC soluble (BA 10) RB (Bmax)	10 D (on drug)/10	NS		WB	NS		CN
PKC particulate (BA 10) RB (Bmax)	10 D (on drug)/10	NS					

Marker/Method/Author	Sample/ Control	p	Effect size	Consortium method	Consortium p	Effect Size	Replication status
Coull, 2000							
PKCa (BA 9)	10 D (2/10=B)/10	NS		WB	NS		CN
WB							
PKCe (BA 9)	10 D (2/10=B)/10	NS			membrane bound an		
WB					soluble		
PKCa (BA 10)	10 D (2/10=B)/10	NS					
WB							
PKCe (BA 10)	10 D (2/10=B)/10	NS					
WB							
Hrdina, 1998							
GLIAL MARKERS:							
GFAP (BA 10)							
WB	14 S/12	NS		WB	0.005	-1.470	NC
NB	14 S/12	NS					
Karson, 1999							
GFAP (BA 9)	5 S/4	NS (> 0.05)	-0.365	WB	0.005	-1.470	PC
WB							
GFAP (BA 10)	6 S/6	NS (> 0.05)	-0.272				
WB							
Perrone-Bizzozero, 1996							
RECEPTORS:							
AMPA	16 S (6/16=drug free)/9	NS (Layer 2/3)	NA				
RB		NS (Layer 4/5)	NA				
Healy, 1998							

Marker/Method/Author	Sample/ Control	p	Effect size	Consortium method	Consortium p	Effect Size	Replication status
AMPA-GluR1 (BA 9)	16 S (6/16=drug			ISH (BA 46)	0.017 (Layer 1)	0.921	NC
ISH	free)/9	NS (Layer 2/3)	NA		0.012 (Layer 2)	1.000	NC
Healy, 1998					0.019 (Layer 3)	0.875	NC
		NS (Layer 4/5)	NA		0.064 (Layer 4)	0.705	NC
					0.089 (Layer 5)	0.650	NC
					NS (Layer 6)		
AMPA-GluR1 (BA 46)							
RB	16 S (6/16=drug	NS (Layer 2/3)	NA				
	free)/9	NS (Layer 4/5)	NA				
ISH	16 S (6/16=drug			ISH (BA 46)	0.017 (Layer 1)	0.921	NC
Healy, 1998	free)/9	NS (Layer 2/3)	NA		0.012 (Layer 2)	1.000	NC
					0.019 (Layer 3)	0.875	NC
		NS (Layer 4/5)	NA		0.064 (Layer 4)	0.705	NC
					0.089 (Layer 5)	0.650	NC
					NS (Layer 6)		
AMPA-GluR1 (BA 9)	21 S/9	< .001	-0.622	ISH (BA 46)	0.017 (Layer 1)	0.921	NC
PCR					0.012 (Layer 2)	1.000	NC
Sokolov, 1998					0.019 (Layer 3)	0.875	NC
					0.064 (Layer 4)	0.705	NC
					0.089 (Layer 5)	0.650	NC
					NS (Layer 6)		
AMPA-GluR2 (BA 9)							
RB	16 S (6/16=drug	NS (Layer 2/3)	NA				
	free)/9	NS (Layer 4/5)	NA				
ISH	16 S (6/16=drug			ISH (BA 46)	NS (Layer 1)		
Healy, 1998	free)/9	NS (Layer 2/3)	NA		NS (Layer 2)		CN
					NS (Layer 3)		
		NS (Layer 4/5)	NA		NS (Layer 4)		CN
					NS (Layer 5)		

Marker/Method/Author	Sample/ Control	p	Effect size	Consortium method	Consortium p	Effect Size	Replication status
					0.054 (Layer 6)	-0.750	NC
AMPA-GluR2 (BA 46)							
RB	16 S (6/16=drug free)/9	NS (Layer 2/3)	NA				
		NS (Layer 4/5)	NA				
ISH	16 S (6/16=drug free)/9			ISH (BA 46)	NS (Layer 1)		
Healy, 1998		NS (Layer 2/3)	NA		NS (Layer 2)		CN
					NS (Layer 3)		
		NS (Layer 4/5)	NA		NS (Layer 4)		CN
					NS (Layer 5)		
					0.054 (Layer 6)	-0.750	NC
AMPA-GluR3 (BA 9)							
RB	16 S (6/16=drug free)/9	NS (Layer 2/3)	NA				
		NS (Layer 4/5)	NA				
ISH	16 S (6/16=drug free)/9			ISH (BA 46)	NS (Layer 1)		
Healy, 1998		NS (Layer 2/3)	NA		NS (Layer 2)		CN
					NS (Layer 3)		
		NS (Layer 4/5)	NA		NS (Layer 4)		CN
					NS (Layer 5)		
					NS (Layer 6)		
AMPA-GluR3 (BA 46)							
RB	16 S (6/16=drug free)/9	NS (Layer 2/3)	NA				
		NS (Layer 4/5)	NA				
ISH	16 S (6/16=drug free)/9			ISH (BA 46)	NS (Layer 1)		
Healy, 1998		NS (Layer 2/3)	NA		NS (Layer 2)		CN
					NS (Layer 3)		
		NS (Layer 4/5)	NA		NS (Layer 4)		CN
					NS (Layer 5)		
					NS (Layer 6)		

Marker/Method/Author	Sample/ Control	p	Effect size	Consortium method	Consortium p	Effect Size	Replication status
AMPA-GluR3 (BA 9) ISH	7 S/10	NS	NA	ISH (BA 46)			CN
AMPA-GluR3 (BA 10) ISH	7 S/10	NS	NA		NS (Layer 1)		
Ohnuma, 1998					NS (Layer 2)		
					NS (Layer 3)		
					NS (Layer 4)		
					NS (Layer 5)		
					NS (Layer 6)		
AMPA-GluR4 (BA 9)							
RB	16 S (6/16=drug free)/9	NS (Layer 2/3)	NA				
		NS (Layer 4/5)	NA				
ISH	16 S (6/16=drug free)/9			ISH (BA 46)	NS (Layer 1)		
Healy, 1998		NS (Layer 2/3)	NA		NS (Layer 2)		CN
					NS (Layer 3)		
		NS (Layer 4/5)	NA		NS (Layer 4)		CN
					NS (Layer 5)		
					NS (Layer 6)		
AMPA-GluR4 (BA 46)							
RB	16 S (6/16=drug free)/9	NS (Layer 2/3)	NA				
		NS (Layer 4/5)	NA				
ISH	16 S (6/16=drug free)/9	NS (Layer 2/3)	NA				
Healy, 1998		NS (Layer 4/5)	NA				
Dopamine D1 RB	7 S/8	NS (superficial layer)	0.879	ISH	0.005 (Layer 1)	1.080	NC
					NS (Layer 2)	0.250	CN
Knable, 1996		NS (intermediate layer)	0.762		NS (Layer 3)	-0.333	CN
					NS (Layer 4)	0.005	CN
		NS (deep layer)	0.326		NS (Layer 5)	0.167	CN
					NS (Layer 6)	0.071	CN

Marker/Method/Author	Sample/ Control	p	Effect size	Consortium method	Consortium p	Effect Size	Replication status
Dopamine D2	6 D/9	NS	NA	ISH	0.090 (Layer 1)	0.643	NC
WB					NS (Layer 2)		
Jorgensen, 1985					NS (Layer 3)		
					NS (Layer 4)		
					NS (Layer 5)		
					NS (Layer 6)		
Dopamine D2	11 S/9	< .01	1.92	ISH	NS (Layer 1)		NC
RB (Bmax)					NS (Layer 2)		
Mita, 1986					NS (Layer 3)		
					NS (Layer 4)		
					NS (Layer 5)		
					NS (Layer 6)		
Dopamine D4							CN
PCR	8 S/9	NS		ISH	NS (Layer 1)		
Mulcrone, 1996					NS (Layer 2)		
PCR	11 S/19	NS			NS (Layer 3)		
Roberts, 1996					NS (Layer 4)		
					NS (Layer 5)		
					NS (Layer 6)		
Kainate-GluR5 (BA 9)	7 S/10			ISH	NS (Layer 1)		
ISH					NS (Layer 2)		
Ohnuma, 1998		NS (Layer 3)	NA		NS (Layer 3)		CN
					NS (Layer 4)		
		NS (Layer 5)	NA		NS (Layer 5)		CN
					NS (Layer 6)		
Kainate-GluR5 (BA 10)	7 S/10				NS (Layer 1)		
ISH	7 S/10				NS (Layer 2)		
Ohnuma, 1998		0.065 (Layer 3)	NA		NS (Layer 3)		NC

Marker/Method/Author	Sample/ Control	p	Effect size	Consortium method	Consortium p	Effect Size	Replication status
		0.065 (Layer 5)	NA		NS (Layer 4) NS (Layer 5) NS (Layer 6)		NC
Kainate-GluR7 (BA 9) PCR Sokolov, 1998	21 S/9	0.013	-0.673	ISH	NS (Layer 1) NS (Layer 2) NS (Layer 3) NS (Layer 4) NS (Layer 5) NS (Layer 6)		NC
Kainate (^{3}H-Kainate) (BA 9,10,46) RB	14 S/10	< .01	1.48	RB	NS (Layer 1) 0.083 (Layer 2) 0.057 (Layer 3) NS (Layer 4) NS (Layer 5)	 0.636 0.800	 PC PC
Kainate (^{3}H-Kainate) (BA 8) RB Toru, 1988		< .01	1.46				
Kainate-KA1 (BA 9) PCR Sokolov, 1998	21 S/9	0.003	-1.34	ISH	NS (Layer 1) NS (Layer 2) NS (Layer 3) NS (Layer 4) NS (Layer 5) NS (Layer 6)		NC
Neurotensin RB Lahti, 1998	5 S (drug-free)/5 5 S (on drug)/5	< .05 < .05	-1.98 -2.45	RB	NS		NC
NMDA (^{3}H-CGP39653) (BA 10 RB	22 D/22	< .05	-0.83	RB	0.072 (inner) 0.072 (middle)	-0.683 -0.683	PC

Marker/Method/Author	Sample/ Control	p	Effect size	Consortium method	Consortium p	Effect Size	Replication status
Nowak, 1995					0.072 (outer)	-0.683	
NMDA (^{3}H-MK-801) (BA 9,10,46)	11 S/8	NS		RB	NS (inner)		CN
RB							
NMDA (^{3}H-MK-801) (BA 8)	10 S/7	NS			NS (outer)		
RB							
Ishimaru, 1994							
NMDA (^{3}H-MK-801)	13 S/12	NS		RB	NS (inner)		CN
RB					NS (outer)		
Kornhuber, 1989							
NMDA-NR1 (BA 9)	21 S/9	0.002	-1.34	ISH	0.094 (Layer 1)	0.614	NC
PCR					NS (Layer 2)		
Sokolov, 1998					NS (Layer 3)		
					NS (Layer 4)		
					0.099 (Layer 5)	0.619	NC
					NS (Layer 6)		
NMDA-NR1 (BA 10)	15 S/15	NS					
ISH		NS (Layer 1)			0.094 (Layer 1)	0.614	NC
Akbarian, 1996		NS (Layer 2)			NS (Layer 2)		CN
		NS (Layer 3)			NS (Layer 3)		CN
		NS (Layer 4)			NS (Layer 4)		CN
		NS (Layer 5)			0.099 (Layer 5)	0.619	NC
		NS (Layer6)			NS (Layer 6)		CN
NMDA-NR2A (BA 10)	15 S/15	NS		ISH			
ISH		NS (Layer 1)			0.042 (Layer 1)	0.774	NC
Akbarian, 1996		NS (Layer 2)			0.100 (Layer 2)	0.625	NC
		NS (Layer 3)			NS (Layer 3)		CN
		NS (Layer 4)			0.064 (Layer 4)	0.686	NC
		NS (Layer 5)			0.069 (Layer 5)	0.690	NC

Marker/Method/Author	Sample/ Control	p	Effect size	Consortium method	Consortium p	Effect Size	Replication status
		NS (Layer6)			NS (Layer 6)		CN
NMDA-NR2B (BA 10)	15 S/15	NS		ISH			
ISH		NS (Layer 1)			0.006 (Layer 1)	1.080	NC
Akbarian, 1996		NS (Layer 2)			0.091 (Layer 2)	0.667	NC
		NS (Layer 3)			0.072 (Layer 3)	0.700	NC
		NS (Layer 4)			0.065 (Layer 4)	0.682	NC
		NS (Layer 5)			NS (Layer 5)	0.529	CN
		NS (Layer6)			NS (Layer 6)	0.429	CN
NMDA-NR2C (BA 10)	15 S/15	NS		ISH			
ISH		0.02 (Layer 1)	NA		NS (Layer 1)		NC
Akbarian, 1996		0.02 (Layer 2)	NA		NS (Layer 2)		NC
		0.02 (Layer 3)	NA		NS (Layer 3)		NC
		0.02 (Layer 4)	NA		NS (Layer 4)		NC
		0.02 (Layer 5)	NA		NS (Layer 5)		NC
		0.02 (Layer6)	NA		NS (Layer 6)		NC
NMDA-NR2D (BA 10)	15 S/15	NS		ISH			
ISH		NS (Layer 1)			NS (Layer 1)		CN
Akbarian, 1996		NS (Layer 2)			NS (Layer 2)		CN
		NS (Layer 3)			NS (Layer 3)		CN
		NS (Layer 4)			NS (Layer 4)		CN
		NS (Layer 5)			NS (Layer 5)		CN
		NS (Layer6)			NS (Layer 6)		CN
SYNAPTIC & DEVELOPMENTAL PROTEINS:							
Reelin (BA 10 & 46)							
PCR	14 S/14	<.00001	-2.25	PCR	0.001	-1.690	C
WB	14 S/14	NS					
IH	14 S/14	< .01 (Layer 1)	NA		0.008	-1.130	C

Marker/Method/Author	Sample/ Control	p	Effect size	Consortium method	Consortium p	Effect Size	Replication status
Impagnatiello, 1998		< .01 (Layer 2)	NA				
		NS (Layer 3)	NA				
		NS (Layer 4)	NA				
		NS (Layer 5)	NA				
		NS (Layer 6)	NA				
SNAP 25 (BA 10)							
WB	14/S/12	<.002	-2.00				NC
NB	15 S/13	NS		PCR	NS		CN
Karson, 1999							
SNAP 25 (BA 8)	10 S/13	NS	NA	PCR	NS		NC
WB							
Gabriel, 1997							
SNAP 25 (BA 9)	10 S/6	< .05	0.933	PCR	NS		C (with Karson, 1999)
WB							
SNAP 25 (BA 10)	5 S/5	< .05	-1.82				C (with Karson, 1999)
WB							
Thompson, 1998							
Synaptophysin (BA 9, 46)							
ISH	18 S/18	NS		PCR	NS		CN
IH	18 S/18	NS					
WB	18 S/18	NS					
Eastwood, 2000							
Synaptophysin (BA 9)	10 S/10	< .008	-0.364	PCR	NS		C (with Honer, 1999; Karson, 1999; Perrone-Bizzozero, 1996)
IH							
Synaptophysin (BA 46)	10 S/10	< .001	-0.403				
IH							
Glantz, 1997							

Marker/Method/Author	Sample/ Control	p	Effect size	Consortium method	Consortium p	Effect Size	Replication status
Synaptophysin (BA 9) ISH Glantz, 2000	10 S/10	NS		PCR	NS		CN
Synaptophysin ELISA Honer, 1999	11D/10	0.06	NA	PCR	NS		C (with Glantz, 1997; Karson, 1999; Perrone-Bizzozero, 1996)
	13S/10		NA				
Synaptophysin (BA 10) WB	14 S/12	< .05	-1.00				C (with Glantz, 1997; Honer, 1999; Perrone-Bizzozero, 1996)
NB Karson, 1999	15 S/13	NS		PCR ISH	NS NS (deep) NS (middle) NS (superficial)		CN
Synaptophysin (BA 8) WB Gabriel, 1997	10 S/13	NS	NA	ISH	NS NS (deep) NS (middle) NS (superficial)		CN
Synaptophysin (BA 9) WB	5 S/4	< .05	-1.53				C (with Glantz, 1997; Honer, 1999; Karson, 1999)
Synaptophysin (BA 10) WB Perrone-Bizzozero, 1996	6S/6	< .05	-1.19				

BA = Brodmann's area, C = confirmed, CN = confirmed negative, D = depression, IH = immunohistochmistry, ISH = in situ hybridization, NA = not available, NB = Northern blot, NC = non-confirmed, NS = non-significant (p>0.10), PC = potentially confirmed, PCR = polymerase chain reaction, RB = radioligand binding, WB = Western blot.

From these studies, four markers were considered confirmed in subjects with schizophrenia: GAD 67 RNA (with a potentially confirming observation on protein level as well), reelin RNA, reelin protein by immunohistochemistry, and synaptophysin protein. All of these abnormalities were observed. Three other markers were potentially confirmed: GFAP in schizophrenia, ^{3}H-CGP39653 binding in depression, and ^{3}H-kainate binding in schizophrenia.

Glutamic acid decaboxylase (GAD) is the principal synthetic enzyme for gamma-amino-butyric-acid (GABA), which is the principal inhibitory neurotransmitter in the brain. A loss, or dysfunction, of prefrontal inhibitory interneurons has been proposed as a possible explanation for prefrontal cortical abnormalities in schizophrenia. Decreased density or impaired function of inhibitory interneurons has been suggested by direct cell counting (Benes et al., 1991), decreased expression of calcium-binding proteins (Beasely and Reynolds, 1997), upregulation of GABA-A receptor numbers (Benes et al., 1992; Ohnuma et al., 1999), abnormal arborization of the dendritic tree of prefrontal neurons (Selemon et al., 1995), and reduced GABA transporter expression (Woo et al., 1998; Volk et al., 2001). It has also been argued that downregulation of GAD may occur independently of structural alterations of prefrontal interneurons and may exist as a functional alteration in the protein itself (Guidotti et al., 2000).

Reelin is an extracellular matrix glycoprotein that is secreted from Cajal-Retzius interneurons during development and from other GABA-ergic interneurons during adult life. The secretion of reelin is important for the regulation of genes necessary for synaptic plasticity and morphological changes associated with learning. Mice that are haploinsufficient for reelin exhibit some features that may overlap with some of those reported in schizophrenia (Guidotti et al., 2000a). Abnormal reelin expression in psychiatric diseases may prove to underlie subtle alterations in cortical architechture, as has been shown for other conditions characterized by cortical dysgenesis (Fatemi, 2001).

Messenger RNA levels for synaptophysin, a presynaptic terminal-associated protein, were unchanged in the Stanley Consortium. Negative results have also been observed by other investigators (Gabriel et al., 1997; Karson et al., 1999; Eastwood et al., 2000; Glantz et al., 2000). However, synaptophysin protein levels, which have not yet been measured in the Stanley Consortium, are reported to be decreased in schizophrenia in four studies (Perrone-Bizzozero et al., 1996; Glantz et al., 1997; Gabriel et al., 1997; Karson et al., 1999). In a fifth study, synaptophysin levels were reduced at the trend level (Honer et al., 1999). These results support the notion that prefrontal synaptic complexity may be diminished in schizophrenia, which has been suggested by other methodologies as well (Selemon et al., 1995; Garey et al., 1998).

Interestingly, glial fibrillary acidic protein (GFAP) levels were decreased in all three of the psychiatric groups in the Stanley Consortium.

These results may be congruent with other reports suggesting that there are reductions in glial density in psychiatric disorders (Ongur et al., 1998; Rajkowska et al., 1999). While most investigators have felt that a lack of gliosis in postmortem brain specimens from subjects with schizophrenia is firmly established, and that this lack of gliosis supports a developmental rather than degenerative etiology for neuropathology, more recent findings suggest that a normal gliotic reaction to injury may not be possible in psychiatric disorders. In the study of Perrone-Bizzozero (1996), Western blotting for GFAP was not significantly different between schizophrenia and controls, but the effect size indicates a change in the same direction as that observed in the Stanley Consortium tissue.

CGP-39653 is a radioligand that binds predominantly to the NR2A and NR2B subunits of the NMDA receptor (Christie et al., 2000). Nowak et al. (1995) have previously reported a decrease in CGP-39653 binding in a sample of subjects with depression. In the Stanley Consortium tissue, binding in subjects with depression was decreased at a trend level of significance. Although it is not included in the Table 9, the expression of mRNA for the NR2A receptor subunit was also decreased in subjects with depression at trend levels of significance (Layers I, II, IV, V $p< 0.06$; Layers II, VI $p <0.10$), but messenger RNA for the NR2B was not altered in subjects with depression. An NMDA receptor hypofunction hypothesis has been proposed for schizophrenia, whereby dysfunction of glutamatergic input upon cortical GABA-ergic interneurons, or upon brainstem catecholaminergic neurons, leads to diminished inhibitory input on cortical pyramidal neurons (Olney and Farber, 1995). While this hypothesis is not supported by work performed with Stanley Consortium tissue, other work (Nowak et al., 1993), which suggests that NMDA receptor hypofunction may also be relevant for depression, is more clearly supported.

Toru et al. (1988) have previously reported a decrease in radiolabeled kainic acid binding in the prefrontal cortex of subjects with schizophrenia. In the Stanley Consortium tissue, kainic acid binding was decreased at the trend level of significance in layers II and III from subjects with schizophrenia. Interestingly, mRNA levels for the KA2 subunit of the kainate receptor were decreased in superficial cortical layers of all three psychiatric groups within the Stanley Consortium tissue. Although the function of kainic acid receptors is still incompletely understood, it appears that these receptors may be both pre- and post-synaptic to cortical pyramidal cells, and similarly to NMDA receptors, may be involved with regulation of inhibitory inputs upon pyramidal cells (Frerking and Nicoll, 2000).

Conclusions

In summary, a collection of studies performed on prefrontal cortex from the Stanley Foundation Neuropathology Consortium has identified 14 potential disease markers for further study. Because the study of this brain collection was meant to stimulate research on pathophysiology and potential new routes for treatment for these diseases, the findings reported here are considered exploratory rather than explanatory. However, a number of tentative statements can be made in conclusion:

1. The findings include abnormalities related to a variety of neural systems effecting neuronal maturation, survival and plasticity, glial cells, neurotransmitters, and signal transduction.
2. Schizophrenia has the largest number of abnormalities and overlaps to a considerable degree with bipolar disorder (see Table 7). The preponderance of findings in schizophrenia and bipolar disorder is consistent with clinical ideas concerning the severity and chronicity of these illnesses and argues against the notion that the findings have emerged for purely stochastic reasons.
3. Bipolar disorder is more similar to schizophrenia than it is to depression with these neuropathological markers.
4. The vast majority of the abnormalities represented a decline in function. The directionality of the findings also argues against a purely stochastic explanation for the findings and suggests that there is a widespread failure of gene expression in the major psychiatric illnesses. Such a generalized failure of gene expression may be due to an overriding global insult such as infection, malnutrition, hypoxia, or some other unknown mechanism.
5. None of the findings can be clearly explained by exposure to neuroleptic drugs or illicit drugs, but these factors cannot be absolutely ruled out.
6. Some of the abnormal markers may be partly explained by differential effects of descriptive variables between the groups (see Table 7) or by unforeseen interactions between combinations of descriptive variables (such as sex and brain hemisphere analyzed).

References

Akbarian S, Kim JJ, Potkin SG, Hagman JO, Tafazzoli A, Bunney WE Jr, Jones EG. Gene expression for glutamic acid decarboxylase is reduced without loss of neurons in prefrontal cortex of schizophrenics. Arch Gen Psychiatry 1995;52: 258–266.

Akbarian S, Sucher NJ, Bradley D, Tafazzoli A, Trinh D, Hetrick WP, Potkin SG, Sandman CA, Bunney WE Jr, Jones EG. Selective alterations in gene expression for NMDA receptor subunits in prefrontal cortex of schizophrenics. J Neurosci 1996;16: 19–30.

Bayer TA, Schramm M, Feldmann N, Knable MB, Falkai P. Antidepressant drug exposure is associated with mRNA levels of tyrosine receptor kinase B in major depressive disorder. Prog Neuropsychopharmacol Biol Psychiatry 2000;24: 881–888.

Beasley CL, Reynolds GP. Parvalbumin-immunoreactive neurons are reduced in the prefrontal cortex of schizophrenics. Schizophr Res 1997;24: 349–355.

Benes FM, McSparren J, Bird ED, SanGiovanni JP, Vincent SL. Deficits in small interneurons in prefrontal and cingulated cortices of schizophrenic and schizoaffective patients. Arch Gen Psychiatry 1991;48: 996–1001.

Benes FM, Vincent SL, Alsterberg G, Bird ED, SanGiovanni JP. Increased GABA-A receptor binding in superficial layers of cingulated cortex in schizophrenics. J Neurosci 1992;12: 924–929.

Caberlotto L, Hurd YL. Reduced neuropeptide Y mRNA expression in the prefrontal cortex of subjects with bipolar disorder. Neuroreport 1999;10: 1747–1750.

Christie JM, Jane DE, Monaghan DT. Native N-methyl-D-aspartate receptors containing NR2A and NR2B subunits have pharmacologically distinct competitive antagonist binding sites. J Pharmacol Exp ther 2000;292: 1169–1174.

Cohen, J. Statistical Power Analysis for the Behavioral Sciences. 2nd ed., Lawrence Erlbaum, Hillsdale, NJ, 1998; p 42.

Coull MA, Lowther S, Katona CLE, Horton RW. Altered brain protein kinase C in depression: a postmortem study. Eur Neuropsychopharmacol 2000;10: 283–288.

Daviss SR, Lewis DA. Local circuit neurons of the prefrontal cortex in schizophrenia: selective increase in the density of calbindin-immunoreactive neurons. Psychiatry Res 1995;59: 81–96.

Eastwood SL, Cairns NJ Harrison PJ. Synaptophysin gene expression in schizophrenia: investigation of synaptic pathology in the cerebral cortex. Br J Psychiatry 2000;176: 236–242.

Fatemi SH. Reelin mutations in mouse and man: from reeler mouse to schizophrenia, mood disorders, autism and lissencephaly. Mol Psychiatry, in press.

Frerking M, Nicoll RA. Synaptic kainate receptors. Curr Opin Neurobiol 2000;10: 342–351.

Gabriel SM, Haroutunian V, Powchik P, Honer WG, Davidson M, Davies P, Davis KL. Increased concentrations of presynaptic proteins in the cingulated cortex of subjects with schizophrenia. Arch Gen Psychiatry 1997;54: 559–566.

Garey LJ, Ong WY, Patel TS, Kanani M, Davis A, Mortimer AM, Barnes TR, Hirsch SR. Reduced dendritic spine density on cerebral cortical pyramidal neurons in schizophrenia. J Neurol Neurosurg Psychiatry 1998;65: 446–453.

Glantz LA, Lewis DA. Reduction of synaptophysin immunoreactivity in the prefrontal cortex of subjects with schizophrenia: regional and diagnostic specificity. Arch Gen Psychiatry 1997;54: 943–952.

Glantz LA, Austin MC, Lewis DA. Normal cellular levels of synaptophysin mRNA expression in the prefrontal cortex of subjects with schizophrenia. Biol Psychiatry 2000;48: 389–397.

Goldman-Rakic PS, Selemon LD. Functional and anatomical aspects of prefrontal pathology in schizophrenia. Schizophr Bull 1997;23: 437–458.

Guidotti A, Auta J, Davis JM, Gerevini VD, Dwivedi Y, Grayson DR, Impagnatiello F, Pandey G, Pesold C, Sharma R, Vzonov D, Costa E. Decrease in reelin and glutamic acid decarboxylase 67 (GAD67) expression in schizophrenia and bipolar disorder: a postmortem brain study. Arch Gen Psychiatry 2000;57: 1061–1069.

Guidotti A, Pesold C, Costa E. New neurochemical markers for psychosis: a working hypothesis of their operation. Neurochem Res 2000a;25: 1207–1218.

Healy DJ, Haroutunian V, Powchik P, Davidson M, Davis KL, Watson SJ, Meador-Woodruff JH. AMPA receptor binding and subunit mRNA expression in prefrontal cortex and striatum of elderly schizophrenics. Neuropsychopharmacology 1998;19: 278–286.

Honer WG, Falkai P, Chen C, Arango V, Mann JJ, Dwork AJ. Synaptic and plasticity-associated proteins in anterior frontal cortex in severe mental illness. Neuroscience 1999;91: 1247–1255.

Hrdina P, Faludi G, Li Q, Bendotti C, Tekes K, Sotonyi P, Palkovits M. Growth-associated protein (GAP-43), its mRNA, and protein kinase C (PKC) isoenzymes in brain regions of depressed suicides. Mol Psychiatry 1998;3: 411–418.

Impagnatiello F, Guidotti A, Pesold C, Dwivedi Y, Caruncho H, Pisu MG, Uzunov DP, Smalheiser NR, Davis JM, Pandey GN, Pappas GD, Tueting P, Sharma RP, Costa E. A decrease of reelin expression as a putative vulnerability factor in schizophrenia. Proc Natl Acad Sci USA 1998;95: 15718–15723.

Ishimaru M, Kurumaji A, Toru M. Increases in strychnine-insensitive glycine binding sites in cerebral cortex of chronic schizophrenics: evidence for glutamate hypothesis. Biol Psychiatry 1994;35: 84–95.

Jorgensen OS, Riederer P. Increased synaptic markers in hippocampus of depressed patients. J Neural Transm 1985;64: 55–66.

Karson CN, Mrak RE, Schluterman KO, Sturner WQ, Sheng JG, Griffin WST. Alterations in synaptic proteins and their encoding mRNAs in prefrontal cortex in schizophrenia: a possible neurochemical basis for 'hypofrontality'. Mol Psychiatry 1999;4: 39–45.

Knable MB, Hyde TM, Murray AM, Herman MM, Kleinman JE. A postmortem study of frontal cortical dopamine D1 receptors in schizophrenics, psychiatric controls and normal controls. Biol Psychiatry 1996;40: 1191–1199.

Kornhuber J, Mack-Burkhardt F, Riederer P, Hebenstreit GF, Reynolds GP, Andrews HB, Beckmann H. [^{3}H]MK–801 binding sites in postmortem brain regions of schizophrenic patients. J Neural Transm 1989;77: 231–236.

Kozlovsky N, Belmaker RH, Agam G. Low GSK-3 beta immunoreactivity in postmortem frontal cortex of schizophrenic patients. Am J Psychiatry 2000;157:831–833.

Lahti RA, Cochrane EV, Roberts RC, Conley RR, Tamminga CA. [^{3}H]Neurotensin receptor densities in human postmortem brain tissue obtained from normal and schizophrenic persons: an autoradiographic study. J Neural Transm 1998;105: 507–516.

Lewis DA, Gonzalez-Burgos G. Intrinsic excitatory connections in the prefrontal cortex and the pathophysiology of schizophrenia. Brain Res Bull 2000;52: 309–317.

Mita T, Hanada S, Nishino N, Kuno T, Nakai H, Yamadori T, Mizoi Y, Tanaka C. Decreased serotonin S_2 and increased dopamine D_2 receptors in chronic schizophrenics. Biol Psychiatry 1986;21: 1407–1414.

Mulcrone J, Kerwin RW. No difference in the expression of the D4 gene in postmortem frontal cortex from controls and schizophrenics. Neurosci Lett 1996;219: 163–166.

Nowak G, Paul IA, Popik P, Young A, Skolnick P. Ca2+ antagonists effect an antidepressant-like adaptation of the NMDA receptor complex. Eur J Pharmacol 1993;247: 101–102.

Nowak G, Ordway GA, Paul IA. Alterations in the N-methyl-D-aspartate (NMDA) receptor complex in the frontal cortex of suicide victims. Brain Res 1995;675: 157–164.

Ohnuma T, Augood SJ, Arai H, McKenna PJ, Emson PC. Expression of the human excitatory amino acid transporter 2 and metabotropic glutamate receptors 3 and 5 in the prefrontal cortex from normal individuals and patients with schizophrenia. Mol Brain Res 1998;56: 207–217.

Ohnuma T, Augood SJ, Arai H, McKenna PJ, Emson PC. Measurement of GABAergic parameters in the prefrontal cortex in schizophrenia: focus on GABA content, GABA(A) receptor alpha-1 subunit messenger RNA and human GABA transporter-1 (HGAT-1) messenger RNA expression. Neuroscience 1999;93: 441–448.

Olney JW, Farber NB. Glutamate receptor dysfunction and schizophrenia. Arch Gen Psychiatry 1995;52: 998–1007.

Ongur D, Drevets WC, Price JL. Glial reduction in the subgenual prefrontal cortex in mood disorders. Proc Natl Acad Sci USA 1998;95: 13290–13295.

Perrone-Bizzozero NI, Sower AC, Bird ED, Benowitz LI, Ivins KJ, Neve RL. Levels of the growth-associated protein GAP-43 are selectively increased in association cortices in schizophrenia. Proc Natl Acad Sci USA 1996;93: 14182–14187.

Rajkowska G, Miguel-Hidalgo JJ, Wei J, Dilley G, Pittman SD, Meltzer HY, Overholser JC, Roth BL, Stockmeier CA. Morphometric evidence for neuronal and glial prefrontal pathology in major depression. Biol Psychiatry 1999;45: 1085–1098.

Roberts DA, Balderson D, Pickering-Brown SM, Deakin JFW, Owen F. The relative abundance of dopamine D_4 receptor mRNA in postmortem brains of schizophrenics and controls. Schizophr Res 1996;20: 171–174.

Schramm M, Falkai P, Feldmann N, Knable MB, Bayer TA. Reduced tyrosine kinase receptor C mRNA levels in the frontal cortex of patients with schizophrenia. Neurosci Lett 1998;257: 65–68.

Shapiro J, Belmaker RH, Bigeon A, Seher A, Agam A. Scyllo-inositol in postmortem brain of suicide victims, bipolar, unipolar and schizophrenic patients. J Neural Transm, in press.

Selemon LD, Rajkowska G, Goldman-Rakic PS. Abnormally high neuronal density in the schizophrenic cortex: a morphometric analysis of prefrontal area 9 and occipital area 17. Arch Gen Psychiatry 1995;52: 805–818.

Sokolov BP. Expression of NMDAR1, GluR1, GluR7, and KA1 glutamate receptor mRNAs is decreased in frontal cortex of "neuroleptic-free" schizophrenics: evidence on reversible up-regulation by typical neuroleptics. J Neurochem 1998;71: 2454–2464.

Song L, Greendorfer AJ, Bartolucci AA, Jope RS. Phosphoinositide signal transduction system and G-protein levels in postmortem brain from subjects with bipolar disorder, major depression, and schizophrenia. J Psychiatr Res, in press.

Thompson PM, Sower AC, Perrone-Bizzozero NI. Altered levels of the synaptosomal associated protein SNAP-25 in schizophrenia. Biol Psychiatry 1998;43: 239–243.

Toru M, Watanabe S, Shibuya H, Nishikawa T, Noda K, Mitsushio H, Ichikawa H, Kurumaji A, Takashima M, Mataga N, Ogawa A. Neurotransmitters, receptors and neuropeptides in postmortem brains of chronic schizophrenic patients. Acta Psychiatr Scand 1988;78: 121–137.

Torrey EF, Webster M, Knable M, Johnston N, Yolken RH. The Stanley Foundation Brain Collection and Neuropathology Consortium. Schizophr Res 2000;44: 151–155.

Uranova NA, Vostrikov VM, Olovskaya DD, Rachmanova VI. Oligodendroglial density in the prefrontal cortex area 9 in schizophrenia and mood disorders: A study of the Stanley Foundation Neuropathology Consortium. Submitted manuscript.

Volk DW, Austin MC, Pierri JN, Sampson AR, Lewis DA. Decreased glutamic acid decarboxylase$_{67}$ messenger RNA expression in a subset of prefrontal cortical gamma-aminobutyric acid neurons in subjects with schizophrenia. Arch Gen Psychiatry 2000;57: 237–245.

Volk DW, Austin MC, Pierri JN, Sampson AR, Lewis DA. GABA transporter–1 mRNA in the prefrontal cortex in schizophrenia: decreased expression in a subset of neurons. Am J Psychiatry 2001;158: 256–265.

Woo T, Miller JL, Lewis DA. Schizophrenia and the parvalbumin-containing class of cortical local circuit neurons. Am J Psychiatry 1997;154: 1013–1015.

Woo T-U, Whitehead RE, Melchitzky DS, Lewis DA. A subclass of prefrontal gamma-aminobutyric acid axon terminals are selectively altered in schizophrenia. Proc Natl Acad Sci USA 1998;95: 5341–5346.

Xing G, Post R. Reduced Nurr1 (NOT1) and NGFI-B (TR3) expression in the prefrontal cortex of schizophrenic and unipolar patients. Submitted manuscript.

8 MACROANATOMICAL FINDINGS IN POSTMORTEM BRAIN TISSUE FROM SCHIZOPHRENIC PATIENTS

Peter Falkai, MD

Abstract

The macroscopic study of the postmortem brain in schizophrenia has revealed changes in area, volume and shape measures in several cortical and subcortical regions. There seems to be an overall subtle reduction of whole brain volume (about 3%), accompanied by increased area/volume of the ventricular system and a more regionalised volume loss in the temporal lobe, especially the hippocampus and entorhinal cortex (about 5 – 10%). Looking at other cortical regions the frontal and parietal lobe reveal changes which are more subtle compared to the temporal cortex. Subcortically the thalamus demonstrates volume reduction (about 10 – 15%) comparable to limbic system structures. Beside this the cerebellum and the basal ganglia show volume/area changes awaiting replication. Although the introduction of structural imaging has been very helpful in replicating and enlarging on these macroscopic findings, imaging will never replace postmortem studies because of its limited resolution and refusing access to the microscopic neurobiology of schizophrenia.

Usefulness of Macroanatomical Findings in Schizophrenia

In 1897 Alois Alzheimer was one of the first to publish a paper describing neuronal abnormalities in the neocortex of patients suffering from schizophrenia (Alzheimer 1897). Subsequent to this paper a wealth of qualitative findings were published until 1951, when during the first World Congress of Neuropathology the experts could not agree on a common morphological substrate for schizophrenia. After that schizophrenia was well regarded as a functional illness lacking a structural basis. In the late seventies some groups started to research this area again using quantitative, statistically based methods. This revieval of postmortem-studies was enabled by the introduction of imaging, like Computertomography (CT) and later Magnetic Resonance Imaging (MRI), supporting the notion of schizophrenia being a brain disorder (e.g. Johnstone et al. 1976). Up to now this kind of macroscopic postmortem research demonstrated changes in several brain regions focussing on temporo-limbic structures.

Confounding Variables

Critical questions concerning these changes in area, volume or shape are whether they are part of the disease process or just a consequence of chronic illness and accompanying life style. At least concerning the volume decrease of the amygdala-hippocampus complex, internal segment of the globus pallidus and substantia nigra we know from studies on recently treated patients with schizophrenia, that these changes are not due to any somatic treatment including neuroleptics (e.g. Bogerts et al. 1985). Concerning volume reduction of the thalamus on the other hand (Bogerts et al. 1985, 1990) one can not be absolutely sure. Concerning cortical regions there are no published data allowing a comparision between treated and untreated schizophrenia. Still, it is very important that postmortem-studies in schizophrenia are only based on brain tissue, which has been examined by neuropathologists in detail excluding control and schizophrenic cases with degenerative or vascular changes. However one has to be cautious not to exclude every pathology in the control subjects, while not using the same criteria in the schizophrenic cases. Futhermore the clinical documentation has to be good enough, that any experienced psychiatrist is able to replicate the clinical diagnosis based on international criteria, e.g. DSM-IV.

Meaning

From studies examining the clinical symptoms caused by brain tumors or head trauma (e.g. Hillboom 1954) we know that lesions of the frontal and temporal lobe will more likely cause psychotic symptoms compared to other brain regions. Furthermore about 25% of people with Alzheimer's Disease and 1 –2 % of patients with Temporal Lobe Epilepsy will develop psychotic symptoms in the course of their illness, which points to the temporal lobe as a key structure in the pathophysiology of psychotic illness. More recently animal models with temporal lobe, specifically entorhinal lesions (Lipska and Weinberger 1994, Talamini et la. 1998) have been able to mimic some behavioral disturbances in schizophrenia and replicate some of morphological findings outside the temporal lobe involving cortical thinning, disorganized cortical layering and abnormal temporal asymmetries (Talamini et al. 1998). Finally the reeler-mouse, a mutant lacking the reelin gene, reveals cytoarchitectonic abnormalities seen in schizophrenia (for review see Falkai et al. 2000). A mutation of the reelin gene was found to be causative for lissencephaly in two recently described pedigrees (Hong et al. 2000). Most interestingly the family members suffering from lissencephaly showed enlarged ventricles and temporal horns as well as reduced volumes of hippocampi bilaterally, traits frequently seen in schizophrenia.

In summary temporal lobe lesions demonstrated with macroscopic postmortem-studies in schizophrenia are central to the disease process and unlikely a consequence of treatment or other confounding variables.

Limitations and Promises

Postmortem studies on human tissue have several limitations of which some are mentioned: 1) Artifacts due to cause of death, postmortem-delay and histological procedures. 2) The clinical information is limited and additional information can usually not be aquired. 3) Standardised clinical ratings or neuropsychological work-up is usually not possible, with the exception of the few prospective studies available.

On the other hand imaging studies lack the limitations mentioned above, but the resolution of them is limited to the macroscopic level. Therefore trying to understand the underlying neuronal basis of area/volume/shape changes in schizophrenia is linked to postmortem-studies. In the future both research strategies should be combined by undertaking imaging on postmortem tissue and then performing histological studies.

In summary macroscopic postmortem-studies are still valuable despite of the advances of structural imaging made in recent years. They add to our knowledge with their superior resolution and the possibility to perform more detailed measurements e.g. on structures like fiber tracts (see cerebral asymmetries in this chapter).

Whole Brain Measures

Several planimetric postmortem studies of the whole cortex have been performed, some reporting significant reduction of cortical volume (12 per cent) and central grey matter (6 per cent) and others reporting no difference in volumes of cortex, white matter and whole hemispheres between schizophrenics and controls. Other general brain parameters measured have shown reduced brain length, brain weight, and increased ventricular area/volume (see table 1).

In summary there seems to be evidence from postmortem-studies that they are changes relating to the entire brain in schizophrenia, like a generalized reduction of the gray matter and a significant increase of the ventricular system.

Table 1: Gross morphometric cortical findings in schizophrenia

Region/Parameter	Finding
General	
Brain length	(↓)
Brain weight	(↓)
Ventricular area/volume	↑
Cortex thickness	(↓)
Temporal lobe	
Lobar area/volume	-
Hippocampal area/volume	-
Parahippocampal area/volume	(↓)
Parahippocampal cortical thickness	(↓)
Amygdala area/volume-	
Sylvian fissure length, planum temporal volume	↓
Sulcogyral pattern	(Abnormal)
Frontal, parietal and occipital lobes	
Cingulate cortical thickness	-
Insula area/volume	-
Corpus callosum thickness	(↑)
Internal capsule area/volume	-

In comparison to controls:
↓ = reduced
↑ = increased
- = no difference
( ) = finding not or only partially replicated
Adapted from Arnold S.E. and Trojanowski. J.Q. Recent advances in defining the neuropathology of schizophrenia. Acta Neuropathologica 1996; 92; 217-231.

Regional Measures

Cortical Structures

Temporal Lobe. Since the first report of reduced tissue volume in temporolimbic structures of schizophrenics (Bogerts et al. 1983), numerous quantitative or qualitative anatomical postmortem studies on temporo-limbic structures of schizophrenics have been published. Of these studies the majority found subtle structural changes (15-20 per cent mean volume reduction) in at least one of the investigated areas, whereas only a few yielded entirely negative results (for review see Dwork 1997). The findings comprise reduced volumes (Falkai and Bogerts 1986) or cross-sectional areas (Altshuler et al. 1987) of the hippocampus, amygdala (Bogerts et al.

1985), and parahippocampal gyrus (Falkai et al. 1988, Brown et al. 1986), which were later corroborated by my morphometric MRI studies. Other findings in limbic brain regions are left temporal horn enlargement (Crow et al. 1989), white matter reductions in parahippocampal gyrus (Colter et al. 1987) or hippocampus, and an increased incidence of a cavum septi pellucidi (Degreef et al. 1992).

Currently there is a growing literature examining the temporal lobe as well as other structures with respect to disturbed laterality in schizophrenia. Therefore a separate paragraph is dedicated to this subject.

Frontal Lobe. There is up to now no published postmortem data on the frontal lobe in schizophrenia. However from the MRI literature it can be deducted, that there is a subtle volume reduction present, but the degree of change is more moderate compared to the temporal lobe. A recent postmortem-study has raised the question whether the gyrification pattern is altered in the frontal lobe (Vogeley et al. 2000). Using the Gyrification Index by Zilles et al (1988) (GI) significant hypergyrification was again found in the right frontal lobe in patients suffering from schizophrenia compared to control cases (Vogeley et al, 2000, see Figures 1a and 1b).

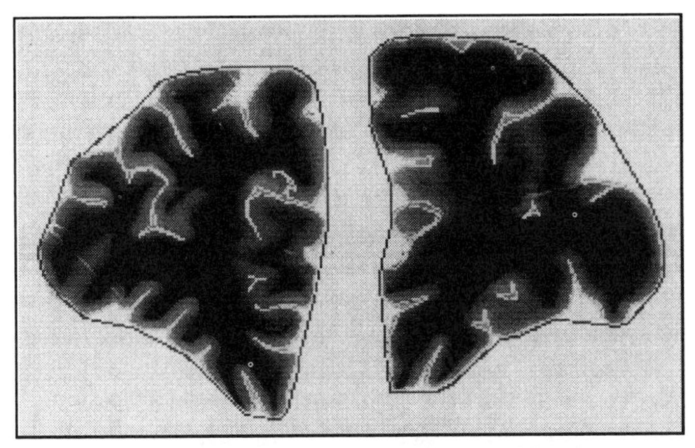

Figure 1a: Determining frontal lobe gyrification using the Gyrification Index (GI)

$$GI = \frac{\text{Outer Contour}}{\text{Inner Contour}}$$

Zilles et al., Anat Embryol 179, 174 -179, 1988

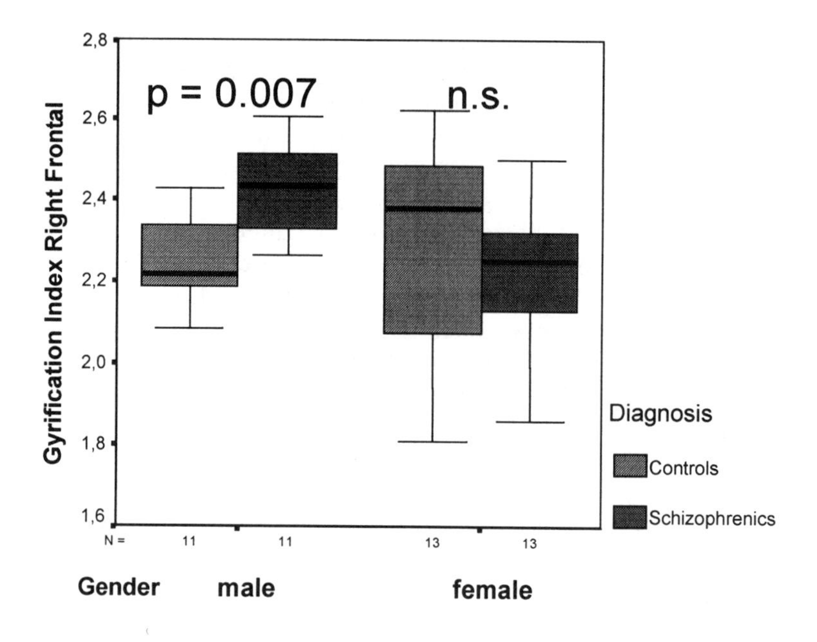

Figure 1b: Hypergyrification in the right frontal lobe in schizophrenia

Vogeley et al., Am J Psychiatry 157(1), 34-39, 2000

Following up on this finding the GI was determined in schizophrenic patients and their healthy siblings. Significant hypergyrification was found in the frontal lobe again (Vogeley et al. 2001).

The complexity of the cortical cytoarchitectonics has avoided the performance of quantitative anatomical studies of the cortex in schizophrenia. The introduction of stereology and the Grey Level Index (GLI) methodology (e.g. Kawasaki et al. 2000) has yielded some interesting findings concerning volumetric changes. It seems to be that at least in Brodmann Area 9 there is a lack of neuronal loss, but decrease of neuropil leading to reduced gray matter volume in schizophrenia (Selemon et al. 1995).

Cortical Structures. Unchanged volumes of the striatum and external pallidum but a subtle volume decrease in the internal pallidal segment were found in brains from the pre-neuroleptic era. Pallidal volume reduction was due to a reduction in the catatonic subgroup (Bogerts et al. 1986). The initial findings have to be pursued, as longitudinal MRI studies suggest that enlargement of basal ganglia can be seen in schizophrenia as a consequence of treatment with classical neuroleptics, which can be reversed by the use of atypical substances (Chakos et al. 1994).

After initially finding no volumetric changes in the thalamic nuclei, subsequently the area/volume of the whole thalamus, but especially the mediodorsal nucleus and anteroventral thalamic nucleus were found to be decreased (Bogerts et al. 1990, Danos et al. 1998).

Changes in area measurements of the corpus callosum were described in some studies. The findings, however, are inconsistent; there are reports of increased as well as of decreased midline areas. More consistent are reports of shape abnormalities, in that the sex difference in anterior and posterior callosal thickness in normal controls seems to be reversed in schizophrenics and the mean curvature in the corpus callosum is bent upwards (for review see Woodruff et al. 1995). Only a limited number of postmortem-studies were publihed after the first reports of cerebellar atrophy in schizophrenia. Reduced size of anterior vermis or reduced density of Purkinje cells were not replicated in subsequent studies (for review see Supprian et al. 2000).

Findings of decreased volume of the substantia nigra (Bogerts et al. 1983) and the periventricular gray matter (Lesch et al. 1984) as well as no volumetric change in the locus coeruleus (Jeste et al. 1989) await replication.

In summary, looking at regionalized changes macroscopic morphometric postmortem studies demonstrate volume/area reduction of temporolimbic nuclei in schizophrenia (see Table 2). Findings on other regions such as the basal ganglia, thalamus, cerebellum and brainstem await replication.

Table 2: Gross morphometric subcortical findings in schizophrenia

Region/Parameter	**Finding**
Basal ganglia	
Globus pallidum area/volume	(↓)
Nucleus accumbens area/volume	↓
Caudate-putamen area/volume	↑
Thalamus	
Mediodorsal nucleus area/volume	↓
Whole and various nuclei area/volume	-
Cerebellum	
Anterior vermis area	↓
Brainstem	
Substantia nigra volume	↓
Locus coerluleus volume	-
Periventricular gray volume	↓

In comparison to controls:
↓ = reduced
↑ = increased
- = no difference
( ) = finding not or only partially replicated

Cerebral Asymmetries

One of the first was Crichton-Brown (1874) who proposed that schizophrenia has morphological changes in the neocortex, which are more pronounced on the left than on the right hand side. However it was Crow et al. (1989) determining the volumes of the ventricular system in postmortem brains of schizophrenic patients and controls finding a left lateralised increased of the temporal horn in schizophrenia. From this it was concluded that disturbed asymmetry is central to schizophrenia. This hypothesis has stimulated some research concentrating on those structures, which are asymmetric in the human brain.

Frontal and Occipital Lobes

One of the most prominent asymmetries are the so called frontal and occipital lobes. The left frontal lobe being bigger than the right one and the right occipital lobe being bigger and longer than the left counterpart. CT (e.g. Falkai et al. 1995a) and MRI (Bilder et al. 1994) have demonstrated loss of this asymmetry pattern in schizophrenia.

Measuring the length from the frontal pole to the central sulcus dorsally over the external surface of the postmortem brain on both hemispheres, female controls had a left-greater-than-right asymmetry, and the male controls had a right-greater-than-left asymmetry. This pattern was reversed in schizophrenia. The converse effect was observed on a similar measure of the occipito-parietal lobes. Further, within the patient group, the frontal lobe asymmetry was related to age of onset such that leftward asymmetrical brains were associated with a later age of onset than rightward asymmetrical brains (Highley et al. 1998a).

Temporal Lobe

Measuring the volume of cortex underlying the Planum temporale as loss of asymmetry was found in schizophrenia (Falkai et al. 1995b). This was due to a reduction of the length and not mean width of the left Planum temporale expecially in male schizophrenia. An another study could replicate this finding describing a significant left-sided reduction in the superior temporal gyrus with schizophrenia, the total volume of temporal lobe grey and white matter being significantly reduced as well. Although being more marked on the left than the right, the lateralisation for these grey and white matter measures attained formal statistical significance only for the superior temporal gyrus (Highley et al. 1999a).

Along these lines the same group found in postmortem brains that the degree of gyral folding was significantly increased in schizophrenia, but the orientation of the sulci was not changed. In addition the temporal lobes were significantly shortened in schizophrenia. These two structural changes were interpreted to refelct an alteration of the cortico-cortical connectivity of the brain in schizophrenia (Highley et al. 1998b).

Focussing on temporo-basal regions the same group determined the volume of the gray matter of the parahippocampal and fusiform gyri. In relation to the comparison subjects, the schizophrenic patients had lower volumes of the parahippocampal and fusiform gyri on the left side. For both structures a left-greater-than-right volume asymmetry was present in the comparison subjects, but this asymmetry was reversed in the schizophrenic patients. A sex difference was present with respect to age at onset – degree of anomaly of asymmetry for both gyri increased with age at onset in men but not in women (McDonald et al. 2000).

Fibre Tracts

Hemispheres are closely connected with fibre tracts. The more asymmetric a structure is (Rosen et al. 1989), the less fibres connect the two regions via the corpus callosum. In order to make use of this fact the fibre composition of the corpus callosum, anterior commissure and fornix were examined postmortem in schizophrenia.

Looking at the corpus callosum it was found, that amongst controls, females had greater density than males; in patients with schizophrenia this difference was reversed. A reduction in the total number of fibres in all regions of the corpus callosum except the rostrum was observed in female patients. These findings were interpreted as evidence for subtle and gender-dependent alteration in the forebrain commissures relating to the deviations in asymmetry seen in other studies (Highley et al. 1999b). Examining the anterior commissure its cross-sectional area was unaffected, but female not male patients revealing a significant reduction in fiber densities (Highley et al. 1999 b). Finally the density of fibres of the left fornix was increased in male schizophrenics, while the total fiber number was not differently affected by gender or diagnosis. These findings were interpreted as evidence against a primary limbic encephalopathy in schizophrenia (Chance et al. 1999).

In summary there is some evidence from postmortem-studies that fronto-occipital as well as temporal asymmetries are changed in schizophrenia. Most interestingly the changes in the temporal lobe seem to focus on the parahippocampal/fusiform and the superior temporal gyri, which is supported by recent metanalysis of the MRI literature (Lawrie and Abukmeil 1998, Wright et al. 2000).

Conclusions

Macroscopic postmortem examinations together with brain imaging studies have helped considerably to restore the concept of schizophrenia being a brain disorder with an underlying morphological basis. Major findings are a subtle reduction of the whole brain volume accompanied by a significant increase of the ventricular system, especially lateral ventricles.

There is hardly any region of the brain where some morphological abnormality has not been described for schizophrenia. Looking at cortical regions macroscopic changes like volume/area/shape changes are replicated for temporal lobe structures, like the hippocampus, parahippocampal gyrus and superior temporal gyrus. Concerning subcortical structures there is some evidence for volume reduction of the thalamus, however changes of corpus callosum, basal ganglia and brain stem await replication. Recent studies dedicated to asymmetrical brain structures have yielded interesting results supporting the notion a disturbed lateralisation in schizophrenia, but they need replication, too. Matching these results with recent metanalyses of structural MRI studies it seems to be, that schizophrenia is indeed a generalised brain disorder (whole brain volume reduction and ventricular enlargement) with replicated structural changes in the temporal lobe (hippocampus, parahippocampal gyrus and superior temporal gyrus). Compared to structural imaging macroscopic postmortem-studies have a superior resolution and offer the possibility to study the brain differently, e.g. quantify fibre tracts. Finally they are the gateway to microscopic examinations, which are the morphological basis to the neurobiology of schizophrenia. Therefore postmortem-studies in general and their macroscopic variant is still needed to understand the neural substrate of severe mental illness.

Acknowledgements

Supported by the Wada and Theodore Stanley Foundation.

References

Altshuler LL, Conrad A, Kovelman JA, Scheibel A. Hippocampal pyramidal cell orientation in schizophrenia. A controlled neurohistologic study of the Yakovlev collection. Arch Gen Psychiatry 1987; 44: 1094-1098.

Alzheimer A. Beiträge zur pathologischen Anatomie der Hirnrinde und zur anatomischen Grundlage einiger Psychosen. Mschr. Psychiat. Neurol. 1897; 2: 82-120.

Bilder RM, Houwei W, Bogerts B, Degreef G, Ashtari M, Alvir JMJ et al. Absence of regional hemispheric asymmetries in first-episode schizophrenia. Am J Psychiatry 1994; 151: 1437-1447.

Bogerts B, Häntsch J, Herzer M. A morphometric study of the dopamine containing cell groups in the mesencephalon of normals, Parkinson patients and schizophrenics. Biol Psychiatry 1983; 18: 951-960.

Bogerts B, Meertz E, Schönfeld-Bausch R. Morphometrische Untersuchungen an Gehirnen Schizophrener. 2. Kongr. Dtsch. Ges. Biol. Psychiat. Hrsg. von Hopf A. Biologische Psychiatrie. Springer Berlin 1984; 227-233.

Bogerts B, Meertz E, Schönfeldt-Bausch R. Basal ganglia and limbic system pathology in schizophrenia. A morphometric study of brain volume and shrinkage. Arch Gen Psychiatry 1985; 42: 784-791.

Bogerts B, Falkai P, Tutsch J. cell numbers in the pallidum and hippocampus of schizophrenics. In Shagass C. et al Biol Psychiatry Elsevier-North Holland Amsterdam 1986; 1178-1180.

Bogerts B, Ashtari M, Degreef G, Alvir JMJ, Bilder RM, Liebermann JA. Reduced temporal limbic structure volumes on magnetic resonance images in first-episode schizophrenia. Psychiat Res Neuroimaging 1990; 35: 1-13.

Brown R, Colter N, Corsellis JAN et al. Postmortem evidence of structural brain changes in schizophrenia. Arch Gen Psychiatry 1986; 43: 36-42.

Chakos MH, Liebermann JA, Bilder RM, Borenstein M, Lerner G, Bogerts B, Wu H, Kinon B, Ashtari M. Increase in caudate nuclei volumes of first-episode schizophrenic patients taking antipsychotic drugs. Am J Psychiatry 1994; 151: 1430-1436.

Chance SA, Highley JR, Esiri M.M, Crow TJ. Fiber Content of the Fornix in Schizophrenia: Lack of Evidence for a Primary Limbic Encephalopathy. Am J Psychiatry 1999; 156: 1720-1724.

Colter N, Battal S, Crow TJ, Johnstone EC, Brown R, Bruton C. White matter reduction in the parahippocampal gyrus of patients with schizophrenia (letter). Arch Gen Psychiatry 1987; 44: 1023.

Crichton-Brown J. On the weight of the brain and ist component parts in the insane. Brain 1874; 2: 42-67.

Crow TJ, Ball J, Bloom SR, et al. Schizophrenia as an anomaly of development of cerebral asymmetry. A postmortem study and a proposal concerning the genetic basis of the disease. Arch Gen Psychiatry 1989; 46: 1145-1150.

Danos P, Baumann B, Bernstein HG, Franz M, Stauch R, Northoff G, Krell D, Falkai P, Bogerts B. Schizophrenia and anteroventral thalamic nucleus: selective decrease of parvalbumin-immuno reactive thalamocortical projection neurons. Psychiatry Res 1998; 82(1): 1-10.

Degreef G, Bogerts B, Falkai P, Greve B, Lantos G, Ashtari M, Liebermann J. Increased prevalence of the cavuum septum pellucidum in MRI scans and postmortem brains of schizophrenic patients. Psychiatry Res 1992; 45: 1-13.

Dwork AJ. Postmortem studies of the hippocampal formation in schizophrenia. Schizophr Bull 1997; 23: 385-402.

Falkai P, Bogerts B. Cell loss in the hippocampus of schizophrenics. Europ Arch Psychiat neurol Sci 1986; 236: 154-161.

Falkai P, Bogerts B, Rozumek M. Limbic pathology in schizophrenia: the entorhinal region-a morphometric study. Biol Psychiatry 1988; 24: 515-521.

Falkai P, Bogerts B, Schneider T, Greve B, Pfeiffer U, Pilz K, Gonsiorzcyk C, Majtenyi C, Ovary I. Disturbed planum temporale asymmetry in schizophrenia: a quantitative postmortem study. Schizophr Res 1995; 14: 161-176.

Falkai P, Schneider T, Greve B, Klieser E, Bogerts B. Reduced frontal and occipital lobe assymetry on the CT scans of schizophrenic patients: ist specificity and clinical significance. J Neural Transm Gen Sect 1995; 99: 63-77.

Falkai P, Schneider-Axmann T, Honer WG. Entorhinal cortex pre- alpha cell clusters in Schizophrenia: quantitative evidence of a developmental abnormality. Biol. Psychiatry 2000; 47(11): 937-43.

Highley JR, Esiri MM, McDonald B, Cooper SJ, Crow TJ. Temporal-lobe length is reduced, and gyral folding is increased in schizophrenia: a postmortem study. Schizophr Res 1998; 34: 1-12.

Highley JR, Esiri MM, McDonald B, Cortina-Borja M, Cooper SJ, Herron BM, Crow TJ. Anomalies of cerebral asymmetry in schizophrenia interact with gender and age of onset: a postmortem study. Schizophr Res 1998; 34: 13-25.

Highley J.R., Esiri M.M., McDonald B., Roberts H.C., Walker M.A., Crow T.J. The Size and Fibre Composition of the Anterior Commissure with Respect to Gender and Schizophrenia. Biol Psychiatry 1999; 45; 1120-1127.

Highley RJ, Esiri MM, McDonald B, Cortina-Borja M, Herron BM, Crow TJ. The size and fibre composition of the corpus callosum with respect to gender and schizophrenia: a postmortem study. Brain 1999; 122: 99-110.

Hillboom E. Schizophrenia-like psychoses after brain trauma. Acta psychiat neurol scand 1954; 60: 36-47.

Hong SE, Shugart YY, Huang DT, Shahwan SA, Grant PE, Hourihane JO, Martin ND, Walsh CA. Autosomal recessive lissencephaly with cerebellar hypoplasia is associated with human RELN mutations. Nat Genet 2000; 26(1): 93-96.

Jeste DV, Lohr JB. Hippocampal pathologic findings in schizophrenia. Arch Gen Psychiatry 1989; 46: 1019-1024.

Johnstone EC, Crow TJ, Frith CD, Husband J, Kreel L. Cerebral ventricular size and cognitive impairment in chronic schizophrenia. Lancet 1976; 2(7992): 924-926.

Kawasaki Y, Vogeley K, Jung V, Tepest R, Hütte H, Schleicher A, Falkai P. Automated image analysis of disturbed cytoarchitecture in Brodmann area 10 in schizophrenia: a postmortem study. Prog Neuropsychopharmacol. Biol Psychiatrie 2000; 24: 1093-1104.

Lawrie SM, Abukmeil SS. Brain abnormality in schizophrenia: a systematic and quantitative review of volumetric magnetic resonance imaging studies. Br J Psychiatry 1998; 172: 110-120.

Lesch A, Bogerts B. The diencephalon in schizophrenia: Evidence for reduced thickness of the periventricular grey matter. Europ Arch Psychiat neurol Sci 1984; 234: 212-219.

Lipska BK, Weinberger DR. Delayed effects of neonatal hippocampal damage on haloperidol-induced catalepsy and apomorphine-induced stereotypic behaviors in the rat. Brain Res Dev Brain Res 1993; 75(2): 213-222.

McDonald B, Highley JR, Walker MA, Herron BM, Cooper SJ, Esiri MM, Crow TJ. Anomalous Asymmetry of Fusiform and Parahippocampal Gyrus Gray Matter in Schizophrenia: A postmortem Study. Am J Psychiatry 2000; 157: 40-47.

Rosen GD, Sherman GF, Galaburda AM. Interhemispheric connections differ between symmetrical and asymmetrical brain regions. Neurosci 1989; 33:525-533.

Selemon LD, Rajkowska G, Goldman-Rakic PS. Abnormally high neuronal density in the schizophrenic cortex: a morphometric analysis of prefrontal area 9 and occipital area 17. Arch Gen Psychiatry 1995; 52: 805-820.

Supprian T, Ulmar G, Bauer M, Schüler M, Püschel K, Retz-Junginger P, Schmitt HP, Heinsen H. Cerebellar vermis area in schizophrenic patients-a postmortem study. Schizophr Res 2000; 42: 19-28.

Talamini LM, Koch T, Ter Horst GJ, Korf J. Mehylazoxymethanol acetate-induced abnormalities in the entorhinal cortex of the rat; parallels with morphological findings in schizophrenia. Brain Res 1998; 789: 293-306.

Vogeley K, Schneider-Axmann T, Pfeiffer U, Tepest R, Bayer TA, Bogerts B, Honer WG, Falkai P. Disturbed gyrification of the prefrontal region in male schizophrenic patients: A morphometric postmortem study. Am J Psychiatry 2000; 157 (1): 34-39.

Vogeley K, Tepest R, Pfeiffer U, Schneider-Axmann T, Maier W, Honer WG, Falkai P. Right frontal hypergyria differentiation in affected and unaffected siblings from families multiply affected with schizophrenia: a morphometric MRI study. Am J Psychiatry 2001; 158(3): 494-496.

Woodruff PWR, McManus IC, David A.S. A metaanalysis of corpus callosum size in schizophrenia. J Neurol Neurosurg Psychiatry 1995; 58: 457-461.

Wright IC, Rabe-Hesketh S, Woodruff PWR, David AS, Murray RM, Bullmore ET. Meta-Analysis of Regional Brain Volumes in Schizophrenia. Am J Psychiatry 2000; 157:16-25.

9 MICROANATOMICAL FINDINGS IN POSTMORTEM BRAIN TISSUE FROM SUBJECTS WITH SCHIZOPHRENIA: Disturbances in Thalamocortical and Corticocortical Connectivity in Schizophrenia

T. Hashimoto and D. A. Lewis

Abstract

Evidence from clinical studies suggest that the prefrontal cortex and hippocampal formation are sites of dysfunction in schizophrenia. Postmortem investigations have revealed several types of cellular alterations that appear to be common to both of these brain regions. These changes include reductions in cortical volume or thickness, increased neuronal density without a change in total neuronal number, reduced somal size and alterations in the dendritic trees of projection neurons, and decreased markers of presynaptic axon terminals. This chapter reviews the data that support these conclusions and considers these findings in the context of the extent to which they reflect alterations in thalamocortical and corticocortical connectivity in schizophrenia.

Introduction

Since Emil Kraeplin first used the term "dementia praecox" to describe an independent entity among mental disorders based upon its clinical symptoms and chronic deteriorating course, it has been expected that some "organic changes" would be uncovered in the brain that would explain the pathophysiology of this illness now called schizophrenia. Thus, postmortem brain tissue from subjects with schizophrenia became a target of histopathological investigation following Kraeplin's description. However, many positive findings reported in initial studies were not replicated in subsequent investigations and serious questions about limitations in methodology were raised. The initial enthusiasm turned to a skepticism concerning, not only postmortem histopathological studies (Plum, 1972), but also the idea of neuronal or structural alterations in the brain in schizophrenia.

Progress in discovering brain abnormalities in schizophrenia occurred with the development of *in vivo* imaging techniques, such as computed tomography (CT) and magnetic resonance imaging (MRI). During the last 15 years, findings obtained from those imaging techniques have triggered a new era of postmortem histopathological studies. In addition, the application of new methodologies, such as unbiased 3-dimensional (3-D) sampling methods (i.e., stereology) and statistical strategies to control for confounding effects of variables such as age, sex, and postmortem interval (PMI), have enhanced the rigor of postmortem studies. Now, in concert with structural and functional imaging studies, postmortem brain research is beginning to elucidate the pathophysiology of schizophrenia. Specifically, postmortem studies provide the only means to gain insight into the microanatomic correlates of the disease, information that is critical for understanding the molecular, cellular and circuitry bases of the functional brain disturbances in subjects with schizophrenia.

Abnormalities in multiple brain regions and the connections among them, are likely to contribute to, or arise from, the primary pathophysiological mechanisms operative in schizophrenia. However, the greatest number of postmortem studies have focused on the cerebral cortex, especially the prefrontal regions, and the hippocampal formation. Thus, this chapter will review these findings with an emphasis on how they may inform our understanding of neural circuitry disturbances in schizophrenia.

Cerebral Cortex of Subjects with Schizophrenia

General findings

Deficits in higher cognitive functions observed in individuals with schizophrenia appear to be attributable to the dysfunction of association regions of the cerebral cortex where sensory inputs are integrated and executive control of behavior is generated. Structural MRI studies have reported a reduction in cortical gray matter volume by approximately 2-6%, as well as a diffuse enlargement of lateral and third ventricles in subjects with schizophrenia (Staal et al., 2000; Wright et al., 2000; Sigmundsson et al., 2001, for review see Shenton et al., 2001). In postmortem studies, two macro-anatomical stereological analyses on the total volume of the cerebral cortex including the hippocampal formation showed different results: one reported a significant 12% reduction in subjects with schizophrenia (Pakkenberg, 1987), but the other found no difference (Heckers et al., 1991b), though both reported significant enlargement of the ventricles (Pakkenberg, 1987; Heckers et al., 1990a). This discrepancy indicates an inherent difficulty for postmortem studies to assess subtle changes in cortical volume because of possible artifacts arising from brain tissue swelling after death and shrinking during fixation (Heckers et al., 1991b). However, the above MRI findings strongly support the presence of a slight reduction of cortical volume in schizophrenia. A subsequent stereological study (Pakkenberg, 1993) addressing micro-anatomical substrates of this reduced cortical volume unexpectedly revealed increased neuron density with the reduced cortical volume. Thus, total neuron number was found to be unchanged in all four lobes of the cerebral cortex of subjects with schizophrenia. These findings, together with the apparent absence of gliosis throughout the cortical regions (for review see Harrison and Lewis, 2001), suggested that degenerative loss of cortical neurons is not a central feature in schizophrenia.

Among cortical association areas, several regions have been implicated by clinical, psychological and functional imaging studies to have functional relevance to the cognitive deficits and/or psychotic symptoms of schizophrenia. Recent structural MRI studies have also revealed a reduction in gray matter volume in these regions, namely the dorsal and medial prefrontal cortex (PFC) (Staal et al., 2000; Gur et al., 2000a; Hirayasu et al., 2001), the hippocampal formation (Nelson et al., 1998; Velakoulis et al., 1999; Razi et al., 1999; Wright et al., 2000; Gur et al., 2000b; Sigmundsson et al., 2001) and the superior temporal cortex (for review see Shenton et al., 2001). Accordingly, many postmortem histopathological studies have focused on these regions, especially on the PFC and the hippocampal formation. Interestingly, through studies performed by several research groups using different collections of postmortem brains, common abnormalities have emerged in these two most intensively investigated cortical areas over the past ten years. These features include a subtle reduction in cortical volume and thickness, an absence of neuronal loss and gliosis, a reduced size of projection neurons, and alterations in their dendritic morphology.

Prefrontal Cortex

Cortical Thickness and Volume

Recent structural MRI studies have consistently reported a 7-11% decrease in the prefrontal gray matter volume in subjects with schizophrenia, even at the first episode of the disease (Staal et al., 2000; Gur et al., 2000a; Hirayasu et al., 2001). However, despite the statistical significance of these findings, considerable overlap in cortical volume was present between control subjects and subjects with schizophrenia. In postmortem studies, a 3-10% decrease was also reported for volume or thickness of PFC by 3-D or 2-D measurements, but it did not reach statistical significance with relatively large variance across subjects (Akbarian et al., 1995; Selemon et al., 1995, 1998; Woo et al., 1997; Volk et al., 2000; Thune et al., 2001). Significant reduction, on the other hand, was observed in assessments of the relative proportion of each layer to the whole cortical thickness. For example, the relative thickness of layer V in area 9 and of layer II in area 46 were found to be reduced by 7% and 12%, respectively, in subjects with schizophrenia (Selemon et al., 1995, 1998).

Neuronal Number and Density

To determine whether alterations in neuronal number were present in the PFC in schizophrenia, Benes et al. (1986, 1991a) performed the first systematic evaluation of neuronal density on the basis of 2-D measurements, and reported that neuronal density, especially for nonpyramidal neurons, was significantly reduced in the PFC (area 10) of subjects with schizophrenia. However, alterations in neuronal density do not necessarily correlate with a change in total number of neurons since density is also influenced by the volume in which the neurons are distributed. In a 3-D stereological estimation of total neuronal number that assessed reference volume, Pakkenberg (1993) found no change in neuron number in the frontal lobe of subjects with schizophrenia compared to control subjects. Recently, Thune et al. (2001) further assessed total neuron number in the PFC using stereological approaches and reported no change in subjects with schizophrenia. On the other hand, Selemon et al. (1995), using a 3-D measurement technique, demonstrated a 17% increase in neuronal density in area 9 of subjects with schizophrenia, with the largest increase (24%) in layers IV and V. The densities of pyramidal and nonpyramidal cells were equally elevated. The same group subsequently reported a 21% increase in the neuronal density in area 46 of subjects with schizophrenia (Selemon et al., 1998). Based upon these observations, together with a slight decrease in cortical thickness, Selemon et al. (1995, 1998) hypothesized a reduction of neuropil, including dendrites and axons, in the PFC of subjects with schizophrenia. Although not all groups have observed an increase in density of PFC neurons in schizophrenia (Akbarian et al., 1995; Thune et al., 2001), this hypothesis is supported by observations (described below) suggesting aberrant neuronal connections in the PFC of schizophrenic subjects.

Neuronal Somal Size

Benes et al. (1986) first reported that the cross-sectional somal area of large neurons in layers III and VI of area 10 was not significantly different between subjects with schizophrenia and normal controls, although the mean somal area showed a 4-7% reduction in schizophrenia. Rajkowska et al. (1998) also reported a non-significant 4% reduction in overall mean size of neuronal soma in area 9 of subjects with schizophrenia, with a significant 7% decrease in layer III. This reduction in somal size was accompanied by a shift of the normal distribution curve for somal size toward smaller sizes in layers III and V, and by a significant 40% decrease in the density of extra large neurons in layer IIIc. Thus, it appeared that large pyramidal neurons in layer IIIc underwent the most severe atrophic or dystrophic changes in schizophrenia. Pierri et al. (2001) focused on the pyramidal neurons in deep layer III and revealed a significant 9.2% reduction in somal volume by using systematic random sampling and nucleator probes in a large cohort of subjects with schizophrenia. Glantz et al. (2000) also demonstrated a non-significant 4.9-12% reduction in the cross-sectional area of synaptophysin mRNA-positive neurons in layers III to VI in subjects with schizophrenia. In contrast to pyramidal neurons, nonpyramidal neurons were shown to be unchanged in their somal sizes in the PFC of subjects with schizophrenia by other studies (Akbarian et al., 1995; Woo et al., 1997; Volk et al., 2000).

Dendritic Morphology

To date, three studies have examined dendritic morphology in the PFC of subjects with schizophrenia. Garey et al. (1998) utilized the rapid Golgi technique to assess dendritic spine on layer III pyramidal cells in areas 10, 11 and 45. They found a 67% reduction in the spine density in schizophrenic subjects. Glantz and Lewis (2000) determined the spine density on basilar dendrites of Golgi-impregnated pyramidal neurons in superficial and deep layer III in area 46. In subjects with schizophrenia, the density was decreased on deep layer III pyramidal neurons by 23% compared with normal controls, whereas the density on superficial layer III pyramidal neurons showed a smaller and non-significant change. Importantly, the spine density on deep layer III neurons in area 46 did not differ between normal controls and non-schizophrenic psychiatric subjects with a history of antipsychotic medication use. This study also showed an 18% reduction in total length of basilar dendrites of deep layer III pyramidal neurons in subjects with schizophrenia compared to normal controls, although this difference was not specific to the diagnosis of schizophrenia. Kalus et al. (2000) analyzed dendritic arborization of layer III pyramidal cells in area 11 and found a 30% reduction in total dendritic length, a 28% reduction in mean length of each dendrite and a 23% reduction in total number of basilar dendritic segments in subjects with schizophrenia, but no change in apical dendritic morphology.

Since neurons receive synaptic input mainly at dendrites, reductions in dendritic spine density and arborization indicate a reduced number of synaptic inputs on pyramidal neurons, especially on those in PFC deep layer III. Consistent with this interpretation, reduced levels of the presynaptic protein synaptophysin have been found in the PFC of subjects with schizophrenia by a number of research groups (Perrone-Bizzozero et al., 1996; Glantz and Lewis, 1997; Davidsson et al., 1999; Honer et al., 1999; Karson et al., 1999).

Hippocampal Formation: Volume and Size

The results of structural MRI studies, including meta-analyses, indicate that hippocampal volume is significantly reduced by 4-8% in subjects with schizophrenia (Nelson et al., 1998; Velakoulis et al., 1999; Gur et al., 2000b; Wright et al., 2000). In postmortem studies, quantitative analyses of the Vogt collection, which was collected before the introduction of anti-psychotic medication, reported a 31% reduction in the anterior hippocampal volume (Bogerts et al., 1985). These investigators also found, in the same collection, a 9-35% reduction in volume of each subregion of the hippocampus with a comparable decrease in volume of pyramidal and granule cell layers (Falkai and Bogerts, 1986). These observations were, however, followed by reports of smaller or non-significant reductions of hippocampal volume in schizophrenia. In leukotomized subjects with schizophrenia from the Yakovlev collection, Jeste and Lohr (1989) showed that total volume of the left hippocampus in subjects with schizophrenia was non-significantly reduced by 40% and 20% compared with leukotomized controls and normal controls, respectively. They also found a volume reduction varying from 5% to 46% for each hippocampal subregion in subjects with schizophrenia, but statistical significance was detected only for CA4 volume between subjects with schizophrenia and leukotomized controls. Heckers et al. (1990a, 1991b), applying stereological approaches, found only a non-significant tendency for the whole hippocampus of subjects with schizophrenia to be smaller (3-6%) on both sides compared to age- and sex-matched controls with no change in the volume of the pyramidal cell layer. They suggested that errors in the calculation for tissue shrinkage compensation, analysis restricted to the anterior hippocampus, and the inclusion of controls unmatched for gender could have accounted for the large volume reduction observed in the initial study. However, Bogerts et al. (1990) found a significant reduction of bilateral hippocampal volume (18% for male, 13% for female) in a new group of subjects with schizophrenia compared with age- and sex- matched controls, using improved compensation for tissue shrinkage.

The volume of the parahippocampal gyrus, which includes the entorhinal cortex, was suggested to be decreased by 5-25% in schizophrenia in some structural MRI studies (Razi et al., 1999; Wright et al., 2000; Sigmundsson et al., 2001); however, others have found no detectable difference (Hirayasu et al., 1998; Staal et al., 2000). Similarly, several postmortem investigations reported a significant volume reduction of the parahippocampal gyrus, whereas others did not. Brown et al. (1986) reported a significant 11% decrease in the cortical width of the parahippocampal gyrus at a single level just anterior to the mammillary bodies in subjects with schizophrenia compared to subjects with affective disorders. In the studies of the Vogt collection, a 44% reduction in cortical volume of the parahippocampal gyrus (Bogerts et al., 1985) and a 27% reduction in the volume of the entorhinal cortex (Falkai et al., 1988) were reported in subjects with schizophrenia. The significant reduction of cross-sectional area of the parahippocampal gyrus was also demonstrated in other 2-D studies (Colter et al., 1987; Altshuler et al., 1990). However, the stereological study performed by Heckers et al. (1990b) did not detect a difference in volume of the entorhinal cortex between subjects with schizophrenia and their matched controls. They indicated difficulty in evaluating the volume of the entorhinal cortex due to a wide variety of macroscopic configurations of the parahippocampal gyrus across subjects. Recently, Krimer et al. (1997) performed stereological measurement of the volume of each lamina in each subarea of the rostral entorhinal cortex and detected no difference between subjects with schizophrenia and age-, sex- and PMI-matched control subjects.

Neuron Density and Number

Using the Vogt collection, Falkai and Bogerts (1986) determined the density of pyramidal neurons and calculated the total number of pyramidal neurons in the whole hippocampus by multiplying the density with the volume of the hippocampal pyramidal layer. Although the density showed a slight increase, total pyramidal neuron number was significantly decreased by 9-30% in each subregion of the hippocampus in subjects with schizophrenia due to the marked reduction of the volume. Jeste and Lohr (1989) also reported a significant reduction of pyramidal neuron densities in leukotomized subjects with schizophrenia compared with normal controls. This reduction, however, did not reach a significant level when compared with leukotomized controls. In contrast to these non-stereological measurements, the stereological study by Heckers et al., (1991a) demonstrated no differences in neuron density and tissue volume for the subiculum, CA1, CA2/3 and CA4 between subjects with schizophrenia and controls. Therefore, the total neuron number was indicated to be unchanged in all the hippocampal subregions in subjects with schizophrenia. Subsequent 2-D studies reporting unchanged (Arnold et al., 1995) or even slightly elevated (Zaidel et al., 1997a) neuron density are consistent with unchanged total neuron number in the hippocampus of subjects with schizophrenia, taking into account the above observations of non-significant, slight reduction in tissue volume.

In the entorhinal cortex, the initial study using the Vogt collection showed a 37% reduction in total neuron number accompanied with a robust reduction of tissue volume in subjects with schizophrenia (Falkai et al., 1988). However, a stereological study by Krimer et al. (1997) revealed that total neuron number and neuron density in each layer of the cytoarchitechtonically-defined subdivisions of rostral entorhinal cortex did not differ significantly between subjects with schizophrenia and age-, sex- and PMI-matched controls. Arnold et al. (1995) reported a trend for neuron density to be higher in layers II, III and V of the entorhinal cortex for subjects with schizophrenia than for the normal controls. Thus, a consistent decrease in neuron number seems unlikely to be the case in the hippocampal formation, but the reported slight elevation of neuronal density (Arnold et al., 1995; Zaidel et al., 1997a) could be related to a reduction of neuropil in this brain region, as in the PFC.

Neuronal Size

In leukotomized subjects with schizophrenia, Christison et al. (1989) assessed the cross-sectional area of CA1 neurons in hippocampal sections cut at the coronal, horizontal and sagittal planes, and reported no change compared to leukotomized or normal control subjects. In non-leukotomized subjects with schizophrenia, Benes et al. (1991b) reported a significant 13-18% reduction in the cross-sectional area of pyramidal neuron soma through CA1-4 of the posterior hippocampus. In the mid-hippocampus and the posterior part of the entorhinal cortex, Arnold et al. (1995) also measured the cross-sectional areas of neurons. They found smaller neuronal size in all the Ammonic subfields, the subiculum and the entorhinal cortex in subjects with schizophrenia compared to normal controls. Within these regions, the somal area reduction was prominent and statistically significant for pyramidal neurons in CA1 (13%) and the subiculum (21%), and for stellate neurons in layer II of the entorhinal cortex (14%). Zaidel et al. (1997b) also reported an overall 6.8% reduction in somal area across the hippocampal subregions with a statistical significance for left CA1/2 and right CA3. Although no stereological study on somal volume has been performed in the hippocampal formation yet, consistent findings of these 2-D studies suggest a reduced somal size of projection neurons in the hippocampus and the entorhinal cortex.

Dendritic Morphology

Kovelman and Scheibel (1984) first described an abnormality of pyramidal cell dendritic polarity in the anterior to mid hippocampus in subjects with chronic paranoid schizophrenia by using Golgi staining. However, they considered the disturbance in dendritic polarity to be a reflection of the disarray of hippocampal pyramidal cells. Zaidel et al. (1997b) reported that pyramidal cells in the left CA1 and CA3 had a more "elongated" shape in normal control subjects than in subjects with schizophrenia, whereas pyramidal cells in the left subiculum were more "elongated" in subjects with schizophrenia compared to normal controls. The relationship of this observation with dendritic morphology, however, was not clear since they used Nissl-stained sections. Recently, Rosoklija et al. (2000), using rapid Golgi impregnation, demonstrated that arborization of apical dendrites of pyramidal neurons in the subiculum was less extensive in subjects with schizophrenia than in normal controls, whereas there was no difference in basilar dendrite morphologies across subjects. Furthermore, the subicular pyramidal neurons exhibited a marked reduction of spine density on apical dendrites in subjects with schizophrenia; no overlap in these measures was present between subjects with schizophrenia and normal controls. At the molecular level, the expression level of M2, a dendrite specific cytoskeletal protein, was reported to be decreased in the subiculum and entorhinal cortex of subjects with schizophrenia by one group (Arnold et al., 1991b, 1999), and to be increased in CA1-4 and the subiculum by another group (Cotter et al., 1997, 2000), possibly reflecting an inherent difficulty of quantitative assessment of protein levels in immunohistochemical specimens. However, the decreased levels of presynaptic proteins, such as synaptophysin, SNAP-25, the synapsins and the complexins, reported in several studies (Browning et al., 1993; Eastwood and Harrison, 1995; Harrison and Eastwood, 1998; Young et al., 1998), together with the finding of reduced dendritic spines, suggesting a lowered density of synapses on the dendrites on hippocampal neurons in schizophrenia.

Cytoarchitectonic Abnormalities

The characteristic cytoarchitecture of the hippocampal formation has been the focus of a number of studies designed to determine whether alterations in cellular arrangements are present in this region. In the hippocampus, Kovelman and Scheibel (1984) first reported that the orientation of pyramidal neurons was altered in subjects with schizophrenia. This pyramidal cell 'disarray' was most prominent in transitional zones between CA1 and CA2 and between CA1 and subiculum, and was interpreted as a result of a disturbance in neuronal migration during development. Subsequent studies from the same group supported this disorganization of pyramidal cell orientation (Althauser et al., 1987; Conrad et al., 1991). However, several other independent studies, using computerized image analysis of neuronal orientation, have not replicated the observation of neuronal disarray in the hippocampus (Christison et al., 1989; Benes et al., 1991b; Arnold et al., 1995; Zaidel et al., 1997b).

In the entorhinal cortex, Jakob and Beckmann (1986) first described misplaced and aberrantly clustered neurons in layers II and III. They also claimed that this finding was attributable to a developmental disturbance in cell migration. Although Arnold et al. (1991a, 1997) reported similar observations, other studies with a larger number of subjects that examined the intrinsic and individual heterogeneity of the entorhinal cortex (Heinsen et al., 1996; Akil and Lewis, 1997; Krimer et al., 1997; Bernstein et al., 1998) failed to detect such changes. However, Falkai et al. (2000) have recently reported quantitative evidence of abnormally located and smaller sized cluster of neurons in the entorhinal cortex.

Alterations in Neuronal Connections in Schizophrenia: Thalamocortical vs. Corticocortical Circuitry

Through the postmortem studies described above, some common alterations appear to be present in both the PFC and the hippocampal formation of subjects with schizophrenia. These include 1) a tendency for reduction in cortical volume/thickness, but no change in total neuron number, 2) a slight elevation of neuronal density, 3) reduced somal size of projection neurons, 4) alterations in pyramidal neuron dendritic trees, and 5) reductions in markers of presynaptic axon terminals. Thus, these lines of evidence indicate that compromised neuronal connections may represent the major anomaly in the PFC and hippocampus of subjects with schizophrenia. Together, these observations raise the question of which specific neuronal circuits are compromised in schizophrenia. In response to this question, the following sections consider the evidence for alterations in thalamocortical and corticocortical connectivity in schizophrenia.

Thalamocortical Connection

The PFC and the hippocampal formation are reciprocally connected with specific nuclei in the thalamus. Three stereological investigations have revealed a significant reduction in number of neurons in the mediodorsal thalamic (MD) nucleus, the principal source of thalamic projections to the PFC (Giguere and Goldman-Rakic, 1988). Pakkenberg (1990) first demonstrated that total neuronal number in the MD nucleus was decreased by 40% in subjects with schizophrenia. The volume of the MD nucleus, and the number of glial cells, were also reduced by 40% and 45%, respectively, in schizophrenia. On the other hand, no significant change was detected for volume or total neuronal number in the basolateral nucleus of the amygdala, another subcortical origin of excitatory projections to portions of the PFC. Pakkenberg (1992) further suggested that these changes in the MD nucleus were not due to antipsychotic treatment since subjects with schizophrenia never treated with neuroleptics showed a similar reduction in MD nucleus volume compared to treated subjects. Recently, Young et al. (2000) also demonstrated that the total neuronal number and volume of the MD nucleus were significantly reduced by 35% and 24%, respectively, in subjects with schizophrenia.

The MD nucleus consists of the parvocellular, densocellular and magnocellular subnuclei, which innervate different regions of the PFC. Popken et al. (2000) analyzed total neuronal numbers in each MD subnucleus. They found 30.9% and 24.5% reduction in total neuron number in the parvocellular and densocellular subnuclei, respectively, which project to the dorsolateral PFC, whereas they did not detect a significant decrease in the magnocellular subnucleus, which projects to ventromedial and orbital regions of the PFC. They also estimated total neuron number in a control thalamic nucleus, the ventral posterior nucleus, which relays somatosensory inputs, and found no change. Importantly, in each of these three studies, there was virtually no overlap in the range of neuron number in the MD nucleus between normal control subjects and subjects with schizophrenia. Although it remains to be determined whether projection or local circuit or both neuronal populations are affected in the MD nucleus, the magnitude of reduction in neuronal number is almost certain to be associated with disturbances in the projection from the MD nucleus to the PFC.

Are similar alterations present in the thalamic nuclei that project to the hippocampal formation? The anteroventral (AV) and anteromedial (AM) nuclei send projections to the limbic cortices including the hippocampal formation and posterior cingulate cortex (Amaral and Cowan, 1980; Vogt et al., 1987). Danos et al. (1998) demonstrated that the density of neurons expressing parvalbumin (PV), a marker of projection neurons in the thalamus (Jones and Hendry, 1989), was decreased by more than 30% in the AV nucleus of subjects with schizophrenia. Young et al. (2000) also reported a significant 16% reduction of total neuron number and a 17% reduction of volume, which approached significance, in the AV and AM nuclei of subjects with schizophrenia. Although the reported reduction in total neuron number was less in AV and AD nuclei than in the MD nucleus, direct evidence for the reduced projection neurons may be consistent with an anomaly in thalamic projections to the hippocampus in addition to other cortical regions.

These postmortem findings are consistent with and supported by a number of imaging studies reporting that subjects with schizophrenia exhibited a reduced volume (Konick and Friedman, 2001) and activity (Buchsbaum et al., 1996) of the thalamus, especially in the mediodorsal and anterior nuclei (Hazlett et al., 1999; Byne et al., 2001; Gilbert et al., 2001), even in the initial stage of the illness (Gilbert et al., 2001; Ettinger et al., 2001).

At least for the PFC, the alterations in thalamic connectivity are supported by several other lines of evidence. First, the extent of dendritic arborization and the density of dendritic spines were reported to be decreased in layer III pyramidal neurons (Garey et al., 1998; Glantz and Lewis, 2000; Kalus et al., 2000). These changes were specifically observed in the basilar dendrites of deep layer III pyramidal neurons (Glantz and Lewis, 2000; Kalus et al., 2000), whose spines are likely targets of the excitatory projections from the MD nucleus (Giguere and Goldman-Rakic, 1988; Melchitzky et al., 1999). Because dendritic spine density reflects the number of excitatory inputs, these findings are consistent with a decreased excitatory input from the MD nucleus to the PFC. Second, the density of putative PV-positive axon terminals was decreased within deep layer III and IV of the PFC, but unchanged in the superficial layers (Lewis et al., 2001). PV-positive terminals of thalamic projection neurons were shown to form excitatory synapses mainly in deep layer III and IV, whereas the PV-positive terminals of local inhibitory neurons distribute across layers II-IV (Condé et al., 1994; Melchitzky et al., 1999). Thus, the selective reduction of PV-positive terminals in deep layer III and IV suggests the reduction of inputs to the PFC from the MD nucleus.

Corticocortical Connection

Postmortem findings in the cerebral cortex also suggest disturbances in corticocortical connections. The smaller size of layer III pyramidal neurons could reflect reduced axonal as well as dendritic arborization, because in addition to the evidence for a relationship between somal size and the extent of dendrites (Hayes and Lewis, 1993; Jacobs et al., 1997), somal size correlates well with the extent of axonal arbors (Gilbert and Kelly, 1975; Lund et al., 1975). Given that pyramidal cells in PFC layer III provide associational or commissural projections to other cortical regions, as well as intrinsic projections terminating locally within the same area of the PFC (Schwartz and Goldman-Rakic, 1984; Barbas and Pandya, 1989; Levitt et al., 1993; Pucak et al., 1996), the reduced somal size of these pyramidal cells may reflect a decrease in one or more of these types of cortical connections in schizophrenia.

In the hippocampal formation, the reduced size of neurons in layer II of the entorhinal cortex, the subiculum and CA1 could also be related to aberrant corticocortical connections. In the entorhinal cortex, afferent projections from association cortical areas mainly terminate in layer II and, then, stellate neurons in layer II provide the perforant path projection to the hippocampus proper (Insausti et al., 1987; Witter et al., 1989; Suzuki and Amaral, 1994a). The subiculum and CA1 receive not only intrinsic inputs from the hippocampal subregions, but also direct inputs from association cortical areas, and pyramidal neurons in these regions project directly to association cortical areas, as well as to the entorhinal cortex (Rosene and Van Hoesen, 1977; Witter and Amaral, 1991; Yukie, 2000). Based on these anatomical data, the smaller size of projection neurons in these regions may also be consistent with altered cortical connectivity.

Anatomically, the subiculum and the entorhinal cortex are densely and reciprocally interconnected with the PFC via the inferior temporal cortex (Suzuki and Amaral, 1994b). Interestingly, a recent structural MRI study showed, in subjects with schizophrenia, a significant volume reduction of white matter including the uncinate fasciculus (Sigmundsson et al., 2001), through which the PFC and the inferior temporal cortex interact with each other (Gutnikov et al., 1997). Moreover, a supra-regional analysis of structural MRI suggested a significant correlation between volume changes in the frontal and temporal cortices (Wright et al., 1999). Thus, it is possible that the corticocortical connections between the PFC and the hippocampal formation relayed by the inferior temporal cortices are affected in schizophrenia.

A contribution of altered corticocortical connectivity to the pathophysiology of schizophrenia is also suggested by several indirect lines of evidence. For example, the reduction in the presynaptic marker synaptophysin protein (Perrone-Bizzozero et al., 1996; Glantz and Lewis, 1997; Davidsson et al., 1999; Honer et al., 1999; Karson et al., 1999), in the absence of a change in the expression level of synaptophysin mRNA (Karson et al., 1999; Eastwood et al., 2000; Glantz et al., 2000) in the PFC, could reflect a reduction of synaptic inputs from extrinsic sources. However, it seems unlikely that abnormalities only in projections from the MD nucleus, which terminate mainly in middle layers of the cortex, could account for this reduction because the superficial and deep layers of this cortical region exhibit a similar decrease in synaptophysin protein (Glantz and Lewis, 1997). Alternatively, this pattern of decreased synaptophysin immunoreactivity could be due to reduced corticocortical inputs to the PFC in schizophrenia which terminate across all cortical layers. The findings of the preferential decrease of the dendritic spine density on basal dendrites of deep layer III pyramidal neurons might be due to the reduction of thalamic inputs as described above (Glantz and Lewis, 2000). However, the reported 23% reduction in spine density is unlikely to be fully explained by a reduction in thalamic inputs. For example, in cat visual cortex, thalamocortical afferents comprise a small population (<10%) of the total excitatory inputs to the targeted cortical neurons (Ahmed et al., 1994). If this finding can be extrapolated to the human PFC, even a complete loss of thalamocortical afferents would not be sufficient to account for the observed decrease in spine density. Thus, the marked reduction in the spine density may reflect alterations in, not only the thalamic inputs, but also the cortical inputs provided by commissural, associational and/or intrinsic projections.

Changes in GABAergic Local Inhibitory Neurons in the PFC: A Reflection of Altered Thalamocortical and Corticocortical Connections?

In addition to the above indications of altered excitatory connections in the PFC, other lines of evidence reveal disturbances in GABAergic local circuit neurons in the PFC of subjects with schizophrenia. This evidence includes the observations of a reduction in depolarization-induced release of GABA (Sherman et al., 1991), an increase in GABA binding sites (Hanada et al., 1987), and a decrease in the gene expression of GAD_{67}, an enzyme involved in the synthesis of GABA (Akbarian et al., 1995; Guidotti et al., 2000; Volk et al., 2000). Based on these findings, it has been suggested that, in schizophrenia, synthesis and release of GABA is attenuated in local inhibitory neurons. In the primary visual cortex of monkeys, markers of GABA cell function, such as GAD_{67} expression, are regulated by neuronal activity and reduced by a loss of thalamic inputs (Benson et al., 1994). Thus, it has been proposed that decreased PFC GABAergic neurotransmission in schizophrenia could be a secondary response to the reduction in the thalamic projections from the MD nucleus (Jones, 1993).

Recent studies further indicate that markers of reduced GABAergic neurotransmission in the PFC of subjects with schizophrenia are confined to a subset of local inhibitory neurons. Volk et al. (2000) showed a 25-35% reduction in the density of GAD_{67} mRNA-positive neurons in layers III to V, without a change in mRNA expression level in the other GABA neurons. Since no reduction has been demonstrated in the density of total neurons or nonpyramidal neurons in the PFC (Akbarian et al., 1995; Selemon et al., 1995), the reduction in GAD_{67} mRNA-positive cell density unlikely to be due to the loss of inhibitory neurons. Instead, this observation indicates that GAD_{67} mRNA expression was reduced below detection level in a subset of local inhibitory neurons in the PFC.

This observation raised the question of whether a specific population of GABA neurons is affected in schizophrenia. In the primate PFC, separate populations of GABAeric local inhibitory neurons express one of three calcium-binding proteins, calbindin-28, PV and calretinin (Condé et al., 1994). Among these populations, PV-positive chandelier cells are likely to be among the affected subset of local inhibitory neurons in schizophrenia. Chandelier cells provide linear arrays of axon terminals, termed "cartridges", which synapse at the axon initial segment of pyramidal neurons. Immunohistochemical visualization of GABAergic terminals using anti-GABA transporter-1 (GAT-1) antibody revealed a specific 40% decrease in the density of cartridges without a change in the density of total GAT-1-positive terminals in the PFC of subjects with schizophrenia (Woo et al., 1998; Pierri et al., 1999). In addition, the density of GAT-1 mRNA-positive cells was reduced to the same extent as that of GAD_{67} mRNA-positive cells, and GAT-1 mRNA-positive cell density was positively correlated with the density of GAT-1-labled cartridges across subjects with schizophrenia (Volk et al., 2001). These findings suggest that chandelier cells are among the subset of GABA neurons that are affected in schizophrenia. Interestingly, in monkey PFC, approximately 50% of the intrinsic axon collaterals arising from layer III pyramidal cells synapse locally on the dendrites of GABA neurons (Melchitzky et al., 1998), specifically the subpopulation that contains PV, which includes chandelier cells (Melchitzky et al., 2000). Thus, through these intrinsic connections, a decrease in activity of layer III pyramidal cells resulting from reduced afferent inputs from the MD nucleus and/or corticocortical connections could lead to specific alterations in GABAergic neurotransmission in a subset of local inhibitory neurons.

Conclusions

In concert with the development of neuroimaging techniques, postmortem studies have begun to reveal abnormalities in the brains of subjects with schizophrenia. Studies of the PFC and the hippocampal formation have revealed several abnormalities shared by these regions including 1) a slight reduction of tissue volume, but no change in total neuron number, 2) a slight elevation of neuronal density, 3) reduced somal size of projection neurons, 4) a reduction in spine density and dendritic arborization of pyramidal neurons, and 5) decreased levels of presynaptic marker proteins. Together, these observations suggest that compromised neuronal connections may represent the major anomaly in the PFC and hippocampus of subjects with schizophrenia.

The altered neuronal connections may include projections from the thalamus since a marked reduction in neuron number has been detected in the thalamic nuclei that project to the PFC and the hippocampal formation. Altered thalamocortical connectivity is also supported by observations of decreased dendritic spine density in the termination zone of thalamic axons and a reduced density of putative thalamic axon terminals in the PFC. On the other hand, a disturbance in corticocortical connectivity has also been suggested by findings such as the smaller somal size of corticocortical projection neurons and a decrease in presynaptic marker proteins across all cortical layers. Finally, on the basis of studies of neuronal circuitry in nonhuman primates, disturbances in either thalamocortical or corticocortical connections could be related to and/or responsible for the well-documented alterations in a subset of inhibitory local circuit neurons in the PFC of subjects with schizophrenia.

Additional studies in both postmortem tissue and in animal model systems are needed to determine the relative contribution of altered thalamocortical and corticocortical connections to PFC and hippocampal dysfunction in schizophrenia, and to determine whether these circuitry abnormalities reflect the primary disease process, or a consequence of, or compensation for, some other more fundamental brain disturbance (Lewis, 1999).

References

Ahmed B, Anderson JC, Douglas RJ, Martin KAC, Nelson JC. Polyneuronal innervation of spiny stellate neurons in cat visual cortex. J Comp Neurol 1994; 341: 39-49.

Akbarian S, Kim JJ, Potkin SG, Hagman JO, Tafazzoli A, Bunney JrWE, Jones EG. Gene expression for glutamic acid decarboxylase is reduced without loss of neurons in prefrontal cortex of schizophrenics. Arch Gen Psychiatry 1995; 52: 258-266.

Akil M, Lewis DA. The cytoarchitecture of the entorhinal cortex in schizophrenia. Am J Psychiatry 1997; 154: 1010-1012.

Althauser LL, Conrad A, Kovelman JA, Scheibel A. Hippocampal pyramidal cell orientation in schizophrenia. Arch Gen Psychiatry 1987; 44: 1094-1032.

Altshuler LL, Casanova MF, Goldberg TE, Kleinman JE. The hippocampus and parahippocampus in schizophrenic, suicide, and control brains. Arch Gen Psychiatry 1990; 47: 1029-1034.

Amaral DG, Cowan WM. Subcortical afferents to the hippocampal formation in the monkey. J Comp Neurol 1980; 189: 573-591.

Arnold SE, Hyman BT, Van Hoesen GW, Damasio AR. Some cytoarchitectural abnormalities of the entorhinal cortex in schizophrenia. Arch Gen Psychiatry 1991a; 48: 625-632.

Arnold SE, Lee VMY, Gur RE, Trojanowski JQ. Abnormal expression of two microtubule-associated proteins (MAP2 and MAP5) in specific subfields of the hippocampal formation in schizophrenia. Proc Natl Acad Sci USA 1991b; 88: 10850-10854.

Arnold SE, Franz BR, Ruben BA, Gur C, Gur RE, Shapiro RM, Moberg PJ, Trojanowski JQ. Smaller neuron size in schizophrenia in hippocampal subfields that mediate cortical-hippocampal interactions. Am J Psychiatry 1995; 152: 738-748.

Arnold SE, Ruscheinsky DD, Han L-Y. Further evidence of abnormal cytoarchitecture of the entorhinal cortex in schizophrenia using spatial point pattern analyses. Biol Psychiatry 1997; 42: 639-647.

Arnold SE, Han L-Y, Rioux L, Falke E. Abnormal MAP2 neuron representation in subiculum and entorhinal cortex in poor-outcome schizophrenia. Soc Neurosci Abstr 1999; 25: 575.

Barbas H, Pandya DN. Architecture and intrinsic connections of the prefrontal cortex in the rhesus monkey. J Comp Neurol 1989; 286: 353-375.

Benes FM, Davidson J, Bird ED. Quantitative cytoarchitectural studies of the cerebral cortex of schizophrenics. Arch Gen Psychiatry 1986; 43: 31-35.

Benes FM, McSparren J, Bird ED, SanGiovanni JP, Vincent SL. Deficits in small interneurons in prefrontal and cingulate cortices of schizophrenic and schizoaffective patients. Arch Gen Psychiatry 1991a; 48: 996-1001.

Benes FM, Sorensen I, Bird ED. Reduced neuronal size in posterior hippocampus of schizophrenic patients. Schizophr Bull 1991b; 17: 597-608.

Benson DL, Huntsman MM, Jones EG. Activity-dependent changes in GAD and preprotachykinin mRNAs in visual cortex of adult monkeys. Cereb Cortex 1994; 4: 40-51.

Bernstein H-G, Krell D, Baumann B, Danos P, Falkai P, Diekmann S, Henning H, Bogerts B. Morphometric studies of the entorhinal cortex in neuropsychiatric patients and controls: Clusters of heterotopically displaced lamina II neurons are not indicative of schizophrenia. Schizophr Res 1998; 33: 125-132.

Bogerts B, Meertz E, Schonfeldt-Bausch R. Basal ganglia and limbic system pathology in schizophrenia. Arch Gen Psychiatry 1985; 42: 784-791.

Bogerts B, Falkai P, Haupts M, Greve B, Ernst S, Tapernon-Franz U, Heinzmann U. Post-mortem volume meansurements of limbic system and basal ganglia structures in chronic schizophrenics. Schizophr Res 1990; 3: 295-301.

Brown R, Colter N, Corsellis AN, Crow TJ, Frith CD, Jagoe R, Johnstone EC, Marsh L. Postmortem evidence of structural brain changes in schizophrenia. Arch Gen Psychiatry 1986; 43: 36-42.

Browning MD, Dudek EM, Rapier JL, Leonard S, Freedman R. Significant reductions in synapsin but not synaptophysin specific activity in the brains of some schizophrenics. Biol Psychiatry 1993; 34: 529-535.

Buchsbaum MS, Someya T, Teng CY, Abel L, Chin S, Najafi A, Haier RJ, Wu J, Bunney JrWE. PET and MRI of the thalamus in never-medicated patients with schizophrenia. Am J Psychiatry 1996; 153: 191-199.

Byne W, Buchsbaum MS, Kemether E, Hazlett EA, Shinwari A, Mitropoulou V, Siever LJ. Magnetic resonance imaging of the thalamic mediodorsal nucleus and pulvinar in schizophrenia and schizotypal personality disorder. Arch Gen Psychiatry 2001; 58: 133-140.

Christison GW, Casanova MF, Weinberger DR, Rawlings R, Kleinman JE. A quantitative investigation of hippocampal pyramidal cell size, shape, and variability of orientation in schizophrenia. Arch Gen Psychiatry 1989; 46: 1027-1023.

Colter N, Battal S, Crow TJ, Johnstone EC, Brown R, Bruton C. White matter reduction in the parahippocampal gyrus of patients with schizophrenia (Letter). Arch Gen Psychiatry 1987; 44: 1023.

Condé F, Lund JS, Jacobowitz DM, Baimbridge KG, Lewis DA. Local circuit neurons immunoreactive for calretinin, calbindin D-28k, or parvalbumin in monkey prefrontal cortex: Distribution and morphology. J Comp Neurol 1994; 341: 95-116.

Conrad AJ, Abebe T, Austin R, Forsythe S, Scheibel AB. Hippocampal pyramidal cell disarray in schizophrenia as a bilateral phenomenon. Arch Gen Psychiatry 1991; 48: 413-417.
Cotter D, Kerwin R, Doshi B, Martin CS, Everall IP. Alterations in hippocampal non-phosphorylated MAP2 protein expression in schizophrenia. Brain Res 1997; 765: 238-246.
Cotter D, Wilson S, Roberts E, Kerwin R, Everall IP. Increased dendritic MAP2 expression in the hippocampus in schizophrenia. Schizophr Res 2000; 41: 313-346.
Danos P, Baumann B, Bernstein H-G, Franz M, Stauch R, Northoff G, Krell D, Falkai P, Bogerts B. Schizophrenia and anteroventral thalamic nucleus: Selective decrease of parvalbumin-immunoreactive thalamocortical projection neurons. Psychiatry Res : Neuroimaging 1998; 82: 1-10.
Davidsson P, Gottfries J, Bogdanovic N, Ekman R, Karlsson I, Gottfries C-G, Blennow K. The synaptic-vesicle-specific proteins rab3a and synaptophysin are reduced in thalamus and related cortical brain regions in schizophrenic brains. Schizophr Res 1999; 40: 23-29.
Eastwood SL, Harrison PJ. Decreased synaptophysin in the medial temporal lobe in schizophrenia demonstrated using immunoautoradiography. Neuroscience 1995; 69: 339-343.
Eastwood SL, Cairns NJ, Harrison PJ. Synaptophysin gene expression in schizophrenia: Investigation of synaptic pathology in the cerebral cortex. Brit J Psychiatry 2000; 176: 236-242.
Ettinger U, Chitnis XA, Kumari V, Fannon DG, Sumich AL, O'Ceallaigh S, Doku VC, Sharma T. Magnetic resonance imaging of the thalamus in first-episode psychosis. Am J Psychiatry 2001; 158: 116-118.
Falkai P, Bogerts B. Cell loss in the hippocampus of schizophrenics. Eur Arch Psychiatr Neurol Sci 1986; 236: 154-161.
Falkai P, Bogerts B, Rozumek M. Limbic pathology in schizophrenia: The entorhinal region: A morphometric study. Biol Psychiatry 1988; 24: 515-521.
Falkai P, Schneider-Axmann T, Honer WG. Entorhinal cortex pre-alpha cell clusters in schizophrenia: Quantitative evidence of a developmental abnormality. Biol Psychiatry 2000; 47: 937-943.
Garey LJ, Ong WY, Patel TS, Kanani M, Davis A, Mortimer AM, Barnes TRE, Hirsch SR. Reduced dendritic spine density on cerebral cortical pyramidal neurons in schizophrenia. J Neurol Neurosurg Psychiatry 1998; 65: 446-453.
Giguere M, Goldman-Rakic PS. Mediodorsal nucleus: Areal, laminar, and tangential distribution of afferents and efferents in the frontal lobe of rhesus monkeys. J Comp Neurol 1988; 277: 195-213.
Gilbert AR, Rosenberg DR, Harenski K, Spencer S, Sweeney JA, Keshavan MS. Thalamic volumes in patients with first-episode schizophrenia. Am J Psychiatry 2001; 158: 618-624.
Gilbert CD, Kelly JP. The projections of cells in different layers of the cat's visual cortex. J Comp Neurol 1975; 63: 81-106.
Glantz LA, Lewis DA. Reduction of synaptophysin immunoreactivity in the prefrontal cortex of subjects with schizophrenia: Regional and diagnostic specificity. Arch Gen Psychiatry 1997; 54: 943-952.
Glantz LA, Lewis DA. Decreased dendritic spine density on prefrontal cortical pyramidal neurons in schizophrenia. Arch Gen Psychiatry 2000; 57: 65-73.
Glantz LA, Austin MC, Lewis DA. Normal cellular levels of synaptophysin mRNA expression in the prefrontal cortex of subjects with schizophrenia. Biol Psychiatry 2000; 48: 389-397.
Guidotti A, Auta J, Davis JM, Gerevini VD, Dwivedi Y, Grayson DR, Impagnatiello F, Pandey G, Pesold C, Sharma R, Uzunov D, Costa E. Decrease in reelin and glutamic acid decarboxylase$_{67}$ (GAD$_{67}$) expression in schizophrenia and bipolar disorder. Arch Gen Psychiatry 2000; 57: 1061-1069.

Gur RE, Cowell PE, Latshaw A, Turetsky BI, Grossman RI, Arnold SE, Bilker WB, Gur RC. Reduced dorsal and orbital prefrontal gray matter volumes in schizophrenia. Arch Gen Psychiatry 2000a; 57: 761-768.

Gur RE, Turetsky BI, Cowell PE, Finkelman C, Maany V, Grossman RI, Arnold SE, Bilker WB, Gur RC. Temporolimbic volume reductions in schizophrenia. Arch Gen Psychiatry 2000b; 57: 769-775.

Gutnikov SA, Ma YY, Gaffan D. Temporo-frontal disconnection impairs visual-visual paired association learning but not configural learning in Macaca monkeys. Eur J Neurosci 1997; 9: 1524-1529.

Hanada S, Mita T, Nishino N, Tanaka C. [^{3}H]Muscimol binding sites increased in autopsied brains of chronic schizophrenics. Life Sci 1987; 40: 239-266.

Harrison PJ, Eastwood SL. Preferential involvement of excitatory neurons in medial temporal lobe in schizophrenia. Lancet 1998; 352: 1669-1673.

Harrison PJ, Lewis DA. Neuropathology in schizophrenia. In: Hirsch S and Weinberger D R (eds). Schizophrenia. Blackwell Science Ltd., Oxford, 2001; in press.

Hayes TL, Lewis DA. Interhemispheric differences in the dendritic aborization of magnopyramidal neurons of the anterior speech region. Soc Neurosci Abstr 1993; 19: 844.

Hazlett EA, Buchsbaum MS, Byne W, Wei T-C, Spiegel-Cohen J, Geneve C, Linderlehrer R, Mehmet Haznedar M, Shihabuddin L, Siever LJ. Three-dimensional analysis with MRI and PET of the size, shape, and function of the thalamus in the schizophrenia spectrum. Am J Psychiatry 1999; 156: 1190-1199.

Heckers S, Heinsen H, Heinsen YC, Beckmann H. Limbic structures and lateral ventricles in schizophrenia: A quantitative postmortem study. Arch Gen Psychiatry 1990a; 47: 1016-1022.

Heckers S, Heinsen H, Heinsen YC, Beckmann H. Morphometry of the parahippocampal gyrus in schizophrenics and controls. Some anatomical considerations. J Neural Transm 1990b; 80: 151-155.

Heckers S, Heinsen H, Geiger B, Beckmann H. Hippocampal neuron number in schizophrenia. A stereological study. Arch Gen Psychiatry 1991a; 48: 1002-1008.

Heckers S, Heinsen H, Heinsen YC, Beckmann H. Cortex, white matter, and basal ganglia in schizophrenia: A volumetric postmortem study. Biol Psychiatry 1991b; 29: 556-566.

Heinsen H, Gössmann E, Rüb U, Eisnemenger W, Bauer M, Ulmar G, Bethke B, Schüler M, Schmitt H-P, Götz M, Lockeman U, Püschel K. Variability in the human entorhinal region may confound neuropsychiatric diagnoses. Acta Anat 1996; 157: 226-237.

Hirayasu Y, Shenton ME, Salisbury DF, Dickey CC, Fischer IA, Mazzoni P, Kisler T, Arakaki H, Kwon JS, Anderson JE, Yurgelun-Todd D, Tohen M, McCarley RW. Lower left temporal lobe MRI volumes in patients with first-episode schizophrenia compared with psychotic patients with first-episode affective disorder and normal subjects. Am J Psychiatry 1998; 155: 1384-1391.

Hirayasu Y, Tanaka S, Shenton ME, Salisbury DF, DeSantis MA, Levitt JJ, Wible C, Yurgelun-Todd D, Kikinis R, Jolesz FA, McCarley RW. Prefrontal gray matter volume reduction in first episode schizophrenia. Cereb Cortex 2001; 11: 374-381.

Honer WG, Falkai P, Chen C, Arango V, Mann JJ, Dwork AJ. Synaptic and plasticity-associated proteins in anterior frontal cortex in severe mental illness. Neuroscience 1999; 91: 1247-1255.

Insausti R, Amaral DG, Cowan WM. The entorhinal cortex of the monkey: II. Cortical afferents. J Comp Neurol 1987; 264: 356-395.

Jacobs B, Driscoll L, Schall M. Life-span dendritic and spine changes in areas 10 and 18 of human cortex: A quantitative Golgi study. J Comp Neurol 1997; 386: 661-680.

Jakob H, Beckmann H. Prenatal developmental disturbances in the limbic allocortex in schizophrenics. J Neural Transm 1986; 65: 303-326.

Jeste DV, Lohr JB. Hippocampal pathologic findings in schizophrenia: A morphometric study. Arch Gen Psychiatry 1989; 48: 1019-1024.

Jones EG, Hendry SHC. Differential calcium binding protein immunoreactivity distinguishes classes of relay neurons in monkey thalamic nuclei. Eur J Neurosci 1989; 1: 222-246.

Jones EG. GABAergic neurons and their role in cortical plasticity in primates. Cereb Cortex 1993; 3: 361-372.

Kalus P, Müller TJ, Zuschratter W, Senitz D. The dendritic architecture of prefrontal pyramidal neurons in schizophrenic patients. NeuroReport 2000; 11: 3621-3625.

Karson CN, Mrak RE, Schluterman KO, Sturner WQ, Sheng JG, Griffin WST. Alterations in synaptic proteins and their encoding mRNAs in prefrontal cortex in schizophenia: A possible neurochemical basis for 'hypofrontality'. Mol Psychiatry 1999; 4: 39-45.

Konick LC, Friedman L. Meta-analysis of thalamic size in schizophrenia. Biol Psychiatry 2001; 49: 28-38.

Kovelman JA, Scheibel AB. A neurohistological correlate of schizophrenia. Biol Psychiatry 1984; 19: 1601-1602.

Krimer LS, Herman MM, Saunders RC, Boyd JC, Hyde TM, Carter JM, Kleinman JE, Weinberger DR. A qualitative and quantitative analysis of the entorhinal cortex in schizophrenia. Cereb Cortex 1997; 7: 732-739.

Levitt JB, Lewis DA, Yoshioka T, Lund JS. Topography of pyramidal neuron intrinsic connections in macaque monkey prefrontal cortex (areas 9 & 46). J Comp Neurol 1993; 338: 360-376.

Lewis DA. Distributed disturbances in brain structure and funtion in schizophrenia (editorial). Am J Psychiatry 1999; 157: 1-2.

Lewis DA, Cruz DA, Melchitzky DS, Pierri JN. Lamina-specific reductions in parvalbumin-immunoreactive axon terminals in the prefrontal cortex of subjects with schizophrenia: Evidence for decreased projections from the thalamus. Am J Psychiatry 2001; in press.

Lund JS, Lund RD, Hendrickson AE, Bunt AH, Fuchs AF. The origin of efferent pathways from the primary visual cortex, area 17, of the macaque monkey as shown by retrograde transport of horseradish peroxidase. J Comp Neurol 1975; 164: 287-304.

Melchitzky DS, Sesack SR, Pucak ML, Lewis DA. Synaptic targets of pyramidal neurons providing intrinsic horizontal connections in monkey prefrontal cortex. J Comp Neurol 1998; 390: 211-224.

Melchitzky DS, Sesack SR, Lewis DA. Parvalbumin-immunoreactive axon terminals in monkey and human prefrontal cortex: Laminar, regional and target specificity of Type I and Type II synapses. J Comp Neurol 1999; 408: 11-22.

Melchitzky DS, Gonzalez-Burgos G, Barrionuevo G, Lewis DA. Synaptic targets of the intrinsic axon collaterals of supragranular pyramidal neurons in monkey prefrontal cortex. J Comp Neurol 2000; 430: 209-221.

Nelson MD, Saykin AJ, Flashman LA, Riordan HJ. Hippocampal volume reduction in schizophrenia as assessed by magnetic resonance imaging: A meta-analytic study. Arch Gen Psychiatry 1998; 55: 433-440.

Pakkenberg B. Post-mortem study of chronic schizophrenic brains. Br J Psychiatry 1987; 151: 744-752.

Pakkenberg B. Pronounced reduction of total neuron number in mediodorsal thalamic nucleus and nucleus accumbens in schizophrenics. Arch Gen Psychiatry 1990; 47: 1023-1028.

Pakkenberg B. The volume of the mediodorsal thalamic nucleus in treated and untreated schizophrenics. Schizophr Res 1992; 7: 95-100.

Pakkenberg B. Total nerve cell number in neocortex in chronic schizophrenics and controls estimated using optical disectors. Biol Psychiatry 1993; 34: 768-772.

Perrone-Bizzozero NI, Sower AC, Bird ED, Benowitz LI, Ivins KJ, Neve RL. Levels of the growth-associated protein GAP-43 are selectively increased in association cortices in schizophrenia. Proc Natl Acad Sci USA 1996; 93: 14182-14187.

Pierri JN, Chaudry AS, Woo T-U, Lewis DA. Alterations in chandelier neuron axon terminals in the prefrontal cortex of schizophrenic subjects. Am J Psychiatry 1999; 156: 1709-1719.

Pierri JN, Volk CLE, Auh S, Sampson A, Lewis DA. Decreased somal size of deep layer 3 pyramidal neurons in the prefrontal cortex in subjects with schizophrenia. Arch Gen Psychiatry 2000; 58: 466-473.

Plum F. Neuropathological findings. In: Kety SS and Malthysse S M (eds). Prospects for Research on Schizophrenia. MIT Press, Cambridge, MA, 1972; pp 385-388.

Popken GJ, Bunney Jr. WE, Potkin SG, Jones EG. Subnucleus-specific loss of neurons in medial thalamus of schizophrenics. Proc Natl Acad Sci 2000; 97: 9276-9280.

Pucak ML, Levitt JB, Lund JS, Lewis DA. Patterns of intrinsic and associational circuitry in monkey prefrontal cortex. J Comp Neurol 1996; 376: 614-630.

Rajkowska G, Selemon LD, Goldman-Rakic PS. Neuronal and glial somal size in the prefrontal cortex: A postmortem morphometric study of schizophrenia and Huntington disease. Arch Gen Psychiatry 1998; 55: 215-224.

Razi K, Greene KP, Sakuma M, Ge S, Kushner M, DeLisi LE. Reduction of the parahippocampal gyrus and the hippocampus in patients with chronic schizophrenia. Br J Psychiatry 1999; 174: 512-519.

Rosene DL, Van Hoesen GW. Hippocampal efferents reach widespread areas of cerebral cortex and amygdala in the rhesus monkey. Science 1977; 198: 315-317.

Rosoklija G, Toomayan G, Ellis SP, Keilp J, Mann JJ, Latov N, Hays AP, Dwork AJ. Structural abnormalities of subicular dendrites in subjects with schizophrenia and mood disorders: Preliminary findings. Arch Gen Psychiatry 2000; 57: 349-356.

Schwartz ML, Goldman-Rakic PS. Callosal and intrahemispheric connectivity of the prefrontal association cortex in rhesus monkey: Relation between intraparietal and principal sulcal cortex. J Comp Neurol 1984; 226: 403-420.

Selemon LD, Rajkowska G, Goldman-Rakic PS. Abnormally high neuronal density in the schizophrenic cortex: A morphometric analysis of prefrontal area 9 and occipital area 17. Arch Gen Psychiatry 1995; 52: 805-818.

Selemon LD, Rajkowska G, Goldman-Rakic PS. Elevated neuronal density in prefrontal area 46 in brains from schizophrenic patients: Application of a three-dimensional, stereologic counting method. J Comp Neurol 1998; 392: 402-412.

Shenton ME, Dickey CC, Frumin M, McCarley RW. A review of MRI findings in schizophrenia. Schizophr Res 2001; 49: 1-52.

Sherman AD, Davidson AT, Baruah S, Hegwood TS, Waziri R. Evidence of glutamatergic deficiency in schizophrenia. Neurosci Lett 1991; 121: 77-80.

Sigmundsson T, Suckling J, Maier M, Williams SCR, Bullmore ET, Greenwood KE, Fukuda R, Ron MA, Toone BK. Structural abnormalities in frontal, temporal, and limbic regions and interconnecting white matter tracts in schizophrenic patients with prominent negative symptoms. Am J Psychiatry 2001; 158: 234-243.

Staal WG, Pol HEH, Schnack HG, Hoogendoorn MLC, Jellema K, Kahn RS. Structural brain abnormalities in patients with schizophrenia and their healthy sibling. Am J Psychiatry 2000; 157: 416-421.

Suzuki WA, Amaral DG. Topographic organization of the reciprocal connections between the monkey entorhinal cortex and the perirhinal and parahippocampal cortices. J Neurosci 1994a; 14: 1856-1877.

Suzuki WA, Amaral DG. Perirhinal and parahippocampal cortices of the macaque monkey: Cortical afferents. J Comp Neurol 1994b; 350: 497-533.

Thune JJ, Uylings HB, Pakkenberg B. No deficit in total number of neurons in the prefrontal cortex in schizophrenics. J Psychiatr Res 2001; 35: 15-21.

Velakoulis D, Pantelis C, McGorry PD, Dudgeon P, Brewer W, Cook M, Desmond P, Bridle N, Tierney P, Murrie V, Singh B, Copolov D. Hippocampal volume in first-episode psychoses and chronic schizophrenia. A high-resolution magnetic resonance imaging study. Arch Gen Psychiatry 1999; 56: 133-140.

Vogt BA, Pandya DN, Rosene DL. Cingulate cortex of the rhesus monkey: I. Cytoarchitecture and thalamic afferents. J Comp Neurol 1987; 262: 256-270.

Volk DW, Austin MC, Pierri JN, Sampson AR, Lewis DA. Decreased GAD_{67} mRNA expression in a subset of prefrontal cortical GABA neurons in subjects with schizophrenia. Arch Gen Psychiatry 2000; 57: 237-245.

Volk DW, Austin MC, Pierri JN, Sampson AR, Lewis DA. GABA transporter-1 mRNA in the prefrontal cortex in schizophrenia: Decreased expression in a subset of neurons. Am J Psychiatry 2001; 158: 256-265.

Witter MP, Van Hoesen GW, Amaral DG. Topographical organization of the entorhinal projection to the dentate gyrus of the monkey. J Neurosci 1989; 9: 216-228.

Witter MP, Amaral DG. Entorhinal cortex of the monkey: V. Projections to the dentate gyrus, hippocampus and subicular complex. J Comp Neurol 1991; 307: 437-459.

Woo T-U, Miller JL, Lewis DA. Parvalbumin-containing cortical neurons in schizophrenia. Am J Psychiatry 1997; 154: 1013-1015.

Woo T-U, Whitehead RE, Melchitzky DS, Lewis DA. A subclass of prefrontal gamma-aminobutyric acid axon terminals are selectively altered in schizophrenia. Proc Natl Acad Sci USA 1998; 95: 5341-5346.

Wright IC, Sharma T, Ellison ZR, McGuire PK, Friston KJ, Brammer MJ, Murray RM, Bullmore ET. Supra-regional brain systems and the neuropathology of schizophrenia. Cereb Cortex 1999; 9: 366-378.

Wright IC, Rabe-Hesketh S, Woodruff PW, David AS, Murray RM, Bullmore ET. Meta-analysis of regional brain volumes in schizophrenia. Am J Psychiatry 2000; 157: 16-25.

Young CE, Arima K, Xie J, Hu L, Beach TG, Falkai P, Honer WG. SNAP-25 deficit and hippocampal connectivity in schizophrenia. Cereb Cortex 1998; 8: 261-268.

Young KA, Manaye KF, Liang C-L, Hicks PB, German DC. Reduced number of mediodorsal and anterior thalamic neurons in schizophrenia. Biol Psychiatry 2000; 47: 944-953.

Yukie M. Connections between the medial temporal cortex and the CA1 subfield of the hippocampal formation in the Japanese monkey (Macaca fuscata). J Comp Neurol 2000; 423: 282-298.

Zaidel DW, Esiri MM, Harrison PJ. The hippocampus in schizophrenia: Lateralized increase in neuronal density and altered cytoarchitectural asymmetry. Psychol Med 1997a; 27: 703-713.

Zaidel DW, Esiri MM, Harrison PJ. Size, shape, and orientation of neurons in the left and right hippocampus: Investigation of normal asymmetries and alterations in schizophrenia. Am J Psychiatry 1997b; 154: 812-818.

10 IN SITU/HISTOLOGICAL APPROACHES TO NEUROTRANSMITTER-SPECIFIC POSTMORTEM BRAIN STUDIES OF SCHIZOPHRENIA

Susan E. Bachus and Joel E. Kleinman

Abstract

To bridge the gulf between genetic vulnerability and clinical phenomenology we are obliged to understand gene expression and protein function at the cellular level of resolution. In situ postmortem methods, which merge the advantages of neurochemical specification and neuroanatomical localization, can enable this objective. Our ability to visualize and measure proteins by immunocytochemistry, neurotransmitter receptors with quantitative receptor autoradiography, and mRNA levels with in situ hybridization histochemistry, is presented, both in terms of critical methodological considerations, and according to insights these strategies have revealed into the neuropathology of schizophrenia. The greatest challenge now facing us is the synthesis of this wealth of findings into a coherent account of a dysfunctional neuroanatomically and neurochemically specified circuit subserving the pathophysiology of schizophrenia. The potential for multiple-labeling with in situ methods may aid in the reconstruction of this circuit from the connections between the microscopic elements of neuropathology.

Introduction

Plus ça change... Neurohistopathologists were mining postmortem brains in the search for the cause of "dementia praecox" while it was still widely believed to be a "functional disorder" (e.g. Alzheimer, 1913, Southard, 1914), even before the discovery (Loewi, 1921) that the "coin of the realm" of information transfer, within the staggeringly intricate nervous system, is the neurotransmitter molecule. Awareness dawned early of the hazards of a multitude of confounding experimental variables which required careful controls (e.g. Dunlap, 1924). Nonetheless, some of the earliest observations, such as "reduced neuropil", should seem astute, if not prophetic, to us today (see below). It was also perceived early on (e.g. Roth,

1957) that the schizophrenic phenotype reflects a complex interaction between genetic predisposition and environmental influences (see Tsuang, 2000 for a recent review). Yet appreciation of complexity does not necessarily translate to tools to unravel it. Today, having swept across the threshold of the sequencing of the human genome, we face the challenge of the vista of the "proteome": the elucidation of the regulation of gene expression--translation of an individual's DNA template into functional proteins, via mRNA--in order to bridge the gulf between the dichotomous perspectives arising from advances from genetic and clinical approaches to the study of schizophrenia. Each of these, in isolation, fails to address the intervening variables between genotypic endowment and behavioral phenotype. It has been estimated that we understand the function of only 2% of human genes (Hyman, 2000). "Functional genomics" require that we determine with "bottom-up tools" when, where, and in what circumstances "vulnerability genes" are expressed, and the identity and function of their products, at the cellular level of resolution in neurochemically defined neuroanatomical circuits, or what has been dubbed the "neuronal phenotype" (Akil & Watson, 2000). As Freedman (1998) put it, the question "can we find the genes for schizophrenia?" has evolved into that of "can we identify the biology of the genes for schizophrenia?" Recent methodological advances enable us to tackle the problem at the interface between genes, neurochemistry and neuroanatomy, resurrecting postmortem approaches to understanding schizophrenia from Plum's (1972) infamous "graveyard". An especially exciting prospect emerges from the powerful marriage between novel molecular biological techniques and conventional histopathological methods, for "in situ" histochemical evaluation, in slide-mounted thin sections of postmortem brain, of mRNAs (by in situ hybridization histochemistry [ISH]) and proteins (by immunocytochemistry [ICC] and quantitative receptor autoradiography [QRA]). Such hybrids have enabled histopathology to evolve from a real estate mentality fixation on "location, location and location" to a 3 dimensional perspective merging location, at the microscopic and even molecular level, with neurochemical specification and quantitation.

It must first be acknowledged that postmortem research, likened to an "archaeological excavation" (Benes, 1997) of shards with microscopic/ molecular resolution which need to be pieced together, usually only accessible decades after emergence of the disease, has significant limitations. It is most valuable viewed as a complementary strategy within the larger context of a more diverse arsenal. Distinguishing between primary etiology and secondary brain alterations becomes a serious dilemma. Our "most daunting challenge is the plasticity of the brain that may permit variable compensation for altered gene function" (Coyle & Draper, 1996, p. 5). It is especially problematic that treatments used to ameliorate psychosis, whether leucotomy in an earlier era, or various anti-psychotic drugs used today, have dramatic and widespread impact on brain neuroanatomy and neurochemistry. Postmortem studies also tend to be plagued by cohorts that are relatively small and often somewhat aged (superimposing the added challenge of disentangling schizophrenia from pathologies associated with aging), and a host of methodological concerns. The microscopic nature of the "slice" investigated, though a strength, dictates that we turn to genetic, pharmacological, in vivo imaging, neuropsychological and clinical studies for clues as to where in the "big picture" of the brain to focus our scrutiny. A unique characteristic afforded postmortem studies by in situ techniques is their straddling of structural (neuroanatomical) and functional (neurochemical), and cellular and molecular levels of analysis. Integrating leads yielded by this approach with those gained by other strategies will ultimately optimize their value. As each approach suffers methodological limitations, convergent evidence spanning varied efforts is our best hope in battling this tragic affliction.

Our focus here will be on applications of in situ/histological postmortem methods to neurotransmitter-specific alterations in the schizophrenic brain for two reasons. First, consensus has mounted that since no pathognomonic lesion has been identified in schizophrenia, we must levy our efforts at elucidating a dysfunctional network or circuit (e.g. Weinberger, 1991; Lewis, 1997; Benes, 2000). As the neurotransmitter is the functional unit of information processing in the nervous system, neurochemical coding provides the key to decipher the heterogeneity of its intricate circuitry. Of course neuroanatomical connections cannot as yet be established in human brain by anterograde and retrograde tracing methods employed in animal studies. But the "neuronal phenotype" that in situ methods, especially utilizing "double labeling" strategies, can effectively garner such information. For example, (postsynaptic) neurotransmitter specific receptors and (presynaptic) synthetic machinery for neurotransmitter release can be localized to an individual neuron. Reviews of the light shed on understanding schizophrenia by postmortem methods (e.g. Harrison, 1999; Weinberger, 1999) have concluded that the most consistent finding to date appears to be, as presaged by Alzheimer (1913), paucity of complexity or abundance of synaptic connections. We would argue that further specification of *which* (neurochemically defined) synapses are implicated remains a critical question if we hope to delineate the *circuitry* involved in schizophrenia. Second, even should we fail in our search for the "holy grail" of the etiology of schizophrenia, elucidation of neurotransmitter-specific pathology may inspire novel and improved pharmacotherapeutic strategies, which are after all our ultimate quest, short of discovery of a means of eradication of the disease.

In Situ or in Vitro?

Before delving into the wealth of data unearthed by in situ methods in the past few decades we should contrast them with homogenate-based postmortem methods, which overlap substantially with the former in what they can measure. Homogenate-based methods include radioimmunoassay, immunoblotting, or enzyme-linked immunoadsorbent assay of proteins, assay of ligand binding to receptors, and Northern blot analysis or solution hybridization/nuclease protection assay of mRNA.

The obvious forte of in situ methods is their ability to localize abnormalities anatomically, at the level of regional, subregional or laminar, cellular and even subcellular resolution, in combination with neurochemical specificity. At the regional level, study of an intact slice of tissue avoids dissection artifacts from extraneous tissue or differential representation within a region, resulting from difficulties entailed in carving a region out of either soft fresh brain or frozen slabs (with the attendant risk of partial thawing) (Casanova & Kleinman, 1990; Kleinman et al., 1995). Within a region, we increasingly learn that even territory considered until recently to be homogeneous in structure possesses remarkable heterogeneity, to wit, the labyrinthine "striosomal" organization of neostriatum (Graybiel & Ragsdale, 1978), which only became apparent upon probing for neurochemical markers, respected in contemporary in situ studies (e.g. Hurd et al., 1997). There is considerable differentiation of involvement of cortical laminae in most findings in cortex. To be sure, in some cases there is a similar trend across laminae which may reflect failure to reach statistical significance in some layers due to "noise" from a host of methodological factors. Yet in others there appears to be a very clear cut dichotomy between laminae that are grossly abnormal in schizophrenia and others with values overlapping nearly perfectly with the normal cases (e.g. Volk et al., 2000 & 2001). Again, this provides information relevant to mapping circuits. There is evidence of abnormal neurotensin receptor binding localized to layer ii of entorhinal cortex (Wolf et al., 1995) and abnormal mRNA for cholecystokinin (CCK) in layer vi (Bachus et al., 1997), each of which project to the CA3 subfield in the primate hippocampal formation (Witter & Amaral, 1991) suggestive of a faulty "link" in a larger circuit. Similarly, hippocampal subfields vary in their involvement across a broad range of indices, with CA3/4 particularly implicated (see Weinberger, 1999). At the cellular level of resolution, there is potential to more specifically implicate subpopulations of neurotransmitter-defined neurons in schizophrenia, especially with double labeling in situ strategies. Neurons positive for co-modulators or other factors that characterize some GABA neurons, such as calbindin and calretinin (Daviss & Lewis, 1995), parvalbumin (Woo et al., 1997) or nitric oxide synthase (Akbarian et al., 1993a&b, 1996a), or glutamate neurons, such as CCK (Bachus et al., 1997), may aid us in honing in more closely on the neuropathological substrate in schizophrenia.

While tissue on the order of 1-5 grams is typically required for homogenate assays, that chunk of tissue may be sectioned into hundreds of thin (~ 10 μ) sections sufficing for numerous in situ studies. Conversely, homogenate studies are generally more sensitive than in situ methods. An exception would be for markers localized in sparse cells, such that their signal would be "swamped" in homogenates, in which case in situ methods would provide superior detection (ISH can detect fewer than 5 hybrids per cell [Young, 1990]). Though the protection assay enjoys the reputation of being more quantitative than ISH for estimates of mRNA, it has been shown to yield different values under apparently identical conditions (Young, 1990), whereas ISH has been found to have an inter-assay co-efficient of variation of less than 6% (Guilot & Rahier, 1995) and to compare quite favorably to in vitro methods in quantitative ability (Young, 1989a). Reverse transcriptase-polymerase chain reaction (RT-PCR) has the greatest sensitivity, though there is some concern with variability due to the exponential nature of the reaction.

There are functional aspects of gene expression, such as enzymatic activity, receptor pharmacokinetics, functional linkage of receptors to second messenger systems, calcium-dependent stimulus evoked neurotransmitter release, in vitro translation, transcription rate and neurotransmitter signaling, that can only be measured in homogenates. That said, there are protein or mRNA markers of receptor-activated second messenger systems, such as binding by phorbal ester to protein kinase C or forskolin to adenylate cyclase (e.g. Kerwin & Beats, 1990; Opeskin et al., 1996; Dean et al., 1997b) that can be studied in situ.

Unique methodological factors can threaten the validity of each of these approaches, for example, variability in mRNA extraction or purification in homogenate studies, or in thickness of tissue section or nuclear track emulsion in studies in situ. Some functional assays that require homogenates are more vulnerable to long freezer storage time. Leonard et al. (1993) found that after 5 years of freezer storage, mRNA could still be isolated, but was compromised in assays of biological activity such as in vitro expression or directional cloning. Occasionally it is impossible for a donated brain to be processed quickly enough that cellular morphology can be preserved, negating much of the value of in situ methods, yet it is still viable for homogenate studies. Some brain banks are prepared for alternative processing for this eventuality (e.g. Vonsattel et al., 1995). Ultimately, though the fine anatomical resolution of in situ methods is critical for functional genomics, their value is increased by the availability of alternative methods, such as those using homogenates, to provide convergent evidence. Ideally, alternative methods will be used on the same cohort (e.g. Huntsman et al., 1998) to rule out extraneous variables should discrepancies arise.

Protein or mRNA?

For either homogenate or in situ studies of gene expression, the question can be raised as to the relative merits of studying mRNAs or their protein products. The neuronal phenotype incorporates all of the complex interactive effects between multiple genes, developmental processes, the internal milieu and experience with the external environment. Since mRNA is only expressed in cells actively synthesizing protein, its levels do not simply reflect inheritance but may reveal functional changes in regulation of gene expression. Granted, it has not been universally proven that mRNA levels parallel protein levels. Cases of apparent "mismatch" may be due to methodological limitations, differential susceptibility to confounding variables, or real functional differences. Synthesis of a neurotransmitter may be high yet matched by release and/or turnover, in which case dynamic shifts in regulation might be more accurately portrayed by measurements of mRNA. Alternatively, post-translational changes in proteins, altered protein level caused by abnormal translation efficiency or degradation or the proportion of mRNA that is translationally active, or scavenging of proteins by neurons not equipped with machinery to synthesize them, might be better tracked by direct measurements of the proteins themselves (Wiesner & Zak, 1991). Altered neurotransmitter subunit composition, which might be tracked via subunit mRNAs, can produce shifts in receptor function that do not influence ligand binding and thus are not apparent in binding studies. Another important consideration is that while some mRNAs are found in dendrites or terminals (Steward, 1997), most are predominantly localized in the soma, while terminal-associated proteins may be situated some distance away.

It has been generally established for many mRNAs that their levels vary in a meaningful way with a range of physiological and pharmacological manipulations, supporting the value of their role as functional markers (Lewis et al., 1986; Tecott et al., 1994). Though there is tremendous variability in stability of both mRNAs and proteins, mRNAs are generally more stable in postmortem brain, perhaps because the endonucleases that degrade them are quickly inactivated. The first flurry of neurochemical postmortem studies generated a storm of failures to replicate and ensuing controversy, in large part due to the postmortem vulnerability of enzymes such as glutamate decarboxylase (GAD) (Bird et al., 1977; Perry et al., 1978; Crow et al., 1978a&b; Bird et al., 1978; Cross et al., 1979). Yet as flawed as these early studies may have been, GAD, now studied via its mRNA, remains a suspect.

Once again, our best bet is a double pronged attack by studying both in concert. Studies have found reassuringly parallel evidence from mRNA and protein measurements (e.g. for complexins I and II in hippocampus [Harrison & Eastwood, 1998] and NMDA glutamate receptors in thalamus

[Ibrahim et al., 2000]). Apparent discrepancies between ligand binding and mRNA levels (e.g. for serotonin 1A receptors in dorsolateral prefrontal cortex and serotonin 2A receptors in parahippocampal gyrus [Burnet et al., 1996], thalamic AMPA and kainate glutamate receptors [Ibrahim et al., 2000] or striatal M1 cholinergic receptors [Crook et al., 1999; Dean et al., 1996, 2000]) should be followed by attempts at replication and thoughtful speculation as to possible explanations, such as presynaptic (terminal) localization of receptors, or post-translational protein processing, rather than assuming that the findings are spurious.

Neuroanatomical Specificity

A crucial question for in situ studies is "where to look?" Historically the answer has been heavily driven by pharmacological advances. The cornerstone of the "dopamine hypothesis" is the correlation between D2 dopamine receptor binding efficacy and therapeutic potency across antipsychotic drugs (Creese et al., 1976, Seeman et al., 1976), along with psychotomimetic effects of dopaminergic stimulants. Initial excitement over elevated dopamine receptors found in postmortem studies was quickly tempered by contradictory results (see Bachus & Kleinman, 1996) and concern that chronic antipsychotic treatment can itself increase dopamine receptors (Burt et al., 1997). Antipsychotics fail to ameliorate cognitive symptoms of schizophrenia (Goldberg et al., 1991), suggesting that dopamine hyperfunction is not the whole story (see Carlsson, 1995 and Willner, 1997). The bottom line is that though recent in vivo imaging (e.g. Laruelle et al., 1996) and genetic (e.g. Egan et al, 2001) studies do continue to implicate dopamine in schizophrenia, decades of concentrated effort in postmortem research have failed to uncover a dopaminergic hyperfunction "smoking gun".

Ironically, while part of the enthusiasm over the delineation of the cortical dopamine terminal projection field centered on its apparent confinement to "limbic cortex" (Hökfelt et al., 1974) it has subsequently been found to be far more dispersed in primate cortex (Berger et al., 1991; Huntley et al., 1992). Yet the focus on the limbic system in schizophrenia, which predated (Stevens, 1973, Torrey & Peterson, 1974) the dopamine hypothesis, has reclaimed center stage. Abnormalities in nigral neurons in schizophrenia appear to be localized to the medial, "limbic" portion (Bogerts et al., 1983) and thalamic abnormalities appear to be restricted to "limbic nuclei" (Ibrahim et al., 2000). In vivo imaging, despite its poor anatomical resolution, enjoys the ability to visualize the brain in toto. In vivo multislice proton magnetic resonance spectroscopic examination of N-acetyl-aspartate, a marker of neuronal integrity, found abnormalities *only* in hippocampus and dorsolateral prefrontal cortex, components of the limbic circuit as redefined by Nauta (1971), from a broad anatomical survey (Bertolino et al, 1996). Cognitive deficits in schizophrenia (e.g. Fey, 1951;

Saykin et al., 1991) have also been linked to these regions (Milner, 1963; Gruzelier et al., 1988), and a range of syndromes produced by temporal lobe lesions mimic psychosis (Davison, 1983). The hippocampal formation has emerged as the locus of the most consistent and convergent evidence, from a wide variety of electroencephalographic, in vivo imaging (gross morphological and functional), neuropsychological, and postmortem studies (Dwork, 1997; Weinberger, 1999; Arnold, 2000).

A clue springs from the emphasis on a dysfunctional circuit in schizophrenia: Glutamate is responsible for signaling in links throughout the corticolimbic circuit, from cortical efferent pyramidal cells (Conti et al., 1988) to midbrain dopamine neurons (Sulzer et al., 1998). Drugs that antagonize NMDA glutamate receptor function are psychotomimetic, especially mimicking cognitive deficits not elicited by dopaminergic stimulants, while those that potentiate glutamate function show therapeutic promise, especially against symptoms refractory to classical antipsychotics (Hirsch et al., 1997). Ironically, though initial interest in CCK in schizophrenia was due to its colocalization with dopamine, they appear to be more segregated in primate brain (see Bachus et al., 1997). Localization of reduced CCK mRNA in entorhinal cortical pyramidal neurons in schizophrenia (Bachus et al., 1997) further implicates limbic cortical glutamate, rather than midbrain dopamine. Newly discovered candidates, including gaseous molecules synthesized by glia, continue to stretch the envelope of the definition of neurotransmitter. D-serine, apparently an endogenous ligand at the NMDA receptor complex (Snyder & Ferris, 2000), will be intriguing to study.

An additional challenge in elucidating neuroanatomical specificity with in situ approaches to the neuropathology of schizophrenia will be to establish where there are NOT abnormalities, lest we fall victim to the "searching under the streetlamp" phenomenon. Studies often include a "negative control" region, expected not to differ, such as primary visual cortex, yet abnormalities have been detected here as well (e.g. Burnet et al., 1996). Thus the microscopic resolution of in situ methods will prove to also be a hurdle, yet one well worth the endeavor. Double dissociation between diagnostic and anatomical specificity may at times aid in this (e.g. abnormal CCK mRNA in entorhinal cortex in schizophrenia and in dorsolateral prefrontal cortex in suicide [Bachus et al., 1997]).

A wealth of new insights into schizophrenia have been yielded by in situ methods. Findings prior to 1997 are reviewed elsewhere (Bachus & Kleinman, 1996; Harrison, 1999). We will emphasize findings from the past few years that have been replicated or offer convergent evidence. Ultimately what is needed is careful evaluation of the fit of any potentially important finding within the larger perspective of results from other postmortem and in vivo approaches, but this undertaking exceeds our scope here. We will concentrate instead on consideration of methodological issues that should help to guide strides toward that goal.

General Methodological Considerations

The quantitative ability afforded by the most sophisticated molecular biological method is worthless without careful controls for a multitude of potentially confounding variables. These have been detailed by Casanova & Kleinman (1990), Kleinman et al. (1995), Kittell et al. (1999), Harrison & Kleinman (2000), and Everall & Harrison [2001]).

Diagnostic specificity is especially daunting in postmortem research, and must at times be reconstructed post hoc from scanty medical records, police reports, and family interviews. It would be a mistake to settle for diagnostic designations from coroners' autopsy reports. Hill et al. (1996) investigated the validity of such labels from case histories. 31% failed to meet criterion for schizophrenia by any of 5 diagnostic assessment systems, and only 21% were confirmed by all 5. Issues concerning choice of diagnostic criteria for research purposes have been reviewed by Kendell (1982). It may eventually prove feasible to explore differential neuropathology for finer grained distinctions between subsyndromes or symptom types: first passes have been attempted, albeit with painfully small sample sizes (Virgo et al., 1995; Bachus et al., 1997). More detailed clinical records may be compiled in prospective studies (Arnold et al., 1995), though trade-off between more detailed characterization but restricted tissue availability has been noted (Wagman, 1992).

Organic cerebral disorders may co-exist with psychosis, especially in those without genetic loading for schizophrenia (Davison, 1983). A study that found a 45% rate of neuropathology among patients clinically diagnosed by treatment teams as schizophrenic, blindly retrospectively reclassified every case with significant pathology to an organic sydrome (Healy et al., 1996). Thus, evidence for organic disease in schizophrenia may instead speak more to the need for diagnostic accuracy. It is crucial that any exclusion criteria (Kleinman et al., 1995) be applied equally in both schizophrenic and control groups. Caution must be exercised that stringency not be relaxed for schizophrenic brains due to their scarcity.

In addition to a host of demographic factors that must be considered in this research (see Everall & Harrison, 2001) there is the need to consider secondary effects on brain function of consequences of decades of living with schizophrenia, such as social isolation, impoverishment, institutionalization, and poor health care and nutrition. It may well be that only another group of mentally ill patients can begin to approximate such conditions. Abnormalities found in schizophrenics have been seen in other patient populations as well, such as reductions in GABA neuron density in hippocampus of bipolar patients (Benes et al., 1998). The danger inherent in including such a control group is that other diseases are bound to be characterized by neuropathology of their own, which may overlap to some extent with that of schizophrenia.

Treatments of schizophrenia may have dramatic impact on the brain. Antipsychotics have been linked to effects as subtle as size and orientation of hippocampal pyramidal cells (Benes et al., 1991b). Though strategies are being developed to apply new methods to archival material, effects of leucotomy are a concern in pre-neuroleptic era cohorts, and diagnostic standards and records in that era differed from those of today. Studies have attempted to circumvent this problem by including patients off drug for long periods before death, but chronic exposure to these drugs may leave enduring effects, and patients off medication may be refractory to treatment and less representative. Attempts to correlate biological effects with lifetime exposure to drugs may fail to account for this variable if the relationship is not linear. Other patient populations treated with these drugs may not have received comparable regimens, and doubtless have neuropathology of their own. Some studies have investigated effects of these drugs in animals, ideally in subhuman primates (e.g. Pierri et al., 1999; Volk et al., 2000). Even apart from differences between monkey and human, there may be differential effects in normal and diseased brain. Our best hope is convergent evidence from a multiplicity of strategies.

Peri- and postmortem factors are influential in postmortem neurochemistry. As Wyatt et al. (1978) put it, it can be "unclear whether the apparent differences between schizophrenics and controls were present when they were alive or occurred after death." Plum's relegation of neuropathology of schizophrenia to the graveyard was due to rapid degradation of catecholamines postmortem. Bird et al. (1978) conceded that they were misled in their initial observations of postmortem GAD by differences in agonal state. These concerns are further compounded by the social isolation of many patients, such that terminal health care may be poor and death not quickly discovered. Institutionalized patients may be more severely ill and of questionable representativeness. These factors are notoriously difficult to assess. PMI will have drastically different impact depending on ambient conditions. Time to refrigeration might be a more valuable index. PMI effects may be positive or negative, and nonlinear. Agonal state has tremendously variable effects (Perry et al., 1982), which may be estimated via tissue pH (Everall & Harrison, 2001). To further complicate matters, PMI and agonal state effects may be interactive (Perry, unpub., cited in Dodd et al., 1988), and pH correlated with age (Harrison et al., 1995).

Various sources of postmortem tissue, including medical examiners' offices (MEOs), Veterans Administration and state psychiatric hospitals, hospices, brain banks, and patient advocacy groups, each contribute respective methodological limitations to research design. An important advantage of MEOs is that control cases can be provided by the same source and receive identical treatment at autopsy. Psychiatric histories are more detailed from hospitals. These sources have been threatened by recent trends: away from autopsies, shifts to outpatient care, more stringent requirements for informed consent. Attendant problems such as long PMIs may only be alleviated by improved education of patients and their families, advocacy groups, and the general public, about the critical need for donations (Wagman, 1992). Efforts have been forged at large scale standardization and consensus regarding minimal acceptable diagnostic criteria and conditions for postmortem tissue, e.g. by the National Institute of Mental Health in the USA (Wagman, 1992) and the European Dementia and Schizophrenia Network (Riederer et al., 1995).

Numerous artifacts can be inadvertently caused by mis- or differential handling of tissue postmortem (Casanova & Kleinman, 1990). Partial thawing after freezing, e.g. during dissection, is more detrimental to mRNA than an equivalent period at room temperature before freezing (Ross et al., 1992). The merger between histology and biochemistry, common to in situ methods generates a tension between competing demands of preservation of morphology, optimized by fixation, and biochemical reactivity, impeded by fixation. In some cases (most for ISH, but more precarious in ICC), frozen slide-mounted sections can be lightly formalin-fixed for a few minutes just before assay. This varies across proteins and mRNAs and optimal conditions must be established in each case. For slides kept frozen at -80° C, even for several years, freezer time appears to have minimal effect for many measures--again, this must be confirmed for each assay. Fixation can introduce variable swelling and shrinkage, and duration should be equated. Effects of seemingly identical fixation, embedding and histological procedures on volume may still vary between individuals and brain regions (Quester & Schröder, 1997). An alternative is fixing frozen blocks during thawing, under cryoprotection, just prior to sectioning (Jones et al., 1992).

Anatomical factors require close attention in specimen preparation. Slightly different orientation of planes of section can markedly influence 2 dimensional cell counts (Casanova & Kleinman, 1990). Matching of sections in key regions, such as entorhinal and dorsolateral prefrontal cortex, should be done at the cytoarchitectonic level. Early observations of abnormalities in entorhinal cortical cytoarchitecture that lent impetus to the neurodevelopmental hypothesis of schizophrenia were not consistently replicated, perhaps due to bias introduced by matching of levels by hippocampal landmarks, which do differ in schizophrenia (see Bachus & Kleinman, 1996 for review). To the excellent review of the "stereological" solution to this dilemma by Everall & Harrison (2001), we would add the concern that has been raised lest stereology become a "methodological straight jacket" (Guillery & Herrup, 1997), and the caution that, should pathology be heterogeneous within the sampling perimeter, which is where in situ methods are most valuable, it may be missed entirely by the emphasis of stereology on the "whole".

Computer assisted microdensitometry and image analysis have played vital roles in the quantitation and analysis of the flood of data that in situ methods have tapped into (e.g. Benes et al., 1989), and appropriate image analysis software is now freely available (Rasband, NIH, Bethesda, MD, USA). A few statistical considerations are somewhat unique to this field: 1) It is critical to equalize any factors of potential concern as nearly as possible whether or not they are correlated with a particular signal. If groups differ for both a subject of investigation and an extraneous variable, analysis of covariance may artificially "mistake" the group difference for a correlation between these variables, obscuring a real group difference in the process. 2) The argument has been made that because the syndrome itself entails such diversity, this heterogeneity should be mined by examining clinical factors that match outliers for biological measures (Stevens, 1997). Post-hoc speculation would then require carefully controlled replication. 3) Because replication is so critical, and the work so laborious, it is important that carefully done negative studies be published. 4) As Lewis (1997) has suggested, it will be especially valuable for multiple assays to be conducted on the same cohort, and inter-relationships between variables explored. Thin sections used by in situ methods facilitate this endeavor. Of course, it must then be noted that these diverse parameters are not completely independent. 5) A great strength of in situ methods is the ability to examine neuroanatomical subregions/laminae in one section. Numerous separate statistical analyses risk type I error for a sole "finding". Yet there may well be neuro-anatomically specific pathology. A problem with multivariate statistical treatment of these data, apart from the fact that the same measurement in different areas does not truly equal different independent variables, is that high correlations between measures, as is typical, undermine the sensitivity of detection of a real change in any one. The situation may call for latitude, if not ingenuity, in multiple comparisons. As always, of course, replication is essential.

Conventional Histochemistry

Identification of morphological characteristics of neurochemically defined cell populations, such as the glutamatergic cortical pyramidal efferents and GABAergic small interneurons, and localization of clusters of monoaminergic cells in the midbrain, and cholinergic cells in the ventral forebrain, have enabled measurements of numbers and densities of these cells in Nissl-stained histological material. Age is an important variable in some cases, correlated, for example, with melanin density in substantia nigra and locus coeruleus (Kaiya, 1980). Greater tissue thickness in some material, such as the Yakovlev collection, may pose problems of neuronal overlap (Altshuler et al., 1987). Demonstrations of altered cortical dendritic spine density (Garey et al, 1998; Glantz & Lewis, 2000) were made in Golgi stained material. However, to this day we still do not understand what determines which neurons stain with this method, and whether those that do are entirely representative.

Enzyme histochemistry can be also be carried out in situ, and has been used to investigate the laminar distribution of the nitric oxide positive subpopulation of GABA neurons in limbic cortex (Akbarian et al., 1993a&b, 1996a). An abnormal pattern has been interpreted as disrupted neuronal migration and refueled the neurodevelopmental hypothesis. Enzyme histochemistry appears to be especially sensitive to PMI, with many signals completely abolished by 12 hours (Jones et al., 1992). Semi-quantitative fluorescence histochemistry, which does not distinguish between the monoamines, has discerned monoamine neurons in schizophrenia (Olson, 1974), revealing no difference, a disappointment during the heyday of the dopamine hypothesis.

Immunocytochemistry

An early application of ICC in human postmortem brain, with a putative "antibrain" antibody (Heath & Krupp, 1967), focused on schizophrenia. This was not conceived as a methodological advance but inspired by the theory that schizophrenia is an autoimmune disorder.

Among the in situ methods ICC has the finest anatomical resolution, down to the ultrastructural level (Smiley et al., 1992), but is most vulnerable to long PMIs and problematic in matters of quantitation. ICC exploits the ability of the immune system to respond with exquisitely specific antibodies to target inoculated antigens from another species (see Larsson, 1983; Lewis et al., 1986; Honer et al., 2000). Antibodies can be labeled with variously colored fluorescent tags, conjugated enzymes that catalyze reactions yielding visible precipitates, radiolabeled antigens, or particles that can be visualized at the electron microscopic level. Since antibodies have two antigen combining sites, indirect methods, using supamaximal concentration of a secondary "bridging" antibody so that multilayer "sandwiches" are formed, can increase sensitivity on the order of 5-10-fold. Multiple antibodies attaching to different sites can enhance signal. An advantage of using radiolabeled antigens as markers is that specificity is increased since both the tissue antigen and the radiolabeled antigen must bind to the antibody.

The four contradictory requirements in ICC are antigen retention, and preservation of cell morphology, antigenicity and antigen accessibility. Fixation, preferably before sectioning (Hökfelt et al., 1973), the most critical factor in establishing this balance, is essential for preventing diffusion and degradation of many proteins (e.g. Smiley et al., 1992), but can interfere with accessibility and antigenicity. This varies tremendously across antibodies. Prompt fixation at autopsy should be in thin slabs or blocks. Whole brain immersion in fixative may completely abolish immunoreactivity (Casanova & Kleinman, 1990). Frozen blocks can be fixed subsequently (Jones et al., 1992). Alcohols may extract proteins from tissue. Glutaraldehyde, usually necessary for ultrastructural localization, may abolish some signals (Hökfelt et al., 1973). Glutaraldehyde-paraformaldehyde may be optimal for some antibodies (Ravid et al., 1992). Dehydration-paraffin embedded tissue may impair antibodies. Microwave (Shi et al., 1991) or autoclave (Bankfalvi et al., 1994) treatment of fixed sections has been used for "antigen retrieval" prior to ICC, though this introduces another factor in quantitation. Staining, fixation and tissue processing procedures and durations should be identical for comparison's sake. Comparison of neuronal areas can help rule out differential fixation effects (e.g. Akbarian et al., 1993a&b).

Prolonged agonal state can either increase (e.g. Corder et al., 1990) or decrease (e.g. Eastwood et al., 1996) specific immunoreactivities, but generally has relatively less effect on ICC than on mRNA levels (Harrison et al., 1995). ICC results have been shown to be influenced by age (Harrison et al., 1995; Eastwood et al., 1996) and PMI (Eastwood et al., 1996). All these effects vary tremendously across different proteins (e.g. Jones et al., 1992; Harrison et al., 1995) and must be determined on an independent basis. PMI has been shown to variably influence alternatively spliced isoforms of the same protein (Lewis & Akil, 1997). There may even be lamina-specific effects on a particular protein (Lewis & Akil, 1997). Freezer time appears to have minimal effect (e.g. Eastwood et al., 1996). Paraffin-embedded fixed material can be used after virtually indefinite delays. In general, proteins are relatively more vulnerable to PMI (e.g. Jones et al., 1992) compared to mRNA.

Important controls in ICC include tests of specificity and sensitivity. Specificity certified by a supplier is determined by radioimmunoassay and may not hold true in ICC. Uniqueness of the antigenic site will determine specificity. The purest monoclonal antibody may still cross-react with structurally related antigens. Artifactual staining may result from non-immunologic staining by the antibody, tissue autofluorescence, or endogenous enzymes. Specificity of an antibody can be affected by assay pH, buffer strength, temperature, antibody dilution and incubation time (Larsson, 1983). Controls for specificity include comparison of signal for antibodies targeting different parts of the protein, competition or preincubation of the antibody with exogenous protein, or of the tissue with unlabeled antibody or a protease, failure of similar proteins to block the signal, and elimination of the primary and/or secondary antibody or the enzyme substrate from the procedure. Antibodies can vary in nonspecific staining at the same concentration, thus pre-incubation with "non-immune serum" is not an adequate control. Cross-reactivity of antibodies may be minimized by affinity purification of the antiserum or pooling of multiple antibodies against the protein.

As typically practiced, ICC is a qualitative method. Staining intensity above background can be influenced by a number of factors in addition to antigen concentration. For semi-quantitation the kinetics of chemical reactions involved must be controlled, and the quantitative relationship established between the antigen-antibody concentration and the detection modality. Incubation conditions must be selected such that linear relations between antigen, staining and detection hold, levels of substrates are saturating, and diffusion of antibody is prevented. Despite such concerns, good correlations have been found between ICC estimates and the more quantitative homogenate methods (Honer et al., 2000). Immunoautoradiography can be calibrated to standards of known radioactivity, hence is more quantitative, though at the expense of some loss of anatomical resolution due to isotope dispersion. Signal can be visualized at the cellular level by exposing the tissue to nuclear track emulsion, which is then developed, making the tissue itself a photograph of signal as it were. Dipping the slide into liquid emulsion can cause further diffusion of isotope, but an ingenious strategy (Roth et al., 1974), in which a dry cover slip previously coated with emulsion is "hinged" to the slide, enables separation for development after exposure, and then re-apposition, in register with tissue, for cellular analysis. Multiple exposures, of different durations, can be obtained if required because there are grossly different levels of signal in subregions of the section.

Quantitative Receptor (Binding) Autoradiography

Any molecule to which complementary exogenous ligands (drugs) will bind, thus enzymes and neurotransmitter transporters as well as receptors per se, can be quantitated by the tagging with a radioactive label of high affinity ligands (Leysen 1984; Whitehouse, 1985; Kuhar & Unnerstall, 1985; Kuhar et al, 1986; Herkenham, 1988; Palacios & Mengod, 1992). Close on the heels of the development of this method (e.g. Young & Kuhar, 1978) came its application to human postmortem brain (Young & Kuhar, 1979). The forte of QRA is its quantitative rigor, using calibration of film densitometry to standards (which is essential, as film optical density has been shown not to be linearly related to radioactivity [Young et al, 1990]), albeit with some trade-off against anatomical resolution in QRA done on unfixed tissue. However, concern raised about anatomical resolution with QRA because of "mismatch" between localization of terminal structures by ICC and receptor localization by QRA appears to have been unwarranted. Electron microscopic autoradiographic localization of receptors, despite some isotope scatter, indicates that many receptors are simply not situated at synapses (Hamel & Beaudet, 1984).

The "proof" of QRA lies largely in the specificity of the ligand, which can be far more discriminating for receptor subtypes, the open- or closed-state of an ion channel, even second messenger coupling to receptors (Hall et al., 1985), than are neurotransmitters. That said, receptor density may not translate into functional activity (Leysen et al., 1984), and subunits of the receptor complex that modulate function may be distinct from the ligand binding site. Interesting leads have come from our knowledge that some receptor subtypes that can be discriminated pharmacologically are normally linked in co-modulatory complexes. Benes et al. (1997) interpreted a "mismatch" between findings for $GABA_A$ and benzodiazepine receptors in hippocampus in schizophrenia as evidence for their "uncoupling". Controls for specificity include demonstration of high affinity and saturable binding of the ligand, inhibition by competition with unlabeled pharmacological congeners, a regional distribution that matches the innervation pattern, and correlations between binding affinities and potencies in pharmacological and physiological assays. Nonspecific binding to the labeled optical isomer of the ligand is typically subtracted from total binding for an estimate of specific binding. In practice, drugs tend to be "messy", binding to a variety of receptors. A common strategy is to add to the mix cold ligands that bind to the extraneous receptors, but not the receptor of interest, to increase the specificity of the radiolabeled ligand.

Some studies have attempted to "dissect" specific receptors with a "subtractive" methodology, by comparing ligands whose binding profiles differ for one receptor. The dopamine hypothesis was recharged when this approach implicated the D4 dopamine receptor in schizophrenia (see Bachus & Kleinman, 1996). Consternation arose when ligands specific for D4 receptors failed to detect this difference (Lahti et al., 1998). Apparently another distinction between the ligands used in the subtractive method is in binding to sigma receptors (Tang et al., 1997).

Methodological factors can certainly influence results of QRA. In measuring D2 receptor densities by raclopride in the same cohort Dean et al. (1997a) found the full range of possible results, for assay in homogenates (increased), in QRA with a low concentration of ligand with sulpiride displacement (decreased), a high concentration of ligand with sulpiride displacement (no change), or butaclamol displacement (decreased). Moreover there was a differential effect of 5'guanylyl-imidodiphospate in the assay between normals and schizophrenics.

QRA is relatively resilient to peri- and postmortem factors. Binding can still be detected in tissue devoid of metabolic activity (Dodd et al., 1988). PMIs of 72 hours or longer have been found not to alter binding to glutamate (Kornhuber et al., 1988), serotonin (Marcusson et al., 1984; Harrison et al., 1995), or dopamine (Dean et al., 1997b) receptors. Other receptors (e.g. substance P [Burnet & Harrison, 2000] are more sensitive to PMI. There may be regional differences. D2 dopamine receptors vary with PMI in striatum but not substantia nigra (Camps et al., 1989), while muscarinic acetylcholine receptors do in striatum and nucleus basalis but not the rest of substantia innominata (Cortés et al., 1987). PMI may alter subtle effects on receptor binding, e.g. that of zinc in the NMDA glutamate receptor assay (Piggott et al., 1992). In at least one case (NMDA receptors) binding was found to be affected by agonal state (Piggott et al., 1992), but

QRA is less sensitive to agonal state than ISH (Harrison et al., 1995). Long freezer storage can affect binding to many receptors (Kornhuber et al., 1988) and modulation by spermine of NMDA receptor binding (Piggott et al., 1992). Many receptors vary with age (e.g. glutamate [Kornhuber et al., 1988; Piggott et al., 1992]; serotonin [Marcusson et al., 1984; Burnet et al., 1994 & 1996; Dean et al., 1996]; dopamine [Dean et al., 1997a], substance P [Burnet & Harrison, 2000), and acetylcholine [Cortés et al., 1987]), but this can differ by receptor subtype, brain region, the ligand used, or even between normals and schizophrenics (Dean et al., 1997a).

Some methodological quirks of QRA must be taken into account in quantitation. Isotopes vary in the extent to which tissue "quenching" (absorption) of signal, differentially by region, constitutes a problem (e.g. Lidow et al., 1988), with tritium being especially susceptible. Delipidation may reduce this but may in turn leach ligands from the tissue. Correction factors may be calculated, separately for each region. Tritium's advantages include finer resolution and less influence of variable section thickness. Herkenham (1988) has lamented that the most commonly used conditions in QRA (tissue sections stored frozen, incubation without ions, tris-HCl buffers) are the most deleterious, but for the optimal conditions (refrigerator storage of sections) the slides would need to be assayed within a few days of sectioning, hardly feasible in postmortem research on a large cohort. Fixatives interfere with binding, thus QRA is often done on unfixed sections, eliminating problems associated with fixation. Ultrastructural resolution tends to be precluded by isotope scatter, so tissue preservation is somewhat less critical. Binding to low affinity receptors, or with reversible ligands, may have poor resolution due to diffusion of the ligand, which can be reduced by paraformaldehyde vapor fixation and delipidation, but at the cost of some loss of ligand. Development of irreversible ligands poses a more promising solution. Ligand can diffuse in liquid emulsion, but the hinged coverslip solution described for ICC has also been used in QRA.

Both film autoradiographic and emulsion silver grain density values from QRA have been shown to vary in a meaningful fashion in pharmacological studies, comparing well to homogenate binding results (Benes et al., 1989). Corroboration of evidence from QRA might be sought from homogenate binding assays. Reassuring parallels have been found (e.g. Pralong et al., 2000). A mismatch does not necessarily disprove a finding however. Simpson et al. (1996) failed to find the increased 5HT1a serotonin receptor binding in vitro that they observed in situ but reasoned that the lamina-specific change was swamped in the homogenate study, especially since the sensitivity of QRA is orders of magnitude greater than that of in vitro binding (Kuhar et al, 1986).

In Situ Hybridization Histochemistry

In use since the 1960's (Gall & Pardue, 1969), ISH was a latecomer to human postmortem brain research (Mengod et al., 1990) but then quickly added to our armamentarium in studying schizophrenia (Mengod et al., 1991), spawning a wealth of information in the past decade. ISH (Lewis et al., 1993; Young, 1992; Harrison & Pearson, 1990), which has been described as a "shotgun marriage between molecular biology and histology" (Coulton, 1995), due to their competing demands, capitalizes on the formation of stable "hybrids" between complementary strands of nucleic acids to detect mRNAs with labeled complementary probes. ISH is both enriched and complicated by the abundance of potential probes for any particular mRNA. Done with radiolabeled probes, ISH shares many advantages and limitations with QRA, or with nonisotopic probes (labeled with enzyme or fluorochrome reporters, or enzyme- or particle-conjugated, or fluorochromated antibodies), is more comparable to ICC, with similar trade-off between sensitivity/quantitative ability versus anatomical resolution. Nonisotopic biotinylated probes (e.g. Singer et al., 1989) have been used in ultrastructural studies of mRNA localization.

Either radio- or nonisotopic labeling can be used for different types of probe: complementary DNA probes, which suffer several disadvantages and have been largely superseded (Lewis et al., 1985); ribonucleotide ("ribo") probes, long stretches of "antisense" (complementary) RNA that can be labeled throughout their length; or oligodeoxyribonucleotide ("oligo") probes, short single stranded pieces of synthetic DNA, to which a hot "tail" of limited length can be added. Neither ribos nor oligos are intrinsically superior. Given knowledge of the mRNA sequence, oligos are easily manufactured and readily commercially available. Relative to ribos, oligos are more stable, penetrate fixed tissue better, tend to cause lower background label, are less affected by slight mRNA degradation, and less susceptible to RNase contamination, and can be more specific. A single nucleotide difference, as in a single nucleotide polymorphism, can reduce hybridization. More precise controls are possible for specificity of oligos, e.g. by thermal stability (the high temperatures required for ribos would damage tissue structure). Subtly different splice variants (e.g. of the NR1 NMDA glutamate receptor subunit [LeCorre et al., 2000]) have been discriminated by oligos. Ribos, which require cloning, can be labeled to a higher specific activity, thus can be much more sensitive. However, the strategy of using a cocktail of multiple oligos, targeting different stretches of the mRNA, can bring the sensitivity of the assay into the range of ribos, with the added benefit of improved selectivity. The importance of tissue penetration becomes more vital with increased fixation (Moensch et al., 1985), which, together with impairment of Nissl staining by the RNase treatment involved in the ribo assay, means that cell morphology can be better optimized by oligos. The most sophisticated computer programs used to design probes often fail to predict accurately which will label an mRNA well (for example, in human brain tissue some probes bind non-specifically to white matter). The "proof" of the probe lies in the empirical demonstration that it *works*!

Choice of isotope for radiolabeled probes involves the same concerns noted for QRA. ^{33}P has been proposed to combine advantages of ^{32}P and ^{35}S (Durrant et al., 1995). One study (Loiacono & Gundlach, 1999) found this emission not to be detectable by emulsion for cellular analysis.

Controls for specificity are essential in ISH, as false signals can be produced by numerous artifacts, such as contours of differential thickness (e.g. at the edge of a section) or high cell density (e.g. in the dentate gyrus of the hippocampal formation). Some controls borrowed from ICC or QRA turned out to be less useful here: "sense" probes, i.e. with identical, rather than complementary sequence, analogous to the isomeric ligand used for "nonspecific binding" in QRA, frequently produce high signal, perhaps because it is not uncommon to find mirror sequences in the human genome as a consequence of recombination. Signal may also appear with the use of "randomer" probes that contain a soup of sequences with all possible bases at each position, if an mRNA is of very high abundance. Other control probes, easily devised for oligos, are a "mis-sense" probe (the same sequence but with every third based scrambled), or a "reverse-antisense" probe. Dilution of labeled probe by unlabeled probe, or pretreatment of tissue with RNase, inspired by dilution and protease treatment in ICC, are of negligible value in ISH (Young, 1992). Preincubation with unlabeled probe, analogous to ICC, is difficult because not all the buffer can be removed from the tissue. There is no ideal control, the best strategy being a combination. Useful controls include comparison of anatomical distribution with what ICC reveals at the regional and cellular level (unless the protein and mRNA are not co-localized), comparison of distributions for multiple probes targeted against different parts of the sequence, Northern analysis to verify the size of the mRNA, a "melting curve" to establish thermal stability, and a "saturation curve" for probe concentration. Comparing an mRNA of interest to another mRNA by ratios (e.g. Zachrisson et al., 1999), normalization, or differential effect (e.g. Virgo et al., 1995), to control for variability in general mRNA quality, has some potential to mislead due to variable sensitivity of mRNAs to diverse confounding variables.

Though thawing of a block of frozen tissue drastically degrades mRNA (far more than an equivalent delay before freezing) (Ragsdale & Miledi, 1991; Ross et al, 1992), rapid thawing and drying of a thin section onto a slide appears to have negligible effect on mRNA, presumably because enzymatic degradation is quickly limited. ISH in completely unfixed tissue (Dagerlind et al., 1992) has been described, but consensus is that for retention of mRNA in tissue during hybridization and washing, and preservation of cell morphology, some fixation is required. Dipping slide-mounted fresh frozen sections in formalin for a few minutes prior to ISH suffices, while more extensive fixation can prevent probe penetration into the tissue. ISH can be done on formalin-fixed or paraffin-embedded tissue, albeit with some loss of signal (Harrison & Pearson, 1990). Pretreatment with protease has been used to facilitate labeling, but can degrade histological detail. A very exciting potential opportunity for untangling effects of antipsychotic drugs would be a method for performing ISH on archival, pre-neuroleptic era, fixed tissue. A strategy (Lu & Haber, 1992) in which free floating sections are hybridized has approached this goal, but appears limited to a few years of fixation. Another, adopted from ICC, looks promising: Formalin-fixed, paraffin-embedded sections subjected to autoclaving yield signal as high as that from fresh-frozen, post-fixed sections (Oliver et al., 1997). This benefit has been extended to at least several months of fixation, with no effect of fixation time, and shown to improve morphology (Eastwood & Harrison, 1999), albeit with variable results across mRNAs.

Level of mRNA has been found to be relatively insensitive to PMI but tremendously affected by agonal state (Harrison & Pearson, 1990; Harrison et al., 1995; Harrison & Kleinman, 2000; Torrey et al., 2000). PMIs as long as 84 hours have been found not to influence overall polyadenylated mRNA yield, size or biological activity (Perrett et al., 1988). A variety of specific mRNAs were found not to vary with PMIs up to 51 hours (Kingsbury et al., 1995), 66 hours (Mengod et al., 1991), or even 72 hours (Porter et al., 1997; Schramm et al., 1999). In a screen of 51 brains, for which PMIs (up to 46 hours) could be accurately estimated, Vonsattel et al. (1995) concluded that PMI was not a reliable indicator of mRNA levels. There are exceptions (Burnet et al. 1996), considerable variability in the degree to which specific mRNAs are affected by PMI (Burke et al. 1991; Harrison et al., 1995), and regional variability for the affect of PMI on a particular mRNA (Eastwood et al., 1995; Harrison et al., 1995; Porter et al., 1997). Specific mRNAs are vulnerable to a range of peri-mortem factors, including rapidity of death, hypoxia, pyrexia, coma, seizures, dehydration, hypoglycemia, and time of day (Barton et al, 1993; Harrison et al., 1991b & 1994; Morrison-Bogorad et al., 1995; Johnston et al., 1997), and can vary, differentially, in sensitivity to individual factors within the same cohort (Burke et al., 1991; Barton et al., 1993). There can be regional differences in the sensitivity of an mRNA to agonal state (Porter et al., 1997). Tissue pH serves as a good surrogate marker for these effects, even better correlated with a range of mRNA levels than records of the manner of death (Kingsbury et al., 1995). Age (Eastwood et al., 1995; Burnet et al., 1996; Porter et al., 1997) and freezer storage time (Burke et al., 1991; Eastwood et al., 1995; Johnston et al., 1997) have similarly been found to have variable influences on mRNA levels, often differing by region.

Film autoradiography with radiolabeled ISH probes is comparable in quantitative ability to QRA (Lewis et al., 1989; Nunez et al., 1989; Uhl,

1989; Young, 1992), as long as the data are calibrated to a standard curve, with the caveat that it is the amount of mRNA that is *hybridized*, rather than absolute mRNA content, that is quantitated, and that this can be affected by factors such as fixation. But if such factors are carefully controlled the method is certainly sufficiently semi-quantitative for meaningful comparisons. The short half-life of ^{35}S complicates the use of ^{35}S standards, but a conversion factor has been established to interpolate ^{35}S signal to commercially available ^{14}C standards (Miller, 1991). Silver grain density at the cellular level has been empirically demonstrated to be proportional to radioactivity up to a limiting grain density (Lewis et al., 1989), in which case the overlap can be resolved by visual inspection (e.g. Benes et al., 1989). While in some studies cellular analysis has paralleled film quantitation (e.g. Bachus et al., 1977), discordance has been accounted for by speculation that the cells implicated are too sparse for film sensitivity to be adequate, when abnormality was detected at the cellular but not film level (Ohnuma et al., 1998), or that the abnormality is a reduction in number of positive cells, when found at the film but not cellular level (e.g. Volk et al., 2000 & 2001). More controversial is the effort to semi-quantitate nonisotopic ISH, which is more comparable to ICC, and may be preferable for detection, rather than quantitation, at the ultrastructural level.

Synthesis and Future Directions

One take-home message from a review of recent in situ approaches to the study of postmortem brains in schizophrenia is that there is a bewildering assortment of apparently contradictory findings. It is to be hoped that careful consideration of the methodological concerns elaborated above will facilitate the struggle to sift through and reconcile these incongruities. Certainly, some failures to replicate findings may be attributable to type I errors. However, to assume that this is always the case would be a most unfortunate mistake. As laborious as this work is, the seriousness of our quest dictates that we cannot afford to pass up any potential clues. Discrepancies must be mined for leads that could enable us to revise or refine our hypotheses, which would of course demand subsequent replication. In some cases there may be subtle distinctions, such as clinical heterogeneity, that may be associated with biological differences. On the other hand, the danger of future effort being misled by methodological flaws encumbers us with exceptional responsibility to pay the most exacting attention to the considerations detailed above.

The complementary take-home message is that, amazingly, despite all of the pitfalls and uncertainties in this field, a handful of loci of convergent evidence have coalesced, some of them validating some of the earliest hypotheses elaborated. Many are corroborated by a broader range of alternative methodological approaches.

A. The dopamine hypothesis has enjoyed a resurgence, but with an unexpected twist: at least in limbic cortical dopamine, innervation appears to be *hypo-* rather than hyperfunctional. ISH experiments have found D3 and D4 dopamine receptor mRNAs to be reduced (Meador-Woodruff et al., 1997), on the postsynaptic side, adding to the neurochemical specificity of the hypothesis. On the presynaptic side, ICC studies have found reduced tyrosine hydroxylase immunoreactive axons (Akil et al., 1999 & 2000). These aberrations are lamina-specific, adding to the neuroanatomical specificity of the hypothesis. Recent genetic evidence (Egan et al., 2001) adds further support to the hypothesis of reduced cortical dopamine function in schizophrenia.

B. GABA function, first incriminated by the earliest postmortem neurochemical studies (Bird et al., 1977), is increasingly convincingly found to be abnormal in limbic cortex in schizophrenia. Nissl stained histology indicates that small interneuron density is reduced in cingulate cortex and prefrontal cortex (Benes et al., 1991a), while QRA indicates that $GABA_A$ receptor density is reduced in anterior cingulate cortex (Benes et al., 1992b). GABA transporter expression is consistently found to be reduced in prefrontal cortex by ISH (Ohnuma et al., 1999) and ICC (Woo et al., 1998; Pierri et al., 1999; Volk et al., 2001). GAD_{67} mRNA is found consistently by ISH to be reduced in prefrontal cortex (Akbarian et al., 1995; Volk et al., 2000). Increased GABA receptor function suggested by QRA (Benes et al., 1996a &b) and ISH (Ohnuma et al., 1999) might be supposed to reflect receptor supersensitivity secondary to a presynaptic deficit. Homogenate studies, as well, have replicated findings such as reduced GAD_{67} and increased $GABA_A$ receptor subunit expression (Impagnatiello et al., 1998). Perhaps most intriguingly, neurochemically defined subpopulations of GABA neurons appear to be differentially involved (Akbarian et al., 1993a&b, 1996a; Daviss & Lewis, 1995; Kalus et al., 1997; Beasley & Reynolds, 1997; Woo et al., 1997 & 1998; Eastwood & Harrison, 2000), although for parvalbumin positive cells there may be regional differences in the direction of their dysfunction (Kalus et al., 1997; Beasley & Reynolds, 1997). Again, laminar differences in these abnormalities suggest involvement of neuroanatomically specific connections. Interpretation of the abnormal distribution of nitric oxide positive neurons in terms of neuronal migration has lent new fire to the neurodevelopmental hypothesis.

C. Because of glutamate's ubiquitous localization and multifaceted roles within the nervous sytem, its neurotransmitter function can only be investigated indirectly, for example via co-modulators, transporters, or receptors. Concordant with the pharmacological evidence cited above, the limbic cortical glutamate hypofunction hypothesis has been further shored by evidence of abnormal receptor function by ISH (Eastwood et al., 1995; Akbarian et al., 1996b; Ohnuma et al., 1998; Ohnuma et al., 2000a&b; LeCorre et al., 2000; Meador-Woodruff et al., 2001), QIA (Eastwood et al.,

1996), and QRA (Kerwin et al., 1990; Meador-Woodruff et al., 2001; Noga et al., 2001), and reduced co-modulator (Bachus et al., 1997) and transporter function (Ohnuma et al., 1998), by ISH. Increased cell (Benes et al., 1991a) and fiber (Benes et al., 1992a; Longson et al., 1996) density and decreased cell volume (Pierri et al., 2001), might be considered consistent with evidence of reduced neuropil for connections involving these neurons. Homogenate-based studies corroborate much of this evidence (see Bachus et al., 1997 for review). Elucidating roles of various subunits in modulation of receptor function will be essential in understanding the impact of receptor abnormalities.

D. Thalamus, especially its limbic-related nuclei, incriminated by an early stereological investigation of schizophrenia (Pakkenberg, 1990), and also recently by in vivo imaging (Andreasen et al., 1994), is one of the hotspots of postmortem in situ findings, which now lend some neurochemical specificity to this involvement. Abnormal glutamate receptor expression has been implicated by both ISH and QRA (Ibrahim et al., 2000). Reduced cholinergic receptor binding has been found by QRA (Court et al, 1999). Parvalbumin-positive GABA neurons have been found to be reduced in density by ICC (Danos et al., 1998).

E. Elaboration of cortical neuropil, described by Alzheimer (1913), has been consistently confirmed in dorsolateral prefrontal cortex. Pyramidal cell dendritic spine density has been found to be reduced (Garey et al., 1998; Glantz & Lewis, 2000), supporting the inference of reduced neuropil from increased cell density by Selemon et al. (1995). Additional support derives from the in vivo imaging observation of reduced cortical N-acetyl-aspartate signal (Bertolino et al., 1996).

F. The medial temporal lobe, especially the hippocampal formation, has continued to be increasingly implicated, and its pathology increasingly characterized in terms of neurochemical specificity. Abnormalities in κ receptors (Royston et al., 1991), serotonin receptors (Joyce et al., 1993), cholinergic receptors (Freedman et al., 1995; Crook et al., 2000), GABA receptors (Benes et al., 1996a) and neurons (Harrison & Eastwood, 1998; Eastwood & Harrison, 2000), glutamate receptors (Kerwin et al., 1990; Harrison et al., 1991a; Eastwood et al., 1995 & 1996; Porter et al., 1997; Gao et al., 2000), TRH receptors (Lexow et al., 1994), neurotensin receptors (Wolf et al., 1995), CCK receptors (Kerwin et al., 1992) and expression (Bachus et al., 1997), and dopamine innervation (Akil et al., 2000) are all concentrated there. Neuroanatomical specificity is gradually becoming clarified within the hippocampal formation, with CA3/CA4, and interconnected entorhinal cortical layer vi, most consistently implicated (Weinberger, 1999).

The greatest challenge facing us is integrating the embarrassment of riches of multiple microscopic fragments that are slowly emerging as reliably replicated findings into the reconstruction of the dysfunctional macrocircuit. Continued elucidation of the interface between neurochemistry and neuroanatomy, through in situ methods, will allow us to gradually hone in on this goal. It will be reassuring to bolster convincingly replicated postmortem findings with evidence from diverse methods, among in situ methods, between in situ and homogenate-based postmortem methods, and with others not subject to the methodological limitations of postmortem research. Alternatively, discrepancies may render clues enabling us to revise, refine, and characterize more diagnostically, neurochemically, and neuroanatomically specifically our hypotheses. Ultimately, it will take synthesis within the large perspective of convergence from neuropsychological, pharmacological, clinical, genetic, in vivo and postmortem approaches, and even on rare occasions from surgically biopsied tissue (e.g. Ong & Garey, 1993), to accomplish this reconstruction, and ideally, thereby arrive at informed hypotheses for improved treatment strategies. It will prove invaluable to obtain multiple measurements, from diverse strategies, on the same cohort, to exclude extraneous differences between cohorts (e.g. age, diagnostic differences) as reasons for divergent findings. A condition imposed by the Stanley Foundation for receipt of postmortem tissue for research is that data be made available to the Foundation to enable exactly this sort of meta-analysis (Torrey et al., 2000).

We also enjoy potential for applying increasingly sophisticated tools, as they become available, in this endeavor. Some of these "brave new" techniques of the next generation marry in situ methods with sensitive homogenate-based methods. Combined ISH-PCR (Chen & Fuggle, 1993), optimizes anatomical information with ISH and sensitivity with PCR, by performing PCR on tissue sections. Some diffusion of PCR products away from the cells may impair quantitative ability of this method, but it will certainly have greater sensitivity for detection of low abundance mRNAs. A practical limitation is that today's thermocyclers hold restricted numbers and sizes of slides, but future waves of equipment may expand those boundaries. Another promising merger is "laser capture microdissection" of the contents of individual cells in a histological section, for example after they have already been neurochemically defined by immunofluorescence, which can then be subjected to sensitive immunoassay (Simone et al., 2000a), cDNA microarray analysis (Best et al., 2000), or RT-PCR (Murakami et al., 2000). Advances are being made in the ability to quantitate these methods (Simone et al., 2000b).

New homogenate-based techniques which will continue to reveal candidates genes for study in schizophrenia include in vitro translation (Perrett et al., 1992), differential hybridization screening by differential display RT-PCR and RNA fingerprinting by arbitrarily primed-PCR (Yee & Yolken, 1997), proteomic analysis by 2D electrophoresis with mass spectrometric protein sequencing (Johnston-Wilson et al., 2000), cDNA microarray analysis (Mirnics et al., 2000), quantitative competitive RT-PCR (Tallerico et al., 2001), and serial analysis of gene expression/RT-PCR (Sun et al., 2001). New genetic approaches are continually emerging. But elucidation of the *function* of candidate genes, and how their contribution to psychopathology is mediated, will require follow up with functional genomics, which only in situ methods can enlighten.

We need to more fully exploit one of the most unique, and valuable, virtues of in situ methods, the potential for simultaneous double- or even multiple-labeling of diverse biochemical markers on the same section, combined with the two dimensional anatomical information already inherent, for effectively four or more "dimensional" perspectives on neuropathology. Methods already employed include combined alkaline-phosphatase-conjugated and biotinated ICC (Ichimiya et al., 1989), multiple isotopes in ISH (Haase et al., 1985), radiolabeled and nonisotopic ISH probes (Young, 1989b), ICC and fluorescence histochemistry (Hökfelt et al., 1973), ISH and immunofluorescence (Schalling et al., 1989), radiolabeled ISH and ICC (Gendelman et al., 1985), nonisotopic ISH and ICC (van der Loos et al., 1989), ISH and immunofluorescence confocal microscopy (Linares-Cruz et al., 1995), and QRA and fluorescence histochemistry (Roth et al., 1974). It should prove feasible to combine differently colored fluorochromated antibodies in ICC, radioimmunoassay with enzyme-conjugated ICC, immuno-fluorescence with enzyme-conjugated ICC, multiple substrates for enzyme-conjugated ICC that generate different colored precipitates, QRA with a ligand that can bind covalently to fixed tissue combined with nonisotopic ISH, or QRA and nonisotopic ISH in unfixed tissue. Theoretically the possibilities are endless. In practice, careful balancing of competing requirements is entailed and there is some loss of each signal (Baldino et al., 1989; Lewis et al, 1993). Nonetheless, such strategies will continue to parcellate subpopulations of neurochemically defined neurons, and pre- and post-synaptically defined connections, that may be involved in the neuropathology of schizophrenia.

Abnormal cerebral asymmetry for some neurochemical indices has been hinted at (Joyce et al., 1992; Simpson et al., 1996; Zaidel et al., 1997a&b; Bernstein et al., 1998), but rarely explored due to tissue scarcity. Hopefully future studies will include bilateral samples from each brain, and this hypothesis explored more fully. Performance of numerous experiments on thin sections from one sample with in situ methods should relieve some of the pressure for obtaining this material.

There is an urgent need, now that the biological side of the methodology has become so sophisticated, for comparable strides in fine-tuning the resolution of clinical characterization of the cohorts. In addition to obtaining, evaluating, and analyzing more detailed psychiatric historical records, ideally data can be included from neuropsychological tests, in vivo imaging, medication history, evaluation of extrapyramidal side effects, and records of use of nicotine and alcohol. It may only be realistic to attempt to compile such information in prospective studies.

A criticism levied against contemporary postmortem research is that confusion has arisen from the diversity of methods and parameters across studies (Harrison, 1999). We would argue, to the contrary, that now fully revived, the field is due for a period of growing pains as we extend and delineate either the generalizability or specificity--diagnostic, neuroanatomical and neurochemical--of the knowledge gained, expressly through such checks and comparisons by a multitude of methods. Early in postmortem neurochemical research on schizophrenia, Wyatt et al. (1978) chided the reviewers of their paper, who expressed dismay over the "difficulties in doing biochemical studies with autopsied brains". We dare not shrink from grappling with these difficulties. We are optimistic that in situ methods will prove invaluable in this effort.

Acknowledgments

We are grateful to the individuals and their families, whose donations in their time of grief have made this research possible, and to Dr. Richard J. Wyatt, for his generous mentorship.

References

Akbarian S, Bunney WE Jr, Potkin SG, Wigal, SB, Hagman JO, Sandman CA, Jones, EG. Altered distribution of nicotinamide-adenine dinucleotide phosphate-diaphorase cells in frontal lobe of schizophrenics implies disturbances of cortical development. Arch Gen Psychiatry 1993a;50: 169-177.

Akbarian S, Viñuela A, Kim JJ, Potkin SG, Bunney WE Jr, Jones EG. Distorted distribution of nicotinamide-adenine dinucleotide phosphate-diaphorase neurons in temporal lobe of schizophrenics implies anomalous cortical development. Arch Gen Psychiatry 1993b;50: 178-187.

Akbarian S, Kim JJ, Potkin SG, Hagman JO, Tafazzoli A, Bunney WE Jr, Jones EG. Gene expression for glutamic acid decarboxylase is reduced without loss of neurons in prefrontal cortex of schizophrenics. Arch Gen Psychiatry 1995;52: 258-266.

Akbarian S, Kim JJ, Potkin SG, Hetrick WP, Bunney WE Jr, Jones EG. Maldistribution of interstitial neurons in prefrontal white matter of the brains of schizophrenic patients. Arch Gen Psychiatry 1996a;53: 425-436.

Akbarian S, Sucher NJ, Bradley D, Tafazzoli A, Trinh D, Hetrick WP, Potkin SG, Sandman CA, Bunney WE Jr, Jones EG. Selective alterations in gene expression for NMDA receptor subunits in prefrontal cortex of schizophrenics. J Neurosci 1996b;16: 19-30.

Akil H, Watson SJ. Science and the future of psychiatry. Arch Gen Psychiatry 2000;57: 86-87.

Akil M, Pierri JN, Whitehead RE, Edgar CL, Mohila C, Sampson AR, Lewis DA. Lamina-specific alterations in the dopamine innervation of the prefrontal cortex in schizophrenic subjects. Am J Psychiatry 1999;156: 1580-1589.

Akil M, Edgar CL, Pierri JN, Casali S, Lewis DA. Decreased density of tyrosine hydroxylase-immunoreactive axons in the entorhinal cortex of schizophrenic subjects. Biol Psychiatry 2000;47: 361-370.

Altshuler LL, Conrad A, Kovelman JA, et al. Hippocampal pyramidal cell orientation in schizophrenia. A controlled neurohistologic study of the Yakovlev Collection. Arch Gen Psychiatry 1987;44: 1094-1098.
Alzheimer A. Beiträge zur pathologischen Anatomie der Dementia praecox. Allgemeine Zeitschrift für Psychiatrie und Psychisch-Gerichtliche Medizin 1913;7: 810-812.
Andreasen NC, Arndt S, Swayze V II, Cizadlo T, Flaum M, O'Leary D, Ehrhardt JC, Yuh WT. Thalamic abnormalities in schizophrenia visualized through magnetic resonance image averaging. Science 1994;266: 294-8.
Arnold SE. Cellular and molecular neuropathology of the parahippocampal region in schizophrenia. Ann N Y Acad Sci 2000;911: 275-292.
Arnold SE, Gur RE, Shapiro RM, Fisher KR, Moberg PJ, Gibney MR, Gur RC, Blackwell P, Trojanowski JQ. Prospective clinicopathologic studies of schizophrenia: Accrual and assessment of patients. Am J Psychiatry 1995;152: 731-737.
Bachus SE, Kleinman JE. The neuropathology of schizophrenia. J Clin Psychiatry 1996;57[Suppl. 11]: 72-83.
Bachus SE, Hyde TM, Herman MM, Egan MF, Kleinman JE. Abnormal cholecystokinin mRNA levels in entorhinal cortex of schizophrenics. J Psychiatr Res 1997;31: 233- 256.
Baldino F Jr, Chesselet M-F, Lewis ME. High resolution in situ hybridization histochemistry. Methods Enzymol 1989;168: 761-777.
Bankfalvi A, Navabi H, Bier B, Böcker W, Jasani B, Schmid KW. Wet autoclave pretreatment for antigen retrieval in diagnostic immunohistochemistry. J Pathol 1994;174: 223-228.
Barton AJL, Pearson RCA, Najlerahim A, Harrison PJ. Pre- and postmortem influences on brain RNA. J Neurochem 1993;61: 1-11.
Beasley CL, Reynolds GP. Parvalbumin-immunoreactive neurons are reduced in the prefrontal cortex of schizophrenics. Schizophr Res 1997;24: 349-355.
Benes FM. What an archaeological dig can tell us about macro- and microcircuitry in brains of schizophrenia subjects. Schizophr Bull 1997;23: 503-507.
Benes FM. Emerging principles of altered neural circuitry in schizophrenia. Brain Res Rev 2000;31: 251-269.
Benes FM, Vincent SL, SanGiovanni SP. High resolution imaging of receptor binding in analyzing neuropsychiatric diseases. Biotechniques 1989;7: 970-978.
Benes FM, McSparren J, Bird ED, SanGiovanni JP, Vincent SL. Deficits in small interneurons in prefrontal and cingulate cortices of schizophrenic and schizoaffective patients. Arch Gen Psychiatry 1991a;48: 996-1001.
Benes FM, Sorensen I, Bird ED. Reduced neuronal size in posterior hippocampus of schizophrenic patients. Schizophr Bull 1991b;17: 597-608.
Benes FM, Sorensen I, Vincent SL, Bird ED, Sathi M. Increased density of glutamate-immunoreactive vertical processes in superficial laminae in cingulate cortex of schizophrenic brain. Cerebr Cortex 1992a;2: 503-512.
Benes FM, Vincent SL, Alsterberg G, Bird ED, SanGiovanni JP. Increased $GABA_A$ receptor binding in superficial layers of cingulate cortex in schizophrenics. J Neurosci 1992b;12: 924-929.
Benes FM, Khan Y, Vincent SL, Wickramasinghe R. Differences in the subregional and cellular distribution of $GABA_A$ receptor binding in the hippocampal formation of schizophrenic brain. Synapse 1996a;22: 338-349.
Benes FM, Vincent SL, Marie A, Khan Y. Up-regulation of $GABA_A$ receptor binding on neurons of the prefrontal cortex in schizophrenic subjects. Neuroscience 1996b;75: 1021-1031.
Benes FM, Wickramasinghe R, Vincent SL, Khan Y, Todtenkopf M. Uncoupling of $GABA_A$ and benzodiazepine receptor binding activity in the hippocampal formation of schizophrenic brain. Brain Res 1997;755: 121-129.
Benes FM, Kwok EW, Vincent SL, Todtenkopf MS. A reduction of nonpyramidal cells in sector CA2 of schizophrenics and manic depressives. Biol Psychiatry 1998;44: 88-97.

Berger B, Gaspar P, Verney C. Dopaminergic innervation of the cerebral cortex: Unexpected differences between rodents and primates. Trends Neurosci 1991;14: 21-27.

Bernstein H-G, Stanarius A, Baumann B, Henning H, Krell D, Danos P, Falkai P, Bogerts B. Nitric oxide synthase-containing neurons in the human hypothalamus: Reduced number of immunoreactive cells in the paraventricular nucleus of depressive patients and schizophrenics. Neuroscience 1998;83: 867-875.

Bertolino A, Nawroz S, Mattay V, Barnett AS, Duyn JH, Moonen CTW, Frank JA, Tedeschi G, Weinberger DR. Regionally specific pattern of neurochemical pathology in schizophrenia as assessed by multislice proton magnetic resonance spectroscopic imaging. Am J Psychiatry 1996;153: 1554-1563.

Best CJM, Gillespie JW, Englert CR, Swalwell JI, Pfeifer J, Krizman DB, Petricoin EF, Liotta LA, Emmert-Buck MR. New approaches to molecular profiling of tissue samples. Anal Cell Pathol 2000;20: 1-6.

Bird ED, Barnes J, Iversen LL, Spokes EG, Mackay AVP, Shepherd M. Increased brain dopamine and reduced glutamic acid decarboxylase and choline acetyl transferase activity in schizophrenia and related psychoses. Lancet 1977;ii: 1157-1159.

Bird ED, Spokes EG, Barnes J, Mackay AVP, Iversen LL. Glutamic-acid decarboxylase in schizophrenia. Lancet 1978;I: 156, 1978.

Bogerts B, Häntsch J, Herzer M. A morphometric study of the dopamine-containing cell groups in the mesencephalon of normals, Parkinson patients, and schizophrenics. Biol Psychiatry 1983;18: 951-960.

Burke WJ, O'Malley KL, Chung HD, Harmon SK, Miller JP, Berg L. Effect of pre- and postmortem variables on specific mRNA levels in human brain. Mol Brain Res 1991;11: 37-41.

Burnet PWJ, Harrison PJ. Substance P (NK1) receptors in the cingulate cortex in unipo- lar and bipolar mood disorder and schizophrenia. Biol Psychiatry 2000;47: 80-83.

Burnet PWJ, Eastwood SL, Harrison PJ. Detection and quantitation of 5-HT_{1a} and 5- HT_2 receptor mRNAs in human hippocampus using a reverse transcriptase- polymerase chain reaction (RT-PCR) technique and their correlation with binding site densities and age. Neurosci Lett 1994;178: 85-89.

Burnet PWJ, Eastwood SL, Harrison PJ. 5-HT_{1a} and 5-HT_2 receptor mRNAs and binding site densities are differentially altered in schizophrenia. Neuropsychopharmacology 1996;15: 442-455.

Burt DR, Creese I, Snyder SH. Antischizophrenic drugs: chronic treatment elevates dopamine receptor binding in brain. Science 1977;196: 326-328.

Camps M, Cortés R, Gueye B, Probst A, Palacios JM. Dopamine receptors in human brain: Autoradiographic distribution of D_2 sites. Neuroscience 1989;28: 275-290.

Carlsson A. The dopamine theory revisited. In: Hirsch SR, Weinberger DR (eds). Schizophrenia. Blackwell Science, London, 1995; pp 379-400.

Casanova MF, Kleinman JE. The neuropathology of schizophrenia: A critical assessment of research methodologies. Biol Psychiatry 1990;27: 353-362.

Chen RH, Fuggle SV. In situ cDNA polymerase chain reaction: A novel technique for detecting mRNA expression. Am. J. Pathol. 1993;143: 1527-1534.

Conti F, Fabri M, Manzoni T. Immunocytochemical evidence for glutamatergic cortico-cortical connections in monkeys. Brain Res 1988;462: 148-153.

Corder R, Pralong P, Muller AF, Gaillard RC. Regional distribution of neuropeptide Y- like immunoreactivity in human hypothalamus measured by immunoradiometric assay: Possible influence of chronic respiratory failure on tissue levels. Neuroendocrinology 1990;51: 25-30.

Cortés R, Probst A, Palacios JM. Quantitative light microscopic autoradiographic localization of cholinergic muscarinic receptors in the human brain: Forebrain. Neuroscience 1987;20: 65-107.

Coulton G. *In situ* hybridization comes of age. Histochem J 1995;27:1-3.

Court J, Spurden D, Lloyd S, McKeith I, Ballard C, Cairns N, Kerwin R, Perry R, Perry E. Neuronal nicotinic receptors in dementia with Lewy bodies and schizophrenia: α bungarotoxin and nicotine binding in the thalamus. J Neurochem 1999;73: 1590-1597.

Coyle JT, Draper ES. Molecules and mind: A new home for molecular research in psychiatry. Molecular Psychiatry 1996;1: 5-6.

Creese I, Burt DR, Snyder SH. Dopamine receptor binding predicts clinical and pharmacological potencies of antischizophrenic drugs. Science 192:481-483, 1976.

Crook JM, Dean B, Pavey G, Copolov D. The binding of [^{3}H]AF-DX 384, is reduced in the caudate-putamen of subjects with schizophrenia. Life Sci 1999;64: 1761-1771.

Crook JM, Tomaskovic-Crook E, Copolov DL, Dean B. Decreased muscarinic receptor binding in subjects with schizophrenia: A study of the human hippocampal formation. Biol Psychiatry 2000;48: 381-388.

Cross AJ, Crow TJ, Owen F. Gamma-aminobutyric acid in the brain in schizophrenia. 1979;i: 560-561.

Crow TJ, Owen F, Cross AJ, Lofthouse R, Longden AJ. Brain biochemistry in schizophrenia. Lancet 1978a;i: 36-37.

Crow TJ, Owen F , Cross AJ, Lofthouse R, Longden AJ, Joseph MH, Frith CD. Post-mortem handling and brain biochemistry. Lancet 1978b;i: 393-394.

Dagerlind Å, Friberg K, Bean AJ, Hökfelt T. Sensitive mRNA detection using unfixed tissue: Combined radioactive and non-radioactive in situ hybridization histochemistry. Histochemistry 1992;98: 39-49.

Danos P, Baumann B, Bernstein H-G, Franz M, Stauch R, Northoff G, Krell D, Falkai P, Bogerts B. Schizophrenia and anteroventral thalamic nucleus: Selective decrease of parvalbumin-immunoreactive thalamocortical projection neurons. Psychiatry Res Neuroimaging Section 1998;82: 1-10.

Davison K. Schizophrenia-like psychoses associated with organic cerebral disorders: A review. Psychiatric Developments 1983;1: 1-34.

Daviss SR, Lewis DA. Local circuit neurons of the prefrontal cortex in schizophrenia: Selective increase in the density of calbindin-immunoreactive neurons. Psychiatry Res 1995; 59: 81-96.

Dean B, Crook JM, Opeskin K, Hill C, Keks N, Copolov DL. The density of muscarinic M_1 receptors is decreased in the caudate-putamen of subjects with schizophrenia. Molecular Psychiatry 1996;1: 54-58.

Dean B, Pavey G, Opeskin K. [^{3}H]Raclopride binding to brain tissue from subjects with schizophrenia: Methodological aspects. Neuropharmacology 1997a;36: 779-786.

Dean B, Opeskin K, Pavey G, Hill C, Keks N. Changes in protein kinase C and adenylate cyclase in the temporal lobe from subjects with schizophrenia. J Neural Transm 1997b;104: 1371-1381.

Dean B, Crook JM, Pavey G, Opeskin K, Copolov DL. Muscarinic$_1$ and $_2$ receptor mRNA in the human caudate-putamen: No change in m1 mRNA in schizophrenia. Molecular Psychiatry 2000;5: 203-207.

Dodd PR, Hambley JW, Cowburn RF, Hardy JA. A comparison of methodologies for the study of functional transmitter neurochemistry in human brain. J Neurochem 1988;50: 1333-1345.

Dunlap CB. Dementia praecox. Some Preliminary observations on brains from carefully selected cases, and a consideration of certain sources of error. Am J Psychiatry 1924;3: 403-421.

Durrant I, Dacre B, Cunningham M. Evaluation of novel formulations of ^{35}S- and ^{33}P-labelled nucleotides for *in situ* hybridization. Histochem J 1995;27: 89-93.

Dwork AJ. Postmortem studies of the hippocampal formation in schizophrenia. Schizophr Bull 1997;23: 385-402.

Eastwood SL, Harrison PJ. Detection and quantification of hippocampal synaptophysin messenger RNA in schizophrenia using autoclaved, formalin-fixed, paraffin wax-embedded sections. Neuroscience 1999;93: 99-106.

Eastwood SL, Harrison PJ. Hippocampal synaptic pathology in schizophrenia, bipolar disorder and major depression: A study of complexin mRNAs. Molecular Psychiatry 2000;5: 425-432.

Eastwood SL, McDonald B, Burnet PWJ, Beckwith JP, Kerwin RW, Harrison PJ. Decreased expression of mRNAs encoding non-NMDA glutamate receptors GluR1 and GluR2 in medial temporal lobe neurons in schizophrenia. Mol Brain Res 1995;29: 211-223.

Eastwood SL, Kerwin RW, Harrison PJ. Immunoautoradiographic evidence for a loss of α-amino-3-hydroxy-5-methyl-4-isoxazole propionate-preferring non-*N*-methyl-*D*-aspartate glutamate receptors within the medial temporal lobe in schizophrenia. Biol Psychiatry 1996;41: 636-643.

Egan MF, Goldberg TE, Kolachana BS, Callicott JH, Mazzanti CM, Straub RE, Goldman D, Weinberger DR. Effect of COMT Val$^{108/158}$ Met genotype on frontal lobe function and risk for schizophrenia. Proc Natl Acad Sci USA 2001;98: 6917-6922.

Everall I, Harrison PJ. Methodological and stereological considerations in post mortem psychiatric brain research. 2001: This volume.

Fey ET. The performance of young schizophrenics and young normals on the Wisconsin Card Test. J Consult Psychol 1951;15: 311-319.

Fisman M. Brain stem encephalitic lesions and schizophrenia. A report of 3 cases. S Afr Med J 1974;48: 1491-1494.

Freedman R. Biological phenotypes in the genetics of schizophrenia. Biol Psychiatry 1998;44: 939-940.

Freedman R, Hall M, Adler LE, Leonard S. Evidence in postmortem brain tissue for decreased numbers of hippocampal nicotinic receptors in schizophrenia. Biol Psychiatry 1995;38: 22-33.

Gall JG, Pardue ML. Formation and detection of RNA-DNA hybrid molecules in cytological preparations. Proc Natl Acad Sci USA 1969;63: 378-383.

Gao X-M, Sakai K, Roberts RC, Conley RR, Dean B, Tamminga CA. Ionotropic glutamate receptors and expression of *N*-methyl-*D*-aspartate receptor subunits in subregions of human hippocampus: Effects of schizophrenia. Am J Psychiatry 2000;157: 1141-1149.

Garey LJ, Ong WY, Patel TS, Kanani M, Davis A, Mortimer AM, Barnes TRE, Hirsch SR. Reduced dendritic spine density on cerebral cortical pyramidal neurons in schizophrenia. J Neurol Neurosurg Psychiatry 1998;65: 446-453.

Gendelman HE, Moench TR, Narayan O, Griffin DE, Clements JE. A double labeling technique for performing immunocytochemistry and in situ hybridization in virus infected cell cultures and tissues. J Virol Methods 1985;11: 93-103.

Glantz LA, Lewis DA. Deceased dendritic spine density on prefrontal cortical pyramidal neurons in schizophrenia. Arch Gen Psychiatry 2000;57: 65-73.

Goldberg TE, Gold JM, Braff DL. Neuropsychological functioning and time linked information processing in schizophrenia. Review of Psychiatry 1991;10: 60-78.

Graybiel AM, Ragsdale CW Jr. Histochemically distinct compartments in the striatum of human, monkey, and cat demonstrated by acetylthiocholinesterase staining. Proc Natl Acad Sci USA 1978;11: 5723-5726.

Gruzelier J, Seymour K, Wilson L, Jolley A, Hirsch S. Impairments on neuropsychologic tests of temporohippocampal and frontohippocampal functions and word fluency in remitting schizophrenia and affective disorders. Arch Gen Psychiatry 1988;45: 623-629.

Guillery RW, Herrup K. Quantification without pontification: Choosing a method for counting objects in sectioned tissues. J Comp Neurol 1997;386: 2-7.

Haase AT, Walker D, Stowring L, Ventura P, Geballe A, Blum H, Brahic M, Goldberg R, O'Brien K. Detection of two viral genomes in single cells by double-label hybridization in situ and color microradioautography. Science 1985;227: 189-192.

Hall MD, El Mestikawy S, Emerit MB, Pichat L, Hamon M, Gozlan H. [^{3}H]8-Hydroxy-2-(di-*n*-propylamino)tetralin binding to pre- and postsynaptic 5-hydroxytraptamine sites in various regions of the rat brain. J Neurochem 1985;44: 1685-1696.
Hamel E, Beaudet A. Electron microscopic autoradiographic localization of opioid receptors in rat neostriatum. Nature 1984;312: 155-157.
Harrison PJ. The neuropathology of schizophrenia. A critical review of the data and their interpretation. Brain 1999;122: 593-624.
Harrison PJ, Eastwood SL. Preferential involvement of excitatory neurons in medial temporal lobe in schizophrenia. Lancet 1998;352: 1669-1673.
Harrison PJ, Kleinman JE. Methodological issues. In: Harris PJ, Roberts GW (eds). The Neuropathology of Schizophrenia: Progress and Interpretation. Oxford Univ Press, New York, NY, 2000; pp 339-350.
Harrison PJ, Pearson RCA. *In situ* hybridization histochemistry and the study of gene expression in the human brain. Prog Neurobiol 1990;34; 271-312.
Harrison PJ, McLaughlin D, Kerwin RW. Decreased hippocampal expression of a glutamate receptor gene in schizophrenia. Lancet 1991a;337: 450-452.
Harrison PJ, Procter AW, Barton AJL, Lowe SL, Najlerahim A, Bertolucci PHF, Bowen DM, Pearson RCA. Terminal coma affects messenger RNA detection in post mortem human temporal cortex. Mol Brain Res 1991b;9: 161-164.
Harrison PJ, Barton AJL, Procter AW, Bowen DM, Pearson RCA. The effects of Alzheimer's disease, other dementias, and premortem course on β-amyloid precursor protein messenger RNA in frontal cortex. J Neurochem 1994;62: 635-644.
Harrison PJ, Heath PR, Eastwood SL, Burnet PWJ, McDonald B, Pearson RCA. The relative importance of premortem acidosis and postmortem interval for human brain gene expression studies: Selective mRNA vulnerability and comparison with their encoded proteins. Neurosci Lett 1995;200: 151-154.
Healy DJ, Sima AAF, Tapp A, Watson SJ, Meador-Woodruff JH. Frequency of neuropathology in a brain bank from a long-term, domiciliary population. J Psychiatr Res 1996;30: 45-49.
Heath RG, Krupp IM. Schizophrenia as an immunologic disorder. I. Demonstration of antibrain globulins by fluorescent antibody techniques. Arch Gen Psychiatry 1967;16: 1-9.
Herkenham M. Receptor autoradiography: Optimizing anatomical resolution. In: Leslie FM, Altar CA (eds). Receptor Localization: Ligand Autoradiography. Alan R Liss, New York, NY, 1988; pp 37-47.
Hill C, Keks N, Roberts S, Opeskin K, Dean B, MacKinnon A, Copolov D. Problem of diagnosis in postmortem brain studies of schizophrenia. Am J Psychiatry 996;153: 533-537.
Hirsch SR, Das I, Garey LJ, de Belleroche J. A pivotal role for glutamate in the pathogenesis of schizophrenia, and its cognitive dysfunction. Pharmacol Biochem Behav 1997;56: 797-802.
Hökfelt T, Fuxe K, Goldstein M, Joh TH. Immunohistochemical localization of three catecholamine synthesizing enzymes: Aspects on methodology. Histochemie 1973; 33: 231-254.
Hökfelt T, Ljungdahl Å, Fuxe K, Johansson O. Dopamine nerve terminals in the rat limbic cortex: Aspects of the dopamine hypothesis of schizophrenia. Science 1974;184, 177-179.
Honer WG, Young C, Falkai P. Synaptic pathology. In: Harrison PJ, Roberts GW (eds). Oxford Univ Press, New York, NY, 2000; pp 105-136.
Huntley GW, Morrison JH, Prikhozhan A, Sealfon SC. Localization of multiple dopamine receptor subtype mRNAs in human and monkey motor cortex and striatum. Mol Brain Res 1992;15: 181-188.

Huntsman MM, Tran B-V, Potkin SG, Bunney WE Jr, Jones EG. Altered ratios of alternatively spliced long and short γ2 subunit mRNAs of the γ-amino butyrate type A receptor in prefrontal cortex of schizophrenics. Proc Natl Acad Sci USA 1998;95: 15066-15071.

Hurd, Y.L. Herman MM, Hyde TM, Bigelow LB, Weinberger DR, Kleinman JE. Prodynorphin mRNA expression is increased in the patch vs matrix compartment of the caudate nucleus in suicide subjects. Molecular Psychiatry 1997;2: 495-500.

Hyman SE. Genes, gene expression, and behavior. Neurobiology of Disease 2000;7: 528-532.

Ibrahim HM, Hogg AJ, Healy DJ, Haroutunian V, Davis KL, Meador-Woodruff JH. Ionotropic glutamate receptor binding and subunit mRNA expression in thalamic nuclei in schizophrenia. Am J Psychiatry 2000;157: 1811-1823.

Ichimiya Y, Emson PC, Christodoulou C, Gait MJ, Ruth JL. Simultaneous visualization of vasopressin and oxytocin mRNA-containing neurons in the hypothalamus using non-radioactive *in situ* hybridization histochemistry. J Neuroendocrinol 1989;2: 73-75.

Impagnatiello F, Guidotti AR, Resold C, Dwivedi Y, Caruncho H, Pisu MG, Uzunov DP, Smalheiser NR, Davis JM, Pandey GN, Pappas GD, Tueting P, Sharma RP, Costa E. A decrease of reelin expression as a putative vulnerability factor in schizophrenia. Proc Natl Acad Sci 1998;95: 15718-15723.

Johnston NL, Cerevnak J, Shore AD, Torrey EF, Yolken RH, The Stanley Neuropathology Consortium. Multivariate analysis of RNA levels from postmortem human brains as measured by three different methods of RT-PCR. J Neurosci Methods 1997;77: 83-92.

Johnston-Wilson NL, Sims CD, Hofmann J-P, Anderson L, Shore AD, Torrey EF, Yolken RH, The Stanley Neuropathology Consortium. Disease-specific alterations in frontal cortex brain proteins in schizophrenia, bipolar disorder, and major depressive disorder. Molecular Psychiatry 2000;5: 142-149.

Jones EG, Hendry SHC, Liu X-B, Hodgins S, Potkin SG, Tourtellotte WW. A method for fixation of previously fresh-frozen human adult and fetal brains that preserves histological quality and immunoreactivity. J Neurosci Methods 1992;44: 133-144.

Joyce JN, Lexow N, Kim SJ, Artymyshyn R, Senzon S, Lawerence D, Cassanova MF, Kleinman JE, Bird ED, Winokur A. Distribution of beta-adrenergic receptor subtypes in human postmortem brain: Alterations in limbic regions of schizophrenics. Synapse 1992;10: 228-246.

Joyce JN, Shane A, Lexow N, Winokur A, Casanova MF, Kleinman JE. Serotonin uptake sites and serotonin receptors are altered in the limbic system of schizophrenics. Neuropsychopharmacology 1993;8: 315-336.

Kaiya H. Neuromelanin, neuroleptics and schizophrenia. Hypothesis of an interaction between noradrenergic and dopaminergic system. Neuropsychobiology 1980;6: 241-248.

Kalus P, Senitz D, Beckmann H. Altered distribution of parvalbumin-immunoreactive local circuit neurons in the anterior cingulate cortex of schizophrenic patients. Psychiatry Res Neuroimaging Section 1997;75: 49-59.

Kendell RE. The choice of diagnostic criteria for biological research. Arch Gen Psychiatry 1982;39: 1334-1339.

Kerwin RW, Beats BC. Increased forskolin binding in the left parahippocampal gyrus and CA1 region in post mortem schizophrenic brain determined by quantitative autoradiography. Neurosci Lett 1990;118: 164-168.

Kerwin R, Patel S, Meldrum B. Quantitative autoradiographic analysis of glutamate binding sites in the hippocampal formation in normal and schizophrenic brain post mortem. Neuroscience 1990;39: 25-32.

Kerwin R, Robinson P, Stephenson J. Distribution of CCK binding sites in the human hippocampal formation and their alteration in schizophrenia: A postmortem autoradiographic study. Psychol Med 1992;22: 37-43.

Kingsbury AE, Foster OJF, Nisbet AP, Cairns N, Bray L, Eve DJ, Lees AJ, Marsden CD. Tissue pH as an indicator of mRNA preservation in human postmortem brain. Mol Brain Res 1995;28: 311-318.

Kittell DA, Hyde TM, Herman MM, Kleinman JE. The collection of tissue at autopsy: Practical and ethical issues. In: Dean B, Kleinman JE, Hyde TM (eds). Using CNS Tissue in Psychiatric Research: A Practical Guide. Harwood Acad Pub, Amsterdam, 1999; pp 1-18.

Kleinman JE, Hyde TM, Herman MM. Methodological issues in the neuropathology of mental illness. In: Bloom FE, Kupfer DJ (eds). Psychopharmacology: The Fourth Generation of Progress. Raven Press, New York, NY, 1995; pp 859-864.

Kornhuber J, Retz W, Riederer P, Heinsen H, Fritze J. Effect of antemortem and postmortem factors on [^{3}H]glutamate binding in the human brain. Neurosci Lett 1988;93: 312-317.

Kuhar MJ, Unnerstall JR. Quantitative receptor mapping by autoradiography: Some current technical problems. Trends Neurosci 1985;8: 49-53.

Kuhar MJ, DeSouza EB, Unnerstall JR. Neurotransmitter receptor mapping by autoradiography and other methods. Ann Rev Neurosci 1986;9: 27-59.

Lahti RA, Roberts RC, Cochrane EV, Primus RJ, Gallager DW, Conley RR, Tamminga CA. Direct determination of dopamine D_4 receptors in normal and schizophrenic postmortem brain tissue: A [^{3}H]NGD-94-1 study. Molecular Psychiatry 1998;3: 528-533.

Larsson L-I. Methods for immunocytochemistry of neurohormonal peptides. In: Björklund A, Hökfelt T (eds). Handbook of Chemical Neuroanatomy. Vol 1. Methods in Chemical Neuroanatomy. Elsevier Science Pub BV, New York, NY, 1983; pp 147-209.

Laruelle M, Abi-Dargham A, vanDyck CH, Gil R, D'Souza CD, Erdos J, McCance E, Rosenblatt W, Fingado C, Zoghbi SS, Baldwin RM, Seibyl JP, Krystal JH, Charney DS, Innis RB. Single photon emission computerized tomography imaging of amphetamine-induced dopamine release in drug-free schizophrenic subjects. Proc Natl Acad Sci USA 1996;93: 9235-9240.

LeCorre S, Harper CG, Lopez P, Ward P, Catts S. Increased levels of expression of an NMDAR1 splice variant in the superior temporal gyrus in schizophrenia. Neuroreport 2000;11: 983-986.

Leonard S, Logel J, Luthman D, Casanova M, Kirch D, Freedman R. Biological stability of mRNA isolated from human postmortem brain collections. Biol Psychiatry 1993;33: 456-466.

Lewis DA. Schizophrenia and disordered neural circuitry. Schizophr Bull 1997;23: 529-531.

Lewis DA, Akil M. Cortical dopamine in schizophrenia: Strategies for postmortem studies. J Psychiatr Res 1997;31: 175-195.

Lewis ME, Sherman TG, Watson SJ. *In situ* hybridization histochemistry with synthetic oligonucleotides: Strategies and methods. Peptides 1985;6: 75-87.

Lewis ME, Khachaturian H, Schäfer MK-H, Watson SJ. Anatomical approaches to the study of neuropeptides and related mRNA in the central nervous system. In: Martin JB, Barchas JD (eds). Neuropeptides in Neurologic and Psychiatric Disease. Raven Press, New York, NY, 1986; pp 79-109.

Lewis ME, Rogers WT, Krause RG II, Schwaber JS. Quantitation and digital representation of *in situ* hybridization histochemistry. Methods Enzymol 1989;168: 808-821.

Lewis ME, Robbins E, Baldino F Jr. In situ hybridization histochemistry with radioactive and non-radioactive cRNA and DNA probes. In: Sharif NA (ed). Molecular Imaging in Neuroscience. IRL Press, Oxford, 1993, pp 1-22.

Lexow N, Joyce JN, Kim SJ, Phillips J, Casanova MF, Bird ED, Kleinman JE. Alterations in TRH receptors in temporal lobe of schizophrenics: A quantitative autoradiographic study. Synapse 1994;18: 315-327.

Leysen J. Problems in *in vitro* receptor binding studies and identification and role of serotonin receptor sites. Neuropharmacology 1984;23: 247-254.

Lidow MS, Goldman-Rakic PS, Rakic P, Gallager DW. Differential quenching and limits of resolution in autoradiograms of brain tissue labeled with ^{3}H-, ^{125}I- and ^{14}C-compounds. Brain Res 1988;459: 105-119.

Linares-Cruz G, Millot G, DeCremoux P, Vassy J, Olofsson B, Rigaut JP, Calvo F. Combined analysis of *in situ* hybridization, cell cycle and structural markers using reflectance and immunofluorescence confocal microscopy. Histochem J 1995;27: 15-23.

Loewi O. Über humorale Übertragbarkeit der Herznervenwirkung. I. Mitteilung. Pflügers Arch 1921;189: 239-242.

Loiacono RE, Gundlach AL. In situ hybridisation histochemistry: Application to human brain tissue. In: Dean B, Kleinman JE, Hyde TM (eds). Using CNS Tissue in Psychiatric Research: A Practical Guide. Harwood Acad Pub, Amsterdam, 1999; pp 85-105.

Longson D, Deakin JFW, Benes FM. Increased density of entorhinal glutamate-immunoreactive vertical fibers in schizophrenia. J Neural Transm 1996;103: 503-507.

Lu W, Haber SN. In situ hybridization histochemistry: A new method for processing material stored for several years. Brain Res 1992;578: 155-160.

Marcusson J, Oreland L, Winblad B. Effect of age on human brain serotonin (S-1) binding sites. J Neurochem 1984;43: 1699-1705.

Meador-Woodruff JH, Haroutunian V, Powchik P, Davidson M, Davis KL, Watson SJ. Dopamine receptor transcript expression in striatum and prefrontal and occipital cortex. Focal abnormalities in orbitofrontal cortex in schizophrenia. Arch Gen Psychiatry 1997;54: 1089-1095.

Meador-Woodruff JH, Davis KL, Haroutunian V. Abnormal kainate receptor expression in prefrontal cortex in schizophrenia. Neuropsychopharmacology 2001;24: 545-552.

Mengod G, Charli J-L, Palacios JM. The use of *in situ* hybridization histochemistry for the study of neuropeptide gene expression in the human brain. Cell Mol Neurobiol 1990;10: 113-126.

Mengod G, Vivanco MM, Christnacher A, Probst, Palacios JM. Study of pro-opiomelanocortin mRNA expression in human postmortem pituitaries. Mol Brain Res 1991;10: 129-137.

Miller JA. The calibration of ^{35}S or ^{32}P with ^{14}C-labeled brain paste or ^{14}C-plastic standards for quantitative autoradiography using LKB Ultrofilm or Amersham Hyperfilm. Neurosci Lett 1991;121: 211-214.

Milner B. Effects of different brain lesions on card sorting. Arch Neurol 1963;9: 90-100.

Mirnics K, Middleton FA, Marquez A, Lewis DA, Levitt P. Molecular characterization of schizophrenia viewed by microarray analysis of gene expression in prefrontal cortex. Neuron 2000;28: 53-67.

Moench TR, Gendelman HE, Clements JE, Narayan O, Griffin DE. Efficiency of in situ hybridization as a function of probe size and fixation technique. J Virol Methods 1985;11: 119-130.

Morrison-Bogorad M, Zimmerman AL, Pardue S. Heat-shock 70 messenger RNA levels in human brain: correlation with agonal fever. J Neurochem. 1995;64: 235-246.

Murakami H, Liotta L, Star RA. IF-LCM: Laser capture microdissection of immunofluorescently defined cells for mRNA analysis. Kidney Int 2000;58: 1346-1353.

Nauta WJH: The problem of the frontal lobe: A reinterpretation. J Psychiatr Res 1971;8: 167-187.

Noga JT, Hyde TM, Bachus SE, Herman MM, Kleinman JE. AMPA receptor binding in the dorsolateral prefrontal cortex of schizophrenics and controls. Schizophr Res 2001;48: 361-370.

Nunez DJ, Davenport AP, Emson PC, Brown MJ. A quantitative 'in-situ' hybridization method using computer-assisted image analysis. Validation and measurement of atrial-natriuretic-factor mRNA in the rat heart. Biochem J 1989;263: 121-127.

Ohnuma T, Augood SJ, Arai H, McKenna PJ, Emson PC. Expression of the human excitatory amino acid transporter 2 and metabotropic glutamate receptors 3 and 5 in the prefrontal cortex from normal individuals and patients with schizophrenia. Mol Brain Res 1998;56: 207-217.

Ohnuma T, Augood SJ, Arai H, McKenna PJ, Emson PC. Measurement of GABAergic parameters in the prefrontal cortex in schizophrenia: Focus on GABA content, $GABA_A$ receptor α-1 subunit messenger RNA and human GABA transporter-1 (HGAT-1) subunit messenger RNA expression. Neuroscience 1999;93: 441.

Ohnuma T, Tessler S, Arai H, Faull RLM, McKenna PJ, Emson PC. Gene expression of metabotropic glutamate receptor 5 and excitatory amino acid transporter 2 in the schizophrenic hippocampus. Mol Brain Res 2000a;85: 24-31.

Ohnuma T, Kato H, Arai H, Faull RLM, McKenna PJ, Emson PC. Gene expression of PSD95 in prefrontal cortex and hippocampus in schizophrenia. Neuroreport 2000b;11: 3133-3137.

Oliver KR, Wainwright A, Heavens RP, Sirinathsinghji DJS. Retrieval of cellular mRNA in paraffin-embedded human brain using hydrated autoclaving. J Neurosci Methods 1997;77: 169-174.

Olson L. Postmortem fluorescence histochemistry of monoamine neuron systems in the human brain: A new approach in the search for a neuropathology of schizophrenia. J Psychiatr Res 1974;11: 199-203.

Ong WY, Garey LJ. Ultrastructural features of biopsied temporopolar cortex (area 38) in a case of schizophrenia. Schizophr Res 1993;10: 15-27.

Opeskin K, Dean B, Pavey G, Hill C, Keks N, Copolov D. Neither protein kinase C nor adenylate cyclase are altered in the striatum from subjects with schizophrenia. Schizophr Res 1996;22: 159-164.

Pakkenberg B. Pronounced reduction of total neuron number in mediodorsal thalamic nucleus and nucleus accumbens in schizophrenics. Arch Gen Psychiatry 1990;47: 1023-1028.

Palacios JM, Mengod G. Visualization of neurotransmitter receptors and their mRNAs in the human brain. Arzneimittelforschung 1992;42: 189-195.

Perrett CW, Marchbanks RM, Whatley SA. Characterisation of messenger RNA extracted postmortem from the brains of schizophrenic, depressed and control subjects. J Neurol Neurosurg Psychiatry 1988;51: 325-331.

Perry EK, Blessed G, Perry RH, Tomlinson BE. Brain biochemistry in schizophrenia. Lancet 1978;i: 35-36.

Perry EK, Perry RH, Tomlinson BE. The influence of agonal status on some neurochemical activities of postmortem human brain tissue. Neurosci Lett 1982;29: 303-308.

Pierri JN, Chaudry AS, Woo T-UW, Lewis DA. Alterations in chandelier neuron axon terminals in the prefrontal cortex of schizophrenic subjects. Am J Psychiatry 1999;156: 1709-1719.

Pierri JN, Volk CL, Auh S, Sampson A, Lewis DA. Decreased somal size of deep layer 3 pyramidal neurons in the prefrontal cortex of subjects with schizophrenia. Arch Gen Psychiatry 2001;58: 466-473.

Piggott MA, Perry EK, Sahgal A, Perry RH. Examination of parameters influencing [^{3}H] MK-801 binding in postmortem human cortex. J Neurochem 1992:58: 1001-1008.

Plum, F. 1972. Prospects for research on schizophrenia. III. Neurophysiology. Neuropathological findings. Neurosci. Res. Prog. Bull. 10: 384-388.

Porter RHP, Eastwood SL, Harrison PJ. Distribution of kainate receptor subunit mRNAs in human hippocampus, neocortex and cerebellum, and bilateral reduction of hippocampal GluR6 and KA2 transcripts in schizophrenia. Brain Res 1997;751: 217-231.

Pralong D, Tomaskovic-Crook E, Opeskin K, Copolov D, Dean B. $Serotonin_{2A}$ receptors are reduced in the *planum temporale* from subjects with schizophrenia. Schizophr Res 2000;44: 35-45.

Quester R, Schröder R. The shrinkage of the human brain stem during formalin fixation and embedding in paraffin. J Neurosci Methods 1997;75: 81-89.

Ragsdale DS, Miledi R. Expressional potency of mRNAs encoding receptors and voltage-activated channels in the postmortem rat brain. Proc Natl Acad Sci USA 1991;88: 1854-1858.

Ravid R, VanZweiten EJ, Swaab DF. Brain banking and the human hypothalamus—factors to match for, pitfalls and potentials. Prog Brain Res 1992;93: 83-95.

Riederer P, Gsell W, Calza L, Franzek E, Jungkunz G, Jellinger K, Reynolds GP, Crow T, Cruz-Sànchez FF, Beckmann H. Consensus on minimal criteria of clinical and neuropathological diagnosis of schizophrenia and affective disorders for post mortem research. Report from the European Dementia and Schizophrenia Network (BIOMED I). J Neural Transm [Gen Sect] 1995;102: 255-264.

Ross BN, Knowler JT, McCulloch J. On the stability of messenger RNA and ribosomal RNA in the brains of control human subjects and patients with Alzheimer's disease. J Neurochem 1992;58: 1810-1819.

Roth LJ, Diab IM, Watanabe M, Dinerstein RJ. A correlative radioautographic, fluorescent, and histochemical technique for cytopharmacology. Mol Pharmacol 1974;10: 986-998.

Roth M. Interaction of genetic and environmental factors in the causation of schizophrenia. In: Richter D (ed). Schizophrenia: Somatic Aspects. MacMillan, New York, NY, 1957; pp 15-31.

Royston MC, Slater P, Simpson MDC, Deakin JFW. Analysis of laminar distribution of kappa opiate receptor in human cortex: Comparison between schizophrenia and normal. J Neurosci Methods 1991;36: 145-153.

Saykin AJ, Gur RC, Gur RE, Mozley PD, Mozley LH, Resnick SM, Kester B, Stafiniak P. Neuropsychological function in schizophrenia. Arch Gen Psychiatry 1991;48: 618-624.

Schalling M, Friberg K, Bird E, Goldstein M, Schiffmann SN, Mailleux P, Vanderhaeghen J-J, Hökfelt T. Presence of cholecystokinin mRNA in dopamine cells in the ventral mesencephalon of a human with schizophrenia. Acta Physiol Scand 1989;137: 467- 468.

Schramm M, Falkai P, Tepest R, Schneider-Axmann T, Przkora R, Waha A, Pietsch T, Bonte W, Bayer TA. Stability of RNA transcripts in postmortem psychiatric brains. J Neural Transm 1999;106: 329-335.

Seeman P, Lee T, Chau-Wong M, Wong K. Antipsychotic drug doses and neuroleptic/dopamine receptors. Nature 1976;261: 717-719.

Selemon LD, Rajkowska G, Goldman-Rakic PS. Abnormally high neuronal density in the schizophrenic cortex. A morphometric analysis of prefrontal area 9 and occipital area 17. Arch Gen Psychiatry 1995;52: 805-818.

Shi S-R, Key ME, Kalra KL. Antigen retrieval in formalin-fixed, paraffin-embedded tissues: An enhancement method for immunohistochemical staining based on micro-wave oven heating of tissue sections. J Histochem Cytochem 1991;39: 741-748.

Simone NL, Remaley AT, Charboneau L, Petricoin EF III, Glickman JW, Emmert-Buck MR, Fleisher TA, Liotta LA. Sensitive immunoassay of tissue cell proteins procured by laser capture microdissection. Am J Pathol 2000a;156: 445-452.

Simone NL, Paweletz CP, Charboneau L, Petricoin EF III, Liotta LA. Laser capture microdissection: Beyond functional genomics to proteomics. Molecular Diagnosis 2000b;5: 301-307.

Simpson MDC, Lubman DI, Slater P, Deakin JFW. Autoradiography with [^{3}H]8-OH-DPAT reveals increases in 5-HT$_{1A}$ receptors in ventral prefrontal cortex in schizophrenia. Biol Psychiatry 1996;39: 919-928.

Singer RH, Lawrence JB, Silva F, Langevin GL, Pomeroy M, Billings-Gagliardi S. Strategies of ultrastructural visualization of biotinated probes hybridized to messenger RNA in situ. Curr Top Microbiol Immunol 1989;143: 55-55-69.

Smiley JF, Williams SM, Szigeti K, Goldman-Rakic PS. Light and electron microscopic characterization of dopamine-immunoreactive axons in human cerebral cortex. J Comp Neurol 1992;321: 325-335.

Snyder SH, Ferris CD. Novel neurotransmitters and their neuropsychiatric relevance. Am J Psychiatry 2000;157: 1738-1751.

Southard EE. On the topographical distribution of cortex lesions and anomalies in dementia praecox, with some account of their functional significance. Am J Insanity 1914;71: 383-403.

Stevens JR. An anatomy of schizophrenia? Arch Gen Psychiat 29:177, 1973.

Stevens JR. Enough of pooled averages: Been there, done that. Biol Psychiat 1997;41: 633-635.

Steward O. mRNA localization in neurons: A multipurpose mechanism? Neuron 1997;18: 9-12.

Sulzer D, Joyce MP, Lin L, Geldwert D, Haber SN, Hattori T, Rayport S. Dopamine neurons make glutamatergic synapses *in vitro*. J Neurosci 1998;18: 4588-4602.

Sun Y, Zhang L, Johnston NL, Torrey EF, Yolken RH. Serial analysis of gene expression in the frontal cortex of patients with bipolar disorder. Br J Psychiatry 2001;178: S137-S141.

Tallerico T, Novak G, Liu ISC, Ulpian C, Seeman P. Schizophrenia: Elevated mRNA for dopamine $D2_{Longer}$ receptors in frontal cortex. Mol Brain Res 2001;87: 160-165.

Tang SW, Helmeste D, Fang H, Li M , Vu R, Bunney W Jr, Potkin S, Jones EG. Differential labeling of dopamine and sigma sites by [^{3}H]nemonapride and [^{3}H]raclopride in postmortem human brains. Brain Res 1997;765: 7-12.

Tecott LH, Eberwine JH, Barchas JD, Valentino KL. Methodological considerations in the utilization of *in situ* hybridization. In: Eberwine JH, Valentino KL, Barchas JD (eds). In Situ Hybridization In Neurobiology: Advances in Methodology. Oxford Univ Press, New York, NY, 1994; pp 3-23.

Torrey EF, Peterson MR. Schizophrenia and the limbic system. Lancet 1974;ii: 942-946.

Torrey EF, Webster M, Knable M, Johnston N, Yolken RH. The Stanley Foundation brain collection and neuropathology consortium. Schizophr Res 2000;44: 151- 155.

Tsuang M. Schizophrenia: Genes and environment. Biol Psychiatry 2000;47: 210-220.

Uhl GR. *In situ* hybridization: Quantitation using radiolabeled hybridization probes. Methods Enzymol 1989;168: 741-752.

Van der Loos CM, Volkers HH, Rook R, Van den Berg FM, Houthoff H-J. Simultaneous application of in situ DNA hybridization and immunohistochemistry on one tissue section. Histochem J 1989;21: 279-284.

Virgo L, Humphries C, Mortimer A, Barnes T, Hirsch, S, de Belleroche J. Cholecystokinin messenger RNA deficit in frontal and temporal cerebral cortex in schizophrenia. Biol Psychiatry 1995;37: 694-701.

Volk DW, Austin MC, Pierri JN, Sampson AR, Lewis DA. Decreased glutamic acid $decarboxylase_{67}$ messenger RNA expression in a subset of prefrontal cortical γ-aminobutyric acid neurons in subjects with schizophrenia. Arch Gen Psychiatry 2000;57: 237-245.

Volk DW, Austin MC, Pierri JN, Sampson AR, Lewis DA. GABA transporter-1 mRNA in the prefrontal cortex in schizophrenia: Decreased expression in a subset of neurons. Am J Psychiatry 2001;158: 256-265.

Vonsattel JPG, Aizawa H, Ge P, DiFiglia M, McKee AC, MacDonald M, Gusella JF. Landwehrmeyer GB, Bird ED, Richardson EP Jr, Hedley-Whyte ET. An improved approach to prepare human brains for research. J Neuropath Exp Neurol 1995;54: 42-56.

Wagman AMI. Report of a workshop on issues in brain tissue acquisition. Schizophr Bull 1992;18: 149-153.

Weinberger DR. Anteromedial temporal-prefrontal connectivity: A functional neuro-anatomical system implicated in schizophrenia. In: Carroll BJ, Barrett JE, (eds). Psychopathology and the Brain. Raven Press, New York, NY, 1991; pp 25-43.

Weinberger DR. Cell biology of the hippocampal formation in schizophrenia. 1999;45: 395-402.
West MJ. New stereological methods for counting neurons. Neurobiol Aging 1993;14: 275-285.
Whitehouse PJ. Receptor autoradiography: Applications in neuropathology. Trends Neurosci 1985;8: 434-437.
Wiesner RJ, Zak R. Quantitative approaches for studying gene expression. Am J Physiol 1991;260: L179-L188.
Willner P. The dopamine hypothesis of schizophrenia: Current status, future prospects. Int Clin Psychopharmacol 1997;12: 297-308.
Witter MP, Amaral DG. Entorhinal cortex of the monkey: V. Projections to the dentate gyrus, hippocampus, and subicular complex. J Comp Neurol 1991;307: 437-459.
Wolf SS, Hyde TM, Saunders RC, Herman MM, Weinberger DR, Kleinman JE. Autoradiographic characterization of neurotensin receptors in the entorhinal cortex of schizophrenic patients and control subjects. J Neural Transm [Gen Sect] 1995;102: 55-65.
Woo T-U, Miller JD, Lewis DA. Schizophrenia and the parvalbumin-containing class of cortical local circuit neurons. Am J Psychiatry 1997;154: 1013-1015.
Woo T-U, Whitehead RE, Melchitzky DS, Lewis DA. A subclass of prefrontal γ-aminobutyric acid axon terminals are selectively altered in schizophrenia. Proc Natl Acad Sci USA 1998;95: 5341-5346.
Wyatt RJ, Erdelyi E, Schwartz M, Herman M, Barchas JD. Difficulties in comparing catecholamine-related enzymes from the brains of schizophrenics and controls. Biol Psychiatry 1978;13: 317-334.
Yee F, Yolken RH. Identification of differentially expressed RNA transcripts in neuropsychiatric disorders. Biol Psychiatry 1997;41: 759-761.
Young WS III. In situ hybridization histochemical detection of neuropeptide mRNA using DNA and RNA probes. Methods Enzymol 1989a:168: 702-710.
Young WS III. Simultaneous use of digoxigenin- and radiolabeled probes for hybridization histochemistry. Neuropeptides 1989b;13: 271-275.
Young WS III. In situ hybridization histochemistry. In: Björklund A, Hökfelt T, Wouterlood F, van den Pol AN. (eds). Handbook of Chemical Neuroanatomy. Vol 8: Methods for the Analysis of Neuronal Microcircuits and Synaptic Interactions. 1990, Elsevier Science Pub BV, New York, NY; pp 481-511.
Young WS III. *In situ* hybridization with oligodeoxyribonucleotide probes. In: Wilkinson DG (ed). In Situ Hybridization: A Practical Approach. Oxford Univ Press, New York, NY, 1992; pp. 33-44.
Young WS III, Kuhar MJ. Opiate receptor autoradiography: In vitro labeling of tissue slices. In: van Ree JM, Terenius L (eds). Characteristics and Function of Opioids. Elsevier/North-Holland Biomedical Press, Amsterdam; pp 451-452.
Young WS III, Kuhar MJ. Autoradiographic localisation of benzodiazepine receptors in the brains of humans and animals. Nature 1979;280: 393-395.
Zachrisson O, de Belleroche J, Wendt KR, Hirsch S, Lindefors N. Cholecystokinin CCK_B receptor mRNA isoforms: expression in schizophrenic brains. Neuroreport 1999;10: 3265-3268.
Zaidel DW, Esiri MM, Harrison PJ. Size, shape, and orientation of neurons in the left and right hippocampus: Investigation of normal asymmetries and alterations in schizophrenia. Am J Psychiatry 1997a;154: 812-818.
Zaidel DW, Esiri MM, Harrison PJ. The hippocampus in schizophrenia: Lateralized increase in neuronal density and altered cytoarchitectural asymmetry. Psychol Med 1997b;27: 703-713.

11 DEFINING THE ROLE OF SPECIFIC LIMBIC CIRCUITRY IN THE PATHOPHYSIOLOGY OF SCHIZOPHRENIA AND BIPOLAR DISORDER

Francine M. Benes and Sabina Berretta

Abstract

A core component of all corticolimbic circuitry is the GABAergic interneuron. Recent postmortem studies have provided consistent evidence that a defect of GABAergic neurotransmission probably plays a role in both schizophrenia and bipolar disorder. Based on the regional and subregional distribution of changes in GABA cells in layer II of the anterior cingulate regions and sectors CA3 and CA2 of the hippocampal formation, it has been postulated that the basolateral nucleus of the amygdala (BLa), a region that projects preferentially to both of these latter sites, may contribute to these abnormalities by sending an increased flow of excitatory activity. In support of this hypothesis, a "partial" rodent model in which the $GABA_A$ antagonist picrotoxin is injected into the BLa has demonstrated changes in the GABA system that are remarkably similar to those seen in schizophrenia and bipolar disorder. In the years to come, the combined use of studies in rodent, primate and human brain will be useful in identifying how specific phenotypic changes in subclasses of interneurons may have been induced in schizophrenia and bipolar disorder. Such information will undoubtedly provide important new insights into how the integration of GABAergic interneurons with other intrinsic and extrinsic transmitter systems may be altered in neuropsychiatric disease.

Introduction

Schizophrenia presents a special challenge to the field of neuroscience because there are no readily identifiable histopathologic changes that can be used to make a reliable diagnosis (Benes, 1988). Over the past 20 years, however, research efforts have been directed at defining the possible ways in which the circuitry of the limbic system may be altered in patients with this disorder. As a result of intensive efforts using quantitative cytochemical approaches, it now appears likely that schizophrenia probably involves a abnormalities of the glutamate (Tamminga, 1998) and GABA (Benes and Berretta, 2001) systems in key regions of schizophrenic brain. These findings have emphasized the limitations of postmortem research as it is virtually impossible to evaluate the function of intact circuitry after death in human brain. Eventually, brain imaging technology may make it possible to "view" functional changes in specific neural circuits that are adversely affected in the pathophysiology of schizophrenia. In the meantime, however, it seems relevant to consider alternative strategies that might be feasible for current investigations. One approach that offers the hope of more immediate insights into the neurobiology of schizophrenia is the use of carefully-designed animal models that can reproduce some of the functional changes in neural circuits inferred from postmortem findings. Although this so-called 'partial' model approach will not likely produce rats or monkeys that exhibit the clinical syndrome seen in schizophrenia, it can nevertheless help to uncover small pieces of a larger puzzle and in the long run help to build a foundation for a more comprehensive understanding of psychotic disorders.

In the discussion that follows, there will be a review of postmortem findings in the anterior cingulate cortex and hippocampal formation. As the reader will see, this cumulative information has pointed to an abnormality of a specific projection from the amygdala to discrete aspects of these two target sites. Together, the components of this triangulated circuitry comprise the limbic lobe, a phylogenetically older aspect of the cerebral hemispheres that plays a central role in emotion and learning. This circuitry has been used to develop a 'partial' rat model for how interactions of the amygdalo-hippocampal circuitry may be altered in schizophrenia.

Is There Evidence for Neuronal Loss in Limbic Lobe of Schizophrenics?

Over the past 20 years, brain imaging and microscopic studies from several different laboratories have suggested that there may be a reduction in the volume of the parahippocampal gyrus, entorhinal region, hippocampus and amygdala in schizophrenia (for a review see Arnold, 2000). In the hippocampus, not all studies have shown atrophy (Heckers et al., 1990) or neuronal loss (Heckers et al., 1991a), but it seems likely that this finding may occur to a variable degree in different patient populations (Benes, 1997a) It is important to emphasize that when volume loss is present, many different processes, some reversible and others irreversible in nature, could be responsible. On the other hand, when volume loss is not present, this does not preclude the possibility that important histopathologic changes may be present at the microscopic level.

A reduction in neuronal number has been detected in the anterior cingulate cortex (Benes et al., 1986; 1991c) and hippocampus (Benes et al., 1998) of schizophrenics. Other regions also show evidence of neuronal loss and have been reviewed elsewhere (Benes, 1997a). There is mounting evidence that a loss of nonpyramidal neurons may occur in both schizophrenia and affective disorder (Benes et al., 1991c, 1998; Benes, 1998). Not all cell counting studies of schizophrenia, however, have revealed a decrease in the density of neurons (Arnold et al., 1995) (Benes et al., 1991a; Heckers et al., 1991b; Pakkenberg, 1993; Selemon et al., 1995). A variety of factors, including patient samples, tissue processing, regions studied and cell counting methodology could all help to explain the variability of these findings from one study to another (Benes, 1997a). It seems likely, however, that neuronal loss is neither necessary nor sufficient to explain the occurrence of schizophrenia. When present, however, it seems likely that it would play a contributory role in the pathophysiology of this disorder.

A reduction in the density or total numbers of neurons in a brain region is not sufficient to prove that a degenerative process has occurred. It is generally accepted in the field of neuropathology that "typical" neuronal degeneration, like that seen in Huntington's chorea or Alzheimer's disease, is accompanied by a "gliotic reaction." The latter consists of an increase in the number of non-neuronal elements (i.e. glial cells) that are involved in the removal of cellular debris accrued from dying neurons. It is noteworthy that several cell counting studies have systematically evaluated whether the density of glial cells is increased in various regions of schizophrenic brain and have found no evidence for a gliotic reaction (Benes et al., 1986; 1991c; Falkai et al., 1988; Roberts et al., 1986; Selemon et al., 1995). The absence of gliosis in schizophrenia provides a compelling argument that ***necrotic*** cell death (Farber, 1994) does not play a role in the pathophysiology of this disorder. The latter conclusion does not, however, exclude the possibility that an ***apoptotic*** (Farber, 1994) form of neuronal loss may be occurring in this disorder and even play a protective role (Bursch et al., 1992).

Are there any clues that could shed some light on whether apoptosis may play a role in the pathophysiology of schizophrenia? Closer scrutiny of the findings reported in cell counting studies of schizophrenic brain reveals a potentially interesting pattern. In both the anterior cingulate and prefrontal cortex (Benes et al., 1991a), as well as the hippocampal formation (Benes et al., 1998), a preferential decrease in the density of nonpyramidal cells has been reported. This latter neuronal subtype is considered to be a population of local circuit cells that are largely GABAergic in nature (Fairen et al., 1982). Although this change shows a very discrete localization in only some regions, a decrease of inhibitory modulation in a key circuit could nevertheless have a significant impact on information processing (see below).

Evidence for Glutamatergic Dysfunction in Schizophrenia

Early attempts at modelling for postmortem results had suggested that there may be alterations in the arrangements of neuronal clusters in layer II of the anterior cingulate cortex (Benes and Bird, 1987). To test the hypothesis that there might be an increase of vertical axons passing between adjacent clusters in this lamina, an immunocytochemical localization of the phosphorylated epitopes of the 200K neurofilament protein (NFP200K) was used to visualize such fibers in human postmortem cortex. The results demonstrated a 25% higher density of vertical axons in anterior cingulate cortex of schizophrenic subjects when compared to normal controls (Benes et al., 1987); horizontal axons did not show this change. In a subsequent replication study, this same pattern was again found in cingulate cortex, but not the prefrontal area of schizophrenics (Benes, 1993). In yet another replication study, this finding was again observed using an entirely different method in which antibodies against glutamate were employed. In this case, the effect size was much larger and a robust increase (78%) in the density of glutamate-IR vertical axons in layer II and upper portions of layer IIIa was found (Benes et al., 1992a). This latter finding provided further support for the idea that these vertical fibers might be incoming excitatory afferents from other cortical or perhaps even subcortical regions.

This latter increase in the number of glutamatergic inputs to layer II of the anterior cingulate cortex could potentially play a role in the induction of excitotoxicity, particularly if the flow of activity along these axons were also increased (see below). Consistent with this idea, a recent PET scanning study has reported an increase of metabolic activity in relation to auditory hallucinations that occurred in the cingulate gyrus of schizophrenic subjects (Sik et al., 1995). If there is an increase of glutamatergic tone in the cingulate gyrus in this disorder, then a secondary increase would also be expected to occur in the hippocampal formation because this region receives a substantial descending flow of afferent activity via the cingulum bundle. Is there any evidence for an actual increase of excitatory activity in the hippocampal formation of schizophrenics? As with the cingulate cortex, another recent PET scanning study has also demonstrated an elevation in the baseline metabolism of this latter region (Heckers et al., 1998a). Accordingly, both components of the limbic lobe show increased activity in schizophrenia.

Do postmortem studies of the hippocampus also provide corroborative data in support of this latter concept? Neurochemical investigations have suggested that nonNMDA-sensitive glutamate receptors might be decreased in the hippocampal formation of schizophrenics (Harrison et al., 1991; Kerwin et al., 1988; Kerwin et al., 1990). Further support for this idea has come from a recent histochemical study of the

$GluR_{5,6,7}$ subunit of the kainate receptor. A preferential localization of this protein to the apical dendritic shafts of pyramidal cells, particularly in sectors CA3 and CA2 was reported (Good, 1993). Subsequently a blind, quantitative analysis revealed that there was a decrease of $GluR_{5,6,7}$-IR dendrites in sectors CA3 and CA2 in schizophrenics when compared to a group of age and PMI-matched normal controls (Benes et al., 2001). This difference in the schizophrenic group was seen in the stratum radiatum and stratum moleculare, but not the stratum oriens. Distal portions of sector CA1, i.e. those areas closer to the prosubiculum, did not show any difference in the schizophrenics. A matched group of manic depressives did not show a significant change in any of the sectors or laminae examined, whether or not they were treated with neuroleptic drugs. Taken together, these findings are consistent with the idea that the expression of the kainate-sensitive glutamate receptor may be reduced in schizophrenia. Is such a decrease a sign of a primary dysregulation of this receptor protein or possibly to a compensatory down-regulation? The fact that there was little change in the stratum oriens where the basal dendritic branches of pyramidal neurons are found suggests that this change is not distributed uniformly throughout the intracellular space of projections cells. Since the stratum radiatum and stratum moleculare, where the $GluR_{5,6,7}$ subunit expression is decreased, are known to contain both intrinsic and extrinsic glutamatergic inputs, respectively, to sectors CA3, 2 and 1 (Rosene and Van Hoesen, 1987), an increased flow of excitatory activity through these two laminae could be occurring in schizophrenic hippocampus and causing a compensatory down-regulation (see discussion on macrocircuitry below).

Evidence for GABAergic Dysfunction

The observation that nonpyramidal neurons were decreased in density in both the anterior cingulate cortex (Benes et al., 1991c) and hippocampal formation (Benes et al., 1998) has suggested the possibility that these cells might be associated with a decrease of GABAergic function in these regions. Neurochemical evidence in support of this idea has come from studies showing an increase of [^{3}H]muscimol binding (Hanada et al., 1987), a decrease of glutamate decarboxylase (Bird et al., 1979) and a reduction in GABA uptake in the cortex (Simpson et al., 1989) and hippocampus (Reynolds et al., 1990a) of the schizophrenic brain. More recently, an immunocytochemical localization of the calcium binding protein parvalbumin has demonstrated a reduction of cells showing this immunoreactivity in the cingulate cortex of schizophrenics (Beasley and Reynolds, 1997). Since this peptide immunoreactivity appears to be preferentially associated with interneurons believed to be inhibitory basket cells (Conde et al., 1994), these latter results support the hypothesis that GABA cell loss or dysfunction could be a feature of schizophrenia The alternative possibility that the amount of parvalbumin present within a normal number of cells may have simply fallen below the level of detection cannot as yet be excluded. Other studies described elsewhere in this volume have used a similar approach to study neurons with immunoreactivity for calbindin and calretinin in the dorsolateral prefrontal cortex (Davis and Lewis, 1995). There are many different subtypes of nonpyramidal neuron that use GABA as a neurotransmitter and each of these could theoretically show a unique pattern of change in schizophrenic cortex.

If the nonpyramidal cells that are reduced in density in the cingulate cortex (Benes et al., 1991c) and hippocampus (Benes et al., 1998) were GABAergic in nature, a compensatory upregulation of the $GABA_A$ receptor on postsynaptic pyramidal neurons would be expected to occur (Benes et al., 1989). Using a high resolution autoradiographic technique to localize receptor binding on individual neuronal cell bodies, bicuculline-sensitive [^{3}H] muscimol binding was found to be increased by 84% in layer II and 74% in layer III of the cingulate cortex (Benes et al., 1992b). A similar pattern was observed in the dorsolateral prefrontal cortex where the increase was principally noted on pyramidal neurons (Benes et al., 1996a). In the hippocampal formation (Benes, 1996), an increase of this receptor has also been noted, but in this case, the increase was found preferentially in sectors CA4,3 and 2, but not CA1 in schizophrenia. In sectors CA3 and CA2, the most striking increase of the $GABA_A$ receptor binding activity was in the stratum oriens, where potent inhibitory modulation of pyramidal cells is exerted.

How could interneurons be adversely affected in both SZ and manic depression? Apoptosis is a cellular mechanism associated with neuronal death primarily during normal ontogeny (Binder C, 1994; Bursch et al., 1992; Farber, 1994), but oxidative stress is one of its variants that can occur with excessive glutamatergic activity (Coyle and Puttfarcken, 1993). Glucocorticoid hormones, together with increased glutamatergic activity, can result in excitotoxic cell death (Sapolsky, 1992). It is well established that all forms of stress result in an increased release of glucocorticoids, steroid hormones that bind to receptors abundantly present throughout the hippocampus and cortex (Sapolsky, 1992). Of particular relevance to the current discussion is the fact that [^{3}H]corticosterone shows the most extensive nuclear binding in sector CA2 (McEwen, 1982; Stumpf, 1989; where nonpyramidal cells seem to be selectively decreased in schizophrenia (Benes et al., 1998b). Accordingly, increased levels of glucocorticoid hormone associated with the stress response could potentially exert a potent toxic effect on neurons in this sector, particularly if they were simultaneously receiving heightened glutamatergic activity. It is relevant to ask then, whether a high degree of sensitivity to circulating glucocorticoid hormone could potentially predispose sector CA2 to a selective decrease of nonpyramidal cells in schizophrenic and manic depressive subjects exposed to stress (see below). If so, a disturbance of the hippocampal glutamate system acting synergistically to promote oxidative stress would be required for such a change to occur.

Glucocorticoid hormones also have the ability to bind directly to the $GABA_A$ receptor (Sutanto et al., 1989) and increase its activity (Lambert et al., 1987; Majewska et al., 1985). Is it possible that the upregulation of this receptor noted in sectors CA3 and CA2 of schizophrenics (Benes et al., 1996b; 1997a; Benes et al., 1997b) might be related, perhaps in part, to increased glucocorticoid release. This is a potentially important possibility to consider because these hormones are released into the bloodstream in abundant amounts in association with the acute stress response, although they are thought to increase GABAergic inhibitory activity, rather than decrease it (Feldman, 1968; Miller, 1978; Pfaff, 1971). It is theoretically possible that the increase of $GABA_A$ receptor binding noted in the cortex and hippocampus of schizophrenics may not, as previously suggested, be a compensatory change in response to decreased GABAergic activity; but rather, it may represent the sequela of unremitting stress in individuals with chronic psychosis, rather than a compensatory change. It is noteworthy that glucocorticoids also cause an ***uncoupled*** increase of $GABA_A$ binding, without any change in the benzodiazepine site (for a review see Rabow, 1995); a similar uncoupling in the regulation of this receptor complex has been observed in the hippocampus of schizophrenics (Benes et al., 1997b). Since glucocorticoids potentiate excitotoxicity (Sapolsky, 1992) and GABAergic cells in the hippocampus (Pollard, 1994; Zhang et al., 1990) may be susceptible to such injury mediated through kainate receptors (see section on the Glutamate System above), the stress response offers a plausible way of explaining not only the reduced density of nonpyramidal neurons, but also the increased $GABA_A$ receptor binding activity in schizophrenia. The fact that kainate-sensitive glutamate receptors are especially abundant in sectors CA3 and CA2 (Monaghan and Cotman, 1985) and appear to be reduced in the hippocampus of schizophrenics (Kerwin et al., 1988; Kerwin et al., 1990; Harrison et al., 1991; Benes et al., 2001b) gives further credence to this idea.

Another strategy that has been used to study GABAergic cells in the limbic lobe of schizophrenic brain is immunocytochemistry. The 65 kDalton isoform of glutamate decarboxylase (GAD_{65}) is prefentially localized to terminals and provides a potentially sensitive and reliable means of studying the distribution of GABAergic terminals (Kaufman et al., 1991). Using a technique that was adapted to human postmortem tissues, no overall difference in the density of GAD_{65}-immunoreactive terminals was observed, however, in the hippocampus of schizophrenic subjects (Todtenkopf and Benes, 1998a). There was, however, a striking positive correlation between the dose of neuroleptic medication and the density of GABA terminals in sectors CA3 and CA2, particularly in the stratum oriens where the $GABA_A$ receptor showed the most marked increase (Benes et al., 1996b). Antipsychotic medication could theoretically be capable of inducing a trophic sprouting of GABAergic terminals in the hippocampal subfields (see below). The fact that the stratum oriens also showed the largest increase of

$GABA_A$ binding activity in schizophrenics (Benes et al., 1996b) is consistent with the hypothesis that there is a compensatory upregulation of this receptor in schizophrenia. This lamina may, therefore, be a preferential site not only for GABAergic dysfunction in schizophrenia, but also the therapeutic influence of neuroleptic drugs. Networks of GABAergic interneurons within the stratum oriens have recently been shown to play a central role in the generation of theta rhythms and in the coupling of these oscillations with pyramidal cell activity (Csicsvari et al., 1999). Thus, an appreciable defect of GABAergic function in the stratum oriens could potentially contribute significantly a disturbance of electrical activity generated in the hippocampal formation in schizophrenia (see section below).

A Potential Role for Limbic Lobe Interconnections in Schizophrenia

In considering whether there are alterations at this level in schizophrenic brain, it is useful to evaluate structural parameters that may be inferred to represent extrinsic fibers originating in one region and projecting to another (Benes, 1997a). Because the issues for various regions are somewhat different, the discussion in this section is considered separately for the anterior cingulate cortex and the hippocampal formation.

Inferences from Studies of Anterior Cingulate Cortex

As discussed above in the section on Glutamate, three postmortem studies have provided evidence for an increase of excitatory afferents coursing toward the superficial layers of the anterior cingulate cortex may be increased in schizophrenics. Such axons could originate in other cortical regions (for a review see Eccles, 1984), the amygdala (Vogt and Pandya, 1987) or the thalamus (Vogt et al., 1987), since all travel to the superficial layers of this region. In addition, dopaminergic, serotonergic, noradrenergic and cholinergic fibers from the ventral tegmental area (Berger et al., 1974) (Thierry et al., 1973) (Lindvall and Bjorklund, 1984), raphe nuclei (Descarries et al., 1975; Lindvall and Bjorklund, 1984), locus coeruleus (Lindvall and Bjorklund, 1984) (Lindvall and Bjorklund, 1984) and nucleus basalis of Meynert (Mesulam et al., 1983), respectively, project as well to the upper layers of the cingulate cortex.

Are there indirect ways of making inferences regarding the site from which the vertical fibers may have originated? As noted in the discussion above, the fibers visualized were found to be increased in the anterior cingulate but not the dorsolateral prefrontal region (Benes, 1993; Benes et al., 1987; Benes et al., 1992a). A reasonable question to ask is whether the superficial layers of the cingulate region might receive a projection that is not present in the dorsolateral prefrontal area. The basolateral complex of the amygdala presents an intriguing possibility because layer II of the anterior cingulate cortex receives a "massive" projection from this nucleus, while the prefrontal area receives only a sparse innervation (Amaral et al., 1992; Van Hoesen et al., 1993). Since the basolateral complex also sends a substantial projection to the entorhinal cortex, it seemed relevant to ask whether this latter region might also show an increase of glutamate-IR vertical axons in superficial laminae of schizophrenics. A subsequent study demonstrated a 38% increase of glutamate-IR vertical axons in the schizophrenic group (Longson et al., 1996). Taking together these various findings, it seems plausible that there may be an increase of amygdala projections to superficial layers of anterior cingulated cortex (ACCx) and entorhinal cortex (ERCx) in SZ brain. In support of this, there are cell bodies in the basolateral nucleus of the amygdala that show intense glutamate-IR similar to that seen in the vertical axons described above; these latter somata are believed to be cortical projection neurons (McDonald et al., 1989) . Thus, the studies of vertical axons in ACCx and ERCx have suggested the possibility that there may be an increased number of glutamatergic afferents projecting to these regions from the basolateral nuclear complex.

Inferences from Studies of the Hippocampal Formation

For the hippocampus, similar issues can be raised for the decrease of the $GluR_{5,6,7}$ subunit of the kainate receptor (see discussion above in the section on the Glutamate System). Although it is not possible to visualize an actual change in the conduction of impulses along axons in postmortem brain, a decrease of the $GluR_{5,6,7}$ subunit could represent a functional down-regulation secondary to increased glutamatergic activity (Benes et al., 2001b). This change was observed in the stratum radiatum and stratum moleculare, but not the stratum oriens. This latter pattern could occur if there were an increased flow of glutamatergic activity along both intrinsic and extrinsic fibers, respectively, that course through these two laminae. In the stratum moleculare, these fibers would include afferents from the entorhinal cortex and the basolateral nucleus of the amygdala. In the stratum radiatum, on the other hand, they would include projections, such as mossy fibers and Schaffer collaterals that are intrinsic to the hippocampal formation [for an authoritative review, see citation (Rosene and Van Hoesen, 1987).

A Potential Role of Amygdalo-Hippocampal Connectivity in Schizophrenia

The amygdala is a component of the limbic system that plays a pivotal role in the integration of emotional experience and the stress response (LeDoux, 2000). As shown in Table 1, the hypothesis that this region may play a role in the pathophysiology of schizophrenia has been suggested, in part, by the fact that this region sends a massive projection to layer II of the anterior cingulate cortex where several microscopic anomalies have been observed in schizophrenia (Benes and Bird, 1987; Benes et al., 1987; Benes et al., 1991b; Benes et al., 1992a; Benes et al., 1992b; Benes et al., 1997a). Similarly, as seen in Table 2, several postmortem studies that have selectively implicated sectors CA3 and CA2 of the hippocampus have in the pathophysiology of schizophrenia (Benes et al., 1996b; 1997b; 1998; 2001b Todtenkopf and Benes, 1998a; Benes and Todtenkopf, 1999). More direct postmortem evidence for amygdalar involvement comes from the report of a decrease of high affinity GABA uptake in this region of schizophrenic brain (Reynolds et al., 1990a). It is relevant, therefore, to consider more closely the specific ways in which this region might interact with the hippocampal formation, particularly since both of these regions have been found to show volume reduction in brain imaging studies of schizophrenia (for a detailed review, see (Lawrie and Abukmeil, 1998).

Table 1: Laminar Changes in Anterior Cingulate Cortex of Schizophrenics

	I	II	III	V	VI
Nonpyramidal Som.		↓			
$GABA_A$ RBA*		↑↑	↑		
GAD_{65} Term. (CPZ+)**		↑	↑		
$GluR_{5,6,7}$-IR Neurons		↓			
TH-IR* Varicosities******		↑ NPs ↓ PNs			
Klennow + Nuclei*****	↓	↓↓	↓		
Vertical Glut-IR Axons		↑↑	↑↑		

* Receptor Binding Autoradiography

** GAD_{65}-IR terminals were increased only in neuroleptic(CPZ)-treated SZs

*** Tyrosine Hydroxylase Immunoreactivity

**** ACCx-II showed a shift of TH-IR varicosities from PNs to NPs

***** Klennow —positive nuclei = In Situ End-Labeling of Single-Stranded DNA Breaks

Table 2: Subregional Changes in Schizophrenic Hippocampus

	AD*	CA4	CA3	CA2**	CA1
Nonpyramidal N.				↓	
$GABA_A$ RBA***		↑	↑	↑	
GAD_{65} Terminals		↓	↓	↓	
$GluR_{5,6,7}$-IR			↓	↓↓	↓
TH-IR**** Varicosities				↓	
GAD67/65 mRNA			↓	↓	

* AD = Area Dentata

** CA2 has the highest level of nuclear [^{3}H]corticosterone binding + highest expression of kainate receptor protein

*** Receptor Binding Autoradiography

**** Tyrosine Hydroxylase Immunoreactivity

The basolateral subdivision of the amygdala comprises a frontotemporal system that innervates several key components of the corticolimbic system, including the hippocampus formation (Swanson and Petrovich, 1998). As shown in Figure 1 the projections of the amygdala to the hippocampus are rather complex and include the perforant pathway terminations in the stratum moleculare of the area dentata, as well as various other fibers systems that enter the CA subfields either through the stratum oriens (i.e. to CA3) or the stratum moleculare (Pitkanen et al., 2000). The direct projections of the basolateral complex to the CA subfields, together with its indirect influences exerted via the entorhinal region, constitute a compelling network to consider in relation to schizophrenia. Interestingly, blockade of the $GABA_A$ receptor results in marked changes in the regulation of emotional responses mediated by this region (Sanders and Shekhar, 1991, 1995; Shekhar et al., 1999). If, as suggested above, a dysfunction of the GABA system occurs in the amygdala in schizophrenia (see above), an increased outflow of activity from this latter region to the entorhinal region and hippocampus could contribute to the disturbances in affective experience and the heightened response to stress that are observed in this disorder.

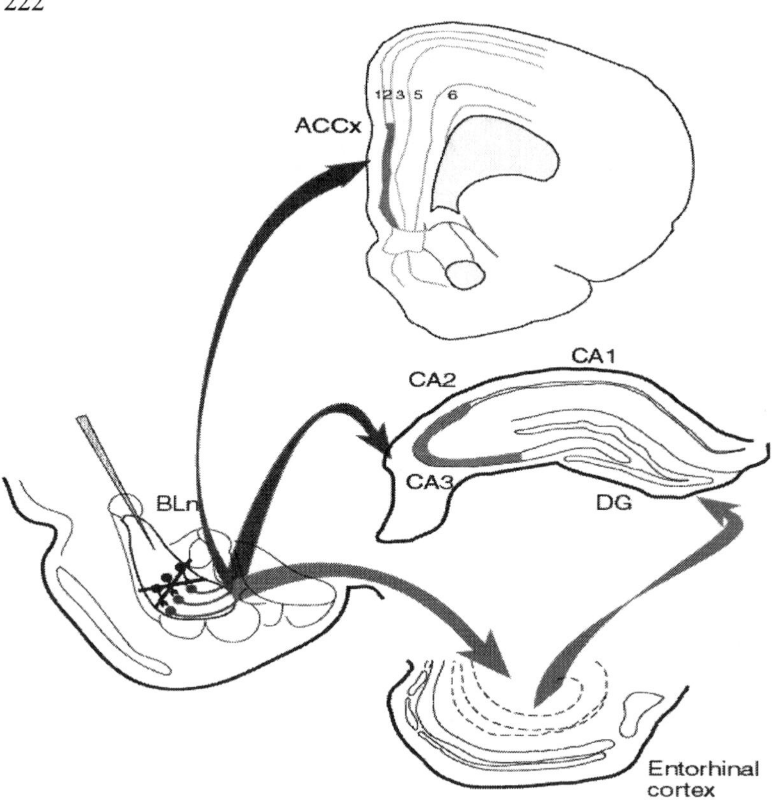

Figure 1. A schematic diagram depicting afferent projections from the basolateral nucleus of the amygdala (BLn) to the entorhinal cortex, hippocampus (HIPP) and anterior cingulate cortex (ACCx). Layer II of ACCx and sectors CA3 and CA2 of the HIPP receive a direct innervation from the BLn. In addition, the BLn also projects to the entorhinal cortex from which it exerts an indirect BLn influence on the flow of activity along the trisynaptic pathway. By delivering picrotoxin, a selective antagonist of the GABAA receptor into the BLn, an increased outflow of excitatory activity to ACCx-II and CA3 and CA2 produces changes in GABAergic interneurons that are remarkably similar in distribution to those seen in schizophrenia and bipolar disorder.

'Partial' Modeling for Postmortem Changes in the Hippocampal GABA System

Is it possible that an increased flow of excitatory activity from the basolateral nucleus of the amygdala could cause changes in the hippocampal GABA system? Bringing together all of these various observations, we have developed a 'partial' rodent model for exploring whether an increased flow of excitatory activity from the amygdala could potentially induce changes in the hippocampal GABA system. Our strategy has been to dissect out, from a complex network of corticolimbic circuitry, one potential source of abnormal afferent activity to the hippocampus and to use selective pharmacological manipulation as a way of inducing changes similar to those reported from the postmortem studies described above. Toward this end, activation of the basolateral nucleus of the amygdala (BLn) has been induced using local intraparenchymal infusion of the $GABA_A$ receptor antagonist picrotoxin in awake, freely moving rats. Changes in the density of GAD_{65}- and GAD_{67}-immunoreactive (IR) somata and terminals have been used as an index of the response of the hippocampal GABA system to amygdalar activation. To our knowledge, this is the first study in the field of schizophrenia research in which a rodent model, other than those addressing neuroleptic effects (Harrison, 1999), has been used to induce microscopic changes similar to those described in our postmortem investigations of schizophrenia.

A study using picrotoxin infusion into the basolateral complex of awake, freely moving rats has demonstrated a marked decrease in the density of GAD_{67}-IR terminals on pyramidal neurons in sectors CA3 and CA2, but not CA1 of drug-treated rats (Berretta et al., 2001). These results are consistent with the hypothesis that exctitatory activity originating in the amygdala could stimulate GABAergic terminals to release GABA (and also GAD) in sectors CA3 and CA2 (see discussion of GABA System in schizophrenia above) where fibers from the basolateral complex terminate (Benes and Berretta, 2000). Since GABA cells are particularly sensitivity to excitotoxic injury mediated through kainate receptors (Zhang et al., 1990), it will be relevant to determine whether chronic stimulation of the BLn, if sustained, can eventually lead to an excitotoxic injury in GABAergic neurons of CA2 where a loss of nonpyramidal cells was observed in schizophrenics (see above).

Direct Amygdalo-Hippocampal and Indirect Amygdalo-Entorhino-Hippocampal Effects

Our anterograde tracing studies confirm previous reports that there is a ***direct*** amygdalo-hippocampal projection (Saunders and Rosene, 1988; Pikkarainen et al., 1999). In recent experiments, injections of biocytin into the BLn were aimed at a caudal portion of the parvicellular subdivision of the BLn using the same stereotaxic coordinates employed for picrotoxin infusion. Anterogradely-labeled fibers were only found in the stratum oriens, radiatum and pyramidale of sectors $CA_{3/2}$ (see also: (Pikkarainen et al., 1999). This distribution suggests that the decrease in GAD-IR terminals in $CA_{3/2}$ reported herein might be mediated by *direct* amygdalo-hippocampal fibers (Fig. 1). Conversely, activation of the ***indirect*** amygdalo-entorhino-hippocampal (perforant) pathway might mediate the increase of GAD_{67}-IR somata found in the DG and CA_4 . Accordingly, multisynaptic effects on GABA cells in CA_3 and CA_2 induced by the *indirect* pathway through mossy fibers, cannot be excluded.

The Amygdala Induces Activation of the GABAergic System of the Hippocampus

In the current study, changes in both GAD_{67}- and GAD_{65}-IR terminals were detected on pyramidal neurons in both CA_3.and CA_2. This observation suggests that GABAergic interneurons affected by the amygdala might be those that make axo-somatic contacts with the neurons giving origin to the Schaffer collaterals. Ultimately those changes would impact on the flow of activity to sector CA_1 and on the resulting hippocampal output. Are the decreased GAD-IR terminals in $CA_{3/2}$ related to the increased IR somata in CA_4/DG? Interestingly, interneurons in the dentate gyrus/hilus region have been shown to have axonal arborizations restricted to the dentate gyrus (Mott et al., 1997). It therefore seems unlikely that the changes found in $CA_{3/2}$ are directly related to those found in the dentate gyrus/hilus unless there is an, as yet unknown, secondary effect mediated by the mossy fiber system. The functions of interneurons in the DG and hilus are not well understood, but their location and axonal distribution suggests a significant role in the modulation of activity flowing from granule cells to the stratum radiatum of CA_3. Accordingly, it seems likely that they would help to gate activity entering the area dentata via the perforant path.

Amygdalo-hippocampal and amygdalo-entorhino-hippocampal afferent systems are presumably excitatory in nature (Colino and Fernandez de Molina, 1986). An increase in the activation of either direct or indirect amygdalar projections to the hippocampus would therefore result in increased excitation of target neurons in this region. In CA_3 and CA_2, it is not known whether GABAergic interneurons are directly contacted by amygdalar fibers or are indirectly activated through hippocampal excitatory neurons; however if there is an activation of GABA cells by those fibers, it would likely result in an increased release of this transmitter from synaptic terminals. Because membrane-associated GAD has been shown to be contained in synaptic vesicles (Nathan et al., 1995; Namchuk et al., 1997; Hsu et al., 1999), an acute and vigorous activation of GABA terminals could theoretically result in a marked depletion of their GAD content. In this regard it is noteworthy that there are fundamental differences in the way the two isoforms of GAD are regulated by neurons (for review see: (Martin and Rimvall, 1993). Overall, the GAD_{65} levels tend to be quite stable, while those of GAD_{67} are rather sensitive to experimental manipulations (Ding et al., 1998). Relevant to the current study is the fact that mRNA for GAD_{67} has been found to be altered in SZ (Akbarian et al., 1995). Although GABA cells in CA_4 and DG express both isoforms, there may be a small subpopulation of interneurons in CA_3 that primarily expresses GAD_{67}, at least under basal conditions (Stone et al., 1999). These differences might potentially account for the more marked changes in GAD_{67}-IR terminals observed in CA_3 and CA_2 in the current study.

In neuron somata, GABA molecules can be diverted into a metabolic shunt and thereby used as an energy source in the setting of increased neuronal activity (for review see: (Martin and Rimvall, 1993). Such a diversion through this "GABA shunt" could theoretically cause a decrease in GABA concentrations in somata, a change that would tend to promote the translation of GAD mRNA into protein transcripts. Such a mechanism could well account for the increase of GAD_{67}-IR somata reported here in DG and hilus. The lack of increased GAD-IR somata in $CA_{3/2}$ is less easily explained unless different cellular mechanims regulating GAD levels occur in various GABA cell subpopulations.

Subregional Distribution of GABA Changes

A strikingly common feature among many hippocampal findings in postmortem schizophrenic brain is their prevalence for sectors CA_4, CA_3 and CA_2, but not in CA_1 (Benes, 1999). This preferential localization is compelling when changes in hippocampal interneurons, and/or markers for the GABAergic system, are considered. For example, a reduction in non-pyramidal neurons was detected exclusively in CA_2 (Benes et al., 1998) and increases in GABA receptor binding were found to be most intense in stratum oriens of CA_3 (Benes et al., 1996c) and GAD_{65}-IR terminals was found to be decreased in stratum pyramidale in CA_4 and CA_3 and in stratum oriens of CA_3 and CA_2 of neuroleptic-free schizophrenics (Todtenkopf and Benes, 1998b). Although the mechanism underlying these alterations in GABA terminals is not known, chronic haloperidol administration has been associated with a marked increase of GABA terminals in rat medial prefrontal cortex (Vincent et al., 1994). It seems likely, however, that differences in connectivity might be responsible, at least in part, for the selectivity of these changes across the hippocampal sectors.

The decrease of GAD-IR terminals in CA_3 and CA_2, but not CA_1, induced in rat hippocampus by amygdalar activation is strikingly similar to that reported by our laboratory in the hippocampus of schizophrenic patients (Todtenkopf and Benes, 1998b). Although the latter finding has not as yet been replicated by another laboratory, it is reasonable to speculate at this juncture that, under pathological conditions in which amygdalar output is increased, significant changes in the intrinsic GABAergic network of the hippocampus, particularly in CA_3 and CA_2 where direct projections terminate, might take place (Benes and Berretta, 2000). Such a mechanism could also potentially contribute to the increase of basal metabolic rate observed in the hippocampus of schizophrenics and account for the poor performance on a memory retrieval task that has been reported in these subjects (Heckers et al., 1998b).

The absence of changes in GAD-IR terminals in CA_4 represents a notable departure from the pattern seen in postmortem studies of SZs. It is noteworthy, however, that excitation conducted along the ***direct*** pathway from the BLn to $CA_{3/2}$ (see above Pikkarainen et al., 1999) would probably be transmitted toward CA_1, rather than CA_4, and would, therefore, be unlikely to induce changes in this latter sector. On the other hand, the BLn also sends an ***indirect*** output to the area dentata via the entorhinal cortex and its perforant path projection (Aggleton, 1986). This latter pathway could potentially contribute to the changes seen in CA_4 in schizophrenia, since activity processed through the area dentata is inevitably transmitted through this sector. In this regard, it is pertinent to note that the use of an acute administration paradigm in the current report contrasts sharply with the chronic course that schizophrenics typically follows. It may well be that chronic amygdalar stimulation would be more likely to involve both the ***direct*** and the ***indirect*** pathways and thereby result in changes in hippocampal GABA cells that more precisely replicate those seen in schizophrenia (Benes and Berretta, 2000).

Implications of the 'Partial' Model for Understanding Pathogenesis

In the past, the use of conventional animal modelling for the study of schizophrenia has proven to be of limited value, as there has typically been a lack of both face and construct validity (Henn and McKinney, 1987). In contrast, the strategy described above overcomes such limitations by providing a form of modelling that approximates more closely the type of pathophysiological changes in GABAergic function that may actually be occurring in the amygdala (Reynolds et al., 1990b) and hippocampus (Benes, 1999) of schizophrenic brain. Accordingly, the importance of the 'partial' model described above lies in its ability to manipulate discrete aspects of a complex circuit with a view toward reproducing specific microscopic changes seen in our recent postmortem studies of schizophrenic brain. In this sense, this model does have construct validity. Moreover, the fact that it is capable of inducing changes in GABA cells that are in many respects similar to those seen in schizophrenia and bipolar disorder suggests that it may also have predictive validity. In this setting, the lack of obvious behavioral changes in picrotoxin-treated rats is not disturbing. Even finite behavioral changes in rodents comparable to aspects of the clinical manifestations of schizophrenia in humans might not be induced by a 'partial' model like the one employed here. In fact, similar changes in the amygdala and hippocampus by themselves might not necessarily resemble, even in humans, the clinical manifestations of schizophrenia, unless they are integrated within a more complex neuropathology.

Another important aspect of this model is its relationship to current hypotheses regarding the etiology of schizophrenia. Both neurodegenerative (Olney and Farber, 1995; Coyle, 1996) and developmental (Benes, 1991; Benes, 1995; Vita and Sacchetti, 1995; Ross and Pearlson, 1996; Weinberger, 1996; Bogerts, 1997; Harrison, 1997; Jones, 1997; Woods, 1998; Ichiki et al., 2000) mechanisms have been proposed. The design of this 'partial' rodent model considers the possibility that changes in schizophrenic hippocampus might be induced during early adulthood, perhaps as a consequence of abnormal GABAergic transmission in the amygdala. The latter could theoretically have been present from birth, but its effects may not be manifest until a later stage when postnatal changes trigger their appearance. This delay could be explained by the protracted time course with which some amygdalar efferent outputs develop (Verwer et al., 1996; Cunningham et al., 2000). Although, the developmental time course of amygdalar projections to the hippocampus is, as yet, not known, it seems plausible that this pathway might also mature during adulthood and influence the nature of the changes seen in schizophrenia.

Conclusions

As discussed above, postmortem studies over the past two decades have begun to demonstrate changes in the GABA and glutamate systems in the limbic lobe of schizophrenic brain. These changes seem to occur preferentially in certain sites, such as layer II of the anterior cingulate cortex and sectors CA3 and CA2 of the hippocampus. By considering the common denominators among these sites, the basolateral nucleus of the amygdala has presented itself as a potentially critical factor as it sends an important projection to both (Figure 1). It is important to emphasize that the use of a 'partial' model, like the one presented above, seeks to explain only a very circumscribed portion of a larger and much more complex network of connections. When viewed in relation to the complex nature of the schizophrenic syndrome, it could be said that the gains from such a strategy might be quite limited. Nevertheless, this study represents an important step forward for postmortem studies of schizophrenia because it provides support for the idea that the changes detected may indeed be related to the pathophysiology of this disorder, rather than to other extraneous factors such as neuroleptic drugs. In addition, however, this 'partial' modelling strategy, when used in a logical, stepwise fashion, can eventually provide important new insights into the pathophysiology of schizophrenia, ones that are defined in terms of precise integrative changes within key regions of the corticolimbic system. Future studies will be directed at understanding how the hippocampal GABAergic system responds to chronic amygdalar stimulation with low dose picrotoxin and how such changes may be reflected in the electrophysiological properties of hippocampal neurons.

Acknowledgments

This work was supported by grants from the National Institutes of Health (MH00423, MH42261, MH31862, MH31152) and the Stanley Foundation.

References

Aggleton JP. A description of the amygdalo-hippocampal interconnections in the macaque monkey. Exp Brain Res 1986;64:515-526.

Akbarian S, Kim JJ, Potkin SG, Hagman JO, Tafazzoli A, Bunney WE, Jr., Jones EG. Gene expression for glutamic acid decarboxylase is reduced without loss of neurons in prefrontal cortex of schizophrenics. Arch Gen Psychiatry 1995; 52:258-266.

Amaral DG, Price JL, Pitkänen A, Carmichael ST. Anatomical organization of the primate amygdaloid complex. In: Amygdala (Aggleton JP, ed), 1992, pp 1-66. New York: Wiley-Liss.

Arnold SE. Hippocampal pathology In: The Neuropathology of Schizophrenia. Harrison PJ, Roberts GW (eds), Oxford Press, Oxford 2000;pp 57-80.

Arnold SE, Franz BR, Gur RC, Gur RE, Shapiro RM, Moberg PJ, Trojanowski JQ. Smaller neuron size in schizophrenia in hippocampal subfields that mediate cortical-hippocampal interactions. Am J Psychiatry 1995; 152:738-748.

Beasley CL, Reynolds GP. Parvalbumin-immunoreactive neurons are reduced in the prefrontal cortex of schizophrenics. Schizophr Res 1997;24:349-355.

Benes FM. Post-mortem structural analyses of schizophrenic brain: study designs and the interpretation of data. Psychiatr Dev 1988;6:213-226.

Benes FM. Evidence for neurodevelopment disturbances in anterior cingulate cortex of post-mortem schizophrenic brain. Schizophr Res 1991;5:187-188.

Benes FM. The relationship of cingulate cortex to schizophrenia. In: Neurobiology of Cingulate Cortex and Limbic Thalamus (Vogt BA, Gabriel M, eds), 1993; pp 581-605. Boston: Birkhäuser, Inc.

Benes FM. A neurodevelopmental approach to the understanding of schizophrenia and other mental disorders. In: Developmental pschychopathology (Cicchetti D, Cohen DJ, eds), 1995; pp 227-253. New York: J. Wiley and Sons.

Benes FM. Is there evidence for neuronal loss in schizophrenia? Int Rev Psychiatry 1997a9:429-436.

Benes FM. What an archaeological dig can tell us about macro- and microcircuitry in brains of schizophrenia subjects. Schizophr Bull 1997a;23:503-507.

Benes FM. Evidence for altered trisynaptic circuitry in schizophrenic hippocampus. Biol Psychiatry 1999;46:589-599.

Benes FM, Bird ED. An analysis of the arrangement of neurons in the cingulate cortex of schizophrenic patients. Arch Gen Psychiatry 1987;44:608-616.

Benes FM, Todtenkopf MS. Effect of age and neuroleptics on tyrosine hydroxylase-IR in sector CA2 of schizophrenic brain. Neuroreport 1999;10:3527-3530.

Benes FM, Berretta S. Amygdalo-Entorhinal Inputs to the Hippocampal Formation in Relation to Schizophrenia. Ann New York Acad Sci 2000; 911:293-304.

Benes FM, Berretta, S. GABAergic interneurons: Implications for understanding schizophrenia and bipolar disorder. Neuropsychopharm 2001; In press.

Benes FM, Todtenkopf MS. Meta-Analysis of nonpyramidal neuron (NP) loss in layer II in anterior cingulate cortex (ACCx-II) from three studies of postortem schizophrenic brain. Soc for Neurosci Abstracts 1998;24:1275.

Benes FM, Davidson B, Bird ED. Quantitative cytoarchitectural studies of the cerebral cortex of schizophrenics. Arch Gen Psychiatry 1986;43:31-35.

Benes FM, Vincent SL, San Giovanni JP. High resolution imaging of receptor binding in analyzing neuropsychiatric diseases. Biotechniques 1989;7:970-972, 974-976, 978.

Benes FM, Sorensen I, Bird ED. Morphometric analyses of the hippocampal formation in schizophrenic brain. Schiz Bull 1991a;17:597-608.

Benes FM, Todtenkopf MS, Taylor JB. Differential distribution of tyrosine hydroxylase fibers on small and large neurons in layer II of anterior cingulate cortex of schizophrenic brain. Synapse 1997a;25:80-92.

Benes FM, Majocha R, Bird ED, Marotta CA. Increased vertical axon numbers in cingulate cortex of schizophrenics. Arch Gen Psychiatry 1987;44:1017-1021.

Benes FM, Vincent SL, Marie A, Khan Y. Up-regulation of GABAA receptor binding on neurons of the prefrontal cortex in schizophrenic subjects. Neuroscience 1996a;75:1021-1031.

Benes FM, Khan Y, Vincent SL, Wickramasinghe R. Differences in the subregional and cellular distribution of GABAA receptor binding in the hippocampal formation of schizophrenic brain. Synapse 1996c;22:338-349.

Benes FM, Kwok EW, Vincent SL, Todtenkopf MS. A reduction of nonpyramidal cells in sector CA2 of schizophrenics and manic depressives. Biol Psychiatry 1998a;44:88-97.

Benes FM, McSparren J, Bird ED, SanGiovanni JP, Vincent SL. Deficits in small interneurons in prefrontal and cingulate cortices of schizophrenic and schizoaffective patients. Arch Gen Psychiatry 1991b;48:996-1001.

Benes FM, Sorensen I, Vincent SL, Bird ED, Sathi M. Increased density of glutamate-immunoreactive vertical processes in superficial laminae in cingulate cortex of schizophrenic brain. Cereb Cortex 1992a;2:503-512.

Benes FM, Vincent SL, Alsterberg G, Bird ED, SanGiovanni JP. Increased GABAA receptor binding in superficial layers of cingulate cortex in schizophrenics. J Neurosci 1992b;12:924-929.

Benes FM, Wickramasinghe R, Vincent SL, Khan Y, Todtenkopf M. Uncoupling of GABA(A) and benzodiazepine receptor binding activity in the hippocampal formation of schizophrenic brain. Brain Res 1997a;755:121-129.

Benes FM, Todtenkopf MS, Kostoulakos P. GluR5,6,7 subunit immunoreactivity on apical pyramidal cell dendrites in hippocampus of schizophrenics and manic depressives. Hippocampus 2001b;In press.

Berger B, Tassin JP, Blanc G, Moyne MA, Thierry AM. Histochemical confirmation for dopaminergic innervation of rat cerebral cortex after the destruction of noradrenergic ascending pathways. Brain Res 1974;81:332-337.

Berretta S, Munno DW, Benes FM. Amygdalar activation alters the hippocampal GABA system: 'partial' modelling for postmortem changes in schizophrenia. J Comp Neurol 2001; 431:129-138.

Binder C AHW. Programmed cell death - many questions still to be answered. Ann Hematol 1994;69:45-55.

Bird ED, Spokes EG, Iversen LL. Increased dopamine concentration in limbic areas of brain from patients dying with schizophrenia. Brain 1979;102:347-360.

Bogerts B. The temporolimbic system theory of positive schizophrenic symptoms. Schizophr Bull 1997;23:423-435.

Bursch W, Oberhammer F, Schulte-Hermann R. Cell death by apoptosis and its protective role against disease. Trends Pharmacol Sci 1992;13:245-251.

Colino A, Fernandez de Molina A. Electrical activity generated in subicular and entorhinal cortices after electrical stimulation of the lateral and basolateral amygdala of the rat. Neuroscience 1986;19:573-580.

Conde F, Lund JS, Jacobowitz DM, Baimbridge KG, Lewis DA. Local circuit neurons immunoreactive for calretinin, calbindin D-28k or parvalbumin in monkey prefrontal cortex: distribution and morphology. J Comp Neurol 1994;341:95-116.

Coyle JT. The glutamatergic dysfunction hypothesis for schizophrenia. Harvard Rev Psychiatry 1996;3:241-253.

Coyle JT, Puttfarcken P. Oxidative stress, glutamate, and neurodegenerative disorders. Science 1993;262:689-695.

Csicsvari J, Hirase H, Czurko A, Mamiya A, Buzsaki G. Oscillatory coupling of hippocampal pyramidal cells and interneurons in the behaving rat. J Neurosci 1999;19:274-287.

Cunningham MG, Bhattacharya S, Benes FM. Post-natal ingrowth of amygdalo-cortical afferents in rat brain continues into adulthood. Soc for Neurosci Abstracts 2000;26.

Davis SR, Lewis DA. Local circuit neurons of the prefrontal cortex in schizophrenia: selective increase in the density of calbindin-immunoreactive neurons. Psychiatry Res 1995;59:81-96.

Descarries L, Beaudet A, Watkins KC. Serotonin nerve terminals in adult rat neocortex. Brain Res 1975;100:563-588.

Ding R, Asada H, Obata K. Changes in extracellular glutamate and GABA levels in the hippocampal CA3 and CA1 areas and the induction of glutamic acid decarboxylase-67 in dentate granule cells of rats treated with kainic acid. Brain Res 1998;800:105-113.

Eccles JC. The cerebral neocortex. A theory of its operation. In: Cerbral Cortex. Functional Properties of Cortical Cells (Jones EG, Peter A, eds), 1984;pp 1-48. New York: Plenum Press.

Fairen A, DeFelipe J, Regidor J. Nonpyramidal neurons. In: Cerebral Cortex (A. P, Jones EG, eds), 1982;pp 201-253. New York: Plenum Press.

Falkai P, Bogerts B, Rozumek M. Limbic pathology in schizophrenia: The entorhinal region -- a morphometric study. Biol Psychiat 1988;24:515-521.

Farber E. Ideas in pathology: Programmed cell death: Necrosis versus apoptosis. Mod Pathology 1994;7:605-609.

Feldman W, Robinson, S. Electrical activity of the brain in adrenalectomized rats with implanted electrodes. J Neurol Sci 1968;6:1-8.

Good PF, Huntley GW, Rogers SW, Heinemann SF, Morrison H. Organization and quantitative analysis of kainate receptor subunit GluR5-7 immunoreactivity in monkey hippocampus. Brain Res 1993;624:347-353.

Hanada S, Mita T, Nishinok N, Tankaka C. 3H-Muscimol binding sites increased in autopsied brains of chronic schizophrenics. Life Sci 1987;40:259-266.

Harrison PH. Schizophrenia: a disorder of neurodevelopment? Current Opinion Neurobiol 1997;7:285-289.

Harrison PJ. The neuropathological effects of antipsychotic drugs. Schizophr Res 1999;40:87-99.

Harrison PJ, McLaughlin D, Kerwin RW. Decreased hippocampal expression of a glutamate receptor gene in schizophrenia. Lancet 1991;337:450-452.

Heckers S, Heinsen H, Heinsen YC, Beckmann H. Limbic structures and lateral ventricle in schizophrenia. A quantitative postmortem study. Arch Gen Psychiatry 1990;47:1016-1022.

Heckers S, Heinsen H, Heinsen Y, Beckmann H. Cortex, white matter, and basal ganglia in schizophrenia: a volumetric postmortem study. Biol Psychiatry 1991a;29:556-566.

Heckers S, Heinsen H, Geiger B, Beckmann H. Hippocampal neuron number in schizophrenia. Arch Gen Psychiatry 1991b;48:1002-1008.

Heckers S, Rauch SL, Goff D, Savage CR, Schacter DL, Fischman,A.J., Alpert NM. Impaired recruitment of the hippocampus during conscious recollection in schizophrenia. Nature 1998a;1:318-323.

Heckers S, Rauch SL, Goff D, Savage CR, Schacter DL, Fischman AJ, Alpert NM. Impaired recruitment of the hippocampus during conscious recollection in schizophrenia. Nature Neurosci 1998b;1:318-323.

Henn FA, McKinney WT. Animals models in psychiatry. In: Psychopharmacology: the third generation in progress (Meltzer HY, ed), 1987;pp 687-695. New York: Raven Press.

Hsu CC, Thomas C, Chen W, Davis KM, Foos T, Chen JL, Wu E, Floor E, Schloss JV, Wu JY. Role of synaptic vesicle proton gradient and protein phosphorylation on ATP-mediated activation of membrane-associated brain glutamate decarboxylase. J Biol Chem 1999;274:24366-24371.

Ichiki M, Kunugi H, Takei N, Murray RM, Baba H, Arai H, Oshima I, Okagami K, Sato T, Hirose T, Nanko S. Intra-uterine physical growth in schizophrenia: evidence confirming excess of premature birth. Psychol Med 2000;30:597-604.

Jones EG. Cortical development and thalamic pathology in schizophrenia. Schizophr Bull 1997;23:483-501.

Kaufman DL, Houser CR, Tobin AJ. Two forms of the γ-aminobutyric acid synthetic enzyme glutamate decarboxylase have distinct intraneuronal distributions and cofactor interactions. J Neurochem 1991;56:720-723.

Kerwin R, Patel S, Meldrum B. Quantitative autoradiographic analysis of glutamate binding sites in the hippocampal formation in normal and schizophrenic brain post mortem. Neuroscience 1990;39:25-32.

Kerwin RW, Patel S, Meldrum B, Czudek C, Reynolds GP. Asymmetrical loss of glutamate receptor subtype in left hippocampus in schizophrenia. Lancet 1988;1:583-584.

Lambert JJ, Peters JA, Cottrell GA. Actions of synthetic and endogenous steroids on the GABAA receptor. Trends Pharmacol Sci 1987; 8:224-227.

Lawrie SM, Abukmeil SS. Brain abnormality in schizophrenia. A systematic and quantitative review of volumetric magnetic resonance imaging studies. Br J Psychiatry 1998;172:110-120.

LeDoux JE. Emotion circuits in the brain. Annu Rev Neurosci 2000;23:155-184.

Lindvall O, Bjorklund A. General organization of cortical monoamine systems. In: Monoamine Innervation of Cerebral Cortex (L. D, Reader TR, Jasper HH, eds), 1984;pp 9-40. New York: Alan R. Liss.

Longson D, Deakin JF, Benes FM. Increased density of entorhinal glutamate-immunoreactive vertical fibers in schizophrenia. J Neural Transm 1996;103:503-507.

Majewska MD, Bisserbe JC, Eskay LR. Glucocorticoids are modulators of GABAA receptors in brain. Brain Res 1985;339:178-182.

Martin DL, Rimvall K. Regulation of gamma-aminobutyric acid synthesis in the brain. J Neurochem 1993;60:395-407.

McDonald AJ, Beitz AJ, Larson AA, Kuriyama R, Sellitto C, Madi JE. Co-localization of glutamate and tubulin in putative excitatory neurons of the hippocampus and amygdala: an immunohistochemical study using monoclonal antibodies. Neuroscience 1989;30:405-421.

McEwen B. Glucocorticoids and hippocampus: Receptors in search of a function. In Adrenal Actions on Brain Ganten D and Pfaff E (eds) Springer-Verlag, Berlin, 1982; pp 1-22.

Mesulam MM, Mufson EJ, Levey AI, Wainer BH. Cholinergic innervation of cortex by the basal forebrain: Cytochemistry and cortical connections of the septal area, diagonal band nuclei, nucleus basals (Substantia innominata) and hypothalamus in the rhesus monkey. J Comp Neurol 1883;214:140-191.

Miller AL, Chaptal, C, McEwen, BS Beck JRE. Modulation of high affinity GABA uptake into hippocampal synaptosomes by glucocorticoids. Psychoneuroendocrinol 1978;3:155-164.

Monaghan DT, Cotman CW. Distribution of N-methyl-D-aspartate-sensitive L-[3H]glutamate- binding sites in rat brain. Journal of Neuroscience 1985;5:2909-2919.

Mott DD, Turner DA, Okazaki MM, Lewis DV. Interneurons of the dentate-hilus border of the rat dentate gyrus: morphological and electrophysiological heterogeneity. J Neurosci 1997;17:3990-4005.

Namchuk M, Lindsay L, Turck CW, Kanaani J, Baekkeskov S. Phosphorylation of serine residues 3, 6, 10, and 13 distinguishes membrane anchored from soluble glutamic acid decarboxylase 65 and is restricted to glutamic acid decarboxylase 65alpha. J Biol Chem 1997;272:1548-1557.

Nathan B, Floor E, Kuo CY, Wu JY. Synaptic vesicle-associated glutamate decarboxylase: identification and relationship to insulin-dependent diabetes mellitus. J Neurosci Res 1995;40:134-137.

Olney JW, Farber NB. Glutamate receptor dysfunction and schizophrenia. Arch Gen Psychiatry 1995;52:998-1007.

Pakkenberg B. Total nerve cell number in neocortex in chronic schizophrenics and controls estimated using optical disectors. Biol Psychiatry 1993;34:768-772.

Pfaff DW, Silva MTA, Weiss JM. Telemeterred recording of hormone effects on hippocampal neurons. Science 1971;172:394-395.

Pikkarainen M, Ronkko S, Savander V, Insausti R, Pitkanen A. Projections from the lateral, basal, and accessory basal nuclei of the amygdala to the hippocampal formation in rat. J Comp Neurol 1999;403:229-260.

Pitkanen A, Pikkarainen M, Nurminen N, Ylinen A. Reciprocal connections between the amygdala and the hippocampal formation, perirhinal cortex, and postrhinal cortex in rat. A review. Ann N Y Acad Sci 2000;911:369-391.

Pollard H, Chariaut-Marlangue C, Cantagrel S, Represa A, Robain O, Moreau J, Ben-Ari Y. Kainate-induced apoptotic cell death in hippocampal neurons. Neuroscience 1994;63:7-18.

Rabow R, SJ, Farb DH. From ion currents to genomic analysis: Recent advances in GABAA receptor research. Synapse 1995;21:189-274.

Reynolds GP, Czudek C, Andrews H. Deficit and hemispheric asymmetry of GABA uptake sites in the hippocampus in schizophrenia. Biol Psychiatry 1990a;27:1038-1044.

Reynolds GP, Czudek C, Andrews HB. Deficit and hemispheric asymmetry of GABA uptake sites in the hippocampus in schizophrenia. Biol Psychiatry 190b;27:1038-1044.

Roberts GW, Colter N, Lofthouse R, Bogerts B, Zec M, Crow TJ. Gliosis in schizophrenia. A survey. Biol Psychiat 1986;39:1043-1050.

Rosene DL, Van Hoesen GW. The hippocompal formation of the primate brain. In: Cerebral Cortex (Peters A, Jones EG, eds), 1987;pp 345-456. New York: Plenum Press.

Ross CA, Pearlson GD. Schizophrenia, the heteromodal association neocortex and development: potential for a neurogenetic approach. Trends Neurosci 1996;19:171-176.

Sanders SK, Shekhar A. Blockade of GABAA receptors in the region of the anterior basolateral amygdala of rats elicits increases in heart rate and blood pressure. Brain Res 1991;567:101-110.

Sanders SK, Shekhar A. Regulation of anxiety by GABAA receptors in the rat amygdala. Pharmacol Biochem Behav 1995;52:701-706.

Sapolsky RM. Stress, the aging brain, and the mechanisms of neuron death. Cambridge, Mass: MIT Press 1992.

Saunders RC, Rosene DL. A comparison of the efferents of the amygdala and the hippocampal formation in the rhesus monkey: I. Convergence in the entorhinal, prorhinal, and perirhinal cortices. J Comp Neurol 1988;271:153-184.

Selemon LD, Rajkowska G, Goldman-Rakic PS. Abnormally high neuronal density in the schizophrenic cortex. A morphometric analysis of prefrontal area 9 and occipital area 17. Arch Gen Psychiatry 1995;52:805-820.

Shekhar A, Sajdyk TS, Keim SR, Yoder KK, Sanders SK. Role of the basolateral amygdala in panic disorder. Ann N Y Acad Sci 1999;877:747-750.

Sik A, Penttonen M, Ylinen A, Buzsáki G. Hippocampal CA1 interneurons: An *in vivo* intracellular labeling study. J Neurosci 1995;15:6651-6665.

Simpson MD, Slater P, Deakin JF, Royston MC, Skan WJ. Reduced GABA uptake sites in the temporal lobe in schizophrenia. Neurosci Lett 1989;107:211-215.

Stone DJ, Walsh J, Benes FM. Localization of cells preferentially expressing GAD(67) with negligible GAD(65) transcripts in the rat hippocampus. A double in situ hybridization study. Brain Res Mol Brain Res 1999;71:201-209.

Stumpf WE, Heiss C, Sar M, Duncan GE, Draver C. Dexamethasone and corticosterone receptor sites. Histochem 1989;92:201-210.

Sutanto W, Handelmann G, de Bree F, de Kloet ER. Multifaceted interaction of corticosteroids with the intracellular receptors and with membrane GABAA receptor complex in the rat brain. J Neuroendocrinol 1989;1:243-247.

Swanson LW, Petrovich GD. What is the amygdala? Trends Neurosci 1998;21:323-331.

Tamminga CA. Schizophrenia and glutamatergic transmission. Crit Rev Neurobiol 1998;12:21-36.

Thierry AM, Blanc G, Sobel A, Stinus L, Glowinski J. Dopaminergic terminals in the rat cortex. Science 1973;182:499-501.

Todtenkopf MS, Benes FM. Distribution of glutamate decarboxylase65 immunoreactive puncta on pyramidal and nonpyramidal neurons in hippocampus of schizophrenic brain. Synapse 1998a;29:323-332.

Todtenkopf MS, Benes FM. Distribution of glutamate decarboxylase 65 immunoreactive puncta on pyramidal and nonpyramidal neurons in hippocampus of schizophrenic brain. Synapse 1998b;29:323-332.

Van Hoesen GW, Morecraft RJ, Vogt BA. Connections of the monkey cingulate cortex. In: Neurobiology of Cingulate Cortex and Limbic Thalamus (Vogt BA, Gabriel M, eds), 1993;pp 249-284. Birkhauser: Boston.

Verwer RW, Van Vulpen EH, Van Uum JF. Postnatal development of amygdaloid projections to the prefrontal cortex in the rat studied with retrograde and anterograde tracers. J Comp Neurol 1996;376:75-96.

Vincent SL, Adamec E, Sorensen I, Benes FM. The effects of chronic haloperidol administration on GABA- immunoreactive axon terminals in rat medial prefrontal cortex. Synapse 1994;17:26-35.

Vita A, Sacchetti E. Developmental brain abnormalities in schizophrenia: contributions of genetic and perinatal factors. Arch Gen Psychiatry 1995;52:157-159.

Vogt BA, Pandya DN. Cingulate cortex of the rhesus monkey: II. Cortical afferents. J Comp Neurol 1987;262:271-289.

Vogt BA, Pandya DN, Rosene DL. Cingulate cortex of the rhesus monkey: I. Cytoarchitecture and thalamic afferents. J Comp Neurol 1987;262:256-270.

Weinberger DR. On the plausibility of "The neurodevelopmental hypothesis" of schizophrenia. Neuropsychopharmacology 1996;14:1S-11S.

Woods BT. Is schizophrenia a progressive neurodevelopmental disorder? Toward a unitary pathogenetic mechanism. Am J Psychiatry 1998;155:1661-1670.

Zhang WQ, Rogers BC, Tandon P, Hudson PM, Sobotka TJ, Hong JS, Tilson HA. Systemic administration of kainic acid increases GABA levels in perfusate from the hippocampus of rats in vivo. Neurotoxicol 1990;11:593-600.

12 DORSOLATERAL PREFRONTAL CORTICAL PARALLEL CIRCUIT IN SCHIZOPHRENIA: Postmortem Abnormalities

Blynn G. Bunney, William E. Bunney, Richard Stein, and Steven G. Potkin

Abstract

Emerging evidence from postmortem studies in schizophrenics implicate disturbances in the dorsolateral prefrontal parallel circuit, specifically in the dorsolateral prefrontal cortex (DLPFC) and mediodorsal nucleus (MD) of the thalamus. Cognitive deficits in schizophrenia include impairments in working memory that are accompanied by test-related decreases in DLPFC activation. The DLPFC is part of the dorsolateral prefrontal cortical parallel circuit, one of three cognitive-affective circuitries. Postmortem evidence for abnormalities in the DLPFC of schizophrenics include abnormal cell settling patterns of interstitial neurons in the white matter, reductions in GAD mRNA, and decreases in γ-2 mRNA of the γ-2 GABA-A receptor subunit as compared to matched controls. The DLPFC has substantial projections to the MD nucleus. Studies in schizophrenics reveal dramatic reductions in neuronal number of the MD with evidence of defects in the MD thalamic subnuclei (i.e., densocellular and parvocellular subnuclei) that project to the DLPFC. Some neuropathological abnormalities in the DLPFC and MD nucleus of schizophrenics could be attributed to disturbances in early cortical development. The neurodevelopmental hypothesis of schizophrenia is supported by the finding that three populations of cortical subplate markers (nicotinamide-adenine dinucleotide phosphate diaphorase, NADPH-d, microtubular-associated protein-2, MAP-2 and a monoclonal antibody that represents a non-phosphorylated neurofilament protein, SMI-32, a 160-200kDA protein), are abnormally distributed in the interstitial white matter of the DLPFC in patients. Disturbances in early neurodevelopment could occur during a period of genetic/environmental vulnerability, most likely in the second trimester, when large numbers of neurons are migrating from the ventricular zone to their target destinations in the cortex. These disturbances might be reflected in abnormal functioning of neuronal circuitries such as the dorsolateral prefrontal parallel circuit containing the DLPFC and MD nucleus of the thalamus.

The dorsolateral prefrontal cortical circuit is one of three parallel circuits in the brain that relates to cognitive-affective function. Figure 1 represents an adaptation of this circuit described by Alexander, DeLong and Strick (1986) and includes the dorsolateral prefrontal cortex (DLPFC), the dorsolateral caudate nucleus of the striatum, the lateral dorsal medial nucleus in globus pallidus, the posterolateral nucleus of the substantia nigra, and the mediodorsal nucleus of the thalamus. Using a neurotropic virus tracer capable of labeling multisynaptic neuronal connections (McIntyre B-strain of herpes simplex virus type 1 - HSV1), Middleton and Strick (2000) identified areas 9 and 46 (two nuclei within the DLPFC) as receiving input from the substantia nigra and globus pallidus via the mediodorsal nucleus of the thalamus. The thalamus is composed of eleven identified nuclei. One of these, the mediodorsal nucleus, projects to the DLPFC. Within the mediodorsal nucleus, three subnuclei have been identified. These include the magnocellular, parvocellular and the densocellular nuclei. The latter two subnuclei project to the middle layers of the DLPFC. Specific cellular and molecular abnormalities in the subnuclei and the DLPFC in schizophrenic brains are reviewed below. These abnormalities clearly could be associated with some of the prominent prefrontal cognitive deficits observed in schizophrenia. To date, as reviewed below, an impressive body of research implicates the DLPFC.

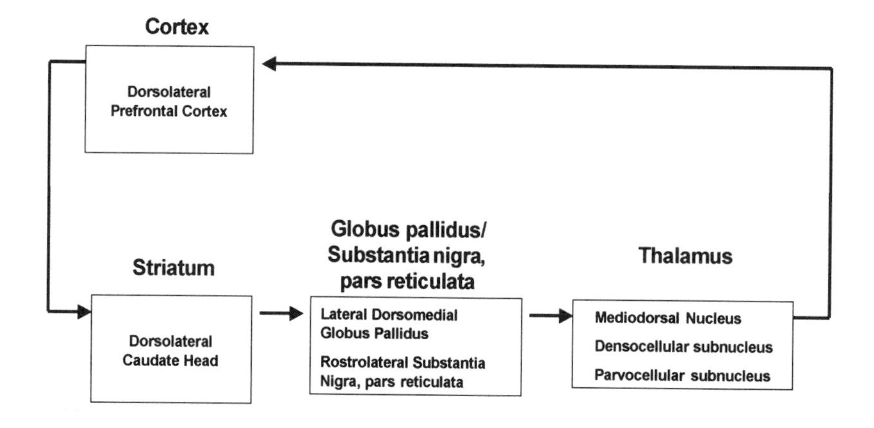

Figure 1. **Dorsolateral prefrontal cortical parallel circuit in schizophrenia**

(Adapted from Alexander, DeLong and Strick, 1986)

Cognitive Deficits Associated with Prefrontal Cortical Function

Abnormalities in DLPFC function associated with schizophrenia appear to be related to deficits in working memory (see review, Weinberger & Lipska, 1995). Schizophrenics perform poorly on tests requiring working memory skills such as the Wisconsin Card Sort (WCS) (Weinberger & Lipska, 1995) and the N-back Working Memory tests (Callicott et al., 2000). The WCS test, for example, involves abstract reasoning and problem-solving skills, and patients tend to make more preservative errors and attain less categorizations than controls (Goldberg et al., 1995). Poor performance on the WCS test is observed even after accounting for the effects of medication or chronicity of the illness and is seen in first-episode, neuroleptic-naïve patients (Berman et al., 1992; Catafau et al., 1994).

A series of studies of monozygotic twins discordant for schizophrenia provide an important body of data linking working memory deficits with abnormalities in test-related DLPFC function. Compared to their well twin counterparts schizophrenic twins perform worse on the WCS (Berman et al., 1992) and have reductions in test-related DLPFC activation (Berman et al., 1992), findings similar to other cohorts of schizophrenics (Goldberg et al., 1995).

Functional neuroimaging studies including 133-xenon regional cerebral blood flow (rCBF), SPECT, fMRI, oxygen-15-PET and magnetic resonance spectroscopy (MRS) reveal a relationship between task difficulty and DLPFC function. In healthy controls, there is an initial activation in test-related DLPFC blood flow that increases as a function of task difficulty. In schizophrenics, the initial activation of the DLPFC is normal; however, as test demands increase, DLPFC activity in the schizophrenic patients falls significantly below that of the controls (Weinberger & Lipska, 1995; Fletcher et al., 1998; Callicott et al., 1998; Carter et al., 1998). Graded working memory tests clearly demonstrate the necessity of sufficient cognitive demands to differentiate schizophrenics from normals in terms of DLPFC activation (Fletcher et al., 1998). Additionally, the specificity of cognitive demands influences DLPFC activity. For example, Ragland et al., (1998) administered both the Paired Associate Recognition Task (PART, a task testing declarative memory) and the WCS test to schizophrenics and controls. Although patients scored poorly on both tests, DLPFC performance was impaired only during the WCS test. Barch et al., (2001) suggest that tasks (such as the WCS) requiring context representations (e.g., instructions, specific prior stimuli or sentence processing) specifically affect DLPFC activation and that context processing is impaired in schizophrenics.

Deficits in DLPFC metabolic activation in schizophrenia are substantiated in other experimental paradigms. Magnetic resonance spectroscopy (^{31}P-MRS and 1H-MRS) studies of N-acetylaspartate (NAA), an interneuronal marker of neuronal functional integrity (Bertolino et al.,

2001; Ohnuma et al., 2000) show lower NAA activity peaks in the DLPFC of schizophrenics compared to controls (Hinsberger et al., 1997; Bertolino et al., 1998a;b). Other studies suggest deficit schizophrenics are more likely to have reductions in NAA in DLPFC and medial prefrontal cortex) as compared to non-deficit patients (Callicott et al., 2000; Delamillieure et al., 2000). Overall, NAA activity is lower in unmedicated patients and is reported in both acute and chronic schizophrenics (Bertolino et al, 1998a; 2000a;b). Antipsychotics increase NAA peaks, particularly in the DLPFC as compared to other brain regions (Bertolino et al., 2001).

Postmortem Findings in Prefrontal Cortical Brain Regions

Morphological investigations of the schizophrenic brain provide evidence of anomalies in many schizophrenic patients compatible with disturbances in neuronal connectivity. Even slight changes in brain structure, including the deletion of particular cell types, and changes in the number and distribution of neurons, may impact on the functioning of neuronal circuits. A substantial body of data collected from the prefrontal cortical tissue of schizophrenics provides evidence of alterations in neuronal density, neuronal cell settling patterns, and in gene expression of transmitter-related molecules (see review, Bunney & Bunney, 2000). Notably absent, in most studies, are neuropathological changes associated with degenerative processes (Rajkowska et al., 1998; Selemon & Goldman-Rakic, 1999; Marenco & Weinberger, 2000). These data provide compelling evidence that a subgroup of schizophrenics might have a primary disturbance in early neurodevelopmental processes.

Neuronal density changes in the DLPFC (BA 9 and 46) of the schizophrenic brain include reductions in neuronal densities in layer VI (Benes et al., 1996), decreases in interneuronal densities in layer II and to a lesser extent in layer I (Benes et al., 1991). Using a 3D counting method, Selemon et al., (1995; 1998) observed increases in overall neuronal density in Layer VI of area 9 (Selemon et al., 1995) and in Layers II, III, IV, VI in area 46 (Selemon et al., 1998). The reduced neuropil hypothesis posited by Selemon and Goldman-Rakic (1999) suggests that disturbances in prefrontal cognitive functioning in schizophrenia are mediated by a process that involves the loss of neuronal processes but not cells. This is consistent with other data showing no neuronal increase or loss in prefrontal cortex of schizophrenics (Thune et al., 2001; Akbarian et al., 1995b). However, these findings do not implicate functional abnormalities, as changes in gene expression (e.g., down-regulation of GAD67 mRNA) can occur in the absence of corresponding alterations in neuronal cell density (Akbarian et al., 1995b).

Layer III pyramidal neurons are thought to play a significant role in corticocortical and thalamocortical connectivity (Glantz & Lewis, 2000). Glantz and Lewis (2000) quantified the density of Golgi-impregnated pyramidal neurons in the superficial and deep portions of layer III and reported a significant decrease in dendritic spine densities only for the deep layers of layer III in schizophrenics compared to controls and non-schizophrenic psychiatric patients. These findings support the hypothesis that the number of cortical and/or thalamic excitatory inputs to these pyramidal neurons is altered in schizophrenia.

Cortical Subplate

Disturbances in neural development during gestation may influence neuronal proliferation, brain connectivity and/or neurotransmitter expression. The failure of connections to form or stabilize properly as a consequence of developmental disturbances could lead to activity-dependent reductions or enhancements of gene expression for molecules involved in neural transmission. The neurodevelopmental hypothesis of schizophrenia suggests a disturbance in cortical formation, and its relevant processes underlie some of the cytoarchitectural defects associated with the illness. As demonstrated in subhuman primates, the cortical subplate plays a critical role in the proper formation of neuronal circuitries (Kostovic & Rakic, 1980; Allendoerfer & Shatz, 1994). The cerebral cortex is formed when neurons migrate from the ventricular walls to the cortical plate, a process that occurs primarily during the second trimester of development. As the neurons migrate, they pass through an intermediate zone below the cortical plate to form a subplate at the junction of the white matter and cortical plate. These early maturing cells appear to serve as a guide or "cellular scaffold" between the developing axons and the target cells in cortical layers. Once the neurons enter the cortical subplate, they receive afferents and send efferents to the developing cortical layers. The axons leaving the subplate migrate first to the deeper layers (e.g. layers II and III) such that the cortex is formed 'inside-out' with the superficial cortical layers being formed last (Allendoerfer & Shatz, 1994; Kostovic & Rakic, 1980). The majority of the subplate cells undergo programmed cell death during late fetal and early postnatal life. However, tens of thousands of subplate cells remain in the interstitial white matter, adjacent to the cortex in the adult brain (Kostovic & Rakic, 1980). The remnant cortical subplate cells could serve as markers of abnormal neuronal development in the schizophrenic brain. Data from animal studies show that lesions to the cortical subplate significantly disrupt neuronal target processes (Allendoerfer & Shatz, 1986). Similar processes in schizophrenics could account for abnormal cell settling patterns reported in the prefrontal cortex and other areas of brain including the hippocampus (Jakob and Beckmann, 1986).

Cytoarchitectural Alterations in Dorsolateral Prefrontal Cortex

The neurodevelopmental hypothesis of schizophrenia suggests that disturbances in programmed cell death or disturbances in neuronal migration during the development of the cerebral cortex could lead to disturbances in cortical connectivity and abnormalities in brain function leading to symptoms associated with schizophrenia. Akbarian et al., (1993a) quantified NADPH-d labeled cells, a subgroup of cortical subplate remnant cells, in prefrontal cortex and underlying white matter of schizophrenics and controls matched for age, gender, and autolysis time. Using immunohistochemistry, densitometry and quantified cell counting methods, observed that NADPH-d stained cells in control brain tissue were distributed most densely in the white matter immediately deep to layer VI of the cortex where they remained from the subplate. In schizophrenics, however, a significant increase in NADPH-d labeled neurons was found in the superficial white matter and in the overlying cortex accompanied by a decrease in the number of neurons in the white matter deeper than 3 mm from the cortex. Confounding effects of neuroleptic treatment are not likely given that the neuroleptics would have to affect the neurochemically and functionally homogenous neurons differently in the cortical gray and white matter. Furthermore, in one of the most severe cases of neuronal disturbance, the patient had not been treated with neuroleptics (Akbarian et al., 1993a).

Since NADPH-d stained cells represent only a small subset of remnant subplate cells, studies were undertaken with two additional markers to evaluate a larger proportion of these interstitial neurons (Akbarian et al., 1996). These were MAP-2, representing the largest fraction of subplate neurons, and SMI-32. Together, the NADPH-d labeled cells, the MAP-2 and SMI-32 neurons represent approximately 85% of the remaining subplate cells. A maldistribution in all three populations of subplate cells was observed in the middle frontal gyrus of schizophrenics compared to matched controls. For each neuronal subpopulation there was an increase in cell density toward the deeper white matter. Thirty-five percent of the 20 schizophrenic brains in the study, but none of the 20 control group brains, showed a maldistribution in at least two of the three marker subpopulations. In a separate study, Anderson et al., 1996) observed that one in five schizophrenic brains had a disturbance in MAP-2 neurons, similar to those reported in the Akbarian et al., (1996) study. Recently, findings in E. Costa's laboratory replicated the Akbarian et al., (1996) findings with NADPH-d in two separate samples (personal communication, 2001).

Other evidence for disorders of early developmental processes is work by Kalus et al., (1997) showing the abnormal distribution of Cajal-Retzius cells labeled with non-phosphorylated neurofilaments (SMI-311) in schizophrenics. Schizophrenics had more SMI-311 labeled cells in the middle and lower third of Layer I of the prefrontal cortex as compared to controls. Since Cajal-Retzius cells play a key role in neuronal migration, these alterations are suggestive of a disturbance in early neurodevelopment.

Abnormalities of Neurotransmitter-Related Molecules in the DLPFC of Schizophrenic Patients

Abnormalities in neurotransmitter-related molecules provide clues to the functional integrity of the neuronal circuit. Emerging data, compatible with alterations in the DLPFC and its relevant circuitry suggest a wide range of defects in schizophrenia. GABA, the major inhibitory neurotransmitter and NMDA, the major excitatory neurotransmitter in cortex, and their associated mRNAs are leading candidates for potential defects in the schizophrenic brain.

GABA

A substantial number of studies document alterations in the major inhibitory neurotransmitter of the brain, GABA, in schizophrenia (see review, Lewis, 2000). GABA-related changes include decreased GABA uptake and release (Sherman et al., 1991; Simpson et al., 1998; Reynolds et al., 1990) and reduced density of GABA axon terminals (Woo et al., 1998) in the schizophrenic brain as compared to controls (see review, Lewis, 2000). Benes et al., (1991) reported a decreased number of small interneurons (most likely GABA-ergic) in prefrontal cortical areas. In a separate investigation, Daviss and Lewis (1995) observed dramatic increases (50-70%) in a subpopulation of calbinin-labeled GABA neurons in layers III, V, and VI in schizophrenics compared to controls matched for age-, gender and autolysis time.

GABA-A receptors are preferentially increased in prefrontal cortical tissue of schizophrenics compared to controls (Benes et al., 1996). Analysis of the expression patterns of six polypeptide subunits of the GABA-A receptor ($\alpha1$, $\alpha2$, $\alpha5$, $\beta1$, $\beta2$, $\gamma2$) revealed non-significant reductions in the $\gamma2$ subunit in the DLPFC in schizophrenics (↓ 28%) compared to controls matched for age-, gender- and autolysis- time (Akbarian et al., 1995a). Due to the relatively large variance in expression patterns of the $\gamma2$ subunits, a follow-up study examined the relative proportions of the long ($\gamma2L$) and short ($\gamma2S$) alternatively spliced forms of the $\gamma2$ receptor subunit mRNA by in situ hybridization and reverse transcription-PCR amplification methods. Instead of the 50:50 relationship proportion between the two mRNAs, schizophrenics showed a relative 30-40% increase in the long I ($\gamma2L$) relative to the short ($\gamma2S$) I. These results, i.e., a net overrepresentation of the functionally less active $\gamma2L$ I, suggest functionally less active GABA-A receptors in the patients as compared to controls (Huntsman et al., 1998).

Another line of evidence for GABA-ergic abnormalities comes from studies of the GABA membrane transporter, GAT-1 (Volk et al., 2001). GAT-1 protein and GAT mRNA were analyzed in DLPFC tissue in matched pairs of schizophrenics and controls. Schizophrenics had significant reductions in GAT-1 protein levels in labeled cells in layers I-V (↓ 21-33%) that were unchanged in layer VI of the prefrontal cortex. There were no significant differences in the relative density of neurons in any of the cortical layers, nor was there a difference in somal size. These data provide evidence that GABA synthesis and reuptake are altered at the level of gene expression in a subset of prefrontal cells (Volk et al., 2001).

GAD

One of the key enzymes in the synthesis of GABA, glutamic acid decarboxylase (GAD), is localized to inhibitory neurons and is regulated in an activity-dependent manner in the cerebral cortex of non-human primates. Blocking optic nerve conduction in adult monkeys produces reversible decreases in immunoreactive GABA and GAD levels (Hendry et al., 1986; 1988) and decreases in GAD mRNA (Hendry et al., 1994). Work by Shatz (for review, see Penn & Shatz, 1999) shows that endogenous and sensory-driven neural activity induces an activity-dependent remodeling of neuronal circuitries. A subhypothesis proposes that neurons that are more active will stimulate more neuronal connections, while less active cells will tend to lose connections. Thus, abnormalities in activity-dependent GABA-ergic systems in schizophrenia could reflect a disruption of the formation of neuronal connections in the schizophrenic brain (Penn & Shatz, 1999).

Significant differences in GAD67 (67 Kd isoform of GAD) mRNA in schizophrenic and control brains were reported by three independent laboratories. First, Akbarian et al., (1995b) measured GAD67 mRNA levels and quantified the number and laminar distribution of GAD67 mRNA-expressing neurons. Research methodologies included in situ hybridization-histochemistry, densitrometry and cell counting methods. CamIIK mRNA (type II calcium-calmodulin-dependent protein kinase) was used as a control. Results showed that schizophrenics, compared to age-, gender, and postmortem interval (PMI) -matched controls had decreased GAD67 mRNA levels in neurons of layer I (↓ 40%), layer II (↓ 48%) and in layers III-VI ((↓ 30%). No significant differences were found in CamIIK mRNA levels, nor were there significant changes in prefrontal gray and white matter volumes. In addition, no differences in the total number of neurons nor in small neurons between patients and controls were observed. Thus, reduced gene expression for GAD in the absence of significant cell loss might relate to functional hypofrontality in the DLPFC of schizophrenics. Volk et al., (2000) reported remarkably similar decreases in GAD67 mRNA, most significantly in layers III-V (↓ 25-35%) without loss of neurons in the prefrontal cortex in schizophrenics compared to matched controls. Finally, Guidotti et al., (2000) reported significant reductions in GAD67 (mRNA and protein levels) in prefrontal cortex of schizophrenic patients. The down-regulation of GAD67 mRNA was shown to be independent of medication effects (Volk et al., 2000; Guidotti et al., 2000) and neuronal damage (Guidotti et al., 2000).

NMDA

One potentially important clue to schizophrenia is the ability of NMDA non-competitive antagonists such as phencyclidine (PCP) and its related compounds to induce schizophrenic-like psychosis in normals. First described by Luby et al., (1959) in a group of medical residents who ingested PCP as part of a research protocol, the psychotomimetic effects of the drug-induced symptoms were sometimes indistinguishable from symptoms of schizophrenia. While healthy controls may become psychotic for 1-2 hours following PCP ingestion, schizophrenics can develop a psychosis lasting as long as 6 weeks (Luby et al.,1959). These observations, in part, later led to the initiation of research for a possible defect in schizophrenia associated with NMDA receptor. Almost every PCP-like compound administered to man is associated with psychosis (see review, Bunney et al., 1995). In a series of studies, ketamine, a PCP-like compound, was administered in subanesthetic doses (0.1-0.5 mg) to healthy volunteers and schizophrenic patients. In healthy controls, ketamine produced schizophrenic-like positive and negative symptoms, altered perception and impaired performance on cognitive tests (Krystal et al., 1994) that were not accounted for by changes in attention (Malhotra et al., 1996). In schizophrenics, ketamine intensified (Tamminga et al., 1995a) and prolonged psychoses lasting from 8-24 hours (Tamminga et al., 1995b). These were described as a discrete activation of psychotic symptoms with striking similarities to symptoms of previous psychotic episodes (Lahti et al., 1995). Ketamine-induced effects were not blocked by pretreatment with haloperidol in patients (Tamminga et al., 1995b) or in controls (Krystal et al., 1999), except for reductions in impairments on executive cognitive function (including the WCST) (Krystal et al., 1999). In contrast, pretreatment with lamotrigine, a drug that inhibits glutamate release, ameliorated more of the ketamine-related symptoms (Anand et al., 2000). Measurements of metabolic activity in healthy controls with oxygen-15 PET and FDG PET showed ketamine to activate prefrontal regions (Breier et al., 1997). In both patients and controls, Tamminga et al., (2000) reported that ketamine elevated cerebral blood flow in the anterior cingulate and the right middle frontal cortex while decreasing activation of the cerebellum and lingual gyrus.

Glutamatergic abnormalities, specifically related to NMDA function, have been reported in schizophrenia (see Bunney et al., 1995; Heresco-Levy et al., 1995). NMDA is part of a group of excitatory amino acid ionotropic receptors including kainate and AMPA. These sites are further subdivided into NMDAR1 (8 splice variants) and NMDAR2 (A-D); AMPA (GluR1-4); kainate (low affinity: GluR5-7 and kainate high affinity

(KA1-KA2) receptors. The NMDA receptor is comprised of a heteromeric assembly of NR1 and NR2 subunits, where the NR1 subunits are capable of binding with a variety of NR2 subunits (A-D). Measurements of the expression patterns of the mRNA of five NMDA receptor subunits (NR1/NR2A-D) were analyzed with in situ hybridization techniques in prefrontal, parieto-temporal and cerebellar cortex in brain tissue of schizophrenics, non-schizophrenic controls (on and off neuroleptics), and matched normal controls (Akbarian et al., 1996b). Significant alterations in schizophrenics were seen only in prefrontal regions. There was a 53% increase in expression of NR2D subunit mRNAs and a shift in the relative proportions of the mRNAs for the NR2 subunit family. No significant differences were observed in other brain regions studied nor were there differences between neuroleptic-treated or untreated controls.

The NR2D subunit is developmentally regulated (Monyer et al., 1994) and may be a necessary component for the induction of specific neuronal connections (Kleinschmidt et al 1987; Schlaggar et al., 1993; Li et al., 1994). Functionally, culture studies show that the NR1/NR2D receptor assemblies have a prolonged rate of decay of glutamate-induced ion currents and a lowered threshold for voltage-dependent Mg^{2+} blockade compared to NR1/2A and NR2B subunits (Kleinschmidt et al., 1987). Furthermore, the NR1/NR2D polypeptide kinetic properties may play a role in effective postsynaptic depolarization under conditions when presynaptic activity is reduced (Kleinschmidt et al., 1987) resulting in more "hyperexcitable" NR1/2D receptors in the prefrontal cortex of schizophrenics. It has been hypothesized that the resulting increased NR1/2D "excitability" may be a compensatory response to a generalized reduction in prefrontal neuronal activity (i.e., hypofrontality) in schizophrenia (Akbarian et al., 1996b).

Additional evidence implicating prefrontal glutamatergic disturbances comes from work in other laboratories. Meador-Woodruff et al., (2001) observed increases in GluR7 mRNA and decreases in KA2 mRNA in prefrontal cortex in schizophrenics compared to controls. Ohnuma et al., (2000) reported significant reductions in PSD95 mRNA, a postsynaptic density protein that binds to the NMDA receptor, in DLPFC (BA 9) but not in the hippocampus of schizophrenics compared to controls.

Akbarian et al., (1995c) conducted a study of AMPA receptors in prefrontal cortical brain tissue of schizophrenics and controls. Reverse-transcription PCR was used to amplify cDNA derived from unedited GluR2 mRNA but not from edited GluR2 mRNA. Only modest increases in AMPA were observed in schizophrenia, consistent with non-significant changes in prefrontal AMPA receptors in elderly schizophrenics (Healy et al., 1996).

Neurodevelopmental Changes: An Overview

There is a growing body of data to suggest that disturbances in neuronal development during gestation can increase the risk of schizophrenia in genetically predisposed individuals. Some of the most convincing data comes from epidemiological studies of mothers exposed to stressors during pregnancy, particularly in the second trimester. Environmental risk factors as reviewed by McDonald & Murray, (2000) include early environmental events such as obstetric complications, maternal infection, city of birth and late winter/spring births. They conclude, however, that familial/genetic factors carry, by far, the greatest risk for schizophrenia, while the effect sizes of environmental factors are modest. A current hypothesis proposes that disruption of fetal development, due to expression of a genetic defect, is more likely if the environmental disruption (e.g., maternal infection) occurs during a critical period of development, such as the second trimester (e.g., Mednick et al., 1994).

In the second trimester almost all neurons have been generated and are migrating from the ventricular zone to the developing cortex and forming connections. In subhuman primates, genetic factors have been shown to interfere with these processes resulting in defective neuronal migration, positioning and connectivity (Rakic, 1974). As illustrated by Komuro & Rakic (1993), it is possible that interference with neurotransmitters such as NMDA during critical developmental periods can adversely modulate the forming of proper neuronal connections.

Studies of influenza epidemics show that offspring of pregnant mothers exposed to influenza during the second trimester are at increased risk for of schizophrenia (Mednick et al., 1994). Other factors including severe psychological stress (Huttanen & Niskana, 1978; Van Os et al., 1998), obstetrical complications (Geddes et al., 1999), and malnutrition (Susser et al., 1996) during fetal development increase the risk for this illness. Influenza may also predispose pregnant mothers to other risk factors. Wright et al., (1995) studied 121 schizophrenics patients whose mothers reported having influenza during the second trimester. Influenza-exposed mothers were almost five times as likely to experience at least one obstetric complication and lower birth weights than were patients whose mothers were not exposed during the second trimester.

A research question raised by Weinberger and others (Wolf & Weinberger, 1996; Weinberger & Lipska, 1996) is if early cortical maldevelopment underlies some symptoms of schizophrenia, why are the prominent symptoms of schizophrenia not apparent until adulthood? Archival data derived from films and medical records indicate that dyskinesia in infancy (Walker et al., 1994), and increases in negative facial emotion as early as the first year of life (Walker et al., 1993) are early predictors of the illness. Walker et al., (1996) analyzed home-movies of 29 schizophrenic patients and 28 of their healthy siblings that covered a time period from infancy through 15 years of age. Facial expression of emotion and neuromotor function were coded on a frame-by-frame basis by trained observers blind to the diagnostic status of the subjects. The raters identified neuromotor abnormalities and increases in negative facial emotion as two factors in children who later became schizophrenic (Walker et al., 1996). Although some of these signs are subtle, they may be important consequences of developmental abnormalities that eventually produce psychotic symptoms.

Postmortem Findings in the Mediodorsal Nucleus of the Thalamus

The importance of the mediodorsal nucleus of the thalamus in schizophrenia is based, in part, on the substantial connections of two of its subnuclei, the parvocellular and densocellular nuclei with the DLPFC as well as its connections to other structures within the DLPFC circuitry (Middleton & Strick, 2000). Three independent studies (Popken et al., 2000; Young et al., 2000; Jones et al., 1998) have confirmed Pakkenberg's original observation (Pakkenberg & Gundersen, 1989; Pakkenberg, 1990;1992;1993) of a highly significant decrease in neuronal number in the mediodorsal nucleus of the thalamus. Popken et al., (2000) reported that the neuronal numbers were reduced in the specific subnuclei of the mediodorsal thalamus that project to the middle layers of the DLPFC. A challenge for future research is to determine whether the initial primary defect occurs in the thalamus or in the DLPFC.

References

Anand A, Charney DS, Oren DA, Berman RM, Hu XS, Cappiello A, Krystal JH. Attenuation of the neuropsychiatric effects of ketamine: support for hypoglutamatergic effects of N-methyl-D-aspartate receptor antagonists. Arch Gen Psychiatry 2000;57: 270-276.

Anderson SA, Volk DW, Lewis DA. Increased density of microtubule associated protein 2-immunoreactive neurons in the prefrontal white matter of schizophrenic subjects. Schiz Res 1996;19: 111-119.

Akbarian S, Bunney WE Jr, Potkin SG, Wigal SB, Hagman JO, Sandman CA, Jones EG. Altered distribution of nicotinamide-adenine-dinucleotide phosphate-diaphorase cells in frontal lobe of schizophrenics implies disturbances of cortical development. Arch Gen Psychiatry 1993a;50:169-177.

Akbarian S, Vinuela A, Kim JJ, Potkin SG, Bunney WE Jr, Jones EG. Distorted distribution of nicotinamide-adenine dinucleotide phosphate-diaphorase neurons in temporal lobe of schizophrenics implies anomalous cortical development. Arch Gen Psychiatry 1993b;50:178-187.

Akbarian S, Huntsman MM, Kim JJ, Tafazzoli A, Potkin SG, Bunney, WE Jr, Jones EG. $GABA_A$ receptor subunit gene expression in human prefrontal cortex: Comparison of schizophrenics and controls. Cerebral Cortex 1995a;5: 550-560.

Akbarian S, Kim JJ, Potkin SG, Hagman JO, Tafazzoli A, Bunney WE Jr, Jones, EG. Gene expression for glutamic acid decarboxylase is reduced without loss of neurons in prefrontal cortex of schizophrenics. Arch Gen Psychiatry 1995b;52: 258-266.

Akbarian S, Smith MA, Jones EG. Editing for an AMPA receptor subunit mRNA in prefrontal cortex and striatum in Alzheimer's disease, Huntington's disease and schizophrenia. Brain Res 1995c; 699:297-304.

Akbarian S, Kim JJ, Potkin SG, Hetrick WP, Bunney WE Jr, Jones EG. Maldistribution of interstitial neurons in prefrontal white matter of schizophrenics. Arch Gen Psychiatry 1996a;53 178-187.

Akbarian S, Sucher NJ, Bradley D, Tafazzoli A, Trinh D., Hetrick, WP, Potkin SG, Sandman CA, Bunney WE Jr, Jones E.G. Selective alterations in gene expression for NMDA receptor subunits in prefrontal cortex of schizophrenics. Journal of Neurosci. 1996;16: 19-30.

Alexander GE, DeLong MR, Strick, PL. Parallel organization of functionally segregated circuits linking basal ganglia and cortex. Ann Rev Neurosci 1986; 9: 357-381.

Allendoerfer KL, Shatz CJ. The subplate, a transient neocortical structure: its role on the development of connections between thalamus and cortex. Ann Rev Neurosci 1986; 9: 357-381.

Barch DM, Carter CS, Craver TS, Sabb FW, MacDonald A, Noll D, Cohen JD. Selective deficits in prefrontal cortex function in medication-naïve patients with schizophrenia. Arch Gen Psychiatry 2001; 58: 280-288.

Benes FM, McSparren J, Bird ED, SanGiovanni JP, Vincent SL. Deficits in small interneurons in prefrontal and cingulate cortices of schizophrenic and schizoaffective patients. Arch Gen Psychiatry 1991;48: 996-1001.

Benes FM, Vincent SL, Marie A, Khan Y. Up-regulation of GABA-A receptor binding on neurons of the prefrontal cortex in schizophrenic subjects. Neuroscience 1996; 75:1021-31.

Berman KF, Torrey EF, Daniel DG, Weinberger DR. Regional cerebral blood flow in monozygotic twins discordant and concordant for schizophrenia. Arch Gen Psychiatry 1992; 49: 927-934.

Bertolino A, Breier A, Callicott JH, Adler C, Mattay VS, Shapiro M, Frank JA, Pickar D, Weinberger DR. The relationship between dorsolateral prefrontal neuronal N-acetylaspartate and evoked release of striatal dopamine in schizophrenia. Neuropsychopharmacology 2000b 22: 125-132.

Bertolino A, Callicott JH, Elman I, Mattay VS, Tedeschi G, Frank JA, Breier A, Weinberger DR. Regionally specific neuronal pathology in untreated patients with schizophrenia: a proton magnetic resonance spectroscope imaging study. Biol. Psychiatry 1998; 43: 641-648.

Bertolino A, Callicott JH, Mattay VS Weidenhammer KM, Rakow R, Egan MF, Weinberger DR. The effect of treatment with antipsychotic drugs in brain N-acetylaspartate measures in patients with schizophrenia. Biol Psychiatry 2001: 39-46.

Bertolino A, Esposito G, Callicott JH, Mattay VS, Van Horn JD, Frank JA, Berman KF, Weinberger DR. Specific relationship between prefrontal neuronal N-acetylaspartate and activation of the working memory cortical network in schizophrenia. Am J Psychiatry 2000a; 157: 26-33.

Bertolino A, Kumra S, Callicott JH, Mattay VS, Lestz RM, Jacobsen L, Barnett IS, Duyn JH, Frank JA, Rapoport JL, Weinberger DR. Common pattern of cortical pathology in childhood-onset and adult-onset schizophrenia as identified by proton magnetic resonance spectroscopic imaging. Am J Psychiatry 1998;155; 1376-1383.

Bredt DS, Glatt CE, Hwang PM, Fotuhi M, Dawson TM, Snyder SH. Nitric oxide synthase protein and mRNA are discretely localized in neuronal populations of the mammalian CNS together with NADPH diaphorase. Neuron 1991; 7: 2811-2814.

Breier A, Malhotra AK, Pinals DA, Weisenfeld NI, Pickar D. Association of ketamine-induced psychosis with focal activation of the prefrontal cortex in healthy volunteers. Am J Psychiatry 1997; 6: 805-811.

Bunney BG, Bunney WE, Carlsson A. Schizophrenia and glutamate. In: Bloom FE, Kupfer et al. (eds.) Psychopharmacology: The Fourth Generation of Progress 1995; pp 1205-1214. New York: Raven Press.

Bunney WE, Bunney BG. Evidence for a compromised dorsolateral prefrontal cortical parallel circuit in schizophrenia. Brain Res Rev 2000; 31: 138-46.

Callicott JH, Bertolino A, Mattay VS, Langheim FJ, Duyn J, Coppola R, Goldberg TE, Weinberger DR. Physiological dysfunction of the dorsolateral prefrontal cortex in schizophrenia revisited. Cerebral Cortex 2000; 10: 1078-1092.

Callicott JH, Ramsey NF, Tallent K, Bertolino A, Knable MB, Coppola R, Goldberg T, van Gelderen P, Mattay VS, Frank JA, Moonen CT, Weinberger DR. Functional magnetic resonance imaging brain mapping in psychiatry: methodological issues illustrated in a study of working memory in schizophrenia. Neuropsychopharmacology 1998; 18:186-196.

Carter CS, Perlstein W, Ganguli R, Brar J, Mintun M, Cohen JD. Functional hypofrontality and working memory dysfunction in schizophrenia. Am J Psychiatry 1998; 155: 1285-1287.

Catafau AM, Parellada E, Lomeana FJ, Bernardo M, Pavaia J, Ros D, Setoain, J, Gonzalez-Monclaus, E. Prefrontal and temporal blood flow in schizophrenia: resting and activation technetium-99m-HMPAO SPECT patterns in young neuroleptic-naïve patients with acute disease. J Nuclear Med 1994; 35: 935-941.

Daviss SR, Lewis DA. Local circuit neurons of the prefrontal cortex in schizophrenia selective increase in the density of calbindin-immunoreactive neurons. Psychiatry Res 1995; 59: 81-96.

Delamillieure P, Fernandez J, Constans JM, Brazo P, Benali K, Abadie P, Vasse T, Thibaut F, Courtheoux P, Petit M, Dollfus S. Proton magnetic resonance spectroscopy of the medial prefrontal cortex in patients with deficit schizophrenia: preliminary report. Am J Psychiatry 2000; 157: 641-643.

Fletcher, P.C., McKenna, P.J., Frith, C.D., Grasby, P.M., Friston, K.J., Dolan, R.J., Brain activations in schizophrenia during a graded memory task studied with functional neuroimaging. Arch Gen Psychiatry 1998; 55: 1001-1008.

Geddes JR, Verdoux H, Takei N, Lawrie SM, Bovet P, Eagles JM, Heun R, McCreadie RG, McNeil TF, O'Callaghan E, Stober G, Willinger U, Murray RM. Schizophrenia and complications of pregnancy and labor: an individual patient data meta-analysis. Schizophrenia Bull 1999; 25: 413-423.

Glantz LA, Lewis DA. Decreased dendritic spine density on prefrontal cortical pyramidal neurons in schizophrenia. Arch Gen Psychiatry 2000; 57: 65-73.

Goldberg TE, Torrey EF, Gold JM, Bigelow LB, Ragland RD, Taylor E, Weinberger DR. Genetic risk of neuropsychological impairment in schizophrenia: a study of monozygotic twins discordant and concordant for the disorder. Schizophrenia Res 1995; 17:77-84.

Goldman-Rakic PS. Working memory dysfunction in schizophrenia. J Neuropsychiatry Clin Neurosci 1994; 6:348-357.

Guidotti A, Auta J, Davis JM, Gerevini VD, Dwivedi Y, Grayson DR, Impagnatiello F, Pandey G, Pesold C, Sharma R, Uzunov D, Costa E. Decrease in reelin and glutamic acid decarboxylase67 (GAD67) expression in schizophrenia and bipolar disorder: A postmortem brain study. Arch Gen Psychiatry 2000; 57: 1061-1069.

Healy DJ, Haroutunian V, Powchik P, Davidson M, Davis KL, Watson SJ, Meador-Woodruff JH. AMPA receptor binding and subunit mRNA expression in prefrontal cortex and striatum of elderly schizophrenics. Neuropsychopharmacology 1998: 278-286.

Hendry SHC, Jones EG. Reduction in number of immunostained GABAergic neurons in deprived-eye dominance columns of monkey area 17. Nature 1986; 320: 750-753.

Hendry, SHC, Jones E.G. Activity-dependent regulation of GABA expression in the visual cortex of adult monkeys. Neuron 1988; 1: 701-712.

Hendry SHC, Huntsman MM, Vinuela A, Mohler H, deBlas AL, Jones EG. $GABA_A$ receptor subunit immunoreactivity in primate visual cortex: distribution in macaques and humans and regulation by visual input in adulthood. J Neurosci 1994; 14: 2382-2401.

Hinsberger AD, Williamson PC, Carr TJ, Stanley JA, Drost DJ, Densmore MM, MacFabe GC, Montemurro DG. Magnetic resonance imaging volumetric and phosphorus 31 magnetic resonance spectroscopy measurements in schizophrenia. J Psych Neurosci 1997; 22; 111-117.

Huntsman MM, Tran BV, Potkin SG, Bunney WE Jr, Jones EG. Altered ratios of alternatively spliced long and short gamma2 subunit mRNAs of the gamma-amino butyrate type A receptor in prefrontal cortex of schizophrenics. Proc Natl Acad Sci 1998; 95: 15066-15071.

Huttunen MO, Niskana P. Prenatal loss of father and psychiatric disorders. Arch Gen Psychiatry 1978; 35:429-431.

Jakob, H., Beckmann, H. Prenatal developmental disturbances in the limbic allocortex in schizophrenics. J Neural Transm 1986; 65:303-326.

Kalus P, Senitz D, Beckmann H. Cortical layer I changes in schizophrenia: a marker for impaired brain development? J Neural Transm 1997; 104: 549-559.

Kleinschmidt A, Bear MF, Singer W. Blockade of NMDA receptors disrupts experience-dependent modifications in kitten striate cortex. Science 1987; 238: 355-358.

Kostovic I, Rakic P. Cytology and time of origin of interstitial neurons in the white matter in infant and adult human and monkey telencephalon. J Neurocytol 1980; 9: 219-242.

Krystal JH, Karper LP, Seibyl JP, Freeman GK, Delaney R, Bremner JD, Heninger GR, Bowers MB Jr, Charney DS. Subanesthetic effects of the noncompetitive NMDA antagonist, ketamine, in humans. Psychotomimetic, perceptual, cognitive and neuroendocrine responses. Arch Gen Psychiatry 1994; 51: 199-214.

Krystal JH, D'Souza DC, Karper LP, Bennett A, Abi-Dargham A, Abi-Saab D, Cassello K, Bowers MB Jr, Vegso, Heninger GR, Charney DS. Psychopharmacology 1999; 145:193-204.

Lahti AC, Koffel B, LaPorte D, Tamminga CA. Subanesthetic doses of ketamine psychosis in schizophrenia. Neuropsychopharmacology 1995; 13: 9-19.

Lahti AC, Holcomb HH, Medoff DR, Tamminga CA. Ketamine activates psychosis and alters limbic blood flow in schizophrenia. Neuroreport 1995; 6: 869-872.

Lewis DA. GABAergic local circuit neurons and prefrontal cortical dysfunction in schizophrenia. Brain Res Rev 2000; 31: 270-276.

Li Y, Erzurumulu RS, Chen C, Jhaveri S, Tonegawa, S. Whisker-related neuronal patterns fail to develop in the trigeminal brainstem nuclei of NMDAR1 knockout mice. Cell 1994; 76: 427-437.

Luby ED, Cohen BD, Rosenbaum G, Gottleib JS, Kelly R. Study of a new schizophrenomimetic drug-Sernyl. Arch Neurol Psychiatry 1959; 81: 363-369.

Malhotra AK, Pinals DA, Adler CM, Elman I, Clifton A, Pickar D, Breier A. Ketamine-induced exacerbation of psychotic symptoms and cognitive impairment in neuroleptic-free schizophrenics. Neuropsychopharmacology 1997; 17: 141-150.

Marenco S, Weinberger DR. The neurodevelopmental hypothesis of schizophrenia: following a trail of evidence from cradle to grave. Dev Psychopathol 2000; 12: 501-527.

McDonald C, Murray RM. Early and late environmental risk factors for schizophrenia. Brain Res Reviews 2000;31: 130-137.

Meador-Woodruff JH, Davis, KL, Harotounian V. Abnormal kainate receptor expression in prefrontal cortex of schizophrenia. Neuropsychopharmacology 2001; 24: 545-552.

Mednick SA, Huttunen MO, Machon RA. Prenatal influenza infections and adult schizophrenia. Schiz Bulletin 1994; 20: (1994) 263-267.

Mednick SA, Machon RA, Huttunen MO. Adult schizophrenia following prenatal exposure to an influenza epidemic. Arch. Gen. Psychiatry 1988; 45: 189-192.

Middleton FA, Strick PL. Basal ganglia output and cognition: evidence from anatomical, behavioral and clinical studies. Brain and Cognition 2000; 42: 183-200.

Monyer H, Burnashev N, Laurie DJ, Sakmann B, Seeburg PH. Developmental and regional expression in the rat brain and functional properties of four NMDA receptors. Neuron 1994; 12: 529-540.

Ohnuma T, Kato H, Arai H, Faull RL, McKenna PJ, Emson PC. Gene expression of PSD95 in prefrontal cortex and hippocampus in schizophrenia. Neuroreport 2000; 11: 3133-3137.

Pakkenberg B, Gundersen HJ. New stereological method for obtaining unbiased and efficient estimates of total nerve cell number in human brain areas. Exemplified by the mediodorsal thalamic nucleus in schizophrenics. APMIS 1989; 97: 677-681

Pakkenberg B. Leucotomized schizophrenics lose neurons in the mediodorsal thalamic nucleus. Neuropathol Appl Neurobiol 1993: 19: 373-380.

Pakkenberg B. Pronounced reduction of total neuron number in mediodorsal thalamic nucleus and nucleus accumbens in schizophrenics. Arch Gen Psychiatry 1990; 47: 1023-1028.

Pakkenberg B. The volume of the mediodorsal thalamic nucleus in treated and untreated schizophrenics. Schizophr Res 1992; 7: 95-100.

Penn AA, Shatz CC. Brain waves and brain wiring: the role of endogenous and sensory-driven neural activity in development. Pediatric Research 1999; 45: 447-458.

Popken GJ, Bunney WE Jr, Potkin SG, Jones EG. Subnucleus-specific loss of neurons in medial thalamus of schizophrenics. Proc Natl Acad Sci 2000; 97: 9276-9280.

Ragland JD, Gur RC, Glahn DC, Censits DM, Smith RJ, Lazareth MG, Alavi A, Gur RE. Frontotemporal cerebral blood flow change during executive and declarative memory tasks in schizophrenia: a positron emission tomography study. Neuropsychology 1998; 12: 399-413.

Rajkowska G, Selemon LD, Goldman-Rakic PS. Neuronal and glial somal size in the prefrontal cortex: a postmortem morphometric study of schizophrenia and Huntington disease. Arch Gen Psychiatry 1998; 55: 215-224.

Reynolds GP, Czudek C, Andrews HB. Deficit and hemispheric asymmetry of GABA uptake sites in the hippocampus in schizophrenia. Biol Psychiatry 1990; 27: 1038-1044.

Schlaggar BL, Fox K, O'Leary DDM, Postsynaptic control of plasticity in developing somatosensory cortex. Nature 1993; 364: 623-626.

Selemon LD, Goldman-Rakic P. The reduced neuropil hypothesis: a circuit based model of schizophrenia. Biol Psychiatry 1999; 45: 17-25.

Selemon LD, Rajkowska G, Goldman-Rakic PS. Abnormally high neuronal density in the schizophrenic cortex. A morphometric analysis of prefrontal area 9 and occipital area 17. Arch Gen Psychiatry 1995; 52: 805-818.

Sherman AD, Davidson AT, Baruah S, Hegwood TS, Waziri R. Evidence of glutamatergic deficiency in schizophrenia. Neurosci Lett 1991; 121: 77-80.

Simpson MD, Slater P, Deakin JF. Gamma-aminobutyric acid uptake binding sites in frontal and temporal lobes in schizophrenia. Biol Psychiatry 1998; 44: 423-427.

Susser E, Neugebauer R, Hoek H, Lin S, Labovitz D, Gorman J. Schizophrenia after prenatal famine: further evidence. Arch. Gen. Psychiatry 1996; 53: 25-31.

Tamminga CA, Vogel M, Gao X-M, Lahti AC, Holcomb HH. The limbic cortex in schizophrenia: focus on the anterior cingulate. Brain Res Rev 2000;31: 364-370.

Thune JJ, Uylings HB, Pakkenberg B. No deficit in total number of neurons in the prefrontal cortex in schizophrenics. J Psychiatry Res 2001; 35: 15-21.

Van Os J, Selten JP. Prenatal exposure to maternal stress and subsequent schizophrenia. The May 1940 invasion of The Netherlands Brit J Psychiatry 1998; 172: 324-326.

Volk DW, Austin MC, Pierri JN, Sampson ARE, Lewis DR. Decreased glutamic acid decarboxylase67 messenger RNA expression in a subset of prefrontal cortical gamma-aminobutyric acid neurons in subjects with schizophrenia. Arch Gen Psychiatry 2000; 57:237-245.

Volk D, Austin M, Pierri J, Sampson A, Lewis D. GABA transporter-1 mRNA in the prefrontal cortex in schizophrenia: decreased expression in a subset of neurons. Am J Psychiatry 2001; 158: 256-265.

Walker E, Lewine RJ. Prediction of adult-onset schizophrenia from childhood home movies of the patients. Am J Psychiatry 1990; 147: 1052-1056.

Walker E, Savoie T, Davis D. Neuromotor precursors of schizophrenia. Schizophr Bull 1994; 20: 453-480.

Walker E, Grimes K, Davis D, Smith A. Childhood precursors of schizophrenia: Facial expressions of emotion. Am J Psychiatry 1993: 150: 1654-1660.

Walker, E., Lewine, R.J., Neumann, C. Childhood behavioral characteristics and adult brain morphology in schizophrenia. Schizophr Res 1996; 22: 93-101.

Weinberger, D.R., Berman, K.F., Zec, R.F. Physiological dysfunction of dorsolateral prefrontal cortex in schizophrenia, I: Regional cerebral blood flow evidence. Arch Gen Psychiatry 43(1986):114-124.

Weinberger DR, Lipska BK. Cortical maldevelopment, anti-psychotic drugs and schizophrenia: a search for common ground. Schizophr Res 1995; 16:87-110.

Wright P, Takei N, Rifkin L, Murray RM. Maternal influenza, obstetric complications and schizophrenia. Am J Psychiatry 1995; 152: 1714-1720.

Woo TU, Whitehead RE, Melchitzky DS, Lewis DA. A subclass of prefrontal gamma-aminobutyric acid axon terminals are selectively altered in schizophrenia. Proc Natl Acad Sci USA 1998; 95: 341-5346.

Young KA, Manaye KF, Liang C, Hicks PB, German DC. Reduced number of mediodorsal and anterior thalamic neurons in schizophrenia. Biol Psychiatry 2000; 47: 944-953.

13 POST MORTEM STUDIES OF THE HIPPOCAMPAL FORMATION IN SCHIZOPHRENIA

Andrew J. Dwork

Abstract

In recent years, dramatic abnormalities have been found in the hippocampal formation in schizophrenia. These include diminished levels of dendritic spines, reelin, and the 67 kd isoform of glutamic acid decarboxylase, and increased levels of brain derived neurotrophic factor. These findings are not limited to the hippocampal formation and are consistent with excessive synaptic pruning. So far, however, there is little to indicate when the process began.

Introduction

Several laboratories, from the mid-1980's through the mid-1990's, reported anatomical abnormalities of the size, neuronal density, neuronal orientation, and cytoarchitectural organization of the hippocampal formation in autopsy brains from individuals with schizophrenia. Unaccompanied by gliosis, these changes were frequently interpreted as indicative of developmental abnormalities in the hippocampus. However, the reports contained many mutual contradictions, leading to questions of whether any of the described abnormalities could be observed reproducibly. The only alteration observed consistently was mild enlargement of the temporal horns of the lateral ventricles, especially on the left side. Methodological issues in these earlier studies have been reviewed in detail (Dwork, 1997). Towards the end of this period, newly developed stereological techniques overcame many of these shortcomings, and well-designed studies failed to confirm previously reported abnormalities.

Anatomical Studies

Heckers et al. (1991) and Benes et al. (1998) found normal volume, neuronal cell density, and total neuron number in each subfield of the cornu amonis (CA1-CA4). Arnold et al (1995) found normal cell density and orientation in all subfields of the cornu amonis, in the subiculum, and in the entorhinal cortex. Two careful studies of the entorhinal cortex (Akil et al., 1997; Krimer et al., 1997) likewise failed to show any abnormalities in schizophrenia. Heinsen et al. (1996) (in a study of 450 brains) and Bernstein et al. (1998) both concluded that in the entorhinal cortex, cytoarchitectural "abnormalities" of the type described in schizophrenia (Falkai et al., 1988; Arnold et al., 1991a) are equally common in nonpsychiatric cases. Thus cytoarchitectonic abnormalities, altered neuronal orientation, and altered numbers of pyramidal cells in the hippocampus could probably be ruled out as characteristic features of schizophrenia.

More subtle abnormalities may yet be present. An interesting study of the entorhinal cortex (Arnold et al., 1997) used a mathematical model to analyze the positions of neurons and measured statistically significant differences that suggested that the schizophrenia cases had more widely spaced neurons in layer II, and a more clustered distribution of neurons in layer III, with no differences in layer V. Falkai et al (2000, and his chapter in this book) measured the distances of entorhinal layer 2 cell clusters from the pial surface and gray-white junction, and found them slightly but significantly deeper (by ~3% of the cortical thickness) in schizophrenia. (The report does not state whether these measurements were made blind to diagnosis.) These studies point to the difficulty, apparent in earlier work, of quantitatively evaluating cytoarchitectural patterns. The application of different measurements to different data sets creates serious problems for evaluating reproducibility, but the earlier, rather dramatic reports of cytoarchitectural abnormalities in the entorhinal cortex (Jakob et al., 1986; Jakob et al., 1989; Arnold et al., 1991a) were apparently normal variations that were not recognized as abnormalities in properly blinded studies. It might be of interest to apply position analysis to the material studied previously. It should also be considered that the earlier studies may have included a number of truly peculiar cases, which might have shared distinct clinical features. Schizophrenia is likely heterogeneous in etiology as well as in clinical and neuropathological features, and correlations among these three domains may emerge if sought.

The failure to confirm cytoarchitectural abnormalities in the hippocampal formation weakens the case for a developmental abnormality in neuronal migration as the underlying defect in schizophrenia. Furthermore, as pointed out clearly by Harrison (1999), even if cytoarchitectural abnormalities were present, absence of gliosis does not provide definitive proof of an early neurodevelopmental origin. The developmental timing of glial responsiveness has not been well studied and may be regionally variable. Gliosis does not always occur or persist after postnatal neural injury, does not accompany apoptosis, and may not occur in the context of a process producing very subtle cytoarchitectural abnormalities, regardless of the age at which such a process takes place.

Neuronal and Dendritic Size

Several studies (Benes et al., 1991; Arnold et al., 1995; Zaidel et al., 1997) have shown that pyramidal cells in several subfields of the hippocampus are modestly (~15%) but significantly smaller in schizophrenia than in nonpsychiatric subjects. The smaller somal sizes suggest that these cells are supporting smaller dendritic arbors and fewer synapses, as discussed below. However, a subsequent study from one of these laboratories failed to confirm these findings (Benes et al., 1998). The discrepancy may be explained by laterality. Zaidel et al. (1997) evaluated each side separately, and found that the schizophrenia subjects had significantly smaller neurons than nonpsychiatric subjects in CA1 and CA2 on the left, and in CA3 on the right. Arnold et al. (1995) sampled bilaterally and apparently combined the two sides for their data analysis, and Benes et al. (1991; 1998) do not state, in either report, which side the sections were taken from, or whether one side was consistently sampled. In a subsequent analysis, Zaidel (1999) found reduced regional variability of pyramidal cell size among different subfields on the right side of the hippocampus in schizophrenia, with similar values to controls on the left side.

Although the correlation between somal size and dendritic extent is hypothetical, there is also direct evidence for diminished dendritic extent in the hippocampal formation. Using Golgi impregnation (Rosoklija et al., 2000), we found a significant reduction in the apical dendritic arbors of left subicular pyramidal cells in schizophrenia. Basilar dendrites, and the dendrites of pyramidal cells in the adjacent temporal neocortex (fusiform gyrus) were unaffected. The subicular pyramidal cells relay the hippocampal output from CA1; which they receive through the alveus, to cortical and subcortical structures. This intrinsic input probably synapses predominantly with basilar dendrites (Finch et al., 1983), while modulatory inputs from the transentorhinal cortex, amygdala, locus ceruleus, and ventral tegmental area (Amaral et al., 1990) are more likely to synapse with apical dendrites (Kosel et al., 1983; Verney et al., 1985; Aggleton, 1985). This arrangement suggests that the functional effect of a lessened apical dendritic arborization is likely not to be an interruption of hippocampal output, but rather an abnormality of its modulation.

Other evidence for subicular dendritic abnormalities is provided by studies of microtubule-associated protein (MAP)-2, a dendritic protein that stabilizes microtubular structure. Arnold et al. (1991b) reported a nearly total loss of MAP-2 immunoreactivity in the subiculum of 5 out of 6 schizophrenia subjects, but in no nonpsychiatric cases. We have found this abnormality in 14 of 64 schizophrenia cases studied, and in none of 14 presumed and 17 confirmed nonpsychiatric cases. A similar frequency of low subicular MAP-2 immunoreactivity was present in dementia cases. However, in dementia, low subicular MAP-2 immunoreactivity is associated with gliosis, and therefore probably represents a nonspecific effect of neuronal loss or injury. In schizophrenia, low subicular MAP-2 immunoreactivity is unassociated with gliosis, and therefore probably represents a more specific abnormality.

Our study and that of Arnold et al. (1991b) used different monoclonal antibodies; both were specific for high molecular weight MAP-2 (MAP-2a and MAP-2b, the major forms in adult brain) but independent of phosphorylation state. In contrast, Cotter et al. (2000) observed immunoreactivity that was approximately 50% more extensive (based on total immunoreactive dendritic length) for nonphosphorylated MAP-2 (the predominant form), and approximately 20% more extensive for total MAP-2, with antibodies that recognized both high and low molecular weight MAP-2 (MAP-2c). There are several possible explanations for these discrepancies. First, it is possible that MAP-2c immunoreactivity persists into postnatal life in schizophrenia, as it does in cortical dysplasia (Yamanouchi et al., 1998). Second, there are many factors, besides the amount of MAP-2 present in dendrites, that will contribute to its immunoreactivity. For example, dissociation of MAP-2 from microtubules diminishes its immunoreactivity (Bigot et al., 1991). Conceivably, some alteration of microtubules in schizophrenia could impair the availability of some MAP-2 epitopes, while enhancing the immunoreactivity of others. The issue will be resolved only by studying MAP-2 with a panel of antibodies in the same specimens, and by applying quantitative assays in addition to immunohistochemistry.

Whether over- or under-expressed, MAP-2 could be involved in the pathophysiology of schizophrenia. Activation of NMDA receptors diminishes MAP-2 phosphorylation, favoring the association of MAP-2 with microtubules, in turn altering its immunoreactivity. Conversely, loss of MAP-2, or decreased baseline phosphorylation, could diminish such functional effects of NMDA receptors as are mediated by the dephosphorylation of MAP-2.

Spines and Synapses

Another dramatic change revealed by Golgi stains is a severe (83%) loss of spines on subicular apical dendrites. Spine densities measured in schizophrenia and nonpsychiatric cases did not overlap (Rosoklija et al., 2000), and the difference is readily apparent on inspection (Figure 1). If confirmed, this would be the first anatomical measure with high sensitivity for schizophrenia (and high specificity, when the comparison group is nonpsychiatric; low spine density was also present in some mood disorder cases). Dendritic spines are the main location for excitatory synapses. In keeping with the loss of spines, several studies have reported a loss of presynaptic proteins in the hippocampal formation. Davidson et al (1999), using western blots, reported a statistically significant 30% loss of synaptophysin and Rab3a (low molecualar weight GTP-binding protein-3a) in the left hippocampus. There was limited overlap between groups, with two-thirds of the nonpsychiatric subjects having higher values than any schizophrenia subject. Eastwood and Harrison (1995b), using autoradiographic immunohistochemistry, reported a statistically significant, 23% loss of synaptophysin in the right subiculum, 20% in the right dentate gyrus molecular layer, and smaller, statistically insignificant decreases in these regions and CA1 on the left and in CA3 and CA4 bilaterally. Bilateral in situ hybridization indicated a 25-40% loss of synaptophysin mRNA in subiculum, CA3, CA4, and dentate gyrus, statistically significant in all regions except the dentate gyrus (Eastwood et al., 1995a), and similar results were obtained in a second study by this laboratory (Eastwood et al., 1999). Vawter et al. (1999) using western blots and normalizing for actin, measured 58% less synaptophysin in the hippocampus (side unstated) in schizophrenia. Browning et al. (1993), with western blots on hippocampus from unstated side(s), found a statistically insignificant, 9% decrease in schizophrenia, but a statistically significant 43% decrease of synapsin I. Using ELISA, Young et al. (1998) found a 45% decrease in synaptophysin and a 25% decrease in SNAP-25 in the hippocampus (side

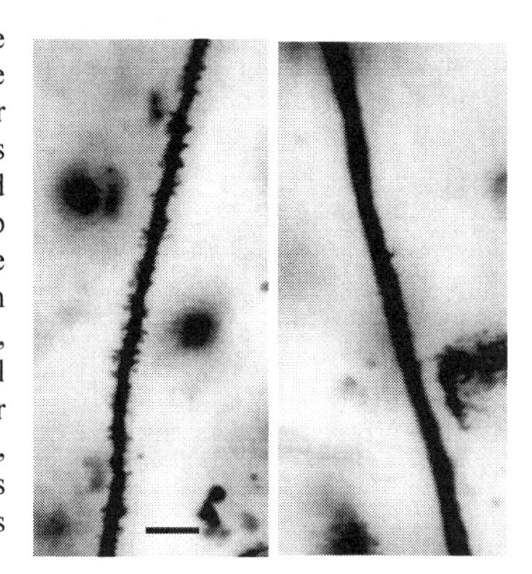

Figure 1 Apical dendrites from subicular pyramidal cells. Left, nonpsychiatric case. Note abundant spines. Right, schizophrenia case.
Bar = 10 microns

unstated), although neither difference was statistically significant. Thus, hippocampal synaptophysin loss of 9% to 58% was reported, with comparable loss of other presynaptic proteins. The only value specifically stated for the synaptophysin in the left subiculum was a decrease of 16%, which was not statistically significant (Eastwood et al., 1999).

The loss of apical dendritic spines was more pronounced than the loss of presynaptic proteins, but spines are not the only sites of synapses, and basilar dendritic spines, while not quantified, were certainly better preserved in our schizophrenia cases than were the spines on the apical dendrites. Furthermore, since dendritic spines are the principal location of excitatory synapses, one might expect a loss of NMDA receptors. However, there is no evidence for decreased NMDA receptor binding in the hippocampus in schizophrenia. Kainate and AMPA receptors may be decreased in CA3 and CA4, but reports are inconsistent (Meador-Woodruff et al., 2000; Gao et al., 2000). This suggests either an up-regulation of NMDA receptors in preserved synapses, or the formation of excitatory synapses on shafts, rather than spines. In addition, some spines that appear to have been lost may in fact have shrunken beyond the resolution of the light microscope, possibly maintaining their original contacts but with vastly altered physiological effects on their neurons. Roberts et al. (1996), using electron microscopy, found that striatal dendritic spines were approximately 30% smaller in schizophrenia subjects than in controls. We are not aware of any such studies in neocortex or hippocampus

Dendritic spines and presynaptic proteins are also lost in the prefrontal cortex in schizophrenia. Garey et al. (1998) reported a two-thirds reduction in spine density in layer 3 temporal (mostly BA 38) and frontal cortex (mostly BA 10 or 11). Several cytoarchitectonic areas were combined in the analyses, apical and basilar dendrites were both included, and no attempt was made to control for distance from the cell body. Only 6 of 13 schizophrenia cases had sufficient tissue to evaluate for senile changes, and 3 of these met the Consortium to Estabilish a Registry for Alzheimer's Disease (CERAD) (Mirra et al., 1991) neuropathological criteria for probable or possible Alzheimer's Disease (although these 3 did not have lower spine counts than the others). These factors could have introduced noise into the data, leading to an overlap between groups that was not present in our study; otherwise, the difference between groups was similar for both studies. Glantz and Lewis (2000) reported a statistically significant 21% loss of spine density on basilar dendrites of pyramidal dendrites in deep layer 3 of the dorsolateral prefrontal cortex (BA 46). The schizophrenia subjects were considerably younger than in the study of Garey et al. (1998) or in ours (Rosoklija et al., 2000). Superficial layer 3 of BA 46 showed a similar but statistically insignificant loss of spine density, while none was seen in layer 3 of primary visual cortex (BA 17). Spine density showed considerably more overlap between groups than in the previous study or ours, which could be related to the anatomical area, or to the age of the subjects, or to the fact that only basilar dendrites were evaluated.

Decreased gray matter volume with preserved neuronal number has been reported in prefrontal cortex (BA 9, BA 10, BA 46), in the cortex as a whole, and in the primary visual cortex (reviewed in Selemon et al., 1999). There is also a report of a 30% reduction in the extent of basilar dendritic arbors of layer 3 pyramidal cells in the prefrontal cortex (BA 11) in schizophrenia (Kalus et al., 2000). These studies suggest that loss of neuropil and of spines may be a widespread phenomenon in schizophrenia. On the other hand, there seems to be more regional specificity in the loss of presynaptic proteins. These were generally found to be decreased in hippocampus and prefrontal cortex, the most frequently studied regions, with negative or less consistent results in most other cortical areas (Table 1).

Reelin, GAD-67 and BDNF

A loss of expression of mRNA for the 67 kd form of glutamic acid decarboxylase (GAD-67) has been described in the left prefrontal cortex (BA 9) in schizophrenia, without a loss of total neurons or of small neurons morphologically typical of GABA-ergic interneurons. The loss was most pronounced (40-50%) in cortical layers I and II (Akbarian et al., 1995). Reelin, a protein synthesized by Cajal-Retzius neurons that plays an important role in developmental migration of neurons, is expressed in non-pyramidal neurons of the cerebral cortex in the adult human (Impagnatiello et al., 1998) and in adult nonhman primates (Rodriguez et al., 2000). The protein is constitutively secreted by these neurons and interacts with integrin receptors on adjacent dendritic spines of pyramidal cells, initiating a cascade that could affect DNA transcription in pyramidal neurons and thereby affect neuronal plasticity (Pesold et al., 1999). In schizophrenia, reelin mRNA was significantly reduced, by ~40% in prefrontal cortex, temporal neocortex, and cerebellum, by 50% in hippocampus, and by ~70% in the caudate nucleus. Western blotting of temporal neocortex revealed significantly decreased immunoreactivity for reelin (by 64%) and GAD-67 (by 72%) in the schizophrenia subjects. Immunohistochemistry on the prefrontal cortex revealed a loss of ~50% of reelin-immunoreactive neurons in layers I and II. These findings were replicated in a subsequent study, which also yielded very similar results in subjects with bipolar disorder, but not with unipolar depression (Guidotti et al., 2000). Fatemi et al. (2000) found decreased numbers of reelin-immunoreactive cells in most sub-regions of the

Table 1: Studies of Synaptic Proteins in Schizophrenia. (Arrows indicate the direction of the observed change in the schizophrenia group; '=' indicates no significant difference)

Study	Synaptophysin protein	Synaptophysin mRNA	Other synaptic proteins	mRNA for other synaptic proteins
Eastwood et al. 2000	= prefrontal (BA 9/46) = superior temporal (BA 22) ?↓ calcarine (BA17)			
Landen et al. 1999	↓ left thalamus =right thalamus		=chromogranin right thalamus =chromogranin left thalamus	
Honer et al. 1999	↓ anterior frontal			
Vawter et al. 1999	↓ hippocampus			
Davidsson et al. 1999	↓ hippocampus ↓cingulate = parietal =frontal =temporal =cerebellum		Rab3a: ↓ hippocampus ↓cingulate ↓ parietal ↓frontal =temporal =cerebellum	
Eastwood and Harrison 1999		↓ hippocampus		
Karson et al. 1999	↓ frontal (BA 10)	= frontal (BA 10)	↓SNAP25 frontal (BA 10)	=SNAP25 frontal (BA 10)
Young et al. 1998	=hippocampus (noon-significant ↓)		↓SNAP25 hippocampus	
Glantz and Lewis 1997	↓ frontal (BA 46) ↓ frontal (BA 9) =calcarine (BA 17)			
Tcherepanov and Sokolov 1997		↑ temporal in young = temporal in old decline with age in schizophrenia, not controls		Synapsin 1A & 1B: ↑ temporal in young = temporal in old decline with age in schizophrenia, not controls
Honer et al.	= cingulate		↑syntaxin	

Study	Synaptophysin protein	Synaptophysin mRNA	Other synaptic proteins	mRNA for other synaptic proteins
1997			cingulate ↑NCAM cingulate	
Gabriel et al. 1997	↑ cingulate =temporal =frontal =parietal		Syntaxin, SNAP25: ↑ cingulate =temporal =frontal =parietal	
Perrone-Bizzozero et al. 1996	↓visual association cortex ↓frontal		GAP-43 ↑visual association ↑frontal	
Eastwood and Harrison 1995	↓throughout medial temporal			
Eastwood et al. 1995	=throughout medial temporal	↓ throughout medial temporal except = in CA1		
Browning et al. 1993	= hippocampus		Hippocampus: =synapsin IIB ↓ synapsin 1	

hippocampus from subjects with schizophrenia, with a total decline of 27% compared with nonpsychiatric subjects. However, the total area in which these cells were counted was 21% smaller in the schizophrenia subjects, so the densities, as well as the densities within each subregion, were similar between the two groups. The significance of this study is therefore unclear. If the difference in area reflects true differences in hippocampal volumes, these are unusual groups of subjects. If sampling was not truly representative, then both groups might have the same hippocampal volume and the same total number and density of reelin-immunoreactive cells. If the difference in size is due to increased shrinkage artifact in the schizophrenia cases, and the pre-shrinkage volumes were the same for both groups, then the number and density of reelin-immunoreactive cells is indeed lower in schizophrenia. The last possibility must be seriously considered, since Impagnatiello et al. (1998) found a 50% reduction of reelin mRNA in the hippocampus in schizophrenia. It would be informative to measure the density of pyramidal neurons in the same sections, which would be increased if tissue shrinkage were excessive. The question is of sufficient interest to warrant new study, with appropriate sampling through the entire extent of the hippocampus in order to obtain an unbiased estimate of the total number of reelin-immunoreactive cells. It should also be remembered, however, that immunohistochemistry , especially with enzyme-linked detection, is not a quantitative procedure. Cells could probably undergo a considerable decline in reelin content without losing their immunoreactivity.

Takahashi et al. (2000) reported hippocampal brain-derived neurotrophic factor (BDNF) levels in schizophrenia over twice as high as those in nonpsychiatric comparison brains. There was very little overlap, with all of the schizophrenia cases above the 90th percentile for control cases. Levels were also significantly increased in the anterior cingulate cortex, although with considerable overlap between groups, but not in prefrontal or occipital cortex. There were no significant differences in nerve growth factor or neurotropin-3. The BDNF receptor, TrkB (receptor tyrosine kinase B), was reduced by ~40% in the hippocampus. The few neuroleptic-withdrawn schizophrenia subjects were typical of the group in all measures, and in rats, a month of haloperidol administration produces either no effect (Takahashi et al., 2000) or a decline (Angelucci et al., 2000) in hippocampal BDNF and TrkB. In the developing brain, over expression of BDNF suppresses the expression of reelin (Ringstedt et al., 1998), but it is not known if this occurs in the adult.

Conclusions

When Feinberg (1982) proposed that schizophrenia might be caused by "a fault in programmed synaptic elimination during adolescence," he was careful to state that it was "not possible to predict whether… the abnormality is due to the elimination of too many or too few synapses or from the wrong ones." Feinberg (1990) later reiterated this position in response to a computer model in which excessive synaptic pruning predicted schizophrenic phenomena (Hoffman et al., 1989), stating, "…one could write other computer programs, containing different assumptions, in which elimination of the 'wrong' synapses or of an insufficient number of synapses would give rise to metabolic abnormalities and behavioral symptoms." More recently, however, cognitive impairment has been recognized as a major component of schizophrenia, and strong correlations between synaptic loss and cognitive impairment have been recognized in Alzheimer's disease (AD), provoking further interest in the hypothesis of a synaptic deficit in schizophrenia. Most post mortem studies lend support to this hypothesis.

A plausible model relating progressive synaptic deficiency to the clinical course of schizophrenia has been proposed (McGlashan et al., 2000). However, neither the clinical symptoms nor the anatomical evidence can tell us how long a brain abnormality was present before the onset of schizophrenia. Anatomical proof of a developmental abnormality is lacking, but there is likewise no evidence to prove a later onset. The most dramatic and consistently described abnormality is a loss of dendritic spines. Spines are extremely plastic; changes in their size and shape have been reported over the course of minutes (Chen et al., 2000). Likewise, synaptic contacts undergo continuous changes. Reelin is an important developmental protein, but it continues to be synthesized in adulthood.

In the past few years, a number of dramatic abnormalities have been observed in post mortem brains: fewer spines, less reelin and GAD67, more BDNF, less TrkB. All are suggestive of altered synaptic connectivity in schizophrenia, but none tells us when this began. Any of these abnormalities could date from intrauterine development, from the onset of schizophrenia, from years later, or from any time in between.

Acknowledgements

Supported by MH60877, MH64168, AG08702, MH46745, MH50727, The National Alliance for Research on Schizophrenia and Depression, the Theodore and Vada Stanley Foundation, and the Lieber Center for Schizophrenia Research at the Department of Psychiatry, College of Physicians and Surgeons of Columbia University.

References

Aggleton, J P, 1985. A description of intra-amygdaloid connections in old world monkeys. Exp Brain Res 1985; 57: 390-399.

Akbarian, S, Kim J J, Potkin S G, Hagman J O, Tafazzoli A, Bunney W E, Jr, and Jones E G. Gene expression for glutamic acid decarboxylase is reduced without loss of neurons in prefrontal cortex of schizophrenics. Arch Gen Psychiatry 1995; 52: 258-266.

Akil M and Lewis D A. Cytoarchitecture of the entorhinal cortex in schizophrenia. Am J Psychiatry 1997; 154: 1010-1012.

Amaral D G and Insausti R. Hippocampal Formation. In Paxinos, G. (Ed). The Human Nervous system. Academic Press, San Diego, CA. 1990; pp. 711-755.

Angelucci F, Mathe A A, and Aloe L. Brain-derived neurotrophic factor and tyrosine kinase receptor TrkB in rat brain are significantly altered after haloperidol and risperidone administration. J Neurosci Res 2000; 60: 783-794.

Arnold S E, Franz B R, Gur R C, Gur R E, Shapiro R M, Moberg P J, and Trojanowski J Q. Smaller neuron size in schizophrenia in hippocampal subfields that mediate cortical-hippocampal interactions. Am J Psychiatry 1995; 152: 738-748.

Arnold SE, Hyman BT, Van Hoesen GW and Damasio A R. 1991a. Some cytoarchitectural abnormalities of the entorhinal cortex in schizophrenia. Arch Gen Psychiatry 1991a; 48: 625-632.

Arnold SE, Lee VM, Gur RE and Trojanowski JQ, 1991b. Abnormal expression of two microtubule-associated proteins (MAP-2 and MAP5) in specific subfields of the hippocampal formation in schizophrenia. Proc Natl Acad Sci USA 1991b; 88: 10850-10854.

Arnold S E, Ruscheinsky DD and Han LY. Further evidence of abnormal cytoarchitecture of the entorhinal cortex in schizophrenia using spatial point pattern analyses. Biol Psychiatry 1997; 42: 639-647.

Benes FM, Kwok EW, Vincent SL and Todtenkopf MS. A reduction of nonpyramidal cells in sector CA2 of schizophrenics and manic depressives. Biol Psychiatry 1998; 44: 88-97.

Benes FM, Sorensen I and Bird ED. Reduced neuronal size in posterior hippocampus of schizophrenic patients. Schizophr Bull 1991; 17: 597-608.

Bernstein HG, Krell D, Baumann B, Danos P, Falkai P, Diekmann S, Henning H and Bogerts, B. Morphometric studies of the entorhinal cortex in neuropsychiatric patients and controls: clusters of heterotopically displaced lamina II neurons are not indicative of schizophrenia. Schizophr Res 1998; 33: 125-132.

Bigot D, Matus A and Hunt SP. Reorganization of the cytoskeleyon in rat neurons following stimulation with excitatory amino acids in vitro. Eur J Neurosci 1991; 3: 551-558.

Browning MD, Dudek EM, Rapier JL, Leonard S and Freedman R. Significant reductions in synapsin but not synaptophysin specific activity in the brains of some schizophrenics. Biol Psychiatry 1993; 34: 529-535.

Chen BE, Lendvai B, Nimchinsky EA, Burbach B, Fox K and Svoboda K. Imaging high-resolution structure of GFP-expressing neurons in neocortex in vivo. Learn Mem 2000; 7: 433-441.

Cotter D, Wilson S, Roberts E, Kerwin R and Everall IP. Increased dendritic MAP-2 expression in the hippocampus in schizophrenia. Schizophr Res 2000; 41: 313-323.

Davidsson P, Gottfries J, Bogdanovic N, Ekman R, Karlsson I, Gottfries CG and Blennow K. The synaptic-vesicle-specific proteins rab3a and synaptophysin are reduced in thalamus and related cortical brain regions in schizophrenic brains. Schizophr Re 1999; 40: 23-29.

Dwork AJ. Postmortem studies of the hippocampal formation in schizophrenia. [Review] [60 refs]. Schizophr Bull 1997; 23: 385-402.

Eastwood SL, Burnet PW and Harrison PJ. Altered synaptophysin expression as a marker of synaptic pathology in schizophrenia. Neuroscience 1995a; 66: 309-319.

Eastwood SL, Cairns NJ and Harrison PJ. Synaptophysin gene expression in schizophrenia. Investigation of synaptic pathology in the cerebral cortex. Br J Psychiatry 2000; 176: 236-242.

Eastwood SL and Harrison PJ. Decreased synaptophysin in the medial temporal lobe in schizophrenia demonstrated using immunoautoradiography. Neuroscience 1995b; 69: 339-343.

Eastwood SL and Harrison PJ. Detection and quantification of hippocampal synaptophysin messenger RNA in schizophrenia using autoclaved, formalin-fixed, paraffin wax-embedded sections. Neuroscience 1999; 93: 99-106.

Fatemi SH, Earle JA, McMenomy T. Reduction in reelin immunoreactivity in hippocampus of subjects with schizophrenia, bipolar disorder and major depression. Mol Psychiatry 2000; 5:654-663.

Falkai P, Bogerts B and Rozumek. Limbic pathology in schizophrenia: the entorhinal region--a morphometric study. Biol Psychiatry 1988; 24: 515-521.

Falkai P, Schneider-Axmann T and Honer WG. Entorhinal cortex pre-alpha cell clusters in schizophrenia: quantitative evidence of a developmental abnormality. Biol Psychiatry 2000; 47: 937-943.

Feinberg I. Schizophrenia: caused by a fault in programmed synaptic elimination during adolescence? J Prsychiatr Res 1982; 17: 319-334.

Feinberg I. Cortical pruning and the development of schizophrenia. Schizophr Bull 1990; 16: 567-570.

Finch DM, Nowlin NL and Babb TL. Demonstration of axonal projections of neurons in the rat hippocampus and subiculum by intracellular injection of HRP. Brain Res 1983; 271: 201-216.

Gabriel SM, Haroutunian V, Powchik P, Honer WG, Davidson M and Davies P. Increased concentrations of presynaptic proteins in the cingulate cortex of subjects with schizophrenia. Arch Gen Psychiatry 1997; 54: 559-566.

Gao XM, Sakai K, Roberts RC, Conley RR, Dean B and Tamminga CA. Ionotropic glutamate receptors and expression of N-methyl-D-aspartate receptor subunits in subregions of human hippocampus: effects of schizophrenia. Am J Psychiatry 2000; 157: 1141-1149.

Garey LJ, Ong WY, Patel TS, Kanani M, Davis A, Mortimer AM, Barnes TR and Hirsch SR. Reduced dendritic spine density on cerebral cortical pyramidal neurons in schizophrenia. J Neuro. Neurosurg Psychiatry 1998; 65: 446-453.

Glantz LA and Lewis DA. Reduction of synaptophysin immunoreactivity in the prefrontal cortex of subjects with schizophrenia. Regional and diagnostic specificity. Arch Gen Psychiatry 1997; 54: 943-952.

Glantz LA and Lewis DA. Decreased dendritic spine density on prefrontal cortical pyramidal neurons in schizophrenia. Arch Gen Psychiatry 2000; 57: 65-73.

Guidotti A, Auta J, Davis JM, DiGiorgi Gerevini V, Dwivedi Y, Grayson DR, Impagnatiello F., Pandey G, Pesold C, Sharma R, Uzunov D and Costa E. Decrease in reelin and glutamic acid decarboxylase67 (GAD67) expression in schizophrenia and bipolar disorder: a postmortem brain study. Arch Gen Psychiatry 2000; 57: 1061-1069.

Harrison PJ. The neuropathology of schizophrenia. A critical review of the data and their interpretation. Brain 1999; 122: 593-624.

Heckers S, Heinsen H, Geiger B and Beckmann H. Hippocampal neuron number in schizophrenia. A stereological study. Arch Gen Psychiatry 1991; 48: 1002-1008.

Heinsen H, Gossmann E, Rub U, Eisenmenger W, Bauer M, Ulmar G, Bethke B, Schuler M, Schmitt HP, Gotz M, Lockemann U and Puschel K. Variability in the human entorhinal region may confound neuropsychiatric diagnoses. Acta Anat (Basel) 1996; 157: 226-237.

Hoffman RE and Dobscha SK. Cortical pruning and the development of schizophrenia: a computer model. Schizophr Bull 1989; 15: 477-490.

Honer WG, Falkai P, Chen C, Arango V, Mann JJ and Dwork AJ. Synaptic and plasticity-associated proteins in anterior frontal cortex in severe mental illness. Neuroscience 1999; 91: 1247-1255.

Honer WG, Falkai P, Young C, Wang T, Xie J, Bonner J, Hu L, Boulianne G L, Luo Z and Trimble WS. Cingulate cortex synaptic terminal proteins and neural cell adhesion molecule in schizophrenia. Neuroscience 1997; 78: 99-110.

Impagnatiello F, Guidotti AR, Pesold C, Dwivedi Y, Caruncho H, Pisu MG, Uzunov DP, Smalheiser NR, Davis JM, Pandey GN, Pappas GD, Tueting P, Sharma RP and Costa E. A decrease of reelin expression as a putative vulnerability factor in schizophrenia. Proc Natl Acad Sci USA 1998; 95: 15718-15723.

Jakob H and Beckmann H. Prenatal developmental disturbances in the limbic allocortex in schizophrenics. J Neural Transm 1968; 65: 303-326.

Jakob H and Beckmann H. Gross and histological criteria for developmental disorders in brains of schizophrenics. J R Soc Med 1989; 82: 466-469.

Kalus P, Muller TJ, Zuschratter W and Senitz D. The dendritic architecture of prefrontal pyramidal neurons in schizophrenic patients. Neuroreport 2000; 11: 3621-3625.

Karson CN, Mrak RE, Schluterman KO, Sturner WQ, Sheng JG and Griffin WS. Alterations in synaptic proteins and their encoding mRNAs in prefrontal cortex in schizophrenia: a possible neurochemical basis for 'hypofrontality'. Mol Psychiatry 1999; 4: 39-45.

Kosel KC, Van Hoesen GW and Rosene DL. A direct projection from the perirhinal cortex (area 35) to the subiculum in the rat. Brain Res 1983; 269: 347-351.

Krimer LS, Herman MM, Saunders RC, Boyd JC, Hyde TM, Carter JM, Kleinman JE and Weinberger DR. A qualitative and quantitative analysis of the entorhinal cortex in schizophrenia. Cereb Cortex 1997; 7: 732-739.

Landen M, Davidsson P, Gottfries CG, Grenfeldt B, Stridsberg M and Blennow K. Reduction of the small synaptic vesicle protein synaptophysin but not the large dense core chromogranins in the left thalamus of subjects with schizophrenia. Biol Psychiatry 1999; 46: 1698-1702.

McGlashan TH and Hoffman RE. 2000. Schizophrenia as a disorder of developmentally reduced synaptic connectivity. Arch Gen Psychiatry 2000; 57: 637-648.

Meador-Woodruff JH and Healy DJ. Glutamate receptor expression in schizophrenic brain. Brain Res. Brain Res Rev 2000; 31: 288-294.

Mirra SS, Heyman A, McKeel D, Sumi SM, Crain BJ, Brownlee LM, Vogel FS, Hughes JP, van Belle G and Berg L. The Consortium to Establish a Registry for Alzheimer's Disease (CERAD). Part II. Standardization of the neuropathologic assessment of Alzheimer's disease. Neurology 1991; 41: 479-486.

Perrone-Bizzozero NI, Sower AC, Bird ED, Benowitz LI, Ivins KJ and Neve RL. Levels of the growth-associated protein GAP-43 are selectively increased in association cortices in schizophrenia. Proc Natl Acad Sci USA 1996; 93: 14182-14187.

Pesold C, Liu WS, Guidotti A, Costa E and Caruncho HJ. Cortical bitufted, horizontal, and Martinotti cells preferentially express and secrete reelin into perineuronal nets, nonsynaptically modulating gene expression. Proc Natl Acad Sci USA 1999; 96: 3217-3222.

Ringstedt T, Linnarsson S, Wagner J, Lendahl U, Kokaia Z, Arenas E, Ernfors P and Ibanez CF. BDNF regulates reelin expression and Cajal-Retzius cell development in the cerebral cortex. Neuron 1998; 21: 305-315.

Rodriguez MA, Pesold C, Liu WS, Kriho V, Guidotti A, Pappas GD and Costa E. Colocalization of integrin receptors and reelin in dendritic spine postsynaptic densities of adult nonhuman primate cortex. Proc Natl Acad Sci USA 2000; 97: 3550-3555.

Rosoklija G, Toomayan G, Ellis SP, Keilp J, Mann JJ, Latov N, Hays AP and Dwork AJ. Structural abnormalities of subicular dendrites in subjects with schizophrenia and mood disorders: preliminary findings. Arch Gen Psychiatry 2000; 57: 349-356.

Selemon LD and Goldman-Rakic PS. The reduced neuropil hypothesis: a circuit based model of schizophrenia. Biol Psychiatry 1999; 45: 17-25.

Takahashi M, Shirakawa O, Toyooka K, Kitamura N, Hashimoto T, Maeda K, Koizumi S, Wakabayashi K, Takahashi H, Someya T and Nawa H. Abnormal expression of brain-derived neurotrophic factor and its receptor in the corticolimbic system of schizophrenic patients. Mol Psychiatry 2000; 5: 293-300.

Tcherepanov AA and Sokolov BP. Age-related abnormalities in expression of mRNAs encoding synapsin 1A, synapsin 1B, and synaptophysin in the temporal cortex of schizophrenics. J Neurosci Res 1997; 49: 639-644.

Vawter MP, Howard AL, Hyde, TM, Kleinman JE and Freed WJ. Alterations of hippocampal secreted N-CAM in bipolar disorder and synaptophysin in schizophrenia. Mol Psychiatry 1999; 4: 467-475.

Verney C, Baulac M, Berger B, Alvarez C, Vigny A and Helle KB. Morphological evidence for a dopaminergic terminal field in the hippocampal formation of young and adult rat. Neuroscience 1985; 14: 1039-1052.

Yamanouchi H, Jay V, Otsubo H, Kaga M, Becker LE and Takashima S. Early forms of microtubule-associated protein are strongly expressed in cortical dysplasia. Acta Neuropathol (Berl) 1998; 95: 466-470.

Young CE, Arima K, Xie J, Hu L, Beach TG, Falkai P and Honer WG. SNAP-25 deficit and hippocampal connectivity in schizophrenia. Cereb Cortex 1998; 8: 261-268.

Zaidel DW. Regional differentiation of neuron morphology in human left and right hippocampus: comparing normal to schizophrenia. Int J Psychophysiol 1999; 34: 187-196.

Zaidel DW, Esiri MM and Harrison PJ. Size, shape, and orientation of neurons in the left and right hippocampus: investigation of normal asymmetries and alterations in schizophrenia. Am J Psychiatry 1997; 154: 812-818.

14 GSK-3 AND WNT MARKERS OF NEURODEVELOPMENTAL ABNORMALITIES IN SCHIZOPHRENIA

Nitsan Kozlovsky, RH Belmaker and Galila Agam

Abstract

The Neurodevelopmental Hypothesis of schizophrenia suggests that interaction between genetic and environmental events occurring during critical early periods in neuronal growth may negatively influence the way by which nerve cells are laid down, differentiated and selectively culled by apoptosis. Recent advances offer insights into the regulation of brain development. The Wnt family of genes plays a central role in normal brain development. Activation of the Wnt cascade leads to inactivation of glycogen synthase kinase-3β (GSK-3β), accumulation and activation of β-catenin and expression of genes involved in neuronal development. The possible role of aberrant GSK-3 in the etiology of schizophrenia is discussed.

Neurodevelopment and the Wnt Signalling Cascade

The neurodevelopmental hypothesis suggests that interaction between genetic and environmental events occurring during critical early periods in neuronal growth may negatively influence the way by which nerve cells are laid down, differentiated, selectively culled by apoptosis and remodeled by expansion and retraction of dendrites and synaptic connections (Weinberger 1987; Bloom 1993; Murray 1994; Roberts 1990). These changes begin in utero, are affected by perinatal events around birth, and become fully expressed in early adulthood (Arnold and Trojanowski 1996; Bunney et al 1995; Harrison 1999).

Recent studies offer insights into possible regulatory mechanisms of developmental brain changes. The Wnt family of genes is a major component in these studies. These genes encode a family of cysteine-rich, secreted glycoproteins involved in embryonic developmental processes such as cell adhesion and rearrangement of synapses (Hall et al 2000; Patapoutian and Reichardt 2000).

During embryogenesis extracellularly secreted Wnt molecules interact with cell-membrane receptors of the frizzled family. The interaction is regulated by glycosaminoglycans, molecules for which Wnt proteins have a high affinity and thus serve as co-receptors. This agonist-receptor interaction induces the hyperphosphorylation and activation of the protein Disheveled (Dvl) and thereby its assembly with three other proteins - Axin, Frat-1/GBP (Frequently Rearranged in Advance T-cell lymphomas 1/Glycogen Synthase Kinase-3 Binding Protein) and Glycogen Synthase Kinase-3β (GSK-3β). The four proteins merge into a heterotetramer complex. This assembly leads to the suppression of GSK-3β's activity, which is constitutively active under resting conditions. While active GSK-3β, facilitated by Axin and Adenomatous Polyposis Coli (APC, a scaffold protein coordinating the interaction of β-catenin and GSK-3β) phosphorylates β-catenin, thereby promoting its assembly with phosphorylated APC leading to degradation of β-catenin through the ubiquitin proteolytic pathway. Inactivation of GSK-3 upon activation of the Wnt cascade protects β-catenin against degradation and allows its translocation into the nucleus. In the nucleus, in the absence of Wnt signaling, TCF/LEF-1 (T-Cell Factor/Lymphocyte Enhancer Factor-1) which belongs to the high-mobility-group (HMG) box architectural transcription factors represses the expression of Wnt target genes. In the presence of Wnt signaling, the rise of β-catenin steady-state levels enables its accumulation in the nucleus of embryonic cells where it interacts with TCF and forms a transcriptional complex that activates gene transcription of crucial development regulatory genes ((Moon et al 1997). An additional cascade which possibly interacts with GSK-3β via dishevelled is the Notch pathway that is involved in cell-fate determination (Anderton et al 2000).

Glycogen Synthase Kinase-3

GSK-3, a downstream component of the Wnt signaling cascade, is a ubiquitous serine/threonine kinase initially identified as an enzyme negatively regulating the activity of glycogen synthase (Cohen 1985). It is highly abundant in brain (Pei et al 1999; Woodgett 1990) and highly conserved during evolution (Yu and Yang 1993). It was first purified from skeletal muscle (Woodgett 1991). Two isoforms with 85% sequence homology and 98% identity in the catalytic domains (Woodgett 1990) referred to as GSK-3 α and β encoded by separate genes mapped to chromosomes 19 and 3, respectively (Shaw et al 1998) have been identified in mammals. Recent studies demonstrated that GSK-3β is also involved in the regulation of growth and development. GSK-3β plays a critical role in axis pattern formation by regulating the establishment of dorsoventral polarity (Dominguez et al 1995) and in cell fate determination in the Wnt signalling cascade (Wodarz and Nusse 1998).

But the role of GSK-3 is not restricted to the Wnt system. GSK-3 is a juncture of three signal transduction cascades, the Wnt, the mitogen activated protein kinase (MAPK) and the phosphatidylinositol-3-kinase (PI-3K) (Srivastava and Pandey 1998). Its multiple targets are spread extracellularly and at various subcellular organelles – membranal, cytoskeletal, nuclear and cytosolic and include transcription factors (Plyte et al 1992), regulatory enzymes (Rubinfeld et al 1996) and structural proteins (Hanger et al 1992); (Mandelkow et al 1992). Thus, GSK-3 mediates metabolic, developmental, differentiational and proliferational processes (Siegfried et al 1992; He et al 1995). The molecular basis for inducing distinct cellular responses via different signalling pathways has recently been unraveled (Weston and Davis 2001). GSK-3 phosphorylates two types of protein substrates: those that are prephosphorylated ("primed") before they can dock with a GSK-3 phosphate-binding site and the "nonprimed" substrates that do not require prephosphorylation (Frame et al 2001).

Differences in Wnt Cascade Markers in Postmortem Brain of Schizophrenic Patients

The Wnt family of genes plays a fundamental role in neuronal development through the control of migration and differentiation. There are indications that a mutation in one or more of these genes may lead to abnormal cerebral development in mice (McMahon and Bradley 1990) The role of this pathway in the regulation of neuronal migration during development suggests that alteration of this pathway may be involved in producing the cytoarchitectural defects observed in schizophrenia (Cotter et al 1998).

Both GSK-3α and GSK-3β play a role in neurodevelopment, hypothesized to be altered in schizophrenic patients. Since GSK-3β is a component of the Wnt pathway, its reduced levels (details in the next subtitle) corroborate recent reports of abnormal Wnt cascade markers in schizophrenia. Miyaoka et al; (Miyaoka et al 1999) reported an increase in Wnt-1 containing neurons in the hippocampal pyramidal cell layers CA3 and CA4 of schizophrenic patients. Reduction in β- and γ- catenin in the CA3 and CA4 regions of the hippocampus have also been noted in this disorder (Cotter et al 1998). In the study by Beasley et al; (2001) showing reduced GSK-3β levels in the frontal cortex of schizophrenic patients there was no significant alteration in β-catenin or dishevelled protein levels. Upon Wnt cascade activation β-catenin translocates from the cytosol to the nucleus (Moon et al 1997), but the assay of Moon et al; (1997) uses whole cell homogenates and thus determines only total cell levels. Abnormalities of social interaction and sensorimotor gating have been reported in the mice lacking the dishevelled-1 gene (Lijam et al 1997). In humans, a deletion of this gene characterizes the DiGeorge syndrome. Indeed, approximately 25% of the adults suffering from this syndrome, which presents with learning disabilities, craniofacial abnormalities and congenital heart defects, develop schizophrenia (Pizzuti et al 1996).

The possible involvement of elements of the Wnt signaling pathway genes in bipolar disorder and schizophrenia was also studied by high-resolution radiation hybrid mapping (Rhoads et al 1999) Frizzled 3, expressed in highest levels in the adult brain (Wang et al 1996) is located on chromosome 7q11.23 shown to be associated with the developmental disorder Williams syndrome (Robinson et al 1996). A potential susceptibility locus at 7q11 has recently been reported for schizophrenia (Blouin et al 1998) and linkage disequilibrium mapping has recently shown that NOTCH4 (see section "The Wnt family of genes" above) is highly associated with schizophrenia (Wei and Hemmings 2000).

GSK-3α and β are Altered in Schizophrenic Patients

We have previously reported that GSK-3 is altered in schizophrenic patients' brains. The study was carried out in frozen postmortem brain specimens obtained from the Stanley Foundation Brain Bank (Torrey et al 2000). The 60 samples of frontal and occipital cortex consisted of 15 schizophrenic patients, 15 bipolar patients, 15 unipolar depressed patients and 15 normal controls. The four groups were matched for age, sex, race, postmortem interval and side of the brain. Quality of preservation was monitored by two factors - mRNA degree of degradation and pH. GSK-3β protein levels were quantified by western immunoblotting. To minimize the effect of interblot variability, a calibration standard curve of known amounts of recombinant GSK-3β units was run in each gel and used to derive the absolute value of each sample band run on the same gel. Each sample was analyzed at least three times and at two different protein concentrations, both within the linear range of detection. Fig 3A shows that frontal cortex samples of schizophrenic patients have 41% reduced GSK-3β protein levels compared with normal controls. There was no correlation between GSK-3β levels of all groups and postmortem interval, or age. In the schizophrenic patients' group there was no correlation between GSK-3β levels and their estimated lifetime antipsychotic drugs consumption (Kozlovsky et al 2000).

We next measured the levels of GSK-3β in the postmortem occipital cortex specimens from the same subjects of the four diagnostic groups. There was no difference between their GSK-3β protein levels, suggesting that the reduction in frontal cortex does not represent a general alteration of GSK-3 β protein levels in these patients' brains (Kozlovsky et al 2001). This is consistent with other neuronal specific molecules, such as synaptophysin (Glantz and Lewis 1997) and GAP-43 (Sower et al 1995), also found altered in prefrontal and frontal cortical areas but unchanged in occipital cortex of schizophrenic patients.

Since GSK-3β is unusual among protein kinases in being constitutively active under non-stimulated conditions in the cell and exerting repression on its substrates (Stambolic and Woodgett 1994), the next question raised was whether low frontal cortex GSK-3β levels are accompanied by low enzymatic activity. To answer this question GSK-3 activity was measured in the same frontal cortex specimens that were used for quantifying protein levels. GSK-3 activity was assayed in homogenates by quantifying the phosphorylation of a phospho-CREB peptide, a specific GSK-3 substrate (Wang et al 1994). We utilized the fact that GSK-3 activity is specifically inhibited by lithium to discriminate between other kinases activities and that of GSK-3. Fig-3B shows that mean GSK-3 activity in the frontal cortex of schizophrenic patients was 45% lower than that of normal controls. The other two diagnostic groups showed no difference from the control group (Kozlovsky et al 2000).

There was no correlation between GSK-3 activity and any demographic parameter. Frontal cortex GSK-3 activity and GSK-3β protein levels did not correlate in the specimens of all four diagnostic groups, nor in the specimens of the schizophrenic patients only (40). Since the phospho-CREB substrate does not distinguish the forms of GSK-3, and lithium ions used to discriminate GSK-3 activity from other kinases inhibit both GSK-3α and β, our assay does not distinguish between the activities of the two GSK-3 isoforms. Therefore, it is not unreasonable that no correlation was found between the GSK-3β protein levels and the GSK-3 activity values.

Beasley et al (Beasley et al 2001) supported our finding in frontal cortex samples from the MRC Brain Bank. They found a significant 19% reduction in GSK-3β protein levels in schizophrenic patients compared to control subjects. They however were not able to replicate their own and our finding in the Stanley Foundation Collection, apparently due to subtle differences in the methodology (Beasley et al, submitted). Yang et al. (Yang et al 1995a) measured the levels of GSK-3α in peripheral tissue of 48 schizophrenic patients. They found that both enzymatic activity and protein levels of GSK-3α were considerably reduced in lymphocytes of schizophrenic patients as compared to controls. The activity of GSK-3α was less than 20% of that of normal controls suggesting that patients with schizophrenia may have a dysfunction common to GSK-3α and GSK-3β. It remains to be investigated whether this dramatically lower GSK-3α activity in peripheral cells reflects the central nervous system and, if so, whether it is brain region specific. If the reduction in GSK-3β in frontal cortex of schizophrenic patients will be found to be reflected in peripheral tissue as well, it could serve as a diagnostic marker.

Difference in GSK-3α-Related Markers in Schizophrenia

GSK-3α and some of its substrates were found to be altered in schizophrenia. As described above, GSK-3α protein levels and activity were reported to be reduced in lymphocytes of schizophrenic patients (Yang et al 1995b). Synapsin I, a neuronal protein involved in modulation of neurotransmission, is phosphorylated and thereby inactivated by GSK-3α (Yang et al 1992). Levels of synapsin I were found to be extremely low in postmortem hippocampal specimens of schizophrenic patients compared with age matched normal controls, whereas levels of synaptophysin, another synaptic vesicle protein, were nearly normal (Browning et al 1993). Given that synapsin I is thought to regulate neurotransmitter release, it is possible that its deficit could result in abnormal processing of neuronal information. It remains to be investigated whether GSK-3α mediates this reduction.

It has been suggested that cell recognition molecules (CRMs) such as Neuronal Cell Adhesion Molecule (N-CAM), mediate abnormalities and disturbances in neurodevelopment related to schizophrenia (Landmesser et al 1990). These substances, which belong to the immunoglobulin superfamily, play an important role in neurodevelopmental processes including axonal guidance, synapse stabilization and cell migration (Hemperly et al 1986; McClain and Edelman 1982). NCAM alteration in schizophrenia was proposed as an explanation for disorientation of cells in the hippocampus (Conrad and Scheibel 1987). Indeed, decreased numbers of hippocampal neurons expressing polysialylated NCAM were found (Barbeau et al 1995). Cerebrospinal fluid (CSF) from schizophrenic patients contains two- to three-fold higher NCAM levels (van Kammen et al 1998). Higher levels of 105 to 115 kDa NCAM were also found in the hippocampus and frontal cortex of schizophrenic patients (Vawter et al 1998). NCAM modified function in brain of schizophrenic patients may either stem from its altered protein levels or be a consequence of reduced GSK-3α levels and activity.

The microtubule–associated proteins MAP-2 and MAP-5 (alternatively known as MAP-1B) are major phosphoprotein components of the neuronal cytoskeleton involved in axonal extension and neuronal polarity (Matus and Riederer 1986). Altered distribution of these two proteins in subiculum and entorhinal cortex, and an increased fraction of the nonphosphorylated form of MAP2 in the subiculum were reported in schizophrenia (Arnold et al 1991; Cotter et al 1997). The ability of MAP1B to bind to microtubules for cytoskeletal re-organization depends upon its phosphorylation state which was found to be regulated by GSK-3α and β in extending axons and growth cones (Garcia-Perez et al 1998; Goold et al 1999).

Summary

What could be the consequences of decreased GSK-3 levels and activity in schizophrenic patients' frontal cortex? Given that GSK-3 is a multisubstrate enzyme that unlike most protein kinases is constitutively active under resting conditions and exerts repressing effects on its substrates, decreased GSK-3 activity in schizophrenic patients may result in an accumulation of β-catenin which, in turn, associates with the HMG-box transcription factors Tcf/LEF family promoting the transcription of developmentally regulated genes in an inappropiate timing during neurodevelopment .

Although there is a clear abnormal signal transduction contribution to schizophrenia, apparently of a neurodevelopmental origin, why is obvious early psychopathology absent in children who will be diagnosed with schizophrenia later in life? An etiology involving a "two-hit" process has been suggested to explain this riddle (Manschreck et al 2000; McCarley et al 1999; Impagnatiello et al 1998). The "two-hit" process means that genetic load, adverse embryonic events and perinatal events consist of a neurodevelopmental first hit that leads to vulnerability to schizophrenia. During puberty hormonal events such as altered neurosteroid biosynthesis, along with the demand to turn on adolescent levels of affect and thought, act as a second hit.

References

Anderton BH, Dayanandan R, Killick R, Lovestone S Does dysregulation of the Notch and wingless/Wnt pathways underlie the pathogenesis of Alzheimer's disease? Mol Med Today 2000; 6:54-59.

Arnold SE, Lee VM, Gur RE, Trojanowski JQ Abnormal expression of two microtubule-associated proteins (MAP2 and MAP5) in specific subfields of the hippocampal formation in schizophrenia. Proc Natl Acad Sci U S A 1991; 88:10850-10854.

Arnold SE, Trojanowski JQ Recent advances in defining the neuropathology of schizophrenia. Acta Neuropathol (Berl) 1996; 92:217-231.

Barbeau D, Liang JJ, Robitalille Y, Quirion R, Srivastava LK Decreased expression of the embryonic form of the neural cell adhesion molecule in schizophrenic brains. Proc Natl Acad Sci U S A 1995; 92:2785-2789.

Beasley C, Cotter D, Khan N, et al glycogen synthase kinase-3beta immunoreactivity is reduced in the prefrontal cortex in schizophrenia. . Neurosci Lett 2001; 20:117-120.

Bloom FE Advancing a neurodevelopmental origin for schizophrenia. Arch Gen Psychiatry 1993; 50:224-227.

Blouin JL, Dombroski BA, Nath SK, et al Schizophrenia susceptibility loci on chromosomes 13q32 and 8p21. Nat Genet 1998; 20:70-73.

Browning MD, Dudek EM, Rapier JL, Leonard S, Freedman R Significant reductions in synapsin but not synaptophysin specific activity in the brains of some schizophrenics. Biol Psychiatry 1993; 34:529-535.

Bunney BG, Potkin SG, Bunney WE, Jr. New morphological and neuropathological findings in schizophrenia: a neurodevelopmental perspective. Clin Neurosci 1995; 3:81-88.

Cohen P The role of protein phosphorylation in the hormonal control of enzyme activity. Eur J Biochem 1985; 151:439-448.

Conrad AJ, Scheibel AB Schizophrenia and the hippocampus: the embryological hypothesis extended. Schizophr Bull 1987; 13:577-587.
Cotter D, Kerwin R, al-Sarraji S, et al Abnormalities of Wnt signalling in schizophrenia--evidence for neurodevelopmental abnormality. Neuroreport 1998; 9:1379-1383.
Cotter D, Kerwin R, Doshi B, Martin CS, Everall IP Alterations in hippocampal non-phosphorylated MAP2 protein expression in schizophrenia. Brain Res 1997; 765:238-246.
Dominguez I, Itoh K, Sokol SY Role of glycogen synthase kinase 3 beta as a negative regulator of dorsoventral axis formation in Xenopus embryos. Proc Natl Acad Sci U S A 1995; 92:8498-8502.
Frame S, Cohen P, Biondi RM A common phosphate binding site explains the unique substrate specificity of GSK3 and its inactivation by phosphorylation. Mol Cell 2001; 7:1321-1327.
Garcia-Perez J, Avila J, Diaz-Nido J Implication of cyclin-dependent kinases and glycogen synthase kinase 3 in the phosphorylation of microtubule-associated protein 1B in developing neuronal cells. J Neurosci Res 1998; 52:445-452.
Glantz LA, Lewis DA Reduction of synaptophysin immunoreactivity in the prefrontal cortex of subjects with schizophrenia. Regional and diagnostic specificity [corrected and republished article originally appeared in Arch Gen Psychiatry 1997 Jul;54(7):660-669]. Arch Gen Psychiatry 1997; 54:943-952.
Goold RG, Owen R, Gordon-Weeks PR Glycogen synthase kinase 3beta phosphorylation of microtubule-associated protein 1B regulates the stability of microtubules in growth cones. J Cell Sci 1999; 112:3373-3384.
Hall AC, Lucas FR, Salinas PC Axonal remodeling and synaptic differentiation in the cerebellum is regulated by WNT-7a signaling [see comments]. Cell 2000; 100:525-535.
Hanger DP, Hughes K, Woodgett JR, Brion JP, Anderton BH Glycogen synthase kinase-3 induces Alzheimer's disease-like phosphorylation of tau: generation of paired helical filament epitopes and neuronal localisation of the kinase. Neurosci Lett 1992; 147:58-62.
Harrison PJ The neuropathology of schizophrenia. A critical review of the data and their interpretation. Brain 1999; 122:593-624.
He X, Saint-Jeannet JP, Woodgett JR, Varmus HE, Dawid IB Glycogen synthase kinase-3 and dorsoventral patterning in Xenopus embryos [published erratum appears in Nature 1995 May 18;375(6528):253]. Nature 1995; 374:617-622.
Hemperly JJ, Murray BA, Edelman GM, Cunningham BA Sequence of a cDNA clone encoding the polysialic acid-rich and cytoplasmic domains of the neural cell adhesion molecule N-CAM [published erratum appears in Proc Natl Acad Sci U S A 1988 Mar;85(6):2008]. Proc Natl Acad Sci U S A 1986; 83:3037-3041.
Impagnatiello F, Guidotti AR, Pesold C, et al A decrease of reelin expression as a putative vulnerability factor in schizophrenia. Proc Natl Acad Sci U S A 1998; 95:15718-15723.
Kozlovsky N, Belmaker RH, Agam G Low GSK-3beta immunoreactivity in postmortem frontal cortex of schizophrenic patients. Am J Psychiatry 2000; 157:831-833.
Kozlovsky N, Belmaker RH, Agam G Low GSK-3 activity in frontal cortex of schizophrenic patients. Schizo Res 2001; in press
Landmesser L, Dahm L, Tang JC, Rutishauser U Polysialic acid as a regulator of intramuscular nerve branching during embryonic development. Neuron 1990; 4:655-667.
Lijam N, Paylor R, McDonald MP, et al Social interaction and sensorimotor gating abnormalities in mice lacking Dvl1. Cell 1997; 90:895-905.
Mandelkow EM, Drewes G, Biernat J, et al Glycogen synthase kinase-3 and the Alzheimer-like state of microtubule-associated protein tau. FEBS Lett 1992; 314:315-321.
Manschreck TC, Maher BA, Winzig L, Candela SF, Beaudette S, Boshes R Age disorientation in schizophrenia: an indicator of progressive and severe psychopathology, not institutional isolation. J Neuropsychiatry Clin Neurosci 2000; 12:350-358.
Matus A, Riederer B Microtubule-associated proteins in the developing brain. Ann N Y Acad Sci 1986; 466:167-179.

McCarley RW, Wible CG, Frumin M, et al MRI anatomy of schizophrenia. Biol Psychiatry 1999; 45:1099-1119.
McClain DA, Edelman GM A neural cell adhesion molecule from human brain. Proc Natl Acad Sci U S A 1982; 79:6380-6384.
McMahon AP, Bradley A The Wnt-1 (int-1) proto-oncogene is required for development of a large region of the mouse brain. Cell 1990; 62:1073-1085.
Miyaoka T, Seno H, Ishino H Increased expression of Wnt-1 in schizophrenic brains. Schizophr Res 1999; 38:1-6.
Moon RT, Brown JD, Torres M WNTs modulate cell fate and behavior during vertebrate development. Trends Genet 1997; 13:157-162.
Murray RM Neurodevelopmental schizophrenia: the rediscovery of dementia praecox. Br J Psychiatry Suppl 1994; :6-12.
Patapoutian A, Reichardt LF Roles of wnt proteins in neural development and maintenance [In Process Citation]. Curr Opin Neurobiol 2000; 10:392-399.
Pei JJ, Braak E, Braak H, et al Distribution of active glycogen synthase kinase 3beta (GSK-3beta) in brains staged for Alzheimer disease neurofibrillary changes. J Neuropathol Exp Neurol 1999; 58:1010-1019.
Pizzuti A, Novelli G, Mari A, et al Human homologue sequences to the Drosophila dishevelled segment-polarity gene are deleted in the DiGeorge syndrome. Am J Hum Genet 1996; 58:722-729.
Plyte SE, Hughes K, Nikolakaki E, Pulverer BJ, Woodgett JR Glycogen synthase kinase-3: functions in oncogenesis and development. Biochim Biophys Acta 1992; 1114:147-162.
Rhoads AR, Karkera JD, Detera-Wadleigh SD Radiation hybrid mapping of genes in the lithium-sensitive wnt signaling pathway. Mol Psychiatry 1999; 4:437-442.
Roberts GW Schizophrenia: the cellular biology of a functional psychosis. Trends Neurosci 1990; 13:207-211.
Robinson WP, Waslynka J, Bernasconi F, et al Delineation of 7q11.2 deletions associated with Williams-Beuren syndrome and mapping of a repetitive sequence to within and to either side of the common deletion. Genomics 1996; 34:17-23.
Rubinfeld B, Albert I, Porfiri E, Fiol C, Munemitsu S, Polakis P Binding of GSK3beta to the APC-beta-catenin complex and regulation of complex assembly [see comments]. Science 1996; 272:1023-1056.
Shaw PC, Davies AF, Lau KF, et al Isolation and chromosomal mapping of human glycogen synthase kinase-3 alpha and -3 beta encoding genes. Genome 1998; 41:720-727.
Siegfried E, Chou TB, Perrimon N wingless signaling acts through zeste-white 3, the Drosophila homolog of glycogen synthase kinase-3, to regulate engrailed and establish cell fate. Cell 1992; 71:1167-1179.
Sower AC, Bird ED, Perrone-Bizzozero NI Increased levels of GAP-43 protein in schizophrenic brain tissues demonstrated by a novel immunodetection method. Mol Chem Neuropathol 1995; 24:1-11.
Srivastava AK, Pandey SK Potential mechanism(s) involved in the regulation of glycogen synthesis by insulin. Mol Cell Biochem 1998; 182:135-141.
Stambolic V, Woodgett JR Mitogen inactivation of glycogen synthase kinase-3 beta in intact cells via serine 9 phosphorylation. Biochem J 1994; 303:701-704.
Torrey EF, Webster M, Knable M, Johnston N, Yolken RH The stanley foundation brain collection and neuropathology consortium. Schizophr Res 2000; 44:151-155.
van Kammen DP, Poltorak M, Kelley ME, et al Further studies of elevated cerebrospinal fluid neuronal cell adhesion molecule in schizophrenia. Biol Psychiatry 1998; 43:680-686.
Vawter MP, Cannon-Spoor HE, Hemperly JJ, et al Abnormal expression of cell recognition molecules in schizophrenia. Exp Neurol 1998; 149:424-432.
Wang QM, Roach PJ, Fiol CJ Use of a synthetic peptide as a selective substrate for glycogen synthase kinase 3. Anal Biochem 1994; 220:397-402.
Wang Y, Macke JP, Abella BS, et al A large family of putative transmembrane receptors homologous to the product of the Drosophila tissue polarity gene frizzled. J Biol Chem 1996; 271:4468-4476.

Wei J, Hemmings GP The NOTCH4 locus is associated with susceptibility to schizophrenia. Nat Genet 2000; 25:376-7.

Weinberger DR Implications of normal brain development for the pathogenesis of schizophrenia. Arch Gen Psychiatry 1987; 44:660-669.

Weston CR, Davis RJ Signal transduction: signaling specificity- a complex affair. Science 2001; 292:2439-2440.

Wodarz A, Nusse R Mechanisms of Wnt signaling in development. Annu Rev Cell Dev Biol 1998; 14:59-88.

Woodgett JR Molecular cloning and expression of glycogen synthase kinase-3/factor A. Embo J 1990; 9:2431-2438.

Woodgett JR A common denominator linking glycogen metabolism, nuclear oncogenes and development. Trends Biochem Sci 1991; 16:177-181.

Yang SD, Huang JJ, Huang TJ Protein kinase FA/glycogen synthase kinase 3 alpha predominantly phosphorylates the in vivo sites of Ser502, Ser506, Ser603, and Ser666 in neurofilament. J Neurochem 1995a; 64:1848-1854.

Yang SD, Song JS, Hsieh YT, Liu HW, Chan WH Identification of the ATP.Mg-dependent protein phosphatase activator (FA) as a synapsin I kinase that inhibits cross-linking of synapsin I with brain microtubules. J Protein Chem 1992; 11:539-546.

Yang SD, Yu JS, Lee TT, Yang CC, Ni MH, Yang YY Dysfunction of protein kinase FA/GSK-3 alpha in lymphocytes of patients with schizophrenic disorder. J Cell Biochem 1995b; 59:108-116.

Yu JS, Yang SD Immunological and biochemical study on tissue and subcellular distributions of protein kinase FA (an activating factor of ATP.Mg-dependent protein phosphatase): a simplified and efficient procedure for high quantity purification from brain. J Protein Chem 1993; 12:667-676.

15 MACROANATOMICAL FINDINGS IN POST-MORTEM BRAIN TISSUE

G.D. Pearlson

Abstract

In this chapter, I will discuss gross pathologic changes associated with major affective disorders, specifically brain weight and both overall and regional volume changes, including a number of microscopic studies that assessed volumes of defined structures.

Any attempt to review and summarize existing macroanatomical findings in post-mortem brain tissue in affective disorders immediately confronts several difficulties. The total number of studies on the neuro-histology of mood disorders is remarkably small, even in comparison to other neuropsychiatric disorders such as schizophrenia.

Introduction

Until very recently, the investigation of mood disorders has failed to capture the imagination of neuropathologists. The samples that constitute neuropathological reports in this disorder are thus small and selective. This latter fact is further complicated by the fact that "mood disorders" themselves are a remarkably heterogeneous category; some of the phenomenologic multiplicity of affective disorders is reviewed by Pearlson and Schlaepfer (1999). In a worst-case scenario from the point of view of complexity, a unique pathological picture could accompany each clinical variant. Recent evidence for example suggests that psychotic bipolar disorder may aggregate within families and share vulnerability genes with schizophrenia, unlike non-psychotic bipolar disorder, (e.g. Potash et al, 2001). Conceivably, the neuropathology of psychotic bipolar disorder may resemble that of schizophrenia more than that associated with non-psychotic bipolar disorders. To date, no studies have attempted to elucidate whether indeed clinical subtypes of affective disorders bear any relationships to the underlying pathophysiology, i.e. whether unique subtypes of clinical disorder are represented by unique pathophysiological events or by unique pathophysiological changes. This is understandable given the multiple subtypes of major affective disorders. However, as emphasized by Perry et al (1996) a conjunction of postmortem neuropathological investigations with pre-mortem clinical assessment has proved useful in a parallel endeavor, the delineation of subtypes of dementia. For example, multi-infarct dementia and senile dementia of the Lewy body type were delineated as separate disease entities by this sort of neuropathologic/clinical comparison (Ferrier and Perry 1992).

Feasibly, brain changes associated with mood disorders may occur at a neurochemical and not at a structural level, or comprise evanescent state phenomena occurring only during periods of manifest illness, or present only at an ultrastructural level. Such alterations might well be undetectable on macroanatomical examination. Vawter et al (2000) for example, raises the question of whether neuropathological changes in bipolar disorder are enduring or whether they are phenomena accompanying clinical state changes that do not result in any permanent biological changes observable in the postmortem brain. The fact that many neuropathologic specimens in affective disorders derive from individuals, who have died as a result of suicide, precludes a definitive answer to this question.

Possible treatment effects represent another complicating factor. First, recent magnetic resonance imaging studies suggest that the administration of lithium at usual therapeutic doses may increase volume of cortical gray matter (Moore et al, 2000). Second, use of typical antipsychotic medications can cause enlargement of neurons in the rat basal ganglia and increases in volumes of the striatum in humans (Chakos et al, 1995). Although these medications are most often prescribed in schizophrenia, they are frequently used in psychotic affective disorder and mania. They thus have the potential to act as confounding factors in being associated with anatomic differences in particular subtypes of affective illness.

Other possible problems with postmortem samples in bipolar disorder may arise from events immediately before or after death; given the bias toward brain material from individuals who have committed suicide particular attention should be given to this potential source of bias.

Overall there are multiple potential problems in interpreting macroanatomic pathologic studies in affective disorder and the investigator must tread carefully in planning and drawing conclusions from them.

Neuroimaging Studies Provide Clues as to Where Neuropathologists Should be Looking

Data obtained from neuroimaging investigations in mood disorders showing structural and functional abnormalities generally present significant guides as to where neuropathologists should be looking in the brain. This chapter will retain this logical sequence; first neuroimaging evidence implicating a particular area will be discussed, followed by the corresponding neuropathological reports. Investigators have reported changes in multiple brain regions of affective disorder patients including: total brain volume, ventricular/sulcal enlargement, amygdala/white matter hyperintensities, hippocampus, limbic forebrain, basal ganglia, hypothalamus, prefrontal cortex – dorsal and ventral, orbitofrontal, Areas 9, 25, 24a, 46, 31, cerebellum, subgenual cingulated and posterior cingulated.

Although neuroimaging has the advantage of being conducted in vivo and potentially at the onset of illness when treatment-related confounds are minimal, Baumann et al (1999) note that MRI resolution is still insufficient to detect subtle structural alterations of small brain nuclei such as the hypothalamus or some parts of the basal ganglia. Postmortem morphometry offers the option of delineating and measuring small brain nuclei such as the nucleus accumbens that have different connections and functions from immediately adjacent structures such as the caudate that are difficult to assess separately using neuroimaging. Thus, as noted by Vawter et al (2000) "leads obtained from neuroimaging report fruitful areas to investigate in postmortem studies"

Structural and functional neuroimaging studies suggest abnormalities in bipolar disorder in the prefrontal cortex, limbic system, 3rd ventricle, cerebellum, temporal lobe, basal ganglia and subcortical white matter. The reader is referred to recent reviews of Drevets et al, (1998), Soares and Mann (1997), Pearlson (1999), Pearlson and Schlaepfer (1999) for a more thorough discussion of this topic and its theoretical implications regarding mood control circuits.

Total Brain Weight and Volume

Neuroimaging Studies

Hoge et al (1999) conducted a meta-analysis of brain size in bipolar disorder. From the studied reviewed, they concluded that bipolar disorder was not associated with overall cerebral volume reductions such as those previously reported in patients with schizophrenia. Many studies of elderly depressed patients have also reported no significant differences in whole brain volumes from normal controls (e.g. Sheline et al, 1999; Kumar et al, 1998; Krishnan et al, 1993). However a meta-analysis of 29 studies of mood disorders examining ventricular enlargement or cortical atrophy with CT or MRI reported an increased prevalence of cortical atrophy and ventricular enlargement (Elkis et al, 1995). Similarly, structural neuroimaging studies in late-life depression (e.g. Schweitzer et al, 2001) also show ventricular enlargement and sulcal widening. These findings that seemingly contradict those of the above paragraph may be resolved because either atrophy/hypoplasia is localized and not general, or conceivably that ventricles are enlarged without a reduction in overall brain tissue.

Neuropathologic Studies

Using coroner records of individuals aged 60 and above, Salib and Tadros (2000) explored variations in postmortem brain weight in 142 individuals who had used various methods of fatal self-harm (including suicide) compared to 150 individuals who had died of natural causes. The brain weight of victims of fatal self-harm victims was significantly higher than those who died of natural causes, although brain weights in both groups were within the normal range. Healy et al, (1996) found a high prevalence of unexpected neuropathologic abnormalities in brains derived from a chronic domiciliary population, but not specific lesions associated with affective disorder. Other studies (Leake et al, 1990; Leake et al, 1991; Perrett et al, 1992; Young et al, 1991) found rather lower estimates.

Limbic System

The role of the limbic system in the normal regulation of emotion and drive is well recognized. Central nervous system disorders involving limbic circuits, e.g. rabies and Alzheimer's disease are frequently associated with pathological emotional states.

Neuroimaging Studies

Reduced hippocampal volumes have been found in major depression by the following investigators: O'Brien et al (1994) and Sheline et al (1999), but not by Vakili et al (2000). Structural neuroimaging studies in late-life depression (e.g. Schweitzer et al, 2001) showed reduced volumes of the hippocampus. In bipolar subjects, Altshuler et al (1991) and Swayze et al (1992) reported smaller volumes of the entire temporal lobe. Although earlier MRI studies found smaller amygdala volumes in bipolar illness, two more recent investigations noted amygdala enlargement, (Altshuler et al, 1998; Strakowski et al, 1999).

Neuropathologic Studies

Baumann et al (1999) reported that the brains of patients with major depressive illness showed trends towards reduced left amygdala volume, which were not seen in bipolar disorder subjects. Volumes of other limbic regions including hippocampus, temporal horn, stria terminalis and basal limbic forebrain were unchanged in the sample as were volumes of thalamus and hypothalamus. In major depression, however, the right hypothalamus was smaller than in controls. Sheline et al (1998) reported decreased volume of amygdala core nuclei in recurrent major depression.

Frontal Lobe

As noted by Rajkowska et al previously (1999) and in her chapter in this book several pieces of evidence suggest that the prefrontal cortex may be involved in the neuropathology of major depressive disorder and bipolar manic-depressive disease. Deficits in working memory consistent with abnormalities in hippocampal and dorsolateral prefrontal cortical function are frequently observed in individuals with major depression. Damage to portions of prefrontal cortex can produce states of apathy and amotivation; post-stroke depression is related to anterior frontal lesions of the dominant hemisphere (Robinson et al, 1981).

Neuroimaging Studies

Structural and functional neuroimaging studies suggest that prefrontal cortex is involved in the neuropathology of mood disorders. For example, reductions in gray and/or white matter volume or evidence of atrophy as well as reductions in glucose metabolism and blood flow have all been reported in this disorder (e.g. Cohen et al, 1989; Drevets et al, 1997; Elkis et al, 1996; Swayze et al, 1990). Reductions in regional or total prefrontal cortical volumes are also reported in familial depression (Drevets et al,1997) and in late life depression by Schweitzer et al (2001), Coffey et al (1993) and Kumar et al (1998). Elkis et al (1995) found prefrontal sulcal prominence in relatively young patients with depression.

Neuropathologic Studies

Although changes in microscopic anatomy have been reported, (for example cell loss in the subgenual prefrontal cortex and cell atrophy in the dorsolateral prefrontal cortex and orbitofrontal cortex) the relationships between these cellular changes in mood disorder and volume alterations on gross neuropathology remain unknown. Both Ongur et al (1998) and Rajkowska et al (1999) noted significant reductions in glial cells in subgenual prefrontal and orbitofrontal regions. Rajkowska et al (1999) noted significant decreases in cortical thickness in the rostral and middle parts of the orbitofrontal region and a 6% decrease in cortical thickness in Brodmann Area 9 of bipolar brains (Rajkowska 1997).

Cerebellum

Little published evidence exists for cerebellar disorders being associated frequently with marked mood disturbances, although some recent published reports document depression in inherited cerebellar degenerative disorders (O'Hearn et al, 2001).

Neuroimaging Studies

Only a very small number of studies have attempted to search for possible cerebellar changes in association with mood disorders; Delbello et al (1999) and Jeste et al (1988) reported abnormalities.

Neuropathologic Studies

A literature search revealed no postmortem macroanatomical investigations of the cerebellum in affective disorders.

Basal Ganglia

Degenerative pathologic conditions with primary basal ganglia involvement such as Huntington's chorea, are frequently associated with disorders of mood, both bipolar disorder and (much more frequently) major depression (Folstein 1989).

Neuroimaging Studies

In patients with unipolar depression, MRI measurements show smaller volumes of the caudate and putamen (Husain et al, 1991; Krishnan et al, 1992). Schweitzer et al (2001) showed reduced volumes of the caudate nucleus in elderly depressed patients. One report (Aylward et al, 1994) found larger caudate nuclei in male bipolar patients; one possible explanation for this is the neuroleptic medication-related effect mentioned earlier. As reviewed by Baumann et al (1999), functional imaging data in depression have shown reduced regional bloodflow or metabolism.

Neuropathologic Studies

In a small study Baumann and Bogerts (1999) found that basal ganglia were smaller in patients with depression irrespective of diagnostic polarity, (i.e. both bipolar and unipolar individuals). Baumann et al (1999) showed volume reductions of 32% in the left nucleus accumbens (which is highly reciprocally interconnected with the prefrontal cortex), of 20% in bilateral external pallidum and of 15% in the right putamen in 8 patients with mood disorders. Age and brain weight did not differ between patients and controls.

UBO

Focal areas of increased signal intensity, sometimes termed "unidentified bright objects" (UBO's), are frequently observed on MRI scans. Although these hyperintensities, which are commoner in white matter are themselves non-specific in their association with cerebrovascular or neuropsychiatric illness, and the most obvious risk factor for them is increased age, some researchers have proposed that white matter hyperintensities constitute a risk marker for mood disorders. Although white matter hyperintensities occur in different disorders and arise from a large range of pathological processes, there is increasing evidence that deep white matter lesions may be vascular in origin and represent areas of ischemia and infarction (Everall et al, 1997; Awad et al, 1986; Fazekas et al, 1993). However, as noted by Thomas et al (2001), no neuropathologic studies of white matter hyperintensities have been carried out in affective disorders to confirm whether or not they are indeed vascular in origin.

Neuroimaging Studies

White matter lesions are more common than expected in depressed subjects. Some studies report these changes to be especially pronounced in patients presenting with late-onset depression, where white matter hyperintensities predominate in deep white matter and subcortical gray matter (Rabins et al, 1991; O'Brien et al, 1996, Greenwald et al, 1996). As noted above however, increasing age by itself is associated with the presence of UBO's even in healthy elderly persons. De Groot et al (2000) sampled over 1,000 non-demented elderly adults from a Dutch community. Persons with severe white matter lesions on MRI were three to five times more likely to have depressive symptoms compared with those who had only mild or no white matter lesions. In addition, individuals with severe subcortical (but not periventricular) white matter lesions were more likely to have had a history of depression with onset after age 60 years. Thus the severity of subcortical white matter lesions was related to the presence of depressive symptoms and to a history of late life onset depression. Several additional studies used MRI to investigate deep white matter lesions and showed these to be more frequent in depressed subjects (e.g. Coffey et al 1990; Rabins et al 1991; O'Brien et al 1996). This association of increased white matter lesions with affective disorders, especially late-onset depression, has led some researchers to propose a concept of "vascular depression" i.e. primary cerebrovascular disease manifesting as first-onset depression in mid or late life, (e.g. Alexopoulos et al 1997; Krishnan et al, 1997). However, as noted above these white matter lesions may have multiple pathologic causes and are not by definition vascular in origin. Several neuroimaging studies have found the presence as well as severity of white matter lesions to be related to the presence of depressive and bipolar symptoms.

Neuropathologic Studies

Thomas et al (2001) conducted a postmortem study on 20 patients who had a history of depression compared to controls. He was able to show that the depressed subjects had greater degrees of arteriosclerosis affecting aortic and large brain vessels, although there was no corresponding increase in microvascular disease.

Influences of Glucocorticoids

The pathologic mechanisms leading to brain alterations in bipolar and depressed patients remain conjectural. Various influences of glucocorticoids, stress and their interactions on neurons have been proposed as key mediators, and evidence adduced from both animal and clinical studies. For example, it has been proposed that affective episodes may be associated with stress-induced alterations in brain regions (especially hippocampus) that may be vulnerable to glucocorticoid activity. Sapolsky et al, (1990) showed that adrenal steroids might result in reduced neuron number in portions of the hippocampus in experimental animals. However, volume reductions of total hippocampal volume were not demonstrated in this study.

Reduced hippocampal volumes have been reported by several studies of individuals with major depression (e.g. Bremner et al, 2000; Krishnan et al, 1991; Shah et al, 1998; Sheline et al, 1999). Other researchers (Altschuler 1991) have proposed a similar mechanism in bipolar disorder, i.e. a hypothesis that repeated episodes of pathologic mood change may lead to increasing structural damage to selected brain regions. However, several recent studies emphasize amygdala rather than hippocampal alterations in bipolar patients.

Differences from Schizophrenia

Neuroimaging

We have noted already that Hoge et al's (1999) meta-analysis of brain size concluded that bipolar disorder, unlike schizophrenia was not associated with overall cerebral volume reductions. Pearlson et al (1997) directly compared bipolar to schizophrenic patients and concluded that reduced volume of entorhinal right amygdala and left anterior temporal gyrus characterized schizophrenia, while a smaller left amygdala was found in bipolar patients. Roy et al (1998) reported temporal horn enlargement to be present in both schizophrenia and bipolar disorder.

Neuropathology

Brown et al (1986) conducted a postmortem study of over 200 patients with diagnoses of schizophrenia or affective disorder who had died over a period of 22 years. From this they derived a sample of 41 patients schizophrenia and 29 with affective disorder. Patients with schizophrenia had larger ventricles, particularly in the temporal horn region, and thinner parahippocampal cortices. Baumann and Bogerts (1999) reviewed postmortem neuro-histological and structural imaging studies of schizophrenia contrasted with those mood disorders. They noted that in contrast to the large number of postmortem studies of schizophrenia published in the last 20 years, very few histological studies of affective disorder are available. They also concluded that despite a broad overlap in structural findings between schizophrenia and mood disorders, regions more affected in schizophrenia included heteromodal association cortex, limbic system and structural asymmetry while subtle structural abnormalities in the basal ganglia and particularly in the nucleus accumbens and the hypothalamus may be more typical of mood disorders.

Conclusions

Neuropathologic and neuroimaging studies provide convergent evidence that mood disorders, however varied in clinical expression, may share common abnormalities in frontal lobe, subcortical and limbic circuits that normally play a role in regulating emotions, mood states and drives. These circuits themselves are strongly reciprocally interconnected (e.g. hippocampus to dorsolateral prefrontal cortex, nucleus accumbens with the prefrontal cortex). Evidence supporting this hypothesis of aberrant structure and function of these neural networks has been demonstrated primarily through neuroimaging; as yet neuropathologic studies are fewer in number and more limited in scope, although results from the two approaches are in general agreement. Pathological studies at these levels have the advantage of being able to yield information on functionally separate anatomically contiguous areas such as caudate head and nucleus accumbens that may be impossible to resolve by neuroimaging. Macroanatomical techniques are potentially powerful, certainly underutilized in affective disorders and deserve wider application. Large-scale efforts to organize systematic samples of brains from individuals with affective disorders (e.g. the Stanley Foundation Collection, Torrey et al, 2000) will greatly aid this effort.

References

Alexopoulos GSS, Meyers ., Young RC, Campbell S, Silberweig D, Charlson M. 'Vascular depression' hypothesis. Arch Gen Psychiatry 1997; 54:915-922.

Altshuler LL, Conrad A, Hauser P, Li .M, Guze BH, Denikoff K, et al. Reduction of temporal lobe volume in bipolar disorder: a preliminary report of magnetic resonance imaging. Arch Gen Psychiatry 1991; 48: 482-483.

Altshuler LL. Bipolar disorder: are repeated episodes associated with neuroanatomic and cognitive changes? Biol Psychiatry 1993; 33:563-565.

Altshuler LL, Bartzokis G, Grieder T, Curran J, Mintz J. Amygdala enlargement in bipolar disorder and hippocampal reduction in schizophrenia: an MRI study demonstrating neuroanatomic specificity. Arch Gen Psychiatry 1998; 55:663-664.

Altshuler LL, Bartzokis G, Grieder T, Curran J, Jimenez T, Leight K, Wilkins J, Gerner R, Mintz J. An MRI study of the temporal lobe structures in men with bipolar disorder or schizophrenia. Biol Psychiatry 2000; 48:147-162.

Awad IA, Johnson PC, Spetzler RF, et al. Incidental subcortical lesions identified on magnetic resonance imaging in the elderly: II. postmortem pathological correlations. Stroke 1986; 17:1090-1097.

Aylward EH, Roberts Twillie JV, Barta PE, et al. Basal ganglia volumes and white matter hyperintensities in patients with bipolar disorder. Am J Psychiatry 1994; 151:687-693

Baumann B, Bogerts B. The pathomorphology of schizophrenia and mood disorders: similarities and differences. Schizophr Res 1999; 39:141-148.

Baumann B, Bogerts B. Neuroanatomical studies on bipolar disorder. Br J Psychiatry 2001; 178:s142-s147.

Baumann B, Danos P, Krell D, Diekmann S, Leschinger A, Stauch R, Wurthmann C, Bernstein H-G, Bogerts B. Reduced volume of limbic system-affialiated basal ganglia in mood disorders: preliminary data from a postmortem study. J Neuropsychiatry Clin Neurosci 1999: 11:71-78.

Bremner JD, Narayan M, Anderson ER, Staib LH, Miller HL, Charney DS. Hippocampal volume reduction in major depression. Am J Psychiatry 2000; 157:115-118.

Brown R, Colter N, Corsellis JA, Crow TJ, Frith CD, Jagoe R, Johnstone EC, Marsh L. Postmortem evidence of structural brain changes in shcizophrenia: difference in brain weight, temporal horn area, and parahippocampal gyrus compared with affective disorder. Arch Gen Psychiatry 1986; 43:36-62.

Chakos MH, Lieberman JA, Alvir J, Bilder R, Ashtari M. Caudata nuclei volumes in schizophrenic patients treated with typical antipsychotics or clozapine. Lancet 1995; 345-456-457.

Coffey CE, Figiel GS, Djang WT, Weiner RD. Subcortical hyperintensity on MRI: a comparison of normal and depressed elderly subjects. Am J Psychiatry 1990; 147:187-189

Coffey CE, Wilkinson WE, Weiner RD, Parashos IA, Djang WT, Webb MC, Figiel GS, Spritzer CE. Quantitative cerebral anatomy in depression: a controlled magnetic resonance imaging study. Arch Gen Psychiat 1993; 50:7-16.

Cohen RM, Semple WE, Gross M, Nordahl TE, King AC, Pickar D, Post RM. Evidence for common alterations in cerebral glucose metabolism in major affective disorders and schizophrenia. Neuropsychopharmacology 1989; 2:241-254.

de Groot JC, de Leeuw F-E, Oudkerk M, Hofman A, Jolles J, Breteler MMB. Cerebral white matter lesions and depressive symptoms in elderly adults. Arch Gen Psychiatry 2000; 57: 1071-1076.

de Groot JC, de Leeuw F-E, Oudkerk M, van Gijn J, Hofman A, Jolles J, Breteler M.B. Cerebral white matter lesions and cognitive function: the Rotterdam Scan Study. Ann Neurol 2000; 47:145-151.

de Leeuw F-E, de Groot JC, Oudkerk M, Witterman JC, Hofman A, van Gijn J, Breteler MMB. A follow-up study of blood pressure and cerebral white matter lesions. Ann Neurol 1999; 46:827-833.

DelBello MP, Strakowski SM, Zimmerman ME, Hawkins JM, Sax KW. MRI analysis of the cerebellum in bipolar disorder: a pilot study. Neuropsychopharmacology 1999; 21:63-68.

Drevets WC, Price JL, Simpson JR Jr, Todd RD, Reich T, Vannier M, Raichle ME. Subgenual prefrontal cortex abnormalities in mood disorders. Nature 1997; 386:824-827

Drevets WC, Ongur D, Price JL. Neuroimaging abnormalities in the subgenual prefrontal cortex: implications for the pathophysiology of familial mood disorders. Molec Psychiatry 1998; 3:190-191, 220-226.

Elkis H, Friedman L, Buckley PF, Lee HS, Lys C, Kaufman B, Meltzer HY. Increased prefrontal sulcal prominence in relatively young patients with unipolar major depression. Psychiatry Res 1996; 67:123-134.

Elkis H, Friedman L, Wise A, Meltzer HY. Meta-analyses of studies of ventricular enlargement and cortical sulcal prominence in mood disorders: comparisons with controls or patients with schizophrenia. Arch Gen Psychiat 1995; 52:734-746.

Everall IP, Chong WK, Wilkinson ID, et al. Correlation of MRI and neuropathology in AIDS. J Neurol Neurosurg Psychiatry 1997; 62:92-95.

Fazekas R, Kleinert R, Offenbacher H, et al. Pathologic correlates of incidental MRI white matter signal hyperintensities. Neurology 1993; 43:1683-1639.

Ferrier IN, Perry EK. Editorial: post-mortem studies in affective disorder. Psychological Medicine 1992; 22:835-838.

Folstein SE. Huntington's disease: a disorder of families. Baltimore: The Johns Hopkins University Press, 1989.

Greenwald B.S., Kramer-Ginsberg E., Krishnan R.R., et al. MRI signal hyperintensities in geriatric depression. Am J Psychiatry 1996;153:1212-1215

Healy DJ, Sima AAF, Tapp A, Watson SJ, Meador-Woodruff JH. Frequency of neuropathology in a brain bank from a long-term, domiciliary population. J Psychiat Res 1996; 30:45-49.

Hoge EA, Friedman L, Schulz SC. Meta-analysis of brain size in bipolar disorder. Schizophr Res 1999; 37:177-181.

Husain MM, McDonald WM, Doraiswamy PM, et al. A Magnetic resonance imaging study of putamen nuclei in major depression. Psychiatry Res 1991; 40:95-99.

Jeste DV, Lohr JB, Goodwin FK. Neuroanatomical studies of major affective disorders: a review and suggestions for further research. Br J Psychiatry 1988; 153:444-459.

Krishnan KR, Doraiswamy PM, Figiel GS, Husain MM, Shah SA. Na C., et al. Hippocampal abnormalities in depression. J Neuropsychiatry Clin Neurosci 1991; 3:387-391.

Krishnan KRR, McDonald WM, Escalona PR, Doraiswamy PM., Na C., Husan M.M., Figiel G.S., Koyko O.B., Ellinwood E.H., Nemeroff C.B. Magnetic Resonance Imaging of the caudate nuclei in depression: preliminary observations. Arch Gen Psychiat 1992; 49:553-557.

Krishnan KRR, McDonald WM, Doraiswamy PM, Tupler LA, Hussain M, Boyko OB, Figiel GS, Ellinwood Jr EH. Neuroanatomical substrates of depression in the elderly. Eur Arch Psychiatry Neurosci 1993; 243:41-46.

Krishnan KR, Hays JC, Blazer DG. MRI-defined vascular depression. Am J Psychiatry 1997; 154:497-501.

Kumar A, Jin Z, Bilker W, Udupa J, Gottleib G. Late-onset minor and major depression: early evidence for common neuroanatomical substrates detected by using MRI. Proc Natl Acad Sci USA 1998; 95:7654-7658.

Kumar A, Bilker W, Jin Z, Udupa J, Gottlieb G. Age of onset of depression and quantitative neuroanatomic measures: absence of specific correlates. Psychiatry Res Neuroimaging 1999; 91:101-110.

Leake A, Fairbairn AF, McKeith IG, Ferrier IN. Studies on the serotonin uptake binding site in major depressive disorder and control post-mortem brain: neurochemical and clinical correlates. Psychiatry Res 1991; 39:155-165.

Leake A, Perry EK, Perry RH, Fairbairn AF, Ferrier .N. Cortical concentrations of corticotropin-releasing hormone and its receptor in Alzheimer type dementia and major depression. Biol Psychiatry 1990; 28:603-608.

Mayberg HS, Lewis PJ, Regenold W, et al. Paralimbic hypoperfusion in unipolar depression. J Nucl Med 1994; 35:929-934.

McDonald WM, Tupler LA, Marsteller FA, Figiel G.S, DiSouza S, Nemeroff CB, Krishnan KR. Hyperintense lesions on magnetic resonance images in bipolar disorder. Biol Psychiatry 1999; 45:965-971.

Moore GJ, Bebchuk JM., Wilds IB, Chen G, Manji HK. Pharmacologic increase in human gray matter. Lancet 2000; 356:1241-1242.

O'Brien JT. Desmond P, Ames D, Schweitzer I, Tuckwell V, Tress B. The differentiation of depression from dementia by temporal lobe magnetic resonance imaging. Pscyhol Med 1994; 24:633-640.

O'Brien J, Desmond P, Ames D, et al. A magnetic resonance imaging study of white matter lesions in depression and Alzheimer's disease. Br J Psychiatry 1996; 168:477-485

O'Brien J, Ames D, Schweitzer I. White matter changes in depression and Alzheimer's disease: a review of magnetic resonance imaging studies. Int J Geriatr Psychiat 1996; 11:681-694.

O'Hearn E, Holmes SE, Calvert PC, Ross CA, Margolis RL. SCA-12: tremor with cerebellar and cortical atrophy is associated with a CAG repeat expansion. Neurology 2001; 56:287-289.

Ongur D, Drevets WC, Price JC. Glial reduction in the subgenual prefrontal cortex in mood disorders. Proc Natl Acad Sci USA 1998; 95:13290-13295.

Parashos IA, Tupler LA, Blitchington T, et al. Magnetic-resonance morphometry in patients with major depression. Psychiatry Res 1998; 84:7-15.

Pearlson GD. Structural and functional brain changes in bipolar disorder: a selective review. Schizophr Res 1999; 39:133-140.

Pearlson GD, Barta PE, Powers RE, et al. Medial and superior temporal gyral volumes and cerebral asymmetry in schizophrenia versus bipolar disorder. Biol Psychiatry 1997; 41:1-14.

Pearlson GD, Schlaepfer TE. Brain imaging in mood disorders. In Psychopharmacology: a third generation of progress, F.E. Bloom, D.J. Kupfer, eds. New York: Raven Press, 1995.

Perrett CW, Whatley SA, Ferrier IN, Marchbanks RM. Changes in relative levels of specific brain mRNA species associated with schizophrenia and depression. Brain Res (Molecular) 1992; 12:163-171.

Perry R, Jarros E, Irving D. What is the neuropathological basis of dementia assoiated with Lewy bodies? In Dementia with Lewy bodies, R. H. Perrym I.G. McKeith, E.K. Perry, eds. Cambridge: Cambridge University Press, 1996.

Potash JB, Willour VL, Chiu YF, Simpson SG, MacKinnon DF, Pearlson GD, DePaulo JR Jr, McInnis MG. The familial aggregation of psychotic symptoms in bipolar disorder pedigrees. Am J Psychiatry 2001; 158:1258-1264.

Rabins PV, Pearlson G., Aylward E, Kumar AJ, Dowell K. Cortical magnetic resonance imaging changes in elderly inpatients with major depression. Am J Psychiatry 1991; 148: 617-620.

Rahman S, Li PP, Young LT, Kofman O, Kish SJ, Warsh JJ. Reduced [3H]cyclic AMP binding in postmortem brain from subjects with bipolar affective disorder. J Neurochem 1997; 68:297-304.

Rajkowska G. Morphometric methods for studying the prefrontal cortex in suicide victims and psychiatric patients. Ann NY Acad Sci 1997; 836:253-268.

Rajkowska G, Miguel-Hidalgo JJ, Wei J, Dilley G, Pittman SD, Meltzer HY, et al. Morphometric evidence for neuronal and glial prefrontal cell pathology in major depression. Biol Psychiatry 1999; 45:1085-1098.

Rajkowska G. Postmortem studies in mood disorders indicate altered numbers of neurons and glial cells. Biol Psychiatry 2000; 48:766-777.

Robinson RG, Szetela B. Mood change following left hemispheric brain injury. Ann Neurl 1981; 9:447-453.

Roy PD, Zipursky RB, Saint-Cyr JA, Bury A, Langevin R, Seeman MV. Temporal horn enlargement is present in schizophrenia and bipolar disorder. Biol Psychiatry 1998; 44:418-422.

Salib E, Tadros G. Brain weight in suicide. Br J Psychiatry 2000; 177:257-261.

Sapolsky RM, Uno H, Rebert CS. Finch CE. Hippocampal damage associated with prolonged glucocorticoid exposure in primates. J Neurosci 1990; 10:2897-2902.

Schweitzer I, Tuckwell V, Ames D, O'Brien J. Structural neuroimaging studies in late-life depression: a review. World J Biol Psychiatry 2001; 2:83-88.

Shah PJ, Ebmeier KP, Glabus MF, Goodwin GM. Cortical grey matter reductions associated with treatment-resistant chronic unipolar depression: controlled magnetic resonance imaging study. Br J Psychiatry 1998; 172:527-532.

Sheline YI, Gado MH, Price JL. Amygdala core nuclei volumes are decreased in recureent major depression. Neuroreport 1998; 9:2023-2028.

Sheline YI, Shanghavi M, Mintun MA, Gado, MH. Depression duration but not age predicts hippocampal volume loss in medically healthy women with recurrent major depression. J Neurosci 1999; 19:5034-5043.

Soares JC, Mann JJ. The anatomy of mood disorders: review of structural neuroimaging studies. Biol Psychiatry 1997; 41:86-106.

Strakowski SM, DelBello MP, Sax KW, Zimmerman ME, Shear PK, Hawkins JM, Larson ER. Brain magnetic resonance imaging of structural abnormalities in bipolar disorder. Arch Gen Psychiatry 1999; 56:254-260.

Swayze VW II, Andreasen NC, Alliger RJ, Ehrhardt JC, Yuh WT. Structural brain abnormalities in bipolar affective disorder: ventricular enlargement and focal signal hyperintensities. Arch Gen Psychiatry 1990; 47:1054-1059.

Swayze VW, Andreasen NC, Alliger .J, Yuh WT, Ehrhardt JC. Subcortical and temporal structures in affective disorder and schizophrenia: a magnetic resonance imaging study. Biol Psychiatry 1992; 31:221-240.

Thomas AJ, Ferrier IN, Kalaria RN, Perry RH, Brown A, O'Brien JT. A neuropathological study of vascular factors in late-life depression. J Neurol Neurosurg Psychiatry 2001; 70:83-87.

Torrey EF, Webster M, Knable M, Johnston N, Yolken RH. The Stanley foundation brain collection and neuropathology consortium. Schizopphr Res 2000; 44:151-155.

Trautner RJ, Cummings JL, Read SL, et al. Idiopathic basal ganglia calcification and organic mood disorder. Am J Psychiatry 1988; 145:350-353.

Vakili K, Pillay SS, Lafer B, Fava M, Renshaw PF, Vonello-Cintron CM. Yurgelun-Todd DA. Hippocampal volume in primary unipolar major depression: a magnetic resonance imaging study. Biol Psychiatry 2000; 47:1087-1090.

Vawter MP, Freed WJ, Kleinman JE. Neuropathology of bipolar disorder. Biol Psychiatry 2000; 48:486-504.

Videbech P. MRI findings in patients with affective disorder: a meta-analysis. Acta Psychiatr Scand 1997; 96:157-168.

Young LT, Li PP, Kish SJ, Siu KP, Warsh JJ. Postmortem cerebral cortex G_s α-subunit levels are elevated in bipolar affective disorder. Brain Res 1991; 553:323-326.

16 QUANTITATIVE CYTOARCHITECTONIC FINDINGS IN POSTMORTEM BRAIN TISSUE FROM MOOD DISORDER PATIENTS

Grazyna Rajkowska

Abstract

A considerable body of pharmacological and neurochemical literature has accumulated on affective disorders. Until recently, however, there have been no quantitative neuroanatomical studies of these disorders at the microscopic level. Clinical neuroimaging studies and pre-clinical animal studies, however, strongly suggest that cell atrophy, cell loss or impairments in neuroplasticity and cellular resilience may underlie the neurobiology of primary mood disorders (i.e., major depressive disorder and bipolar manic-depressive disorder).

Recent quantitative cytoarchitectonic studies on postmortem tissues from patients with mood disorders provide direct evidence that mood disorders are characterized by specific changes in the number, density or size of both neurons and glial cells. Although published reports are scarce and based on rather small sample sizes, these studies are surprisingly consistent in revealing previously unrecognized reductions in glial cell number and density as well as alterations in the density and/or size of specific types of cortical neurons in frontal limbic brain regions.

This chapter reviews the current findings from stereological and morphometric postmortem studies on glia and neurons in primary mood disorders. The relevance of cellular changes in mood disorders to dysfunctional monoaminergic and glutamatergic circuits and a possible role of neurotrophic and neuroprotective factors in cell pathology is discussed. A possible link between cellular changes in mood disorders and the action of psychotherapeutic drugs is suggested as well.

Introduction

Although the existence of depression and the diverse symptomatology of affective disorders have been recognized for decades, the precise anatomical basis of these disorders remains unknown. In sharp contrast to the considerable body of pharmacological, neurochemical and neurophysiological literature on affective disorders, until recently there has been a striking lack of quantitatively controlled neuroanatomical studies on these disorders. Mood disorder patients were primarily used as psychiatric "control" subjects for studies in schizophrenia and were not a subject of separate systematic stereological and morphometric studies.

Quantitative cytoarchitectonic studies on primary mood disorder patients, i.e., diagnosed with major depressive disorder (MDD) or bipolar manic depressive disorder (BPD) may be highly relevant to the profuse neurochemical findings suggesting that the disruption of monoaminergic (in addition to other) neurotransmitter pathways may be critical in the pathophysiology of mood disorders. Changes in monoamine receptors, transporters and related second messenger systems reported in neurochemical and pharmacological studies of mood disorders implicate parallel morphological changes in the monoaminergic brainstem centers and their cortical projection areas. Moreover, results of recent clinical and pre-clinical studies investigating the molecular and cellular targets of antidepressants and mood stabilizers provide intriguing possibilities that impairments in neuroplasticity and cellular resilience related to alterations in the signal transduction pathways may underlie the neurobiology of mood disorders. Finally, the results of recent neuroimaging studies showing changes in the volume of some brain structures and reduced levels of markers for neuronal viability and metabolic integrity strongly imply that alterations in the number of cells in addition to biochemical imbalances underlie the neuropathology of mood disorders.

Quantitative cytoarchitectonic studies undertaken on postmortem tissues from mood disorder patients and reports that have been published within the last three years, clearly demonstrate that mood disorders are characterized by specific morphometric changes in both neurons and glial cells. Although published morphometric reports are scarce and based on rather small sample sizes the findings across such studies are surprisingly consistent. They reveal previously unrecognized reductions in glial cell number and density as well as alterations in the density or size, or both, of specific types of cortical neurons.

This chapter reviews the current findings from postmortem studies on glial and neuronal cell counts in primary mood disorders (MDD and BPD). The relevance of cellular changes in mood disorders to dysfunctional monoaminergic and glutamatergic circuits and a possible role of neurotrophic and neuroprotective factors in cell pathology is discussed. A possible link between cellular changes in mood disorders and the action of psychotherapeutic drugs is suggested as well. It is anticipated that the precise anatomical localization of dysfunctional glia and neurons in mood disorders will reveal cortical targets for novel antidepressants and mood stabilizers.

Alterations in Cell Numbers in Mood Disorders

Gross neuroanatomical examination of postmortem brain from MDD or BPD patients does not reveal any visible changes. The widening of cortical sulci and narrowing of gyri reported by some neuroimaging studies in mood disorder patients is observed in very few postmortem specimens from similar populations. Qualitative microscopic examination of postmortem brain from MDD or BPD patients does not reveal any visible neuropathological changes such as focal lesions or the plaques and tangles seen in Alzheimer disease. However, in some cases, microvascular changes in the gray and white matter are noticeable, and are especially prominent in BPD and older MDD subjects (personal observations). Application of non-biased stereological cell counting and detailed morphometric methods reveal changes in the cell number, density and soma body size in specific brain regions of mood disorder patients.

Currently, three independent laboratories have published stereological and morphometric studies using well-characterized and well-matched postmortem samples. These investigations have found consistent reductions in glial cell number and density in frontal limbic cortical regions in MDD and BPD subjects (Cotter et al 2001; Ongur et al 1998; Rajkowska et al 2001; Rajkowska et al 1999a). These alterations in glia are accompanied by more subtle changes in the density and/or size of specific populations of neurons (Cotter et al 2001; Rajkowska et al 2001; Rajkowska et al 1999a). Specific types of neurons are also reduced in the hippocampal formation (Benes et al 1998) and anterior cingulate cortex in BPD (Benes et al 2001) and MDD (Diekmann et al 1998), whereas glial cells have not been systematically studied in hippocampus to date. Available morphometric data from subcortical structures are less numerous and less consistent due to smaller sample size and heterogeneity of studied groups. These studies indicate increases rather than reductions in the neuronal cell number in the hypothalamus (Purba et al 1996; Raadsheer et al 1994) and brainstem nuclei (Baumann et al 1999; Kasir et al 1998; Underwood et al 1999).

In the following two sections we review findings from several laboratories that have reported on stereological findings in postmortem tissue from mood disorder patients.

Glial Pathology

Stereological analyses of the anterior cingulate cortex conducted by two independent laboratories in postmortem tissues from MDD and BPD subjects provide consistent evidence for reductions in a number (Ongur et al 1998) and density (Cotter et al 2001) of glial cells. Striking 24% reductions in number of glial cells were found in the subgenual region of the anterior cingulate cortex (ventral part of Brodmann's area, BA24) in patients with familial MDD (Ongur et al 1998). Similar 41% reductions were observed in the same region in a small subgroup of subjects with familial BPD, as compared to controls (Ongur et al 1998). However, when familial and non-familial subgroups of patients were combined the reductions were not found. The estimation of glial cell number in this study in both MDD and BPD was combined across all six cortical layers. Thus, no information on laminar specificity of glial loss was provided.

Evaluation of glial cell density in individual cortical layers of the supracallosal region of the anterior cingulate cortex (dorsal part of BA24; (Cotter et al 2001) confirmed findings on glial reductions in MDD. However, the 22% reductions in glial cell density found in this study were confined to only one cortical layer, deep layer VI. Similar laminar analysis of glial cells in the same region in BPD subjects resulted in negative findings (Cotter et al 2001). However, in Cotter's et al., study, unlike in that of Ongur et al., the studied BPD group was not subdivided according to the family history, which could account for the negative results.

Earlier morphometric studies of glial cells in both the dorsolateral prefrontal cortex (dlPFC, BA9) and orbitofrontal cortex (ORB, BA 47) in subjects with MDD also reveal marked decreases in overall (11-15%) and laminar (20-30%; layers III-VI) cell packing densities as compared to controls (Rajkowska et al 1999a). The degree of glial pathology in MDD differs among prefrontal regions and specific cortical layers. Glial reductions were most striking in more caudally located regions of the prefrontal cortex, the caudal ORB and dlPFC and very mild in the rostral part of the ORB cortex. These decreases were found predominantly in deeper cortical layers V and VI of the caudal ORB. The latter observation is similar to findings of Cotter and colleagues (Cotter et al 2001) on layer VI - specific reductions in glial cell density in supracallosal cingulate cortex which is located at the caudal border of the frontal lobe. Reductions in the density of glial cells in dlPFC are accompanied by enlargement in the size of glial nuclei (Rajkowska et al 1999a) suggesting some compensatory mechanisms.

Morphometric analysis of the glial cell density and size in the dorsolateral prefrontal cortex (BA area 9) in BPD subjects revealed no significant differences between BPD and an age- and gender-matched control group when data from all layers and all glial size-types were combined (Rajkowska et al 2001). However, more detailed laminar analysis demonstrated lamina-specific reductions in the density of glial cells in BPD. As compared to control subjects, mean glial cell density was significantly reduced by 19% in BPD in deep layer III of the dlPFC whereas (marginally significant) 12% decreases were found in deep layer V of this cortex. When the density of specific glial types (defined on the basis of nucleus size) were analyzed, even more prominent 35-45% reductions in the density of glia (with medium sized nuclei) were found in layers III, V and when all six cortical layers were combined. These reductions in glial density in BPD were accompanied by an enlargement of the mean size of glial nuclei, likely due to 32-100% increases in the density of glial cells with very large nuclei and parallel 31-45% decreases in the density of glial cells with smaller nuclei. Furthermore, in BPD there were also changes in the shape of glial nuclei to a less rounded conformation. Thus, results of Rajkowska et al., (Rajkowska et al 2001; Rajkowska et al 1999a) are somewhat similar to that of Ongur and colleagues (Ongur et al 1998) since in both studies no reductions in glial cell number or density were found between BPD and controls when the data from all layers and all BPD patients were combined. However, when individual layers were analyzed separately (Rajkowska et al) or subjects were sorted into subgroups based on family history (Ongur et al), significant reductions in glial density or number were detected in BPD.

Glial cell pathology in mood disorders may not be universally noted throughout the cerebral cortex. Ongur et al., (Ongur et al 1998) did not find changes in glial cell density or number in the sensorimotor cortex in either MDD or BPD populations. Recent reports suggest a lack of marked glial pathology in the supracallosal part of the anterior cingulate cortex (Cotter et al 2001) in BPD and in the entorhinal cortex in BPD and MDD (Bowley et al 2000).

In summary, it appears that cortical glial pathology in BPD is less pervasive than in MDD since it is detected only when specific layers or specific types of glia are analyzed. Moreover, glial pathology in BPD is found in some but not all cortical regions in which glial pathology is reported in MDD. Therefore, the regional differences in glial pathology are suggested in mood disorders. These regional differences that exist between the two mood disorders are more apparent in ventrally located regions of the frontal limbic cortex (orbitofrontal cortex and subgenual anterior cingulate cortex) and more subtle in dorsally located regions (dlPFC, supracallosal anterior cingulate cortex). These regional differences may be a reflection of different circuits involved in the neuropathology and clinical symptomatology of the two mood disorders and/or a consequence of different therapeutic treatment.

Subcortical Structures. Glial pathology in mood disorders has not been systematically studied in subcortical structures to date. Only one, recent preliminary report suggests that glial pathology might extend to limbic subcortical regions since a significant reduction in glial number are found in the amygdala in MDD and unmedicated BPD subjects (Bowley et al 2000).

Glial Cell Types. The glial cells analyzed in the above mentioned studies do not represent a homogenous population. They are comprised of distinct populations of oligodendrocytes, microglia and astrocytes, whose crucial role in brain function is currently being re-evaluated (Coyle and Schwarcz 2000; Pfrieger and Barres 1997). These three distinct glial cell types cannot be identified in any of the above mentioned studies since they were conducted on tissue stained for Nissl substance. This staining does not distinguish between specific glial cell types as Nissl staining only visualizes morphological features of glial cell bodies and not glial cells processes. On the other hand, recent immunohistochemical examination of an astrocyte marker, glial fibrillary acidic protein (GFAP), in the dlPFC suggests the involvement of astrocytes in overall glial pathology in MDD (Miguel-Hidalgo et al 2000). Although no significant group differences in GFAP-reactive astrocyte cell density were present when the entire group of MDD (young+old) was compared to normal controls, significant reductions in the population of reactive astrocytes were found in a small subgroup of young MDD subjects as compared to young controls and older MDD subjects. Moreover, a significant positive correlation between the number of immunopositive- astrocytes and the density of glia with large cell body size (revealed by Nissl staining) was found in this study. This suggests that the reductions in the population of astrocytes in the subgroup of young MDD subjects account, at least in part, for the global glial deficit identified in this disorder.

Alterations in GFAP in both BPD and MDD are also suggested by a recent proteomic study where four out of eight proteins displaying disease-specific alterations are forms of GFAP (Johnston-Wilson et al 2000). Another type of glial cell, oligodendrocytes also may be involved in the general reductions in glial number reported in BPD and familial MDD. Recent preliminary observations suggest a reduced density of oligodendrocytes in the dlPFC (layer VI) in BPD subjects and in a subgroup of MDD subjects with family history of severe mental disorder (Orlovskaya et al, 2000). Moreover, morphometric dystrophic alterations in oligodendrocytes were detected at the electron microscopic level in a BPD group (Orlovskaya et al 1999). While these results are intriguing, further immunohistochemical and molecular studies are needed to definitively determine which specific glial cell type(s) are compromised in BPD and whether the same or different glial cell types are involved in the pathology reported in MDD.

In summary, glial deficit is a prominent feature of cortical pathology in both mood disorders, BPD and MDD. Reductions in glial number and density are well documented in the frontal limbic cortical areas in MDD. The findings on BPD seems to be more restricted to specific cortical regions or layers. Discrepancies between several studies may be accounted by differences in the regions studied or the family history of targeted subgroups. The functional significance of glial reductions in mood disorders has not yet been established (see below and Chapter by Cotter et al., of this book for possible functional implications).

Neuronal Pathology

Data on alterations in number or density of neuronal cells in mood disorders are less consistent than reports on glial pathology. Though not numerous, several studies find changes in neuronal density, (Benes et al 1998; Benes et al 2001; Diekmann et al 1998; Rajkowska et al 2001; Rajkowska et al 1999a) number (Purba et al 1996; Raadsheer et al 1994; Underwood et al 1999) or size of neuronal cell bodies (Cotter et al 2001; Rajkowska et al 1999a) whereas, other studies fail to find any alterations in neurons in either MDD, BPD subjects, or both (Cotter et al 2001; Klimek et al 1997; Ongur et al 1998).

Cerebral Cortex. Quantitative histopathology of neurons in mood disorders using stereological or morphometric methods has been estimated in three major cortical regions, 1) prefrontal cortex (dorsolateral and orbitofrontal regions), 2) anterior cingulate cortex (subgenual and supracallosal regions), and 3) hippocampus. In studies in which individual cortical layers are examined or specific morphological type of neurons were analyzed significant reductions in neurons are found in MDD and/or BPD subjects as compared to controls (Rajkowska et al 2001; Rajkowska et al 1999a) (Benes et al 1998; Benes et al 2001; Cotter et al 2001; Diekmann et al 1998) .

Neocortex. Morphometric analysis of neuronal density in individual cortical layers of the ORB cortex revealed the most prominent reductions in the rostral part of this region, whereas the caudal part exhibited mostly reductions in the density of glial cells rather than neurons (Rajkowska et al 1999a). These neuronal reductions in the rostral ORB (corresponding to a transitional area between BA10 and rostral part of BA 47) in MDD were confined to upper cortical layers with most dramatic changes in layer II cells (mostly corresponding to non-pyramidal, inhibitory local-circuit-neurons). Interestingly, reductions in the density of immunohistochemically identified population of layer II non-pyramidal neurons containing the calcium binding protein calretinin are reported in the anterior cingulate cortex in subjects with a history of mood disorders (Diekmann et al 1998). The reductions in neuronal densities in the ORB cortex are paralleled by smaller sizes (5-9%) of neuronal somas and significant, 12-15% decreases in cortical thickness observed in rostral and middle ORB cortex in MDD groups (Rajkowska et al 1999a).

Quantitative examination of another prefrontal regions, dlPFC in MDD did not reveal significant reductions in overall or laminar neuronal density when the entire population of the Nissl stained neurons was analyzed (Rajkowska et al 1999a). Nonetheless, when all neurons were subdivided into four separate subtypes based on the size of their cell bodies, significant reductions (22-37%) in the density of neurons with large (20-36 mm in diameter) body size were found in layers II, III and VI of the dlPFC. These large pyramidal neurons, which are decreased in MDD, are thought to be excitatory pyramidal neurons. However, final identification of this cell type has yet to be confirmed by neurochemical methods. The reductions in the density of large neurons were accompanied by 6-27% increases in the density of the smallest neurons (5-11mm in diameter) and reductions in the mean neuronal size in both layers III and VI. The above suggests that either neuronal shrinkage or a developmental deficiency, rather than neuronal loss, accounts for the overall smaller sizes of neuronal soma in those cortical layers (for further discussion on "Cell Loss or Atrophy" see below). Comparable reductions (23%) in the size of neuronal cell bodies have been recently found in layer VI of the supracallosal region of the anterior cingulate cortex in MDD but not BPD (Cotter et al 2001). These reductions in size were not accompanied by significant decreases in neuronal density.

Laminar analysis of neuron density in the dlPFC region in BPD revealed significant 16-22% reductions in the mean density of all neurons in layer III throughout all three sublayers (Rajkowska et al 2001). In addition, significant decreases in the density of large pyramidal neurons were found in layer III (all three sub layers) and in sub layer Va. Measurements of laminar thickness in these groups indicated a 16% increase in the width of layer IIIc in BPD subjects, where glial size was increased as well. In light of the increased width of layer III, one possibility is that an increase in interneuronal neuropil, perhaps including the processes of hypertrophied glial cells, might account for the reduction in neuronal and glial densities in this layer in BPD patients. The cortical width in the dlPFC region, unlike that in rostral parts of ORB region remained unchanged in both MDD and BPD as compared to control subjects (Rajkowska et al 2001; Rajkowska et al 1999a).

Neuronal Cell Types. Taken together, morphometric analyses of cortical neurons in mood disorders suggest that reductions in both inhibitory and excitatory types of neurons in prefrontal and limbic cortical regions could be involved in the neuropathology underlying mood disorders. Laminar localization of neuronal changes can be, in part, attributed to differential distributions of these neuronal types in the cortex. Nonpyramidal GABAergic (inhibitory) neurons are localized mainly in layer II whereas pyramidal glutamatergic (which are excitatory) neurons reside predominantly in cortical layers III and V. Since information processing in the neocortex involves extensive interactions between excitatory pyramidal neurons and inhibitory GABA nonpyramidal interneurons, dysfunction in one type of cortical neuron likely can lead to malformations in the other. Further immunohistochemical and molecular studies are needed to define more precisely the specific class or classes of cortical neurons that are altered in mood disorders and how specific are these changes.

Cortical processing is also intimately regulated by modulatory monoaminergic afferents, such as dopamine fibers originating in the ventral tegmental area, serotonin axons from the dorsal raphe neurons or noradrenaline axons from the locus coeruleus (reviewed in (Rajkowska 2000a). For example, primate and human postmortem studies reveal that dopamine afferents preferentially target pyramidal neurons (synapsing on the dendritic shafts), whereas serotonin axons preferentially target a subset of nonpyramidal neurons (Krimer et al 1997; Smiley and Goldman-Rakic 1996). Thus, cell type- specific and layer-specific alterations in the density and size of cortical neurons found in mood disorders are in line with the pathophysiology of monoaminergic circuits reported in depression (for further discussion see below, "Linking Cellular Changes to Dysfunctional Monoaminergic/Glutamatergic Circuits").

Hippocampus. Hippocampus constitutes the major part of the limbic archicortex (phylogenetically old cortex consisting of only three layers). This structure has been long implicated in the structural neuropathology of depression and the response to stress. For example, in vivo neuroimaging studies indicate that major depression is associated with the hippocampal atrophy (Bremner et al 2000; Shah et al 1998) which correlates with the total duration of depression (Sheline et al 1999). However, a systematic stereological analysis determining whether hippocampal cell loss contributes to hippocampal atrophy in MDD has not been performed yet. The one morphometric study of the human hippocampus in mood disorders reported significant reductions in the density of nonpyramidal neurons in the CA2 region in BPD subjects as compared to normal controls (Benes et al 1998). Similar analyses have not been conducted in subjects with major depression.

Observations on the reduction in volume and neuronal density reported in the hippocampus could be related to the findings from other postmortem studies that demonstrate 1) decreased dendritic spine density on subicular neurons (in a small group of mixed subjects with BPD and MDD, (Rosoklija et al 2000), 2) decreased level of synaptic proteins found in CA4 hippocampal region in BPD (Eastwood and Harrison 2000), and 3) diminished level of N-acetylaspartate (a marker of neuronal integrity) reported by in vivo neuroimaging in the hippocampus in BPD (Bertolino 1999).

Animal studies suggest that degeneration of specific types of hippocampal neurons is related to a neurotoxic effect of stress and the accompanying increases in glucocorticoids (reviewed in Sapolsky 2000).

Chronic stress or the repeated administration of glucocorticoids to the primate and rodent hippocampus result in dendritic atrophy, shrinkage of the neuronal cell body and nuclear pyknosis (Sapolsky 2000; Watanabe et al 1992). It has been proposed that repeated stress during recurrent depressive episodes may inflict cumulative hippocampal injury as reflected in the loss of structural volume which may be central to the development of depression in genetically vulnerable individuals (Duman 2000). Further studies of cell morphometry would need to be performed on postmortem human and animal hippocampus to determine whether the suggested link between cell atrophy, neurotoxic effect of stress and depression does, indeed, exist.

In summary, the layer-specific neuronal changes found in the prefrontal cortex, anterior cingulate cortex and hippocampus in MDD and/or BPD imply the involvement of frontal limbic neural circuits and several neurotransmitter systems in the neuropathology of mood disorders (for further discussion, see (Rajkowska 2000a; Rajkowska 2000b).

Subcortical Structures. Few postmortem studies have attempted to estimate the number of neurons in the following subcortical structures: hypothalamus, dorsal raphe nucleus and locus coeruleus in postmortem brain tissues of mood disorder patients. While the results were obtained from a small number of subjects they indicate that increases, rather than decreases, characterize the subcortical neuronal pathology in MDD and BPD.

Stereological investigation of specific types of hypothalamic neurons revealed increased numbers of arginine-vasopressin (AVP)-immunoreactive neurons, oxytocin (OXT)-expressing neurons, and corticotropin-releasing hormone (CRH) neurons in the paraventricular nucleus in BPD and MDD subjects, compared to normal controls (Purba et al 1996; Raadsheer et al 1994) . Moreover, increases in CRH-mRNA and in the number of CRH neurons co-localizing AVP were also found in depressed patients (Raadsheer et al 1995; Swaab et al 1993). These findings of increases in specific immunoreactive-neurons are consistent with the evidence of activation of the HPA axis in some subsets of depressed patients (Holsboer et al 1992). Increases in the number of CRH, AVP or OXT cells suggest an enhancement in related cell function, which may, in turn, have a modulatory effect on cortical or brainstem neurons. Increased AVP and OXT cell number is related to the enhanced production of these neuropeptides in the HPA axis (Lucassen et al 1994), and, in turn, may be associated with increased activity of neurochemical signaling systems reported in mood disorders (Jope 1999; Manji et al 2000).

An increased number of neurons is also reported in the brainstem of BPD patients. Bilateral increases were found in the total number of pigmented neurons in locus coeruleus in BPD postmortem brains (Baumann et al 1999). However, findings of that study are based on a very small sample size. Another morphometric study, based on a larger sample size, found no changes in the cell number between MDD and control subjects at several levels of locus coeruleus (Klimek et al 1997). Locus coeruleus neurons are the major source of norepinephrine (NE) in the central nervous system. Alterations in NE projections to neocortical, limbic subcortical regions, or both may play a role in the pathophysiology of mood disorders. Interestingly, another brainstem structure, the dorsal raphe, which is the major source of serotonin projections to the cerebral cortex, also is reported to have increased number, density (Underwood et al 1999), and dendritic processes (Kasir et al 1998) of immunopositive-serotonergic neurons in suicide victims with MDD, compared to controls. Another preliminary study reported on increases in dorsal raphe volume, and, changes in the shape of serotoninergic, raphe neurons from MDD subjects relative to controls (Stockmeier et al 1999). Together, these findings suggest that some re-arrangement in the cellular structure of brainstem neurons projecting to the cerebral cortex is found in mood disorders. However, further stereological studies with a larger number of subjects are required to determine whether locus coeruleus changes are specific to BPD, while those in dorsal raphe are specific to MDD and suicide.

In summary, the unique neuronal changes observed in several distinct brain regions, including the prefrontal cortex, anterior cingulate cortex , hippocampus, locus coeruleus, dorsal raphe and hypothalamic nuclei suggest that several networks of interconnected neural circuits and neurotransmitter systems are undoubtedly implicated in the pathophysiology of MDD and BPD. This is not altogether surprising since the behavioral and physiological manifestations of the illness are complex and include cognitive, affective, motoric, and neurovegetative symptomatology, as well as alterations of circadian rhythms and neuroendocrine systems (for a review, see (Mayberg 2000).

The alterations in cell number and density in mood disorders are likely to be related to the disorder itself and not to the age, postmortem delay or the time of tissue fixation. Statistical analyses conducted in all of the above mentioned morphometric studies yielded no significant correlation between cell numbers and any of these confounding variables. It cannot be ruled out however, that some of the cellular alterations in mood disorders are related to prior treatment with antidepressants and lithium (see further discussion below).

Cell Loss Versus Cell Atrophy

Stereological and morphometric studies reviewed above indicate a loss of glial, but not neuronal cells in familial mood disorders in the anterior cingulate cortex, and lamina-specific reductions in the density of both neurons and glia in the prefrontal cortex in MDD and BPD patients. Whether these prominent reductions in cell density represent cell loss or only atrophy of, their cell bodies, processes, or both has not been established. It is not entirely clear whether cell loss accounts for the reductions in cell packing density since density measurements are dependent not only on the total number of cells present, but, also on the total volume in which cells are counted.

For the estimation of total number of neurons or glia in a particular brain region, it is essential that the borders of this region be established so that sampling is confined to the region within these borders (Gundersen et al 1988). Since the entire extent and borders of the studied cortical areas were not available for examination, estimates of total cell number were not possible. However, indirect assumptions could be made. For example, in MDD, the densities of the largest neurons are significantly reduced in ORB and dlPFC regions (Rajkowska et al 1999a). In contrast, the densities of small neurons are increased in those regions. The latter observation suggests that neuronal shrinkage or a developmental deficiency rather than neuronal loss accounts for the overall smaller sizes of neuronal soma in those cortical layers (Fig. 1). However, if neuronal loss had occurred, it is likely that the density of large neurons would have been decreased without associated increases in the density of small neurons, as was demonstrated in Huntington's disease (Rajkowska et al 1998; Fig. 2). In contrast to MDD, in BPD the density of large neurons is decreased in the dlPFC without accompanying increases in the density of smaller neurons (Rajkowska et al 2001). Thus, a neuronal loss rather than a diminution in neuronal size could be suggested in this disorder (Fig. 1). Such neuronal loss may be region-specific since no changes in neuronal density were found in MDD or BPD in the anterior cingulate cortex (Cotter et al 2001; Ongur et al 1998). A third possibility is that neuron loss may occur after initial atrophy/shrinkage of neurons. However, we do not know the time course for these processes under these conditions.

Definitive answers regarding cell loss in prefrontal cortex and other brain regions in mood disorders await additional postmortem studies with unbiased stereological methods in which the total number of specific types of neurons and glia will be estimated.

Functional and Biological Implications of Cytoarchitectonic Changes

Linking Cellular Changes to Neuroimaging Findings

Structural neuroimaging studies in mood disorders provide evidence of modest but intriguing volumetric changes suggestive of cell loss and/or

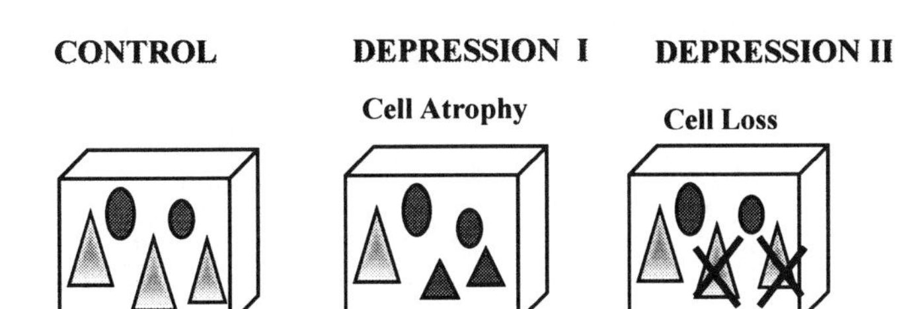

Figure 1. Hypothetical model of cellular alterations in mood disorders. The model reflects two patterns of morphometric changes reported in various cortical regions by postmortem stereological and morphometric studies. The "Depression I" panel illustrates cell atrophy reported in the dlPFC and ORB regions of the prefrontal cortex in MDD (Rajkowska et al 1999a). In these regions, there is a decrease in the density of large neurons (represented by triangles) which is accompanied by parallel increases in the density of small neurons (circles, compare size and number of cells in the "Depressed I" model with the Control drawing). The latter observation suggests that neuronal shrinkage (atrophy) or a developmental deficiency rather than neuronal loss accounts for the overall smaller neuronal sizes in those regions. If neuronal loss had occurred, the density of large neurons would have decreased without associated increases in the density of small neurons as presented in "Depression II". Glial cell loss is reported in the subgenual frontal cortex in MDD and BPD (Ongur et al 1998). Further studies are needed to determine whether loss of neurons occurs in mood disorders.

atrophy. Some, but not all, studies report enlargement of the lateral and third ventricles (Elkis et al 1995) that may be indicative of atrophy of surrounding cortical and subcortical regions. Moderate (7%) reduction in the total volume of the frontal lobe is reported in MDD patients (Coffey et al 1993) whereas, more substantial reductions in volume of gray matter are found in specific regions such as, subgenual part of the anterior cingulate cortex (39-48%, (Drevets et al 1997), hippocampus (Bremner et al 2000; Sheline et al 1999), prefrontal cortex (Lim et al 1999), temporal cortex (Altshuler et al 1991; Hauser et al 1989), and amygdala (Pearlson et al 1997; Sheline et al 1998) in subjects with MDD, BPD, or both. Changes in the volume of cortical regions and subcortical structures are not uniformly restricted to the reductions since increases in the volume of basal ganglia and medial temporal cortex (Pearlson et al 1997) are also observed in mood disorder patients in some neuroimaging studies (for further review on structural neuroimaging see (Soares and Mann 1997)).

Functional neuroimaging studies lend further evidence to physiological abnormalities in cortical and subcortical frontal limbic regions in MDD and BPD. Dysregulation of glucose metabolism, regional cerebral

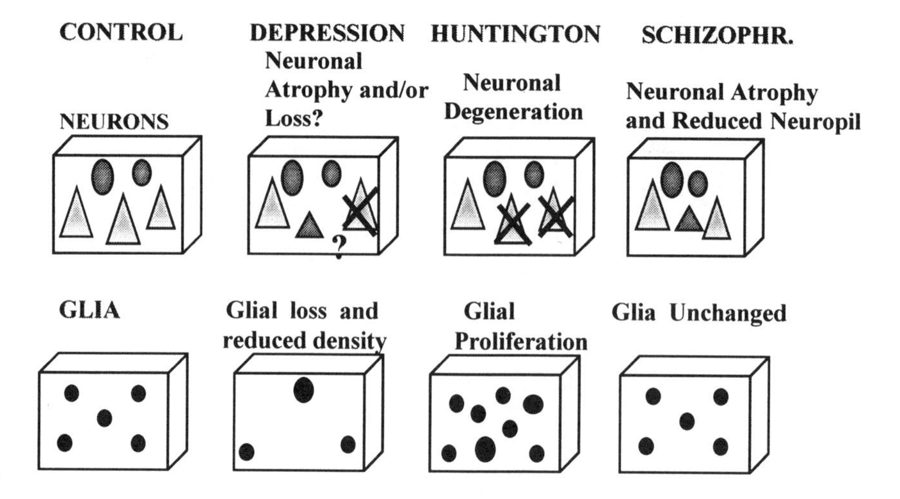

Figure 2. Comparison of the pattern of neuronal and glial changes in the dlPFC between mood disorders (called here "depression"), Huntington disease and schizophrenia. In depression (second column), neuronal atrophy in the dlPFC is suggested by parallel decreases in the density of large cells (triangles) and increases in the density of small cells (circles, see also Fig. 1). In contrast, in Huntington's disease (third column), a well-characterized neurodegenerative disorder, the density of large neurons is decreased without associated increases in the density of small neurons (Rajkowska et al 1998). In the same dlPFC region in schizophrenia (fourth column), increased neuronal density and smaller neuronal sizes suggest neuronal atrophy rather than neuronal loss, and that atrophy may be associated with a reduction in surrounding neuropil (Selemon and Goldman-Rakic 1999). In depression, the reductions in glial densities associated with enlargement of individual glial nuclei are unique for the dlPFC. In Huntington's disease and schizophrenia, opposite patterns of glial changes are reported (increases in glial densities and unchanged glial density, respectively (Rajkowska et al 1998; Selemon et al 1995; Selemon et al 1998). Such striking differences in cell pathology between the three brain disorders may be unique to the dlPFC, as opposed to other frontal regions.

blood flow, and high-energy phosphate metabolism have been observed in the prefrontal and temporal cortex, basal ganglia, and amygdala in mood disorders (Baxter et al 1989; Bench et al 1993; Biver et al 1994; Blumberg et al 1999; Buchsbaum et al 1986; Deicken et al 1995; Drevets 1999; George et al 1993; Kato et al 1995). Neuroimaging studies that examine neurochemical changes in the living brain provide further support for the hypothesis that mood disorders are associated with changes in neuronal viability and function. Recent studies using high resolution magnetic resonance spectroscopy in unmedicated BPD patients have shown decreased N-acetyl-aspartate (NAA) levels bilaterally in the hippocampus (Bertolino et al 1999) and in the dorsolateral prefrontal cortex (Winsberg et al 2000) as compared to healthy controls. In contrast, therapeutic doses of lithium increase levels of NAA in brain of bipolar patients (Moore et al 2000b). These increases are found in a number of regions including frontal cortex, and are localized almost exclusively to gray matter. NAA is regarded as a measure of neuronal, as oppose to glial, viability and function since it is found in mature neurons and not found in mature glial cells, CSF, or blood cells (Tsai and Coyle 1995). Therefore, the changes in NAA levels seen in BPD strongly implicate alterations in cell viability which, itself, may be related to alterations in cell number, cell density, and related volumetric changes.

The prominent reductions in glial cells reported by postmortem studies may also be linked to altered glucose metabolism in neuroimaging studies. Glial cells are involved in glucose metabolism and are considered the primary sites of glucose uptake and phosphorylation during neuronal activity (Tsacopoulos and Magistretti 1996). Glucose uptake may provide much of the observed signal in functional magnetic resonance imaging (fMRI) and positron emission tomography (PET) studies. Astrocytes also promote synapse formation in vitro, and are involved in the development and remodeling of synaptic connections (Pfrieger and Barres 1997; Ullian et al 2001). Thus, reductions in the density of astroglial cells or decreases in the area occupied by astroglial processes could account for a decreased number of functional synapses in mood disorders brains and related neuroimaging changes.

In summary, neuroimaging findings indicate that gross anatomical and metabolic changes are found in the same brain regions in which cytoarchitectonic, postmortem studies report alterations in number, density and size of neurons and glial cells. The affected regions belong to the frontal limbic system and strongly implicate these circuits as an anatomical substrate of mood disorders. On the other hand, the gross anatomical and metabolic changes reported in neuroimaging studies are not very different between mood disorders and other brain disorders such as schizophrenia, Huntington disease or chronic alcohol dependence. For example, dorsolateral prefrontal glucose hypometabolism or volumetric reductions are reported in both schizophrenia and mood disorders (Baxter et al 1989). However, the metabolic changes are reversible with treatment in MDD but not in schizophrenia (reviewed in Mayberg 2000) and the magnitude of structural changes is more pronounced in schizophrenia and psychotic MDD than in nonpsychotic MDD (Jeste et al 1996; Shiraishi et al 1992). The above suggests that similarities and differences between mood disorders and schizophrenia may exist in the regional pattern of cell pathology underlying gross anatomical and metabolic changes. Below, the cell pathology of mood disorders is compared with that of other brain disorders based upon available stereological and morphometric studies.

Comparison Of Cell Pathology Between Mood Disorders And Other Brain Disorders

Reductions in cell density and number found in frontal limbic cortical regions are not restricted to mood disorders. Marked reductions in glial density and atrophic changes in neurons are also reported in some cortical regions of postmortem brain from schizophrenic patients (Cotter et al 2001; Rajkowska et al 1999b; Rajkowska et al 1999c) as well as chronic alcohol abusers (Korbo 1999); Miguel-Hidalgo and Rajkowska, unpublished observations). On the other hand, the cortex from mood disorders subjects exhibits a different pattern of cell pathology from that seen in classic neurodegenerative disorders since glial proliferation associated with neurodegeneration is not observed.

Comparison of Cell Pathology Between Mood Disorders and Schizophrenia. Reductions in glial cell density and atrophic neuronal changes are reported in some cortical regions of postmortem brain from schizophrenic patients. These reductions seem to be region-specific and are not detected by all studies (for further details see, Chapter by Cotter et al., in this book). For example, glial density is reduced in both anterior cingulate (BA 24, (Cotter et al 2001)) and orbitofrontal cortex (BA 47, (Rajkowska et al 1999b)) in postmortem brain of schizophrenics. Accompanying reductions in neuronal size (Cotter et al 2001) and density (Rajkowska et al 1999c) are also noted in these two regions.

In contrast, more dorsally located regions of the prefrontal cortex don't exhibit glial or neuronal reductions. In fact, the pattern of cell pathology in these regions is drastically different (Fig. 2). For example, significant increases in neuronal density, are found in Brodmann's areas 9 and 46 of the dorsolateral prefrontal cortex, while glial density is unchanged in these regions (Rajkowska et al 1998; Selemon et al 1995; Selemon et al 1998). Thus, the two contrasting patterns of cellular changes (decreases and increases) are found in different regions of the frontal cortex in schizophrenia whereas only cell reductions are reported in frontal cortex in mood disorders. Accordingly, cortical cell pathology looks similar between mood disorders and schizophrenia, for regions located ventrally within the frontal lobe (Brodmann's areas 47 and 24), but the pattern of cell pathology is quite different for the dorsolateral prefrontal regions (Brodmann's areas 9 and 46). The above suggests that these two psychiatric disorders have shared and unique features of their cell pathology which may underlie shared and unique clinical symptoms and treatments characterizing depression and schizophrenia. A provocative interpretation of this observation is that anatomic and functional changes in the dlPFC may be related to cognitive dysfunction, whereas anatomic and functional changes in the orbitofrontal cortex and ventral part of the anterior cingulate cortex may be related to depressive/emotional symptoms. Further cytoarchitectural, pharmacological and molecular studies are needed to elucidate the exact morpho-biochemical characteristics of vulnerable neurons and glial cells that may account for unique and overlapping symptoms of mood disorders and schizophrenia.

Cell Pathology in Mood Disorders is Not Classic Neurodegeneration - Comparison to Huntington Disease. A comparison of postmortem cell pathology between mood disorders and neurodegenerative disorders suggests that depression is not a classical neurodegenerative disease. Rather, the data are better associated with impairments of neuroplasticity and cellular resilience. For example, the pattern of neuronal and glial cell pathology found in the dorsolateral prefrontal cortex in MDD and BPD is different than that found in the same region in Huntington's disease, a classic neurodegenerative disorder (Fig. 2). In the dlPFC in Huntington's disease there is a marked loss, compared to normal, of large pyramidal neurons located predominantly in infragranular cortical layers V and VI and in a lesser degree in layer III (Hedreen et al 1991; Rajkowska et al 1998). These neuronal reductions in Huntington's disease are accompanied by prominent gliosis manifested by dramatic increases (50%) in glial cell densities (Selemon et al 1995) and enlargement in glial nuclear sizes (Rajkowska et al 1998). In contrast, in both MDD and BPD the prefrontal cortex is characterized by marked decreases in glial density, enlargement of glial sizes (most prominent in BPD) and reductions in neuronal density predominantly in layer III. Thus, cortex from mood disorders subjects does not exhibit the classic morphometric signature of gliosis associated with neurodegeneration, i.e., glial hypertrophy in conjunction with glial proliferation. Rather, fewer but larger glial cells are present in BPD and MDD, and it could be speculated that these cells can have more elaborate cytoplasmic processes.

The lack of glial proliferation in mood disorders suggests that the glial pathology in these disorders is not a response to ongoing neurodegenerative changes in the cortex. However, if fewer neurons are present in the cortex of depressed patients, the reduction may have occurred prior to the disease onset or may represent a more prolonged or moderate process of degeneration such that full-scale gliosis has not been triggered. It is also possible that some process of neurodegeneration in mood disorders happens later in life. Our recent examination of immunoreactive astroglia in MDD suggests age-related increase in astroglial (Miguel-Hidalgo et al 2000). Further studies will determine whether these increases are related to the progression of depression, the normal process of aging or a combination of both.

Neurotrophic and Neuroprotective Factors and Cell Pathology

Recent biochemical and molecular data suggest involvement of neuroprotective and neurotrophic factors in cell pathology underlying MDD and BPD (reviewed in Duman et al 2000; Manji et al 2000). Reduced neuronal and glial cell density and smaller sizes of neuronal cell bodies observed in postmortem studies in mood disorders may be related to alterations in levels of neuroprotective and/or neurotrophic factors. Cell loss or atrophy may also be related to developmental factors such as diminished amounts of neurotrophic factors, malfunctions in programmed cell death (apoptosis), or both.

Pre-clinical studies suggest that specific target-derived neurotrophic factors, such as brain-derived neurotrophic factor (BDNF), regulate the activity and function of cortical neurons (Rutherford et al 1998; Wang et al 1998). In the neocortex, BDNF is synthesized and secreted in an activity-dependent manner by pyramidal neurons that are innervated by GABA-containing interneurons (Marty et al 1997). BDNF also plays a role in neocortical synaptic plasticity mediated through regulation of excitatory synaptic strengths and the maintenance of the balance of cortical excitation and inhibition (Rutherford et al 1998). Genes for BDNF and its trkB receptors are upregulated in local interneurons adjacent to degenerating apoptotic pyramidal neurons (Wang et al 1998). Apparently, BDNF influences certain interactions between pyramidal and nonpyramidal neurons. Moreover, BDNF-like immunoreactivity is observed in somata and processes of discrete neuronal and glial populations of monkey and human neocortex (Kawamoto et al 1999; Murer et al 1999). Accordingly, it can be proposed that the decreases in neuronal and glial cell density detected in cortical regions of mood disorders subjects may be a response to diminished BDNF.

Experimental data with in situ hybridization histochemistry indicates that the development of cortical neuronal circuits may be related to the expression of BDNF (Huntley et al 1992). Expression of mRNA for BDNF increases during later stages of prefrontal cortical development and continues to increase into adulthood (Friedman et al 1991). Inasmuch as deprivation of neurotrophic factors results in neuron death, any reduction in neurotrophic factors could lead to some degree of neuron death.

Glial deficits found in both mood disorders may also be linked to alterations in levels of neurotrophic factors. The survival of appropriate populations of synaptically connected neurons and supporting glial cells depends on BDNF (Ghosh et al 1994; Ohgoh et al 1998) and other factors, such as glial derived neurotrophic factor (GDNF, Erickson et al 2001;

Ohgoh et al 1998), astroglial neuroprotective protein S-100-beta or fibroblast growth factor (FGF, Mazer et al 1997; Zimmer et al 1995).

In conclusion, it can be proposed that cell alterations found in mood disorders are related to changes in the levels of neurotrophic factors such as BDNF. Since BDNF and other neurotrophic factors are necessary for the survival and function of neurons, sustained reduction of these factors could markedly affect viability of neurons and related glial cells. Future immunohistochemical and molecular measures of BDNF in postmortem brain tissues from depressed and non-psychiatric control patients are needed to test this hypothesis.

Forging A Link Between Cellular Changes And The Action Of Therapeutic Drugs

The question of whether alterations in glial and neuronal number, density and size can be attributed to the effect of therapeutic medications is still open. There have been no systematic studies on the effect of antidepressants and mood stabilizers on cell number and morphology in human brain. On the other hand, molecular studies in human cell cultures and in vivo animal brain suggest that these medications have significant effects on the regulation of gene expression in the central nervous system.

Antidepressant and mood stabilizing medications alter the expression of various neurotrophins, receptors and enzymes involved in neurotransmitter biosynthesis (reviewed in Manji et al 2000). Interestingly, both the second messenger system (e.g., cAMP cascade) and neurotrophins (e.g., BDNF) are reported to be upregulated by antidepressants (Duman et al 2000). Moreover, genes for several endogenous proteins, including neurotrophin receptors are known to be regulated by the signal transduction pathways. For example, several studies have shown that lithium, the major mood stabilizer, modulates an activator protein-1 (AP-1) DNA binding activity. Interestingly, AP-1 DNA binding activity was markedly increased in the frontal cortex and hippocampus of rats treated chronically with mood stabilizers (Manji et al 2000). Neuroprotective effects involving survival promoting processes and proteins like glycogen synthase kinase (GSK)-3 (Agam and Levine, 1998) and Bcl2, have been postulated as a potential component of therapeutic mechanism of lithium in BPD (Bijur et al 2000; Chen et al 1999; Chen and Chuang 1999; Manji et al 2000; Moore et al 2000a). Lithium induces up-regulation of Bcl-2, one of the major neuroprotective proteins in rodent brain and in human neuroblastoma cell culture (Chen and Chuang 1999; Lu et al 1999). In addition, Manji and colleagues found in vivo robust increases in the level of Bcl2 in cortical layers II and III of rat prefrontal cortex after chronic lithium treatment (Chen et al 1999). These are the same cortical layers in which neuronal and glial changes have been found in postmortem BPD studies (Rajkowska et al 2001). The authors interpreted these increases to possibly indicate a neuroprotective effect of lithium. Recent in vivo magnetic resonance spectroscopic studies, conducted by the same group, demonstrated that lithium at therapeutic doses increases brain NAA levels in bipolar patients (Moore et al 2000a). These increases in NAA were found in a number of regions including frontal cortex, and were localized almost exclusively to gray matter. Since NAA is a putative marker of neuronal viability and function, a relative increase in this compound observed in vivo in BPD may reflect a neuroprotective effect of lithium in response to malfunctioning frontal neurons and may be related to increases in the width of specific cortical layer(s) observed postmortem in the dlPFC in BPD subjects (Rajkowska et al 2001). Most BPD subjects used for this cell counting study took lithium for months or years prior to their death. Of relevance to this hypothesis is the observation that the thickness of sub layer IIIc is greater, and, pyramidal cell density tends to be lower in subjects with a long exposure to lithium (Rajkowska et al 2001). Thus, a compensatory increase in dendritic (and/or glial) neuropil and consequent decrease in neuronal density may be a response to medication. Other therapeutic drugs may have similar neuroprotective effects on cortical cells. Treatment with deprenyl, an antidepressant and neuroprotectant drug, enhances performance in cognitive tasks and is linked to increased dendritic tree arborization in primate prefrontal cortex (Shankaranarayana Rao et al 1999). Structural MRI findings of increases in volume of basal ganglia and medial temporal cortex (Pearlson et al 1997) in patients with BPD provide intriguing possibility that these increases are related to the therapeutic effect of mood stabilizers.

A mounting body of data suggests that treatment with antidepressants or mood stabilizers not only regulates neuronal survival but also influences neurogenesis. Pharmacologically induced increases in neurogenesis in adult rodent brain have been recently reported by two independent studies (Chen et al 2000; Malberg et al 2000). Chronic treatment with lithium induces genesis of new neurons in the mouse hippocampus (Chen et al 2000). Similarly, chronic treatment with different classes of antidepressants increases proliferation of hippocampal neurons (Malberg et al 2000). Moreover, there is evidence that treatment with lithium induces increases in the astrocytic protein GFAP in rodents hippocampus (Rocha et al 1998; Rocha and Rodnight 1994) and the neural lobe of the pituitary (Levine et al 2000). Interestingly, chronic exposure to neuroleptics increases rather than decreases glial number and does not change neuronal density in primates prefrontal cortex (Selemon et al 1999).

In summary, it appears that psychiatric drugs have significant effects on the regulation of gene expression in the central nervous system. Increased production of new cells in the hippocampus appears to be a common action of different classes of antidepressants and mood stabilizers such as lithium. However, whether these increases represent a protective or compensatory effect of these medications and the mechanisms underlying the regulation of neurogenesis and glial proliferation has to be further investigated. Other brain regions displaying cell pathology in mood disorders need to be examined as well, given the recent report that neurogenesis also occurs in the adult prefrontal cortex (Gould et al 1999). A precise link between cell loss and atrophy, observed in postmortem human brain, and medication-induced production of new cells, observed in animal brain has yet to be established.

Linking Cellular Changes To Dysfunctional Monoaminergic And Glutamatergic Circuits

The neurotrophins and monoamine neurotransmitters appear to play related roles in stress, depression and therapies for treating depression. Exposure to stress has been associated with changes in monoamines in the prefrontal cortex and hippocampus (Stanford 1993; Thierry et al 1986). Moreover, deficiencies in the monoaminergic neurotransmitter systems have been hypothesized as a critical factors in the pathophysiology of depression (Prange 1964; Schildkraut 1965). In particular, the serotonin and norepinephrine systems would be the main contributors, since nearly all clinically effective antidepressant medications affect neurotransmission mediated by these monoamines (reviewed in Charney 1998). Other neurotransmitters, such as excitatory amino acids, are also implicated in the pathophysiology of depression (Auer et al 2000; Nowak et al 1995). Accordingly, some of the postmortem cellular changes detected in MDD and BPD brains could be related to dysfunctional monoaminergic and other neurotransmitter circuits. Laminar specificity of cellular alterations in cortical areas further suggests the involvement of specific neurotransmitter circuits in the morphopathology of mood disorders.

Monoaminergic Circuits. Results of recent neurochemical studies in depressed patients and suicide victims revealed changes in monoamine receptors, transporters and related second messenger systems in brainstem structures and in the prefrontal cortex (see, chapter by Stockmeier in this book). This implies that monoaminergic axons, receptors, and the cortical targets of these axons, form dysfunctional circuits in major depression.

Serotonin. Cortical cell pathology observed in MDD and BPD may be related to the biochemical and structural changes detected in studies of the serotonergic and noradrenergic neurons located in the brainstem. For example, alterations in the cell composition of upper cortical layers II/III reported in MDD in the rostral ORB cortex seem to be congruent with dysregulation of the serotonin system. Serotonergic axons were shown in monkey prefrontal cortex to synapse predominantly on interneurons of layers II-IV (Smiley and Goldman-Rakic 1996). Moreover, layer II in normal human prefrontal cortex can be distinguished from other cortical layers by a particularly dense lamina of serotonergic receptors (Arango et al 1995; Pazos et al 1987). Finally, the reports in suicide victims (presumably depressed) of an upregulation in serotonin-1A and serotonin-2A receptors and downregulation of serotonin transporters predominantly in layer II and in layers III-IV of the ventrolateral prefrontal cortex (Arango et al 1995) suggest a neuropharmacological link to altered neuronal sizes in upper cortical layers of this region. On the other hand, alterations in the pharmacology and /or morphometry of dorsal raphe neurons, which synthesize serotonin and are the origin of ascending serotonin projections to the cortex, might also be linked to prefrontal cell pathology. Postmortem receptor studies of the dorsal raphe revealed elevated number of serotonin 1A autoreceptors in the dorsal and ventrolateral raphe subnuclei from suicide victims diagnosed with major depression as compared to matched healthy controls (Stockmeier et al 1998). Recent observations from the same laboratory found an increase in raphe volume and changes in the shape of cell bodies of raphe neurons in depression (Stockmeier et al 1999). Volume enlargement of raphe nuclei could be interpreted as a consequence of either increased number of serotonin raphe neurons or enlargement or reshaping of their dendritic trees. Interestingly, increases in the density and number of tryptophan hydroxylase (TrpOH)-immunoreactive neurons (Underwood et al 1999) and neuronal processes (Kasir et al 1998) were reported recently in the dorsal raphe of suicide victims with major depression.

These morphometric and pharmacological observations imply that serotonergic axons, along with the receptors and cortical neurons that are recipients of these axons, may be involved in the pathophysiology of serotonergic neurotransmission reported in depression, and consequently, may be a target of antidepressant medications. As originally described (Blier and de Montigny 1994), chronic antidepressant treatments are thought to enhance serotonergic neurotransmission in one or more sites along serotonin neurons.

Noradrenaline. Potential disruption of the neurotransmission of noradrenergic (NE) neurons has also been implicated in the pathophysiology of depression by a number of neurochemical and pharmacological studies (Anand and Charney 2000). Norepinephrine, synthesized in neurons of the pontine locus coeruleus, is released in the neocortical areas and modulates the activity of prefrontal and cingulate neurons. Animal track-tracing studies confirm that these neocortical neurons are innervated by abundant NE afferents coming from locus coeruleus (Lewis 1992). In primate brain these fibers are particularly dense in lower cortical layers V and VI (Lewis 1992).

In postmortem studies reviewed here, neurons and glia of the same layers in dorsolateral prefrontal, orbitofrontal and anterior cingulate regions had diminished packing densities and smaller sizes suggesting altered functioning of the NE system.

Dopamine. Cellular pathology of upper cortical layers II and III described in postmortem brains from MDD and BPD subjects might also be related to the dopamine dysfunction reported in mood disorders. Animal studies reveal that primate's frontal cortex is robustly innervated by dopamine afferents (Lewis et al 1992; Williams and Goldman-Rakic 1993). These afferents preferentially target spines and distal dendrites of pyramidal neurons in upper cortical layers, layer II and the upper part of layer III (Krimer et al 1997; Smiley and Goldman-Rakic 1993), however, certain subsets of nonpyramidal interneurons also get some dopaminergic input (Sesack et al 1998). Thus, reductions in the density and size of frontal neurons found in mood disorders may be related to the diminished mesolimbic dopaminergic input to the cerebral cortex. Evidence from clinical and animal pharmacological studies suggests that decreased dopamine neurotransmission plays a role in the brain pathology of depression (reviewed in Willner 1995).

Glutamatergic Circuits. In postmortem studies reviewed here, significant reductions in the density and size of large (presumably pyramidal) neurons are reported in layers III, V and VI of the prefrontal cortex (Rajkowska et al 2001; Rajkowska et al 1999a) as well as in layer VI of the anterior cingulate cortex (Cotter et al 2001). From animal studies it is known that these neuronal populations give rise to glutamatergic excitatory projections to associational cortical areas and related neostriatal projections (Selemon and Goldman-Rakic 1985). Hence glutamate, the presumed neurotransmitter of these deeper cortical neurons, also may be involved in the neuropathology of depression. Of relevance are studies of cortical metabolic activity in patients with BPD and MDD indicating reductions not only in the frontal cortex but also the neostriatum (Buchsbaum et al 1986). Moreover, depression has been associated with abnormal activation of a network of areas that receive projections from the dorsolateral prefrontal cortex, including the amygdala, anterior cingulate, and temporal cortices (George et al 1993). It is noteworthy that all these structures are reported as sites of cell pathology in mood disorders. Finally, reduced glutamate transmission has recently been reported in the anterior cingulate cortex of depressed patients (Auer et al 2000) and alterations in the glutamatergic recognition site and its coupling to the NMDA receptor complex are reported in the prefrontal cortex from depressed suicide victims (Nowak et al 1995).

Glia Pathology and Neurotransmitter Circuits. Mounting evidence suggests that glial pathology in addition to changes in neurons is relevant to dysfunctional neurotransmitter circuits in mood disorders.

Reductions in glial number and density, in addition to changes in their size and shape, might be related to the dysfunction of monoaminergic and gluatamatergic systems reported extensively in depression. For example, astrocytes express virtually all of the receptor systems and ion channels found in neurons including transporter systems which regulate both the concentration of neurotransmitters at synapses, and the availability of neurotransmitters for release (reviewed by Cotter et al in this volume). Glia may also play a role in serotonin, norepinephrine or dopamine neurotransmission via postsynaptic receptors distributed on glial cell bodies and processes (Griffith and Sutin 1996; Khan et al 2001; Shimizu et al 1996; Whitaker-Azmitia et al 1993). Astroglia are also important in regulating NMDA receptor activity (Wolosker et al 1999). Finally, astroglia are the primary sites of glutamate uptake by glial transporters (Tanaka et al 1997). Thus, astroglia regulate the levels of extracellular glutamate and have been shown in vitro to protect neurons from cell death (Rothstein et al 1996).

In summary, atrophy or loss of vulnerable neurons and related glial cells may be caused by dysfunctional monoaminergic input-circuits, which may also give rise to dysfunctional glutamatergic circuits. Recognition and precise anatomical localization of dysfunctional glial types and their receptors may offer a new cortical model to guide the development of antidepressant and mood stabilizing medications.

Neuron-Glia Interactions: A Proposed Model Of The Sequence Of Cellular Changes In Mood Disorders

The cellular changes described above indicate that both types of brain cells, neurons and glia, are abnormal in mood disorders. Alterations in one type of cell could be a consequence of changes in the other. For example, lack of sufficient glial support (e.g. less structural support due to a reduced number of glial cells or a shortage in glucose, the energy substrate provided to neurons by glia), may initiate neuronal pathology. Reduced glial cell numbers and density could be responsible for reduced neuronal size, reduced levels of synaptic proteins, and abnormalities of cortical neurotransmission. Alternatively, depression-induced reductions in activity of neurons with altered morphology may require less glial support, which may be reflected in a reduced number or density of glial cells. The question also remains whether depressed patients are genetically predisposed to the cellular changes detected postmortem and had smaller neurons and less glia, or both, from birth, or, whether the cellular changes are a consequence of the progression of the disease. Alternatively, those individuals genetically predisposed to the greatest histopathological alterations may exhibit a greater vulnerability to depression. Cell loss or atrophy may also be related to developmental factors such as diminished amounts of neurotrophic factors, malfunctions in programmed cell death (apoptosis), or both.

The sequence of cellular changes in mood disorders may be better understood if the brains from adolescents, young adults and aging adults are studied. It is unknown at this stage whether cellular changes in the brains of depressed adults can also be found in depressed adolescents. There are no reports on morphometry in postmortem tissues from adolescent patients. On the other hand, preliminary MRI data in 8-16 year-olds with mania, as compared to young normal controls, suggest that structural alterations already exist in young manic patients (Botteron et al 1995). Another recent neuroimaging study reveals that depression in adolescents is associated with alterations in orbitofrontal cytosolic choline metabolism (Steingard et al 2000). Reductions in the density and size of neurons and glia found in adults could be a contributing factor to the pathophysiology of depression or could worsen as the illness progresses. It is interesting to speculate that the prefrontal cortex or hippocampus from adolescents with depression might exhibit primarily glial pathology due to genetic factors interacting with early adverse environmental events, which over time (and with recurrence of depressive episodes) leads to neuronal pathology (Fig. 3). Pathology of both neurons and glia may worsen with increasing duration of illness and recurrent depressive episodes and that pathology may eventually lead to cell loss.

The hypothesis that glial changes precede neuronal changes is supported by our recent observations on the distribution of glial fibrillary acid protein (GFAP), an immunoreactive marker for astroglial cells, in the dlPFC in MDD (Miguel-Hidalgo et al 2000)., The significant positive correlation between age and GFAP immunoreactivity (i.e., the GFAP area fraction and packing density of GFAP-immunoreactive glia) in MDD suggests that glial pathology in younger subjects might be different from that in older subjects.

In younger adults (23-45 years) with MDD, the GFAP area fraction is smaller than the smallest value of the controls while there is a tendency for a larger GFAP area fraction in older (46-86 years) MDD subjects as compared to age matched controls. Thus, GFAP-immunoreactive astroglia maybe differentially involved in the pathology of MDD at the early and late stages of this disorder.

In summary, future studies will be necessary to establish whether cellular changes found in postmortem brains in mood disorders worsen with repetitive episodes of depression and whether these changes are reversible. It also remains to be demonstrated whether the use of therapeutic medications can lead to glial proliferation in cases when neuronal changes cannot be reversed.

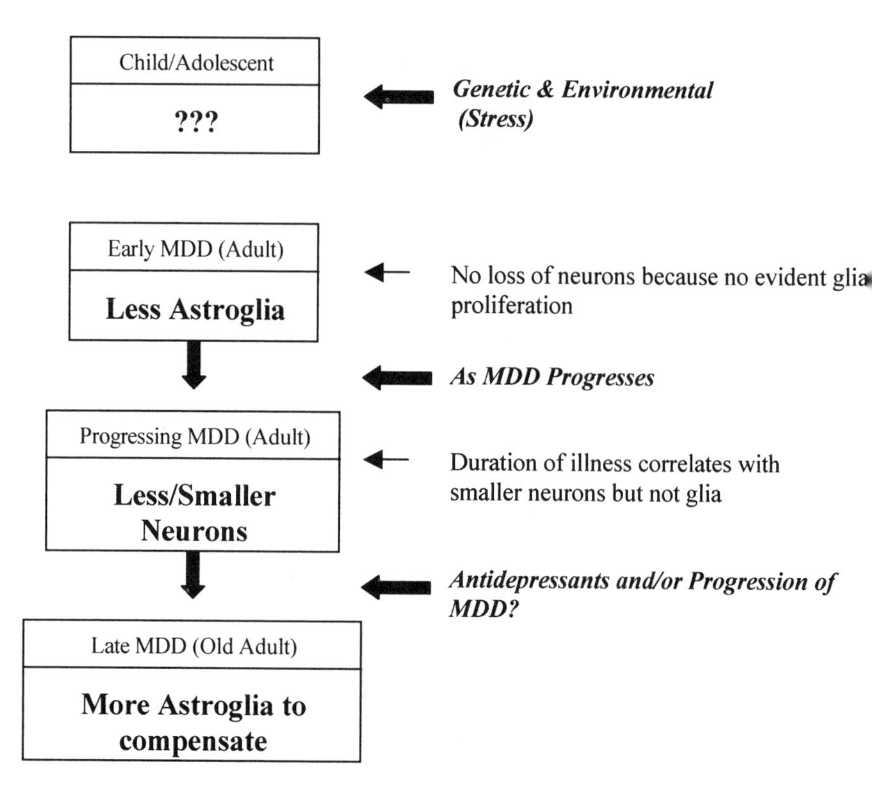

Figure 3. Hypothetical sequence of cellular changes in mood disorders. We propose that the prefrontal cortex and hippocampus from young adults with depression will exhibit primarily glial pathology (reductions in the number or density of cells) due to genetic and/or early adverse environmental events such as stress. Decreased levels of GFAP-immunoreactive astroglial cells are reported in young adults with major depression (Miguel-Hidalgo et al 2000). Over time, a lack of sufficient glial support may lead to recurrence of depressive episodes and neuronal pathology. Reduced densities of neurons and smaller sizes of neuronal cell bodies are correlated with the duration of depression. Progressing neuronal pathology (in combination with medication effect) may in turn stimulate glial activity and eventually lead to increased level of immunoreactive astroglial cells. Increased levels of GFAP-immunoreactive astroglia are observed in older individuals with major depression (Miguel-Hidalgo et al 2000).

(Reprinted by permission of Elsevier Science from "Postmortem Studies in Mood Ddisorders Indicate Altered Numbers of Neurons and Glial Cells" by G. Rajkowska, Biol Psychiatry 2000;48;766-777.)

Conclusions

Findings reviewed here incontrovertibly support the usefulness of postmortem studies in elucidating the microanatomical substrate of depression. Unbiased, stereological counts and measurements of neurons and glial cells in postmortem brains from mood disorder patients established that major depressive disorder and bipolar illness are brain diseases with unique etiological features of brain cytoarchitecture. Co-localization of cellular changes detected in postmortem tissues with in vivo neuroimaging findings is another confirmation of the usefulness of postmortem studies for establishing the anatomical substrate of mood disorders. Both, postmortem and neuroimaging studies revealed changes in the same frontal limbic brain regions.

Moreover, the precise region-specific and layer-specific pattern of changes in cortical neurons and glial architecture observed in mood disorders well coincides with the hypothesis of dysfunction of specific monoaminergic and glutamatergic neurotransmitter systems in these disorders. Disruption of specific serotoninergic, noradrenergic, dopaminergic and glutamatergic pathways may be related to changes in the density, number and size of neurons and glial cells.

In conclusion, available stereological and morphometric postmortem studies suggest that while MDD and BPD are clearly not neurodegenerative disorders, they are associated with impairments of cellular neuroplasticity and resilience. It remains to be fully elucidated to what extent these findings represent neurodevelopmental abnormalities, disease progression, the sequellae of the biochemical changes (for example, in glucocorticoid levels) accompanying repeated affective episodes, per se, or treatment with therapeutic medications. Furthermore, it remains to be determined whether the cellular changes observed postmortem in mood disorders can be reversed by antidepressant and mood stabilizing medications, which alter genomic levels of neurotrophins, receptors and enzymes involved in neurotransmitter biosynthesis.

Acknowledgements

The work reviewed here was supported by grants from a National Alliance for Research on Schizophrenia and Depression (NARSAD) and National Institute of Mental Health (MH55872, MH45488) and the American Foundation for Suicide Prevention.

References

Agam G, Levine J. Glycogen synthase kinase-3 - a new target for lithium's effects in bipolar patients? (Editorial). Human Psychopharmacol. Clin Exp 1998;13: 463-465

Altshuler LL, Conrad A, Hauser P, et al. Reduction of temporal lobe volume in bipolar disorder: a preliminary report of magnetic resonance imaging [letter]. Arch Gen Psychiatry 1991;48: 482-3.

Anand A, Charney DS. Norepinephrine dysfunction in depression. J Clin Psychiatry 2000;61: 16-24.

Arango V, Underwoood MD, Gubbi AV, Mann JJ. Localized alterations in pre- and postsynaptic serotonin binding sites in the ventrolateral prefrontal cortex of suicide victims. Brain Res 1995;688: 121-133.

Auer DP, Putz B, Kraft E, Lipinski B, Schill J, Holsboer F. Reduced glutamate in the anterior cingulate cortex in depression: an in vivo proton magnetic resonance spectroscopy study. Biol Psychiatry 2000;47: 305-13.

Baumann B, Danos P, Krell D, et al. Unipolar-bipolar dichotomy of mood disorders is supported by noradrenergic brainstem system morphology. J Affect Disord 1999;54: 217-24.

Baxter LR, Schwartz JM, Phelps ME, et al. Reduction of prefrontal cortex glucose metabolism common to three types of depression. Arch Gen Psychiatry 1989;46: 243-250.

Bench CJ, Friston KJ, Brown RG, Frackowiak RSJ, Dolan RJ. Regional cerebral blood flow in depression measured by positron emission tomography: the relationship with clinical dimensions. Psychol Med 1993;23: 579-590.

Benes FM, Kwok EW, Vincent SL, Todtenkopf MS. A reduction of nonpyramidal cells in sector CA2 of schizophrenics and manic depressives. Biol Psychiatry 1998;44: 88-97.

Benes FM, Vincent SL, Todtenkopf MS. The Density of Pyramidal and Nonpyramidal Neurons in Anterior Cingulate Cortex of Schizophrenic and Bipolar Subjects. Biol Psychiatry 2001;in press.

Bertolino A. Dysregulation of dopamine and pathology of prefrontal neurons: neuroimaging studies in schizophrenia and related animal models. Epidemiol Psichiatr Soc 1999;8: 248-54.

Bertolino A, Frye M, Callicott JH, et al. Neuronal pathology in the hippocampal area of patients with bipolar disorder. Biol Psychiatry 1999;45: 135S.

Bijur GN, De Sarno P, Jope RS. Glycogen synthase kinase-3beta facilitates staurosporine- and heat shock-induced apoptosis. Protection by lithium. J Biol Chem 2000;275: 7583-90.

Biver F, Goldman S, Delvenne V, et al. Frontal and parietal metabolic disturbances in unipolar depression. Biol Psychiatry 1994;36: 381-388.

Blier P, de Montigny C. Current advances and trends in the treatment of depression [see comments]. Trends Pharmacol Sci 1994;15: 220-6.

Blumberg HP, Stern E, Ricketts S, et al. Rostral and orbital prefrontal cortex dysfunction in the manic state of bipolar disorder. Am J Psychiatry 1999;156: 1986-8.

Botteron KN, Vannier MW, Geller B, Todd RD, Lee BC. Preliminary study of magnetic resonance imaging characteristics in 8- to 16-year-olds with mania. J Am Acad Child Adolesc Psychiatry 1995;34: 742-9.

Bowley MP, Drevets WC, Ongur D, Price JL. Glial chnages in the amygdala and entorhinal cortex in mood disorders. Soc Neurosci Abstr 2000;26: 2313.

Bremner JD, Narayan M, Anderson ER, Staib LH, Miller HL, Charney DS. Hippocampal volume reduction in major depression. Am J Psychiatry 2000;157: 115-8.

Buchsbaum MS, Wu J, DeLisi LE, et al. Frontal cortex and basal ganglia metabolic rates assessed by positron emission tomography with [18F]2-deoxyglucose in affective illness. J Affect Disord 1986;10: 137-52.

Chen G, Rajkowska G, Du F, Seraji-Bozorgzad N, Manji HK. Enhancement of hippocampal neurogenesis by lithium. J Neurochem 2000;75: 1729-34.

Chen G, Zeng WZ, Yuan PX, et al. The mood-stabilizing agents lithium and valproate robustly increase the levels of the neuroprotective protein bcl-2 in the CNS. J Neurochem 1999;72: 879-82.

Chen RW, Chuang DM. Long term lithium treatment suppresses p53 and Bax expression but increases Bcl-2 expression. A prominent role in neuroprotection against excitotoxicity. J Biol Chem 1999;274: 6039-42.

Coffey CE, Wilkinson WE, Weiner RD, et al. Quantitative cerebral anatomy in depression. A controlled magnetic resonance imaging study. Arch Gen Psychiatry 1993;50: 7-16.

Coffman JA, Bornstein RA, Olson SC, Schwarzkopf SB, Nasrallah HA. Cognitive impairment and cerebral structure by MRI in bipolar disorder. Biol Psychiatry 1990;27: 1188-96.

Cotter D, Mackay D, Landau S, Kerwin R, Everall I. Reduced glial cell density and neuronal size in the anterior cingulate cortex in major depressive disorder. Arch Gen Psychiatry 2001;58: 545-553 .

Coyle JT, Schwarcz R. Mind glue: implications of glial cell biology for psychiatry. Arch Gen Psychiatry 2000;57: 90-3.

Deicken RF, Fein G, Weiner MW. Abnormal frontal lobe phosphorous metabolism in bipolar disorder. Am J Psychiatry 1995;152: 915-8.

Diekmann S, Baumann B, Schmidt U, Bogerts B. Significant reduction of calretinin-IR neurons in layer II in the anterior cingulate cortex in subjects with affective disorders. Soc Neurosci Abstr 1998;24: 386.5.

Drevets W, Price J, Simpson JR J, et al. Subgenual prefrontal cortex abnormalities in mood disorders. Nature 1997;386: 824-7.

Drevets WC. Prefrontal cortical-amygdalar metabolism in major depression. Ann N Y Acad Sci 1999;877: 614-37.

Duman RS, Malberg J, Nakagawa S, D'Sa C. Neuronal plasticity and survival in mood disorders. Biol Psychiatry 2000;48: 732-9.

Eastwood SL, Harrison PJ. Hippocampal synaptic pathology in schizophrenia, bipolar disorder and major depression: a study of complexin mRNAs. Mol Psychiatry 2000;5: 425-32.

Elkis H, Friedman L, Wise A, Meltzer HY. Meta-analyses of studies of ventricular enlargement and cortical sulcal prominence in mood disorders. Arch Gen Psy 1995;52: 735-746.

Erickson JT, Brosenitsch TA, Katz DM. Brain-Derived Neurotrophic Factor and Glial Cell Line-Derived Neurotrophic Factor Are Required Simultaneously for Survival of Dopaminergic Primary Sensory Neurons In Vivo. J Neurosci 2001;21: 581-589.

Friedman WJ, Olson L, Persson H. Cells that express brain-derived neurotrophic factor mRNA in the developing postnatal brain. Eur J Neurosc 1991;3: 688-697.

George MS, Ketter TA, Post RM. SPECT and PET imaging in mood disorders. J Clin Psychiatry 1993;54 Suppl: 6-13.

Ghosh A, Carnahan J, Greenberg ME. Requirement for BDNF in activity-dependent survival of cortical neurons. Science 1994;263: 1618-23.

Gould E, Reeves AJ, Graziano MS, Gross CG. Neurogenesis in the neocortex of adult primates. Science 1999;286: 548-52.

Gould E, Tanapat P. Stress and hippocampal neurogenesis. Biol Psychiatry 1999;46: 1472-1479.

Griffith R, Sutin J. Reactive astrocyte formation in vivo is regulated by noradrenergic axons. J Comp Neurol 1996;371: 362-75.

Gundersen HJG, Bagger P, Bendtsen TF, et al. The new stereological tools: disector, fractionator, nucleator and point sampled intercepts and their use in pathological research and diagnosis. Acta Pathol Microbiol Immunol Scand 1988;96: 857-881.

Hauser P, Altshuler LL, Berrettini W, Dauphinais ID, Gelernter J, Post RM. Temporal lobe measurement in primary affective disorder by magnetic resonance imaging. J Neuropsychiatry Clin Neurosci 1989;1: 128-34.

Hedreen JC, Peyser CE, Folstein SE, Ross CA. Neuronal loss in layers V and VI of cerebral cortex in Huntington's disease. Neurosci Lett 1991;133: 257-261.

Holsboer F, Spengler D, Heuser I. The role of corticotropin-releasing hormone in the pathogenesis of Cushing's disease, anorexia nervosa, alcoholism, affective disorders and dementia. Prog Brain Res 1992;93: 385-417.

Huntley GW, Benson DL, Jones EG, Isackson PJ. Developmental expression of brain derived neurotrophic factor mRNA by neurons of fetal and adult monkey prefrontal cortex. Brain Res Dev Brain Res 1992;70: 53-63.

Jeste DV, Heaton SC, Paulsen JS, Ercoli L, Harris J, Heaton RK. Clinical and neuropsychological comparison of psychotic depression with nonpsychotic depression and schizophrenia [see comments]. Am J Psychiatry 1996;153: 490-6.

Johnston-Wilson NL, Sims CD, Hofmann JP, et al. Disease-specific alterations in frontal cortex brain proteins in schizophrenia, bipolar disorder, and major depressive disorder. The Stanley Neuropathology Consortium. Mol Psychiatry 2000;5: 142-9.

Jope RS. Anti-bipolar therapy: mechanism of action of lithium. Mol Psychiatry 1999;4: 117-28.

Kasir SA, Underwood MD, Bakalian MJ, Mann JJ, Arango V. 5-HT1A binding in dorsal and median raphe nuclei of suicide victims. Soc Neurosci Abstr 1998;24: 505.3.

Kato T, Shioiri T, Murashita J, et al. Lateralized abnormality of high energy phosphate metabolism in the frontal lobes of patients with bipolar disorder detected by phase-encoded 31P-MRS. Psychol Med 1995;25: 557-66.

Kawamoto Y, Nakamura S, Kawamata T, Akiguchi I, Kimura J. Cellular localization of brain-derived neurotrophic factor-like immunoreactivity in adult monkey brain. Brain Res 1999;821: 341-349.

Khan ZU, Koulen P, Rubinstein M, Grandy DK, Goldman-Rakic PS. An astroglia-linked dopamine D2-receptor action in prefrontal cortex. Proc Natl Acad Sci USA 2001;98: 1964-1969.

Klimek V, Stockmeier C, Overholser J, et al. Reduced levels of norepinephrine transporters in the locus coeruleus in major depression. J Neurosci 1997;17: 8451-8.

Korbo L. Glial cell loss in the hippocampus of alcoholics. Alcohol Clin Exp Res 1999;23: 164-8.

Krimer LS, Jakab RL, Goldman-Rakic PS. Quantitative three-dimensional analysis of the catecholaminergic innervation of identified neurons in the macaque prefrontal cortex. J Neurosci 1997;17: 7450-61.

Levine S, Saltzman A, Klein AW. Proliferation of glial cells in vivo induced in the neural lobe of the rat pituitary by lithium. Cell Prolif 2000;33: 203-7.

Lewis DA. The catecholaminergic innervation of primate prefrontal cortex. J Neural Trans 1992;36: 179-200.

Lim KO, Rosenbloom MJ, Faustman WO, Sullivan EV, Pfefferbaum A. Cortical gray matter deficit in patients with bipolar disorder. Schizophr Res 1999;40: 219-27.

Lu R, Song L, Jope RS. Lithium attenuates p53 levels in human neuroblastoma SH-SY5Y cells. Neuroreport 1999;10: 1123-5.

Lucassen PJ, Salehi A, Pool CW, Gonatas NK, Swaab DF. Activation of vasopressin neurons in aging and Alzheimer's disease. J Neuroendocrinol 1994;6: 673-9.

Malberg JE, Eisch AJ, Nestler EJ, Duman RS. Chronic antidepressant treatment increases neurogenesis in adult rat hippocampus. J Neurosci 2000;20: 9104-10.

Manji HK, Moore GJ, Chen G. Lithium up-regulates the cytoprotective protein Bcl-2 in the CNS in vivo: a role for neurotrophic and neuroprotective effects in manic depressive illness. J Clin Psychiatry 2000a;61: 82-96.

Manji HK, Moore GJ, Rajkowska G, Chen G. Neuroplasticity and cellular resilience in mood disorders. Mol Psychiatry 2000b;5: 578-93

Marty S, Berzaghi Md, Berninger B. Neurotrophins and activity-dependent plasticity of cortical interneurons. Trends Neurosci 1997;20: 198-202.

Mayberg HS. Depression. In Mazziotta JC, Toga AW, Frackowiak RSJ (eds), Brain Mapping. The Disorders: Academic Press, 2000; pp 485-505.

Mazer C, Muneyyirci J, Taheny K, Raio N, Borella A, WhitakerAzmitia P. Serotonin depletion during synaptogenesis leads to decreased synaptic density and learning deficits in the adult rat: A possible model of neurodev elopmental disorders with cognitive deficits. Brain Res 1997;760X: 68X-73X.

Miguel-Hidalgo JJ, Baucom C, Dilley G, et al. Glial fibrillary acidic protein immunoreactivity in the prefrontal cortex distinguishes younger from older adults in major depressive disorder. Biol Psychiatry 2000;48: 861-73.

Moore GJ, Bebchuk JM, Hasanat K, et al. Lithium increases N-acetyl-aspartate in the human brain: in vivo evidence in support of bcl-2's neurotrophic effects? Biol Psychiatry 2000a;48: 1-8.

Moore GJ, Bebchuk JM, Wilds IB, Chen G, Menji HK. Lithium-induced increase in human brain grey matter. Lancet 2000b;356: 1241-2.

Murer MG, Boissiere F, Yan Q, et al. An immunohistochemical study of the distribution of brain-derived neurotrophic factor in the adult human brain, with particular reference to Alzheimer's disease. Neuroscience 1999;88: 1015-32.

Nowak G, Ordway GA, Paul IA. Alterations in the N-methyl-D-aspartate (NMDA) receptor complex in the frontal cortex of suicide victims. Brain Res 1995;675: 157-64.

Ohgoh M, Kimura M, Ogura H, Katayama K, Nishizawa Y. Apoptotic cell death of cultured cerebral cortical neurons induced by withdrawal of astroglial trophic support. Experimental Neurology 1998;149: 51-63.

Ongur D, Drevets WC, Price JL. Glial reduction in the subgenual prefrontal cortex in mood disorders. Proc Natl Acad Sci U S A 1998;95: 13290-5.

Orlovskaya DD, Vikhreva OV, Zimina IS, Denisov DV, Uranova NA. Ultrastructural Dystrophic chnages of oligodendroglial cells in autopsied prefrontal cortex and striatum in schizophrenia: a morphometric study. Schizophr Res. 1999;36: 82-83.

Orlovskaya DD, Vostrikov VM, Rachmanova NA, Uranova NA. Decreased numerical density of oligodendroglial cells in postmortem prefrontal cortex in schizophrenia, bipolar affective disorder and major depression. Schizophr Res. 2000;41: 105.

Pazos A, Probst A, Palacios J. Serotonin receptors in the human brain--III. Autoradiographic mapping of serotonin-1 receptors. Neuroscience 1987;21: 97-122.

Pearlson GD, Barta PE, Powers RE, et al. Ziskind-Somerfeld Research Award 1996. Medial and superior temporal gyral volumes and cerebral asymmetry in schizophrenia versus bipolar disorder. Biol Psychiatry 1997;41: 1-14.

Pfrieger FW, Barres BA. Synaptic efficacy enhanced by glial cells in vitro. Science 1997;277: 1684-7.

Prange AJ. The pharmacology and biochemistry of depression. Dis Nerv Syst 1964;25: 217-221.

Purba JS, Hoogendijk WJ, Hofman MA, Swaab DF. Increased number of vasopressin- and oxytocin-expressing neurons in the paraventricular nucleus of the hypothalamus in depression. Arch Gen Psychiatry 1996;53: 137-43.

Raadsheer FC, Hoogendijk WJ, Stam FC, Tilders FJ, Swaab DF. Increased numbers of corticotropin-releasing hormone expressing neurons in the hypothalamic paraventricular nucleus of depressed patients. Neuroendocrinology 1994;60: 436-44.

Raadsheer FC, van Heerikhuize JJ, Lucassen PJ, Hoogendijk WJ, Tilders FJ, Swaab DF. Corticotropin-releasing hormone mRNA levels in the paraventricular nucleus of patients with Alzheimer's disease and depression. Am J Psychiatry 1995;152: 1372-1376.

Rajkowska G. Histopathology of the prefrontal cortex in major depression: what does it tell us about dysfunctional monoaminergic circuits? Prog Brain Res 2000a;126: 397-412.

Rajkowska G. Postmortem studies in mood disorders indicate altered numbers of neurons and glial cells. Biol Psychiatry 2000b;48: 766-77.

Rajkowska G, Halaris A, Selemon LD. Reductions in neuronal and glial density characterize the dorsolateral prefrontal cortex in bipolar disorder. Biol. Psychiatry 2001;49: 741-752.

Rajkowska G, Miguel-Hidalgo JJ, Wei J, et al. Morphometric evidence for neuronal and glial prefrontal cell pathology in major depression. Biol Psychiatry 1999a;45: 1085-98.

Rajkowska G, Miguel-Hidalgo JJ, Wei J, Stockmeier CA. Reductions in glia distinguish orbitofrontal region from dorsolateral prefrontal cortex in schizophrenia. Soc Neurosci Abstr 1999b;25: 818.

Rajkowska G, Selemon LD, Goldman-Rakic PS. Neuronal and glial somal size in the prefrontal cortex: a postmortem morphometric study of schizophrenia and Huntington disease. Arch Gen Psychiatry 1998;55: 215-24.

Rajkowska G, Wei J, Miguel-Hidalgo JJ, Stockmeier C. Glial and neuronal pathology in rostral orbitofrontal cortex in schizophrenic postmortem brain. Schizophr Res. 1999c;36: 84.

Rocha E, Achaval M, Santos P, Rodnight R. Lithium treatment causes gliosis and modifies the morphology of hippocampal astrocytes in rats. Neuroreport 1998;9: 3971-4.

Rocha E, Rodnight R. Chronic administration of lithium chloride increases immunodetectable glial fibrillary acidic protein in the rat hippocampus. J Neurochem 1994;63: 1582-4.

Rosoklija G, Toomayan G, Ellis SP, et al. Structural abnormalities of subicular dendrites in subjects with schizophrenia and mood disorders: preliminary findings. Arch Gen Psychiatry 2000;57: 349-56.

Rothstein JD, Dykes-Hoberg M, Pardo CA, et al. Knockout of glutamate transporters reveals a major role for astroglial transport in excitotoxicity and clearance of glutamate. Neuron 1996;16: 675-86.

Rutherford LC, Nelson SB, Turrigiano GG. BDNF has opposite effects on the quantal amplitude of pyramidal neuron and interneuron excitatory synapses. Neuron 1998;21: 521-30.

Sapolsky RM. The possibility of neurotoxicity in the hippocampus in major depression: a primer on neuron death. Biol Psychiatry 2000;48: 755-65.

Schildkraut JJ. The cathecolamine hypothesis of affective disorders: a review of suppoting evidence. Am J Psychiat 1965;122: 509-522.

Selemon LD, Goldman-Rakic PS. Longitudinal topography and interdigitation of corticostriatal projections in the rhesus monkey. J Neuroscience 1985;5: 776-794.

Selemon LD, Goldman-Rakic PS. The reduced neuropil hypothesis: a circuit based model of schizophrenia [In Process Citation]. Biol Psychiatry 1999;45: 17-25.

Selemon LD, Lidow MS, Goldman-Rakic PS. Increased volume and glial density in primate prefrontal cortex associated with chronic antipsychotic drug exposure. Biol Psychiatry 1999;46: 161-72.

Selemon LD, Rajkowska G, Goldman-Rakic PS. Abnormally high neuronal density in the schizophrenic cortex: a morphometric analysis of prefrontal area 9 and occipital area 17. Arch Gen Psychiatry 1995;52: 805-818.

Selemon LD, Rajkowska G, Goldman-Rakic PS. Elevated neuronal density in prefrontal area 46 in brains from schizophrenic patients: application of a three-dimensional, stereologic counting method. J Comp Neurol 1998;392: 402-12.

Sesack SR, Hawrylak VA, Melchitzky DS, Lewis DA. Dopamine innervation of a subclass of local circuit neurons in monkey prefrontal cortex: ultrastructural analysis of tyrosine hydroxylase and parvalbumin immunoreactive structures. Cereb Cortex 1998;8: 614-22.

Shankaranarayana Rao BS, Lakshmana MK, Meti BL, Raju TR. Chronic (-) deprenyl administration alters dendritic morphology of layer III pyramidal neurons in the prefrontal cortex of adult Bonnett monkeys. Brain Res 1999;821: 218-223.

Sheline YI, Gado MH, Price JL. Amygdala core nuclei volumes are decreased in recurrent major depression [published erratum appears in Neuroreport 1998 Jul 13;9(10):2436]. Neuroreport 1998;9: 2023-8.

Sheline YI, Sanghavi M, Mintun MA, Gado MH. Depression duration but not age predicts hippocampal volume loss in medically healthy women with recurrent major depression. J Neurosci 1999;19: 5034-43.

Shimizu M, Nishida A, Zensho H, Yamawaki S. Chronic antidepressant exposure enhances 5-hydroxytryptamine7 receptor- mediated cyclic adenosine monophosphate accumulation in rat frontocortical astrocytes. J Pharmacol Exp Ther 1996;279: 1551-8.

Shiraishi H, Koizumi J, Hori M, et al. A computerized tomographic study in patients with delusional and non- delusional depression. Jpn J Psychiat Neurol 1992;46: 99-105.

Smiley JF, Goldman-Rakic PS. Heterogenous targets of dopamine synapses in monkey prefrontal cortex demonstrated by serial section electron microscopy: a laminar analysis using the silver-enhanced diaminobenzidine sulfide (SEDS) immunolabeling technique. Cereb Cortex 1993;3: 223-238.

Smiley JF, Goldman-Rakic PS. Serotonergic axons in monkey prefrontal cerebral cortex synapse predominantly on interneurons as demonstrated by serial section electron microscopy. J Comp Neurol 1996;367: 431-43.

Soares J, Mann J. The anatomy of mood disorders--review of structural neuroimaging studies. Biol Psychiatry 1997;41: 86-106.

Stanford SC. Monoamines in response and adaptation to stress. In Stanford SC, Salmon P (eds), Stress: from synapse to syndrome. London: Academic Press, 1993; pp 282-332.

Steingard RJ, Yurgelun-Todd DA, Hennen J, et al. Increased orbitofrontal cortex levels of choline in depressed adolescents as detected by in vivo proton magnetic resonance spectroscopy. Biol Psychiatry 2000;48: 1053-61.

Stockmeier CA, Dilley GE, Kulnane LS, Miguel-Hidalgo JJ, Rajkowska-Markow G. Morphometric evaluation of the midbrain dorsal raphe nucleus (DR) in suicide victims with major depression (MD). Soc Neurosc Abstr 1999;25: 2098.

Stockmeier CA, Shapiro LA, Dilley GE, Kolli TN, Friedman L, Rajkowska G. Increase in serotonin-1A autoreceptors in the midbrain of suicide victims with major depression-postmortem evidence for decreased serotonin activity. J Neurosci 1998;18: 7394-401.

Swaab DF, Hofman MA, Lucassen PJ, Purba JS, Raadsheer FC, Van de Nes JA. Functional neuroanatomy and neuropathology of the human hypothalamus. Anat Embryol (Berl) 1993;187: 317-30.

Tanaka K, Watase K, Manabe T, et al. Epilepsy and exacerbation of brain injury in mice lacking the glutamate transporter GLT-1. Science 1997;276: 1699-702.

Thierry AM, Le Douarin C, Penit J, Ferron A, Glowinski J. Variation in the ability of neuroleptics to block the inhibitory influence of dopaminergic neurons on the activity of cells in the rat prefrontal cortex. Brain Res Bull 1986;16: 155-60.

Tsacopoulos M, Magistretti PJ. Metabolic coupling between glia and neurons. J Neurosci 1996;16: 877-85.

Tsai G, Coyle JT. N-acetylaspartate in neuropsychiatric disorders. Prog Neurobiol 1995;46: 531-40.

Ullian EM, Sapperstein SK, Christopherson KS, Barres BA. Control of Synapse Number by Glia. Science 2001;291: 657-661.

Underwood MD, Khaibulina AA, Ellis SP, et al. Morphometry of the dorsal raphe nucleus serotonergic neurons in suicide victims. Biol Psychiatry 1999;46: 473-83.

Wang Y, Sheen VL, Macklis JD. Cortical interneurons upregulate neurotrophins in vivo in response to targeted apoptotic degeneration of neighboring pyramidal neurons. Exp Neurol 1998;154: 389-402.

Watanabe Y, Gould E, McEwen BS. Stress induces atrophy of apical dendrites of hippocampal CA3 pyramidal neurons. Brain Res 1992;588: 341-5.

West MJ. New stereological methods for counting neurons. Neurobiol Aging 1993;14: 275-85.

Whitaker-Azmitia P, Clarke C, Azmitia E. Localization of 5-HT1A receptors to astroglial cells in adult rats: implications for neuronal-glial interactions and psychoactive drug mechanism of action. Synapse 1993;14: 201-5.

Williams SM, Goldman-Rakic PS. Characterization of the dopaminergic innervation of the primate frontal cortex using a dopamine-specific antibody. Cereb Cortex 1993;3: 199-222.

Willner P. Dopaminergic mechanisms in depression and mania. Psychopharmacology: The Fourth Generation of Progress (Bloom FE and Kupfer DJ eds), Raven Press, New York 1995; pp 921-931.

Winsberg ME, Sachs N, Tate DL, Adalsteinsson E, Spielman D, Ketter TA. Decreased dorsolateral prefrontal N-acetyl aspartate in bipolar disorder. Biol Psychiatry 2000;47: 475-81.

Wolosker H, Blackshaw S, Snyder SH. Serine racemase: a glial enzyme synthesizing D-serine to regulate glutamate-N-methyl-D-aspartate neurotransmission. Proc Natl Acad Sci USA 1999;96: 13409-14.

Zimmer DB, Cornwall EH, Landar A, Song W. The S100 protein family: history, function, and expression. Brain Res Bull 1995;37: 417-29.

17 PHOSPHOINOSITIDE SIGNAL TRANSDUCTION SYSTEM IN POSTMORTEM HUMAN BRAIN

Richard S. Jope

Abstract

The phosphoinositide signal transduction system is one of the major receptor-coupled second messenger-producing intracellular signaling systems in the brain. At its most basic level, this signaling system consists of receptors, heterotrimeric G-proteins, and phospholipase C, and it contributes a critical component to a cell's regulation of intracellular calcium levels and activities of protein kinases. Phosphoinositide signaling has been assessed in postmortem brain using membrane preparations incubated with selective stimulatory agents. Membranes are incubated with radiolabeled phosphoinositides and receptor agonists to measure receptor-coupled signaling, with activators of G-proteins to measure G-protein-coupled signaling, and with added calcium to directly stimulate phospholipase C activity. These measurements along with a variety of other techniques have proven useful for assessing the activities of the major protein components of the phosphoinositide signaling system in postmortem brain obtained from subjects with a variety of diseases.

Introduction

Effective activation and regulation of signal transduction systems that mediate interneuronal communication are necessary for maintenance of normal cognitive and emotive functions. One of the primary means used by cells to transduce extracellular signals to intracellular responses is via plasma membrane-spanning receptors coupled to intracellular heterotrimeric G-proteins (G-proteins comprised of α-, β-, and γ-subunits). These G-proteins serve to couple membrane receptors to intracellular effectors, which are often enzymes that catalyze the formation of intracellular messengers, commonly referred to as second messengers.

The phosphoinositide second messenger system is one of the major G-protein-linked signal transduction processes that is known to operate in the central nervous system (CNS). Because of its critical role in neuronal communication, there has been much interest in studying the regulation, activation, and function of this signaling system in human brain, and assessing potential changes in its activity that may be associated with a wide variety of diseases. This interest is especially focused on potential changes in the phosphoinositide signaling system that may be associated with affective disorders, based to a large degree on the finding that lithium, the primary therapeutic agent used to treat bipolar disorder, modulates the metabolism of inositol phosphates that are produced by phosphoinositide signaling (Berridge et al., 1982). This discovery was the catalyst for many basic studies of the regulation of phosphoinositide signaling, and also provided the basis for proposals that signaling systems may be dysregulated in affective disorders and other psychiatric diseases, such as major depression and schizophrenia. In addition to psychiatric disorders, much research attention has been focused on signaling activities in the neurodegenerative disorder Alzheimer's disease based on the severe loss of specific neurotransmitter systems, particularly the cholinergic system, and the use of therapeutic agents that rely on functional second messenger systems. Thus, investigations of the phosphoinositide signal transduction system using postmortem brain samples has begun to reveal information about the consequences of a wide variety of diseases of the CNS.

The Phosphoinositide Signal Transduction System

The classical phosphoinositide signal transduction system is comprised of three primary proteins, receptors, two subtypes of G-proteins, Gq and G11 (signified as Gq/11), and phospholipase Cβ. A simplified scheme of the phosphoinositide signal transduction system as it relates to measurements made in membranes prepared from postmortem human brain is shown in Figure 1. Signal transduction is initiated upon receptor activation by an appropriate agonist, such as carbachol which is often used to activate cholinergic receptors. The stimulated receptor then activates a G-protein that is coupled to it by enhancing the exchange of GDP previously bound to the inactive α-subunit of the G-protein for GTP, which activates the G-protein. The active, GTP-bound α-subunit of Gq/11 then activates phospholipase C through mechanisms not completely understood. Active phospholipase C cleaves phosphoinositides (inositol-containing phospholipids) in the plasma membrane to produce second messengers. For example, phospholipase C-mediated cleavage of phosphatidylinositol-4,5-bisphosphate (PIP2) produces two second messengers, diacylglycerol and inositol-1,4,5-trisphosphate, which activate protein kinase C and release intracellular sequestered calcium, respectively. The inositol-1,4,5-trisphosphate is rapidly sequentially dephosphorylated until free inositol is regenerated. The last step of dephosphorylation is mediated by inositol monophosphatase, an enzyme selectively inhibited by lithium (Hallcher and Sherman, 1980). This inhibition by lithium causes a block in inositol phosphate metabolism, resulting in accumulation of inositol monophosphate. This action of lithium has proven very valuable for studies of phosphoinositide signaling because it allows measurements of inositol monophosphate accumulation to be taken as an estimation of the magnitude of phosphoinositide hydrolysis (Berridge et al.,

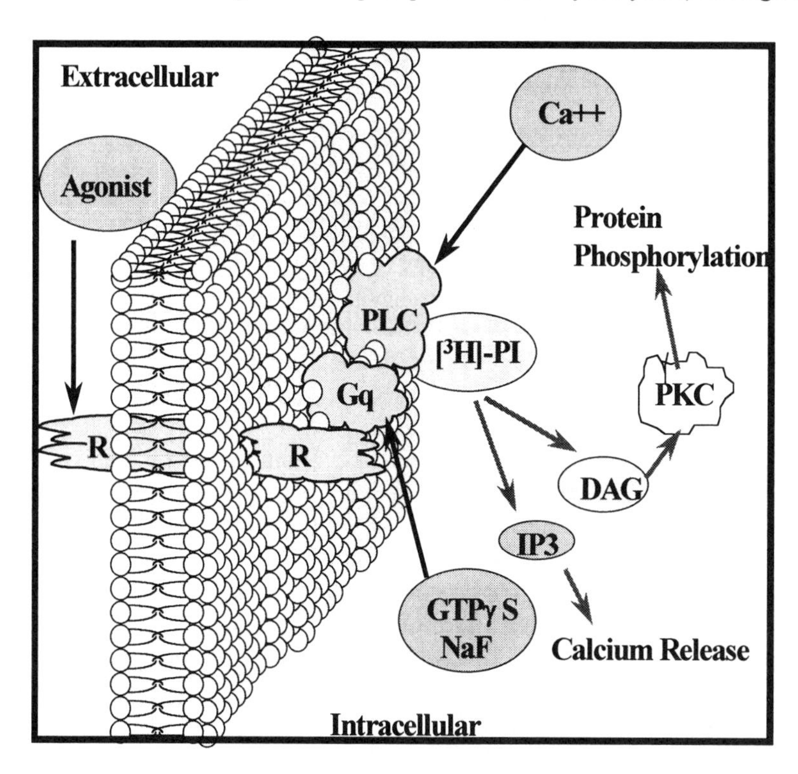

Figure 1. A schematic representation of the phosphoinositide signal transduction is presented arranged to show the components measured using membranes prepared from postmortem brain. [^{3}H]PI is shown as the exogenously added substrate. Addition of calcium directly activates phospholipase C (PLC) to catalyze hydrolysis of [^{3}H]PI. Addition of GTPγS or sodium fluoride directly activates G-proteins which subsequently stimulate phospholipase C activity. Addition of an agonist stimulates membrane-spanning receptors (R) coupled to G-proteins that activate phospholipase C. The two major second messengers produced from phosphoinositide hydrolysis in intact cells are shown: inositol-1,4,5-trisphosphate (IP3), which releases intracellular sequestered calcium, and diacylglycerol (DAG), which activates protein kinase C (PKC).

1982). This action of lithium also raised the intriguing idea that phosphoinositide signaling may be dysfunctional in some manner in disorders in which lithium is therapeutic, i.e., primarily bipolar affective disorder but also depression. In other words, since lithium affects phosphoinositide metabolism, it is logical to propose that the activity of the phosphoinositide signal transduction system may be abnormally regulated in some diseases in which lithium is therapeutic.

Measurements of Phosphoinositide Signaling in Postmortem Human Brain

The ability to measure phosphoinositide signaling in postmortem human brain is based on the experimental foundation established by Claro and colleagues (Claro et al., 1989a; 1989b; Wallace et al., 1991; Wallace and Claro, 1993) who established methods to measure signaling in brain membrane preparations. This was a critical development necessary for extending studies of phosphoinositide signaling to postmortem human brain because brain slices from such specimens are not viable, so membranes constitute the only feasible alternative to measure receptor-coupled second messenger production.

The method for measuring phosphoinositide signaling in specimens of postmortem human brain encompasses incubating preparations of brain membranes with labelled phosphoinositides (Figure 1) followed by measurements of labelled inositol phosphates that are produced as the result of phospholipase C activity. The critical step in the development of this assay was achieving conditions in which there is adequate association of the exogenously added labelled phosphoinositides with receptor-coupled activation of phospholipase C. Both [^{3}H]phosphatidylinositol (PI) and [^{3}H]PIP2 have been used successfully as the labelled substrate. A direct comparison of these two substrates in measurements of phosphoinositide signaling in human brain membranes demonstrated that [^{3}H]PI provided a greater signal-to-noise ratio in the assay than did [^{3}H]PIP2, although both were adequate (Pacheco and Jope, 1997).

Individual components of the phosphoinositide signaling system can be assessed separately in assays using membrane preparations by the use of selective stimulatory agents. The activity of phospholipase Cβ can be measured independently from other signaling components by directly activating the enzyme with calcium and measuring [^{3}H]PI hydrolysis (Jope et al., 1994). Furthermore, the contribution of different subtypes of phospholipase C can be identified and isolated by including in the assay reaction inhibitory antibodies for subtypes of phospholipase C (Pacheco and Jope, 1996). The activity of G-proteins that stimulate phospholipase C can be assessed by directly stimulating the G-proteins with either a stable analog of GTP, such as GTPγS, or with NaF, and measuring [^{3}H]PI hydrolysis (Jope et al., 1994). The specific G-protein subtypes that mediate [^{3}H]PI hydrolysis can be identified by including inhibitory antibodies for specific G-protein-subtypes in the assay. This method was used to demonstrate that nearly 100% of the G-protein-mediated [^{3}H]PI hydrolysis in postmortem human brain membranes is mediated by Gq/11 (Pacheco and Jope, 1996). Finally, receptor-mediated [^{3}H]PI hydrolysis can be assessed by using selective receptor agonists, with parallel confirmatory measurements using selective receptor antagonists. When using agonists, the inclusion of GTPγS in the reaction is necessary to support activation of G-proteins coupling receptors to phospholipase C since endogenous GTP is lost when membranes are prepared. Agonists that have been successfully employed to measure [^{3}H]PI hydrolysis in human brain membranes include carbachol (muscarinic receptors), trans-1-aminocyclopentyl-1,3-dicarboxylic acid (ACPD: glutamatergic metabotropic receptors), SKF 38393 (dopaminergic receptors), serotonin, histamine, and ATP or analogs of ATP (purinergic receptors; Pacheco and Jope, 1996). Notably, norepinephrine has not been used successfully in assays of [^{3}H]PI hydrolysis in human brain membranes for undetermined reasons.

There is a multitude of interindividual differences that could potentially influence comparisons of the phosphoinositide signaling system between groups of individuals. Many of these have been considered previously (Pacheco and Jope, 1996), but few characteristics have been examined experimentally. The responses to most of the agonists listed above have been shown to be stable in membranes prepared from human brain obtained within the postmortem interval range of 5 to 21 hours (Greenwood et al., 1995). However, it is unknown if rapid changes occur during the initial postmortem hours, and longer postmortem intervals lead to deficits in signaling. Activation of [^{3}H]PI hydrolysis by stimulation of phospholipase C or by stimulation of G-proteins also is stable within the postmortem interval range of 5 to 21 hours (Greenwood et al., 1995), and the protein levels of phospholipase C and several G-protein-subtypes are also stable during this postmortem interval (Li et al., 1996). The effects of aging on phosphoinositide signaling and proteins associated with this signaling system also have been reported (Greenwood et al., 1995; Li et al., 1996). Due to changes in phosphoinositide signaling associated with postmortem interval and aging, it is important that these variables be controlled by matching subject characteristics in any studies of this signaling system. Another condition likely to influence these measurements is previous drug history, with prior use of lithium being of particular concern since it is known to affect phosphoinositide signaling in vivo and there is some preliminary evidence that lithium treatment can influence phosphoinositide signaling activities measured in postmortem human brain (Jope et al., 1996). Many other potential influences, such as disease history, characteristics, and subtypes, treatment history and current medications, including the presence of drugs of abuse, agonal state, and many others make studies of any biochemical parameter a daunting task and make almost inevitable substantial interindividual differences which affect statistical analysis. None the less, such studies are immensely important because they open a window into the functioning of the human brain and alterations associated with diseases.

Overall, it has been shown that membranes prepared from postmortem human brain retain the primary protein constituents of the phosphoinositide signal transduction system, and each constituent is activated by appropriate agents. The development of a viable assay of phosphoinositide signaling activity using membranes prepared from postmortem human brain provided the basis for examining this signaling system in specimens from subjects with a variety of disorders.

Phosphoinositide Signaling in Human Disorders of the CNS

The activity of the phosphoinositide signal transduction system has been measured in membrane preparations prepared from subjects with a number of diseases. Upon comparing results from subjects with specific diseases to those from matched control subjects, several significant differences in the activity of the phosphoinositide signal transduction system have been identified, and these are briefly reviewed in the following sections.

Phosphoinositide Signaling in Brain Membranes from Subjects with Bipolar Disorder

The function of the phosphoinositide second messenger system has been assessed in occipital, temporal, and frontal cortex obtained postmortem from ten subjects with bipolar affective disorder and ten matched control subjects by measuring the hydrolysis of [^{3}H]PI incubated with membrane preparations and several different stimulatory agents (Jope et al., 1996). Phospholipase C activity, measured in the presence of 0.1 mM calcium, was not different in bipolar and control specimens. G-protein-mediated [^{3}H]PI hydrolysis was concentration-dependently activated by GTPγS and by NaF. GTPγS-stimulated [^{3}H]PI hydrolysis was 50% lower at all concentrations of GTPγS in occipital cortical membranes prepared from bipolar subjects compared with control subjects, whereas responses to GTPγS in temporal and frontal cortical membranes were similar in bipolar and control specimens. Brain lithium concentrations correlated directly with GTPγS-stimulated [^{3}H]PI hydrolysis in bipolar occipital cortical membranes. Carbachol, histamine, ACPD, serotonin, and ATP in the presence of GTPγS each activated [^{3}H]PI hydrolysis above that obtained with GTPγS alone and these responses were similar in bipolar and control specimens except for the responses to carbachol and serotonin in bipolar occipital cortex which produced deficits equivalent to the deficit detected with GTPγS alone.

Thus, among the three cortical regions examined there was a selective impairment in G-protein-stimulated [^{3}H]PI hydrolysis in occipital cortical membranes from bipolar compared with control subjects, indicative of an impairment in G-protein function. The reason for the selective localization of the deficit is not known, but since only three brain regions were examined, it seems likely that G-protein-mediated phosphoinositide signaling may be impaired in additional brain regions of bipolar subjects which may be more directly associated with the clinical manifestations of the disorder. Studies in the future should measure phosphoinositide signaling in specimens from a larger number of subjects and in a greater variety of brain regions to more completely assess the functional impairment of phosphoinositide signaling associated with bipolar disorder.

Phosphoinositide Signaling in Brain Membranes from Subjects with Major Depression

Pacheco et al (1996) reported the results of a study of the function of the phosphoinositide signal transduction system in postmortem prefrontal cortex areas 8/9 and area 10 from suicide victims with major depression and matched control subjects without psychiatric illness. The hydrolysis of [^{3}H]PI stimulated by phospholipase C, by G-proteins stimulated with GTPγS or with NaF, and by receptor agonists was measured in membrane preparations from thirteen subjects in both groups. Phospholipase C activity was similar in depressed suicide and control subjects in both brain regions. In prefrontal cortex area 10, but not areas 8/9, the GTPγS concentration-dependent stimulation of G-proteins coupled to phosphoinositide hydrolysis was lower by 30% in the depressed suicide group compared to the control group. This decreased G-protein activity was not due to decreased levels of Gq/11 or phospholipase Cβ, which were not different between depressed and control subjects, indicating that differences in the function, not level, of Gq/11 accounted for the decreased signaling in subjects with major depression. Receptor-coupled, G-protein-mediated [^{3}H]PI hydrolysis induced with carbachol, histamine, ACPD, serotonin, or 2-methylthio-ATP stimulated equivalent responses in the two groups of subjects in each brain region.

Overall, these results demonstrated that in the prefrontal cortex of suicide victims with major depression compared with normal control subjects there is a region-selective alteration of G-protein-induced activation of the phosphoinositide signal transduction system. Further investigations with additional subjects and in additional brain regions would more fully clarify the abnormalities of phosphoinositide signaling associated with major depression. Additionally, clarification of the individual roles of suicide and of major depression would be informative.

Phosphoinositide Signaling in Brain Membranes from Subjects with Schizophrenia

Comparisons of the activity of the phosphoinositide signal transduction system were made in membranes prepared from two regions of frontal cortex, areas 8/9 and area 10, from eight schizophrenic and eight matched control subjects (Jope et al., 1998). Also included in this study were specimens from eight alcohol-dependent subjects. The major finding of this study was that G-protein-mediated phosphoinositide signaling, assessed by measuring GTPγS-stimulated [^{3}H]PI hydrolysis, was 50% greater in frontal cortex areas 8/9 from schizophrenic than control or alcohol-dependent subjects. This increased G-protein activity was not due to increased levels of Gq/11 or phospholipase Cβ, which were not different between schizophrenic and control subjects, indicating that differences in the function, not level, of Gq/11 accounted for the increased signaling in schizophrenia. There were no differences among these groups of subjects in the response to GTPγS in frontal cortex area 10. Agonists for dopaminergic, cholinergic, purinergic, serotonergic, histaminergic, and glutamatergic receptors coupled to the phosphoinositide signaling system were tested in this study. Responses to most receptor agonists were similar in all three subject groups in both cortical regions. However, some differences were noted, with the largest difference being a 40% greater response to dopaminergic receptor stimulation in frontal cortex areas 8/9 membranes from subjects with schizophrenia. The large increase in G-protein-activated phosphoinositide signaling in schizophrenia markedly distinguishes this disease from other psychiatric disorders (bipolar disorder and major depression) in which brain regionally selective decreases, rather than increases, in G-protein function linked to phosphoinositide signaling have been detected. Further investigations with a greater number of subjects and in additional brain regions would help to clarify the alterations in phosphoinositide signaling activity that are associated with schizophrenia.

Phosphoinositide Signaling in Brain Membranes from Subjects with Alzheimer's Disease

The most extensive studies of the phosphoinositide signal transduction system associated with any disease have examined changes associated with Alzheimer's disease. The major focus of these studies has been directed towards examining cholinergic muscarinic receptor-coupled phosphoinositide signaling activities. This is because there is a severe deficit of cholinergic activity in Alzheimer's disease, and the major treatment currently available consists of the administration of inhibitors of acetylcholinesterase to facilitate cholinergic neurotransmission. However, such a facilitation would require functional muscarinic receptors to transmit the signal generated by acetylcholinesterase inhibitor-induced increases in acetylcholine. Thus, the basis of this therapeutic intervention requires both the maintenance of muscarinic receptor levels and the maintenance of functional coupling of these receptors to the phosphoinositide signal transduction system. The former has generally been observed, as studies of muscarinic receptors from brain samples from subjects with Alzheimer's disease have provided evidence of the maintenance of near normal muscarinic receptor numbers (Flynn et al., 1991; Warpman et al., 1993). However, as discussed below, these receptors appear to be severely compromised in their ability to couple to G-proteins to generate signals through the phosphoinositide second messenger system.

Several studies investigated the functional coupling of muscarinic receptors in brain samples from subjects with Alzheimer's disease. Using the methods described earlier in this chapter to measure phosphoinositide signaling activity in membranes prepared from postmortem brain, severe deficits were found in muscarinic receptor-mediated phosphoinositide signaling activity in Alzheimer's disease, compared with matched control, brain samples (Jope et al., 1994; Greenwood et al., 1995; Jope et al., 1997). Further examinations of the point of impairment revealed that the activation of G-proteins was deficient in Alzheimer's disease brain samples (Jope et al., 1994; 1997). An example of this deficit is shown in Figure 2. These results show that the G-protein-dependent (i.e., stimulation concentration-dependently by GTPγS) activation of phosphoinositide hydrolysis induced by stimulation of muscarinic receptors with carbachol is much lower in prefrontal cortical membranes prepared from subjects with Alzheimer's disease compared with responses obtained in membranes prepared from matched control subjects. Using a similar experimental approach, Crews et al (1994) also found severely impaired phosphoinositide signaling in Alzheimer's disease brain samples.

The deficit in phosphoinositide signaling in Alzheimer's disease brain was further assessed by examining a variety of cholinergic agonists, by examining several brain regions, and by studying phosphoinositide signaling coupled to other receptors. With each of six different cholinergic receptor agonists, and each of three different acetylcholinesterase inhibitors, the deficit in phosphoinositide signaling in Alzheimer's disease brain was consistently evident (Jope et al., 1997). These results make it appear unlikely that the activity of this signaling system can be enhanced in Alzheimer's disease by varying the agonist or acetylcholinesterase inhibitor that is used. Additionally, the deficit in cholinergic receptor-linked phosphoinositide signaling in

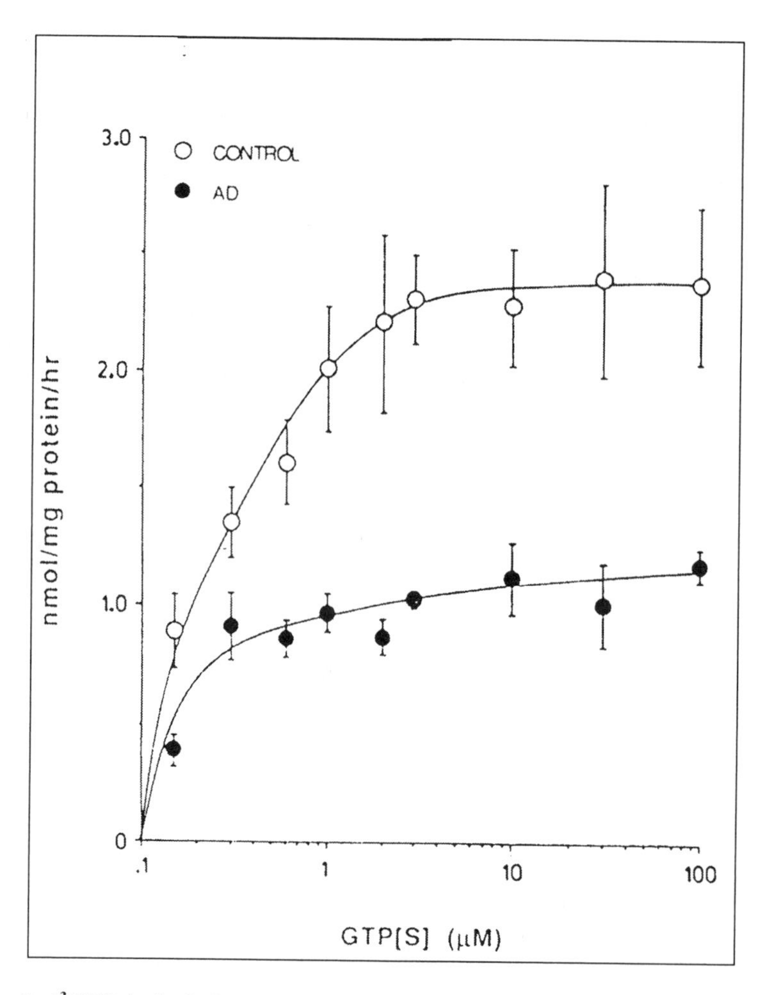

Figure 2. [^{3}H]PI hydrolysis was measured using membranes prepared from postmortem control and Alzheimer's disease (AD) prefrontal cortex by incubation with 1 mM carbachol and varying concentrations of GTPγS (GTP[S]). n = 6 per group

Alzheimer's disease was evident in several brain regions, indicating a widespread deficit in signaling activity (Jope et al., 1997). In contrast to the deficit in phosphoinositide signaling linked to cholinergic receptors, less of a deficit was detected in phosphoinositide signaling linked to glutamatergic metabotropic receptors in Alzheimer's disease brain (Jope et al., 1997), suggesting some selectivity in the impaired signaling. The deficient cholinergic receptor-coupled phosphoinositide signaling activity in Alzheimer's disease likely is a significant contributory factor to the very limited therapeutic success of acetylcholinesterase inhibitors in Alzheimer's disease, since although these drugs may increase acetylcholine levels in the synapse, its ability to stimulate signals through muscarinic receptors is severely impaired. Further studies of phosphoinositide signaling in postmortem human brain membranes may reveal methods to enhance signaling activities which could be applied in combination with acetylcholinesterase inhibitors to increase both acetylcholine and its signaling capability.

Other Approaches to Examine the Phosphoinositide Signaling System in Postmortem Human Brain

In addition to measuring phosphoinositide signaling activity in membranes prepared from postmortem human brain as described in the previous sections of this chapter, several other experimental approaches are also available to examine various aspects of this signal transduction system in postmortem human brain.

A useful method for measuring receptor-coupled activation of G-proteins in postmortem human brain involves measurements of the stimulated binding of labelled GTPγS to heterotrimeric G-proteins. Using this method it has been shown that incubation of membranes prepared from postmortem human brain with receptor agonists causes an increase in the binding of labelled GTPγS to individual G-proteins which are isolated by immunoprecipitation and electrophoresis (Friedman et al., 1993; Wang and Friedman, 1994; Friedman and Wang, 1996; Cowburn et al., 1996). Using this method, significant changes in receptor-mediated activation of Gq/11 associated with bipolar disorder have been reported (Friedman and Wang, 1996). Further applications of this novel method should provide additional valuable information about the function of the phosphoinositide signaling system in human brain.

Individual assessments of each of the three primary components of the phosphoinositide signal transduction system (receptors, Gq/11, phospholipase C) also can be obtained using a variety of other methods. The most common method has been to use classical receptor binding assays to measure the characteristics of receptors that are known to be coupled to this signaling system. Additionally, immunoblotting methods have been widely used to measure the levels of individual components of the phosphoinositide signal transduction system in postmortem human brain and changes associated with a variety of diseases. For example, Mathews et al (1998) reported increased levels of the G-proteins Gq/11 in the brain of subjects with bipolar disorder, and several investigations have examined the levels of these phospholipase C-linked G-proteins in other conditions (Kish et al., 1993; Ozawa et al., 1993; Yang et al., 1998). Investigators have also studied activation of the low Km GTPase activity associated with stimulation of receptors linked to the phosphoinositide signaling system in postmortem human brain samples (Cutler et al., 1994). Recently developed autoradiographic methods also hold great promise for providing information about the functional capacities of G-proteins in postmortem human brain (Rodriguez-Puertas et al., 2000). The third major component of the phosphoinositide signaling system, phospholipase C, also has been examined individually with both immunoblot techniques to estimate levels and by activity measurements (O'Neill et al., 1991). For example, Shimohama and colleagues have reported extensive studies of phospholipase C activity in investigations of the effects of Alzheimer's disease (Shimohama et al., 1992; 1995). Thus, information about each of the individual components of the phosphoinositide signal transduction system can be obtained from postmortem human brain.

In addition to studies of receptors, G-proteins, and phospholipase C, several other components of the phosphoinositide cycle have been examined in postmortem human brain. The levels of inositol, the precursor of PI and thus a critical component of the phosphoinositide signaling system, have been measured in postmortem human brain, with evidence for a decrease associated with suicide victims, bipolar disorder, and schizophrenia (Shimon et al., 1997; 1998). Several investigations examined the level of the inositol trisphosphate receptor in postmortem human brain, and all found significantly decreased levels in Alzheimer's disease brain (Young et al., 1988; Garlind et al., 1995; Haug et al., 1996; Kurumatani et al., 1998). Measurements in postmortem human brain have also encompassed studies of the phosphoinositide synthesizing enzymes (Bothmer and Jolles, 1994; Zubenko et al., 1999), and the inositol phosphate metabolizing enzymes (Atack, 1996). Clearly many individual components of the phosphoinositide signal transduction system are amenable to investigations using postmortem human brain.

Conclusions

The phosphoinositide signal transduction system is one of the major second messenger-producing signaling systems in the central nervous system. However, little is known about the activity and regulation of this critical pathway in human brain. Initial strides have been made towards gaining further information about this system in human brain by using postmortem specimens. Although there are many difficulties encountered when making biochemical measurements using postmortem human brain, a variety of techniques has been applied successfully to gain information both about individual components of the phosphoinositide signaling system and about its functional activity. Application of these methods to groups of subjects with different categories of illnesses has begun to identify specific disease-associated alterations in this signaling system. Further studies using postmortem human brain will undoubtedly contribute further to the understanding of alterations of this signaling system that occur in diseases of the human central nervous system.

Acknowledgements

Research in the author's laboratory has been supported by the National Institutes of Health, the Alzheimer's Association, and the Stanley Association. I am grateful to the many talented colleagues who have made possible the described studies.

References

Atack JR. Inositol monophosphatase, the putative therapeutic target for lithium. Brain Res Rev 1996:183-190.

Berridge MJ, Downes CP, Hanley MR. Lithium amplifies agonist-dependent phosphatidylinositol responses in brain and salivary glands. Biochem J 1982;206: 587-595.

Bothmer J, Jolles J. Phosphoinositide metabolism, aging and Alzheimer's disease. Biochim Biophys Acta 1994;1225:111-124.

Claro E, Garcia A, Picatoste F. Carbachol and histamine stimulation of guanine-nucleotide-dependent phosphoinositide hydrolysis in rat brain cortical membranes. Biochem J 1989a;261: 29-35.

Claro E, Wallace MA, Lee HM, Fain JN. Carbachol in the presence of guanosine 5'-O-(3-thiotriphosphate) stimulates the breakdown of exogenous phosphatidylinositol 4,5-bisphosphate, phosphatidylinositol 4-phosphate, and phosphatidylinositol by rat brain membranes. J Biol Chem 1989b;264: 18288-18295.

Cowburn RF, Wiehager B, Ravid R, Winblad B. Acetylcholine muscarinic M2 receptor stimulated [^{35}S]GTPγS binding shows regional selective changes in Alzheimer's disease postmortem brain. Neurodegeneration 1996;5:19-26.

Crews FT, Kurian P, Freund G. Cholinergic and serotonergic stimulation of phosphoinositide hydrolysis is decreased in Alzheimer's disease. Life Sci 1994;55:1993-2002.

Cutler R, Joseph JA, Yamagami K, Villalobos-Molina R, Roth GS. Area specific alterations in muscarinic stimulated low Km GTPase activity in aging and Alzheimer's disease: implications for altered signal transduction. Brain Res 1994;664:54-60.

Flynn DD, Weinstein DA, Mash DC. Loss of high-affinity agonist binding to M1 muscarinic receptors in Alzheimer's disease: implications for the failure of cholinergic replacement therapies. Ann Neurol 1991;29:256-262.

Friedman E, Butkerait P, Wang HY. Analysis of receptor-stimulated and basal guanine nucleotide binding to membrane G proteins by sodium dodecyl sulfate-polyacrylamide gel electrophoresis. Anal Biochem 1993;214:171-178.

Friedman E, Wang HY. Receptor-mediated activation of G proteins is increased in postmortem brains of bipolar affective disorder subjects. J Neurochem 1996;67:1145-1152.

Garlind A, Cowburn RF, Forsell C, Ravid R, Winblad B, Fowler CJ. Diminished [^{3}H]inositol(1,4,5)P3 but not [^{3}H]inositol(1,3,4,5)P4 binding in Alzheimer's disease brain. Brain Res 1995;681:160-166.

Greenwood AF, Powers RE, Jope RS. Phosphoinositide hydrolysis, Gαq, phospholipase C, and protein kinase C in postmortem human brain: effects of postmortem interval, subject age, and Alzheimer's disease. Neuroscience 1995;69: 125-138.

Hallcher LM, Sherman WR. The effects of lithium ion and other agents on the activity of myo-inositol-1-phosphatase from bovine brain. J Biol Chem 1980;255: 10896-10901.

Haug LS, Ostvold AC, Cowburn RF, Garlind A, Winblad B, Bogdanovich N, Walaas SI. Decreased inositol (1,4,5)-trisphosphate receptor levels in Alzheimer's disease cerebral cortex: selectivity of changes and possible correlation to pathological severity. Neurodegeneration 1996;5:169-176.

Jope RS, Song L, Powers R. [^{3}H]PtdIns hydrolysis in postmortem human brain membranes is mediated by the G-proteins Gq/11 and phospholipase C-β. Biochem J 1994;304: 655-659.

Jope RS, Song L, Li PP, Young LT, Kish SJ, Pacheco MA, Warsh JJ. The phosphoinositide signal transduction system is impaired in bipolar affective disorder brain. J Neurochem 1996;66: 2402-2409.

Jope RS, Song L, Grimes CA, Pacheco MA, Dilley GE, Li X, Meltzer HY, Overholser JC, Stockmeier CA. Selective increases in phosphoinositide signaling activity and G protein levels in postmortem brain from subjects with schizophrenia or alcohol dependence. J Neurochem 1998;70: 763-771.

Kish SJ, Young T, Li PP, Siu KP, Robitaille Y, Ball MJ, Schut L, Warsh JJ. Elevated stimulatory and reduced inhibitory G protein α subunits in cerebellar cortex of patients with dominantly inherited olivopontocerebellar atrophy. J Neurochem 1993;60:1816-1820.

Kurumatani T, Fastbom J, Bonkale WL, Bogdanovic N, Winblad B, Ohm TG, Cowburn RF. Loss of inositol 1,4,5-trisphosphate receptor sites and decreased PKC levels correlate with staging of Alzheimer's disease neurofibrillary pathology. Brain Res 1998;796:209-221.

Li X, Greenwood AF, Powers R, Jope RS. Effects of postmortem interval, age, and Alzheimer's disease on G-proteins in human brain. Neurobiol Aging 1996;17: 115-122.

Mathews R, Li PP, Young LT, Kish SJ, Warsh JJ. Increased G αq/11 immunoreactivity in postmortem occipital cortex from patients with bipolar affective disorder. Biol Psychiatry 1997;41:649-656.

O'Neill C, Fowler CJ, Wiehager B, Alafuzoff I, Winblad B. Assay of a phosphatidylinositol bisphosphate phospholipase C activity in postmortem human brain. Brain Res 1991;543:307-314.

Ozawa H, Katamura Y, Hatta S, Saito T, Katada T, Gsell W, Froelich L, Takahata N, Riederer P. Alterations of guanine nucleotide-binding proteins in postmortem human brain in alcoholics. Brain Res 1993;620:174-179.

Pacheco MA, Jope RS. Phosphoinositide signaling in human brain. Prog Neurobiol 1996;50: 255-273.

Pacheco MA, Stockmeier C, Meltzer HY, Overholser JC, Dilley GE, Jope RS. Alterations in phosphoinositide signaling and G-protein levels in depressed suicide brain. Brain Res 1996;723: 37-45.

Pacheco MA, Jope RS. Comparison of [^{3}H]phosphatidylinositol and [^{3}H]phosphatidylinositol 4,5-bisphosphate hydrolysis in postmortem human brain membranes and characterization of stimulation by dopamine D1 receptors. J Neurochem 1997;69: 639-644.

Rodriguez-Puertas R, Gonzalez-Maeso J, Meana JJ, Pazos A. Autoradiography of receptor-activated G-proteins in postmortem human brain. Neuroscience 2000;96:169-180

Shimohama S, Fujimoto S, Taniguchi T, Kimura J. Phosphatidylinositol-specific phospholipase C activity in the postmortem human brain: no alteration in Alzheimer's disease. Brain Res 1992;579:347-349.

Shimohama S, Perry G, Richey P, Praprotnik D, Takenawa T, Fukami K, Whitehouse PJ, Kimura J. Characterization of the association of phospholipase C-δ with Alzheimer neurofibrillary tangles. Brain Res 1995;669:217-224.

Shimon H, Agam G, Belmaker RH, Hyde TM, Kleinman JE. Reduced frontal cortex inositol levels in postmortem brain of suicide victims and patients with bipolar disorder. Am J Psychiatry 1997;154:1148-1150.

Shimon H, Sobolev Y, Davidson M, Haroutunian V, Belmaker RH, Agam G. Inositol levels are decreased in postmortem brain of schizophrenic patients. Biol Psychiatry 1998;44:428-432.

Wallace MA, Claro E, Carter HR, Fain JN. Phosphoinositide-specific phospholipase C activation in brain cortical membranes. Methods Enzymol 1991;197: 183-190.

Wallace MA, Claro E. Transmembrane signaling through phospholipase C in human cortical membranes. Neurochem Res 1993;18: 139-145.

Wang HY, Friedman E. Receptor-mediated activation of G proteins is reduced in postmortem brains from Alzheimer's disease patients. Neurosci Lett 1994;173:37-39.

Warpman U, Alafuzoff I, Nordberg A. Coupling of muscarinic receptors to GTP proteins in postmortem human brain--alterations in Alzheimer's disease. Neurosci Lett 1993;150:39-43.

Yang CQ, Kitamura N, Nishino N, Shirakawa O, Nakai H. Isotype-specific G protein abnormalities in the left superior temporal cortex and limbic structures of patients with chronic schizophrenia. Biol Psychiatry 1998;43:12-19.

Young LT, Kish SJ, Li PP, Warsh JJ. Decreased brain [^{3}H]inositol 1,4,5-trisphosphate binding in Alzheimer's disease. Neurosci Lett 1988;94:198-202.

Zubenko GS, Stiffler JS, Hughes HB, Martinez AJ. Reductions in brain phosphatidylinositol kinase activities in Alzheimer's disease. Biol Psychiatry 1999;45:731-736.

18 cAMP SIGNAL TRANSDUCTION ABNORMALITIES IN THE PATHOPHYSIOLOGY OF MOOD DISORDERS: CONTRIBUTIONS FROM POSTMORTEM BRAIN STUDIES.

Annisa Chang, Peter P. Li and Jerry J. Warsh

Abstract

During the past decade, considerable advances have been made in second messenger and signal transduction research in mood disorders. In this chapter, evidence derived from human postmortem brain studies is reviewed which supports the emerging view that postreceptor disturbances in G protein-linked cAMP signaling pathways play a major role in the pathophysiology of bipolar (BD) and major depressive disorders (MDD). In addition, the pathophysiological implications of the available evidence are also discussed, within the context of findings from recent neuroimaging and neuropathological studies.

Introduction

Major advances have occurred in the past decade in understanding the molecular mechanisms governing cellular function. A direct outgrowth of this progress has been the recognition of the importance of abnormalities in intracellular signaling systems in the pathophysiology of a number of diseases, among these the mood disorders. It is now evident that a number of signal transduction systems subserve an array of processes involved in propagating, modulating, gating and multiplexing signals in spatially and temporally delimited domains within cells (Jordan et al. 2000). The cyclic adenosine monophosphate (cAMP) signaling pathway is the archetypical intracellular second messenger system first described in hepatocytes (Sutherland and Rall

1958) and later characterized in the central nervous system (CNS) (Greengard 1978). This critical second messenger system quickly became an important focus of investigations into the pathophysiology of mood disorders, particularly given its central role in mediating responses of a number of monoaminergic receptors implicated in the psychobiology of mood disorders and in the therapeutic effects of psychotropic medications (reviewed in Warsh et al. 1988, Hudson et al. 1993). As evidence accrued of extensive crosstalk between different signaling pathways (Jordan et al. 2000), the possibility quickly loomed that disturbances in cAMP-mediated transduction in these disorders could be the result of either a primary, intrinsic abnormality that disrupts neuronal homeostasis, or a secondary event consequent to abnormalities of signal transduction pathways with which the cAMP signal transduction system interacts. In either respect, abnormal cAMP signaling has been recognized as a potentiallyimportant contributing factor in the molecular and cellular pathophysiology of subtypes of the mood disorders.

Biochemical Studies of Mood Disorders

The earliest studies searching for biochemical abnormalities in mood disorders focused intensively on the levels, turnover and functionality of various neurotransmitter systems in the CNS (reviewed in Warsh et al. 1988). However, when the first hints from preclinical and clinical investigations suggested that functional alterations in intracellular signaling mechanisms possibly occur in mood disorders (reviewed in Warsh et al. 1988; Hudson et al. 1993; Warsh and Li 1996), researchers began to focus on postreceptor second messenger signaling processes in an attempt to better understand the pathophysiology of these disorders. Essential to these efforts was the availability of illness-relevant cellular/tissue models in which signal transduction processes and the functionality of signal transducing proteins could be directly examined. Since it is ethically impossible to obtain cellular samples directly from the brain of living, affected individuals, an alternative strategy has been the examination of the proteins that provide the "networking infrastructure" for signal transduction in brain tissue that has been obtained postmortem and frozen for subsequent analyses. The inherent assumption in using this approach is that, while no longer living, the functional state of some proteins is still retained in such preparations. With improved understanding of the mechanisms regulating signal transducing proteins in response to membrane receptor and second messenger activation, it soon became apparent that alterations in their mRNA and protein levels might reflect adaptive changes consequent to abnormal intracellular signaling in specific diseases (Milligan and Wakelam 1992; Morris and Malbon 1999). The pattern of signal transduction changes "frozen" in the postmortem preparation could, therefore, represent the signatures of pathophysiological abnormalities operating in the active state of illness while the individual was alive, and could potentially yield further insights into the cellular and molecular nature of these disorders. It is within this conceptual framework that we review the contributions of postmortem brain research to the understanding of the role of cAMP signaling disturbances in the pathophysiology of mood disorders.

cAMP Signaling Disturbances in Mood Disorders

Following the initial clinical observations of changes in basal plasma cAMP levels (Lykouras et al. 1978) and agonist-stimulated cAMP formation in mononuclear leukocytes (Extein et al. 1979; Pandey et al. 1979) obtained from unmedicated unipolar depressed and BD patients (reviewed in Warsh et al. 1988; Hudson et al. 1993), it gradually became evident that abnormalities might indeed occur beyond the receptor level in mood disorders within the different components of the cAMP-signaling pathway (Hudson et al. 1993; Warsh and Li 1996; Wang et al. 1997). At the present time, several key proteins in this signaling cascade, including guanine nucleotide binding (G) proteins, adenylyl cyclase (AC), cAMP-dependent protein kinase (PKA) and cAMP response element binding protein (CREB) have been implicated.

G Proteins in BD Brain

Immunolabeling studies using specific antisera directed against individual G protein subunits provided the first direct observations implicating postreceptor disturbances in BD brain. Significantly elevated immunoreactive levels of the long spliced variant (52 kDa) of the G_s α-subunit, $G\alpha_{s\text{-}L}$, were reported in frontal, temporal, and occipital cortical regions (Young et al. 1991a; 1993) and moderately increased levels in thalamus (Warsh et al. 2000) of postmortem brain from BD patients compared with non-neurological, non-psychiatric controls matched for age, postmortem interval, and brain pH. Levels of the short spliced variant of $G\alpha_s$ ($G\alpha_{s\text{-}S}$) were also higher in the hippocampus and caudate nucleus of postmortem brain from BD patients compared with controls, but were reduced in occipital cortex (Warsh et al. 2000). In contrast, no difference was observed in $G\alpha_s$ levels in the parietal cortex and cerebellum of BD brain compared with the controls (Young et al. 1993; Warsh et al. 2000). The selective alteration in levels of only one of the spliced variants of $G\alpha_s$ ($G\alpha_{s\text{-}L}$) in BD brain regions examined may reflect the differential regulation of the alternative splicing mechanism(s) for the $G\alpha_s$ gene in affected regions of BD brain. Alternatively, the differential changes in one or the other $G\alpha_s$ subtype could be the result of differences in the turnover of the $G\alpha_s$ isoforms. As subtle differences have been reported between the two spliced variants of $G\alpha_s$ in the interaction with receptor (Seifert et al. 1998) and AC (Walseth et al. 1989), changes in the relative expression levels of the $G\alpha_s$ spliced variants in BD brain could have important consequences for intracellular cAMP signaling.

Interestingly, the increase in $G\alpha_s$ protein levels, noted above, has been replicated by Friedman and Wang (Friedman and Wang 1996), but not by Dowlatshahi and colleagues (Dowlatshahi et al. 1999) who failed to detect any difference in $G\alpha_s$ protein immunolabeling in postmortem BD brain compared with matched controls in the sample collection from the Stanley Foundation Neuropathology Consortium. The basis for the discrepancies between these aforementioned reports is presently unknown, but may be the result of one or more factors such as differences in clinical and demographic characteristics of the patient population from which the brains were obtained postmortem, drug treatments, mode of death, and variations in postmortem handling conditions among the different investigators (Warsh and Li 2000). Thus, additional studies with larger sample size, better-defined subject groups and control of extraneous variables, such as noted, will be necessary to confirm the validity of the $G\alpha_s$ protein changes observed in some but not all studies of postmortem BD brain.

While other G protein subunits, in addition to $G\alpha_s$, are also involved in the regulation of AC activity (reviewed in Chern 2000), no statistically significant differences have been found between BD and control subjects in the protein levels of $G\alpha_{i-1}$, $G\alpha_{i-2}$, $G\alpha_o$, $G\alpha_{olf}$, $G\alpha_z$, $G\beta_{35}$ or $G\beta_{36}$, in any of the autopsied brain regions studied (Young et al. 1993; Friedman and Wang 1996; Warsh et al. 2000). Thus, the changes in $G\alpha_s$ in postmortem BD brain suggest dysregulation occurs in this stimulatory branch of the G protein-mediated cAMP-signaling cascade, in the pathophysiology of BD.

Mechanisms Underlying the $G\alpha_s$ Changes in BD Brain

The exact mechanism(s) underlying the observed changes in $G\alpha_s$ levels in BD brain is not clear. They are not related to differences in age or postmortem delay, as G protein levels are stable with respect to these variables (Young et al. 1991b; Li et al. 1996). The elevations are also not likely to be related to the effects of antemortem lithium therapy since the changes were observed both in BD patients with, and in those without, a history of lithium treatment (Young et al. 1993). Moreover, it has been shown in several animal studies that chronic lithium, carbamazepine and antidepressant administration do not affect cerebral cortical levels of $G\alpha_s$, $G\alpha_{i-1}$, or $G\alpha_{i-2}$ protein (Li et al. 1993; Chen and Rasenick 1995; Emamghoreishi et al. 1996; Dwivedi and Pandey 1997). Possible confounding effects of other psychotropic or nonpsychotropic drugs that the patients may have received antemortem, leading to increased $G\alpha_s$ immunolabeling, cannot be ruled out, however.

At the molecular level, the lack of concomitant changes in $G\alpha_s$ mRNA levels in the same BD cerebral cortical regions in which increased $G\alpha_s$ levels were observed (Young et al. 1996), together with the absence of identifiable mutations in the promoter or coding region of the $G\alpha_s$ gene in BD patients (Ram et al. 1997), argue against the involvement of transcriptional mechanisms as the basis for the changes in $G\alpha_s$ levels observed in BD brain. Instead, disturbances in one or more of the post-translational processes regulating the cellular disposition and degradation of $G\alpha_s$ merit consideration as possible mechanisms that may account for the observed immunolabeling differences in BD brain. In this regard, cholera toxin (CTX)-catalyzed ADP-ribosylation of $G\alpha_s$ (both $G\alpha_{s\text{-}S}$ and $G\alpha_{s\text{-}L}$), a covalent modification that constitutively activates the protein (Toyoshige et al. 1994), has been shown to accelerate $G\alpha_s$ turnover rate (Chang and Bourne 1989; Milligan et al. 1989). Of note, $G\alpha_s$ is a substrate for both endogenous and CTX-catalyzed ADP-ribosylation in autopsied human brain (Andreopoulos et al. 1999). Furthermore, significantly lower endogenous ADP-ribosylation of a novel $G\alpha_s$ isoform has been reported in autopsied temporal cortex of BD patients compared with control subjects (Andreopoulos et al. 1997). Taken together, these latter observations suggest that the increase in $G\alpha_s$ protein levels in key cerebral cortical regions of BD brain may be due in part to alterations in factors regulating the ADP-ribosylation-mediated turnover and degradation of $G\alpha_s$.

Functional Significance of Elevated $G\alpha_s$ in BD

The assessment of receptor-G protein coupling in postmortem brain membranes is problematic because of the relative instability of the functional complex during freezing and storage of the tissue. Nonetheless, there is evidence suggesting that elevations in $G\alpha_s$ levels are associated with altered receptor-G_s protein coupling in membrane preparations from postmortem BD brain. Friedman and Wang (1996) reported significantly higher basal and isoproterenol-stimulated $[^{35}S]GTP\gamma S$ binding to $G\alpha_s$ in the frontal cortex from postmortem BD brain, changes that they attributed to both the elevated $G\alpha_s$ levels and a higher proportion of G_s protein in the heterotrimeric ($\alpha\beta\gamma$) state. It is unlikely that the former alteration is related to changes in β-adrenoceptor (AR) densities as no differences have been found in this parameter in frontal cortex from BD patients compared with matched controls (Young et al. 1994). Interestingly, Andreopoulos et al. (1997) found that the relative proportion of G_s protein in the trimeric state, as assessed by the extent to which it could be ADP-ribosylated in a CTX-dependent manner, was either not different ($G\alpha_{s-L}$) or moderately reduced ($G\alpha_{s-S}$), in BD temporal cortex, in which $G\alpha_{s-L}$ immunolabeling was significantly increased. These contradictory findings suggest that the proportion of G_s protein in the trimeric state is lower in BD temporal cortex compared with matched controls. While differences in experimental techniques (CTX-catalyzed ADP-ribosylation versus immunoprecipitation) or brain regions examined (temporal versus frontal cortex) may account for the discrepant observations, the exact explanation remains to be clarified.

Adenylyl Cyclase Activity in BD brain

The questions of whether and how these $G\alpha_s$ alterations impact on cAMP formation and signaling to downstream target proteins in BD brain have been addressed in several studies in which observations indicating an hyperfunctional cAMP signaling system have been reported. Using assay conditions favoring stimulation by $G\alpha_s$ of the Ca^{2+}/calmodulin (CaM)-insensitive AC isoforms, Young et al. (1993) reported a significant increase in forskolin-stimulated AC activity in temporal and occipital, but not in frontal cortex of postmortem BD brain. The enhanced response, together with the significant correlation found between forskolin-stimulated cAMP response and the levels of $G\alpha_{s-L}$ across these cortical brain regions, supports the notion that the increase in the coupling/activation of AC is a consequence of the elevation in $G\alpha_s$ levels (Young et al. 1993). Alternatively, the enhanced forskolin-stimulated AC activity could be attributed to an increase in the concentrations of specific Ca^{2+}/CaM-insensitive AC subtypes including types II and IV. However, the lack of differences in immunoreactive levels of AC subtypes (I, IV, and V/VI) in BD cerebral cortical regions compared with those of matched control subjects (Reiach et al. 1999) does not support the latter explanation.

Downstream Post-effector Signaling Abnormalities in BD

While the above observations point towards altered G protein-coupled cAMP signaling in BD, it is not yet possible to confirm this through direct measurement of cAMP in living human brain. Moreover, inferences about the state of cAMP signaling can not be drawn based on estimation of its levels in postmortem brain because of the rapid degradation that occurs immediately upon death (Jones and Stavinoha 1979). Therefore, alternative strategies were necessary to search for evidence of cAMP-signaling disturbances in this preparation. One approach was to probe for adaptive changes in cAMP protein targets (e.g. PKA) that would have been expected to occur had the cAMP signaling system been up-regulated in the living state. To this end, our group examined the binding of $[^3H]$cAMP in a number of regions of the postmortem brain from BD patients and controls, reasoning that the level of cAMP binding would reflect the amount of PKA regulatory (R) subunits (Rahman et al. 1997). Significantly lower $[^3H]$cAMP binding was observed in the cytosolic, but not in the particulate fractions of cerebral cortices (frontal, temporal and occipital), cerebellum and thalamus of postmortem brain from BD patients compared with matched controls (Rahman et al. 1997). These observations implied that PKA R subunit abundance was reduced in BD brain, a change that we suggested could have occurred as a result of increased degradation of PKA R subunits induced by persistent, hyperactive stimulation of this transducing enzyme by cAMP. The localization of these alterations to the cytosolic compartment is of some interest given that the translocation of PKA catalytic (C) subunits from the cytosol into the nucleus and their subsequent phosphorylation of transcription factors, such as CREB (see figure 1), is an important mode of regulating gene transcription (Yamamoto et al. 1988). Furthermore, the availability of free C subunits is a stoichiometric function of the abundance of R relative to C subunits, as free R subunits, acting as scavengers, recombine with C subunits forming nascent holoenzyme (Spaulding 1993).

To further characterize the nature of the adaptive changes in the abundance of PKA subunits induced by putative hyperactive cAMP input, Chang et al. (2000) examined PKA R and C subunit abundance in postmortem BD brain by immunoblot assay using PKA subunit-specific antibodies. Significantly higher protein levels of PKA R (mainly RIIβ) and C subunits were found in the cytosolic fraction of temporal cortex from the same postmortem BD brains compared with the same matched controls as reported by Rahman et al. (1997). These immunolabeling results are consonant with observations of significantly higher levels of basal as well as maximally stimulated PKA activities, and a significantly lower activation constant (i.e. EC_{50}) for cAMP in the cytosolic fraction of temporal cortex of the same BD subjects compared with controls (Fields et al. 1999).

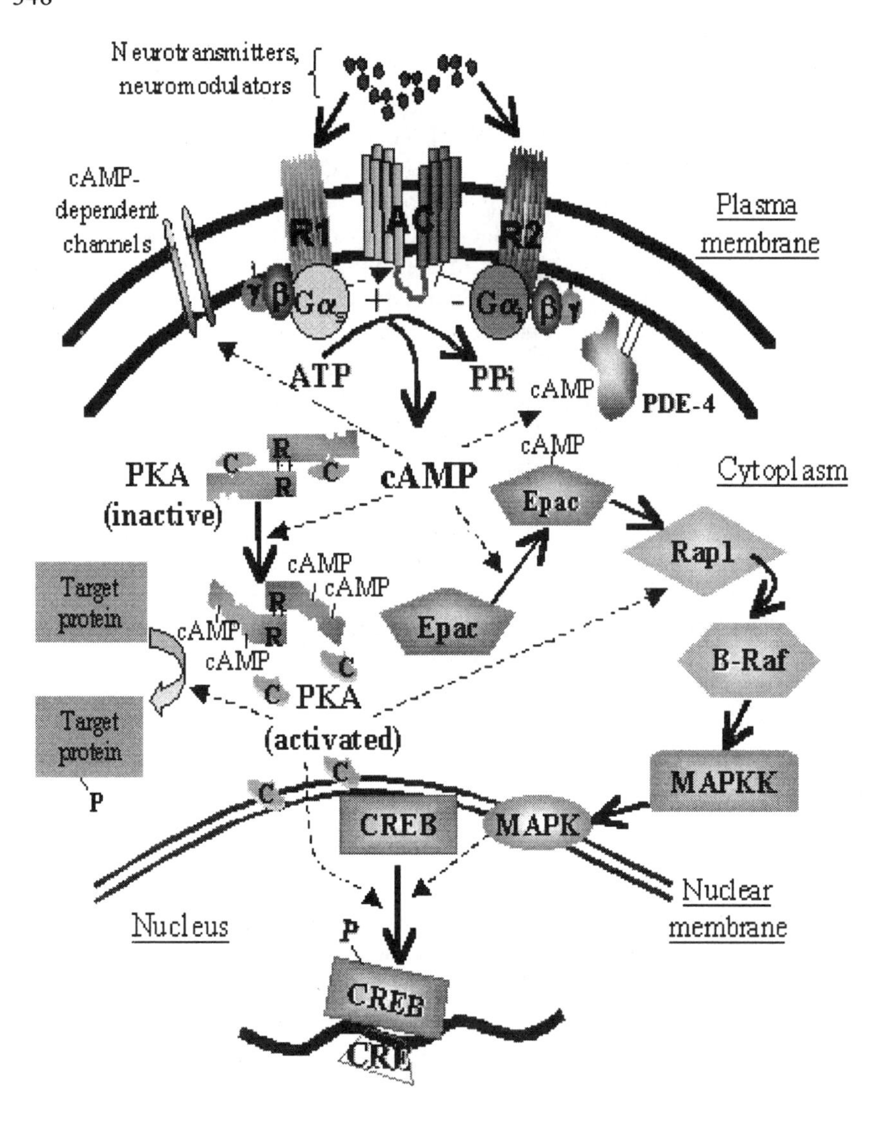

Figure 1. Overview of the cAMP signal transduction pathway. - -▶ denotes stimulatory pathways; - - -┤ denotes inhibitory pathway.

Abbreviations: AC, adenylyl cyclase; C, PKA catalytic subunit; CRE, cAMP response element; CREB, cAMP response element binding protein; Epac, exchange protein directly activated by cAMP; PKA, cAMP dependent protein kinase; MAPK, mitogen-activated protein kinase, MAPKK, MAPK kinase; PDE-4, phosphodiesterase type 4; R, PKA regulatory subunit; R1, receptor 1; R2, receptor 2.

Moreover, the lack of significant differences in PKA subunit immunolabeling in the particulate fraction concurs with the absence of changes in [^{3}H]cAMP binding (Rahman et al. 1997) and PKA activity (Fields et al. 1999; Jensen et al. 2000) in this cellular fraction in BD brain.

While the finding of elevated RIIβ levels in BD brain was unexpected, given the reductions in [^{3}H]cAMP binding reported earlier, on closer scrutiny, several interesting possibilities are evident that may explain these seemingly contradictory results. First, the binding assays employed by Rahman et al. (1997) were conducted at only one ligand concentration. Thus, a decrease in [^{3}H]cAMP binding could signify a reduction in the affinity (i.e. K_d) of PKA for [^{3}H]cAMP rather than a reduction in the levels of cAMP "receptors" (mainly PKA R subunits) if the ligand concentration used was subsaturating. In this regard, the higher PKA RIIβ subunit level found in BD temporal cortex could lead to preferential assembly of PKA-II holoenzyme over PKA-I holoenzyme (Otten and McKnight 1989). Because PKA-II exhibits an almost 10-fold higher K_d (4 – 6 nM) for cAMP than type I (0.1 – 1 nM) (Amieux et al. 1997), a relative shift in PKA-I to PKA-II, as implied by the immunolabeling results, could manifest in reduced [^{3}H]cAMP binding despite higher immunolabeling. Another intriguing possibility is that changes in abundance of other cAMP binding proteins, such as the recently discovered guanine-nucleotide-exchange factors (GEFs) (e.g. Epac - exchange protein directly activated by cAMP) (de Rooij et al. 1998), could result in a fractional reduction affecting net [^{3}H]cAMP binding as reported in postmortem BD brain. The observations that Epac and RIα have similar binding affinities for cAMP (de Rooij et al. 1998) also support such a possibility.

Finally, changes in cAMP signaling could affect a number of downstream protein targets involved in transducing signals entrained in second and/or third messengers into physiological events. Thus, in BD brain, changes should also be expected in protein and/or mRNA levels of other proteins modulated in the cAMP transduction system. A more distal, as well as critical, downstream target in the cAMP pathway is the transcription factor, CREB, which is phosphorylated in response to cAMP-stimulation of PKA (Gonzalez and Montminy 1989). Only one study has examined CREB levels in temporal cortex of postmortem BD brain, however, reporting no significant differences between BD and controls (Dowlatshahi et al. 1999). This failure to detect differences of CREB levels in postmortem BD brain may have been related to the absence of $G\alpha_s$ elevations, or to other confounding extraneous factors in that study, as discussed above. Moreover, total CREB content was measured, not the phosphorylated form, which reflects the transcriptionally active state of the protein (Yamamoto et al. 1988), as the latter is rapidly dephosphorylated and variably detectable in postmortem tissue. Thus, this negative finding does not conclusively rule out the possibility that altered levels and/or activity of phosphorylated CREB occurs in BD brain in association with the upstream disturbances in cAMP signaling implied by the $G\alpha_s$ and PKA changes discussed above.

cAMP Signaling Disturbances in Major Depressive Disorder

As with BD, the number of studies that have examined intracellular signaling in MDD using the postmortem brain strategy is relatively modest. Nevertheless, interesting and important observations have been obtained using this strategy that also implicate disturbances in the receptor-G protein-cAMP-PKA cascade in critical brain regions in depressive disorders, or at least in depressed patients who died of suicide, as these comprise the bulk of subject material for most reports to date. Notably, the nature and pattern of the cAMP-signaling disturbances observed in postmortem MDD brain appear to be quite distinct from those found in studies of BD, suggesting that a number of the molecular disturbances differ between these disorders.

At the receptor level, a number of studies have demonstrated up-regulation of α_2-AR density in prefrontal cortex (PFC) of depressed suicides (Meana et al. 1992; Gonzalez et al. 1994; Ordway et al. 1994). While three α_2-AR subtypes are expressed in human brain, namely α_{2A}, α_{2B}, and α_{2C}, it is the α_{2A}-AR which is selectively elevated in the PFC of depressed suicides (Callado et al. 1998; Garcia-Sevilla et al. 1999). The α_{2A}-AR is coupled to G_i, which, upon stimulation, inhibits AC activity. Furthermore, agonist binding to α_{2A}-AR can also activate specific G protein-coupled receptor kinases (GRK2/3) that are known to phosphorylate and desensitize the agonist-occupied receptor (Jewell-Motz and Liggett 1996). In this regard, GRK-mediated phosphorylation has been shown to desensitize a broad range of G protein-coupled receptors (GPCRs) (Chuang et al. 1996) including noradrenergic (Anand and Charney 2000) and serotonergic receptors (Mann 1999), which have been implicated in depression. Of particular interest in this context is the observation that the level of membrane-associated GRK2 is also increased in the PFC of depressed suicides (Garcia-Sevilla et al. 1999). This is thought to represent an important compensatory mechanism in response to the supersensitive α_{2A}-AR-$G\alpha_{i\text{-}2}$ signaling system (Garcia-Sevilla et al. 1999). The fact that the levels of membrane-associated GRK2 correlate significantly with those of the α_{2A}-AR and $G\alpha_{i\text{-}2}$ in the same brain samples of depressed suicides further supports this idea. Regardless of the exact mechanism, these findings strongly implicate an enhancement in the receptor-mediated inhibitory regulation of cAMP signaling in depressed suicides.

There is also evidence implicating altered G protein levels and function in depressed suicides, though the findings are somewhat conflicting. For example, Garcia-Sevilla et al. (1999) reported up-regulated $G\alpha_{i1/2}$ levels in the PFC of antidepressant-free depressed suicide subjects. Other groups, however, found normal $G\alpha_{i1/2}$ (Cowburn et al. 1994) or reduced $G\alpha_{i2}$ (Pacheco et al. 1996) protein levels in the PFC of depressed suicides. The seemingly contradictory observations could be related to differences in the subregions of PFC examined. In comparison, Pacheco et al. (1996) also reported increased $G\alpha_{s\text{-}S}$ levels in the PFC of depressed suicides. Similarly, the levels of $G\alpha_{s\text{-}S}$ in frontal cortex also showed a tendency to be increased in the violent death suicide and depressed suicide subgroups when compared with controls (Cowburn et al. 1994). These modest increases involved only $G\alpha_{s\text{-}S}$ and not $G\alpha_{s\text{-}L}$ as was reported in a number of cortical regions of postmortem BD brain, as discussed in the previous section.

With regard to G protein functionality, Cowburn et al. (1994) found that the cAMP responses to GTPγS- and forskolin-stimulation were blunted in the frontal cortical region of depressed suicides. Lowther et al. (1996) examined basal, GTPγS- and forskolin-stimulated cAMP formation, as well, but in a more carefully selected group of patients with "firm retrospectively established diagnoses of depression" with and without a history of antidepressant treatment in the three months prior to death. However, only a trend towards lower stimulated cAMP responses was found in frontal and parietal cortices in suicides compared with controls. In addition, these investigators found no significant effect of history of antidepressant treatment or any relationship with the mode of suicide (violent as compared with non-violent means). Cowburn et al. (1994) ascribed the reduced cAMP responses to GTPγS and forskolin to the decrement in basal AC activity, a change also not detected by Lowther et al. (1996) and Reiach et al. (1999). Despite these discrepancies, the results of these studies concur in suggesting that G protein-coupled AC activity may be reduced in suicide victims. The support for a relationship of the reduced cAMP responses with depression per se is more tenuous, however.

The molecular basis for the reduced basal, GTPγS- and forskolin-stimulated cAMP formation is also uncertain. Reduced basal cAMP formation could be related to constitutively increased G_i function or changes in other regulatory inputs (e.g. PKC and/or MAPK) that influence the basal state or levels of AC (Morris and Malbon 1999). In this respect, decreased immunoreactive levels of AC-IV, but not AC-I, V or VI, have been observed in the temporal cortex of depressed suicide subjects compared with controls

(Reiach et al. 1999). In addition, Dwivedi et al. (1999) reported that the B_{max} for [^{3}H]cAMP binding and PKA activity was reduced in postmortem brain of depressed suicide subjects. These observations also support the notion that disturbances occur at loci further down-stream in the cAMP cascade in depressed suicides.

Dysregulation of hypothalamic-pituitary-adrenocortical (HPA) function is thought to play an important role in the vulnerability to depression (reviewed in Holsboer 2001). Moreover, endogenous glucocorticoids can regulate the expression levels of selected isoforms of PKA subunits, the binding of cAMP to PKA and ultimately PKA activity in brain (Dwivedi and Pandey 2000). Therefore, HPA dysregulation that occurs in depression could also affect cAMP signaling at the critical level of PKA functionality, regardless of any effects that these hormonal changes exert on upstream components in this transduction system. Notwithstanding the discrepancies noted above, the accumulating findings suggest that molecular alterations, both at the level of the G protein-AC complex and in PKA regulation, result in blunted cAMP signaling in brain of patients with a history of depression who commit suicide.

As in depressed suicides, blunting of the cAMP signaling pathway has also been implicated in brain of patients with MDD, although there is even less evidence from postmortem brain research in this instance due to the difficulty in separating changes specific to MDD from those related to suicide, and the limited brain material from patients with MDD who died of non-self inflicted causes. Higher photoaffinity GTP labeling levels of $G\alpha_{i/o}$, but not $G\alpha_s$, were found in parietal and temporal cortices of MDD patients (most of whom died of causes other than suicide) compared with controls without any changes in immunoreactive levels of $G\alpha_s$, $G\alpha_{i1-2}$, $G\alpha_o$ or $G\alpha_{q/11}$ (Ozawa et al. 1993). Dowlatshahi et al. (1999) also found no differences in temporal or occipital cortex $G\alpha_s$ and $G\alpha_{i1-2}$ levels between MDD patients who died of suicide and those who did not, and controls. Because photoaffinity GTP labeling reflects the functional activities, and not the quantities, of G proteins, these results suggest that an imbalance between $G\alpha_s$ and $G\alpha_{i/o}$ function, rather than a change in their levels, may be of importance in the pathophysiology of MDD.

Dowlatshahi et al. (1999) also reported significant reductions in CREB concentrations in temporal cortex of untreated MDD patients compared with matched controls implicating downstream cAMP signaling disturbances in the MDD patients studied. Moreover, they noted that a history of antidepressant treatment was associated with significantly higher CREB levels. While these preliminary results are of considerable interest, their findings of lower CREB levels in temporal cortex of depressed subjects

(MDD and BD) who died of suicide (Dowlatshahi et al. 1999) suggest the observed changes may actually be attributable to suicide and are not necessarily specific to MDD. Furthermore, caution is again warranted in interpreting changes in global rather than phosphorylated CREB or its binding to the consensus cAMP response element sequence, which, although more relevant functionally (Yamamoto et al. 1988), can not be reliably measured in postmortem brain tissue.

Pathophysiological Implications of cAMP Signal Transduction Abnormalities

While as reviewed herein, an appealing case can be made supporting the notion that cAMP signaling is disturbed both in BD and MDD (including depressed suicides), the role of these abnormalities in the overall pathophysiological schema of these disorders is poorly understood. However, when considered within the context of observations from in vivo neuroimaging and morphometric studies on postmortem brain of affected individuals, molecular studies of stress, and investigations of antidepressant and mood stabilizer action in cellular and animal models, several relevant pathophysiological themes emerge in which cAMP-mediated signaling disturbances may participate.

Structural brain imaging studies have identified regionally selective reductions in volumes of the basal ganglia, cerebellum, hippocampus, and PFC in MDD and in PFC in familial mood disorders (MDD and BD) (reviewed in Drevets et al. 1997; Soares and Mann 1997; Drevets 2000; Sheline 2000) suggesting underlying cellular changes in these regions in some mood disorder categories. In BD, some consistent neuroanatomical differences, notably subcortical white mater hyperintensities, reduced cerebellar volume and ventricular enlargement (mainly third but possibly left lateral, as well), and possibly PFC (BD with familial mood disorders) also implicate underlying cellular neuropathology (Drevets et al. 1997; Soares and Mann 1997; Drevets 2000; Stoll et al. 2000). Results of functional neuroimaging studies also support disturbances in PFC, amygdala, caudate, and cerebellar vermis in MDD (Drevets 2000), but it is the observations of persistent functional differences in the subgenual anterior cingulate cortex in MDD and BD patients with familial histories of mood disorders (Drevets et al. 1997) which have received considerable interest as recent morphometric studies indicate reduced neuronal and glial densities in this region in these mood disorders (Ongur et al. 1998; Rajkowska 2000). However, differences in the brain regions affected and patterns of cellular morphology between MDD and BD have been suggested. In BD, cell loss has been implicated in the subgenual region with reductions in the density of glial cells and neurons (Ongur et al. 1998; Rajkowska 2000). The cellular morphometric changes in MDD are more complex, with atrophy-like changes and possible cell loss reported in dorsolateral PFC, and orbitofrontal and anterior cingulate cortex (Rajkowska 2000; Cotter et al. 2001) and increased cell numbers in hypothalamus, and dorsal raphe nucleus (Rajkowska 2000). Clearly the nature of the cellular changes and regions affected in BD as compared to MDD suggest that different pathophysiological mechanisms may be at work in these two types of primary mood disorders. These findings complement the suggestions arising from molecular signaling studies of these disorders that different pathophysiological mechanisms operate in BD compared with MDD. Of particular note, the findings suggesting neuronal and glial cell loss in specific limbic regions of BD brain have raised the possibility that cellular viability is affected in a uniquely brain region specific manner in BD (Manji et al. 2000b), although it is unknown whether this applies to BD in general or a subgroup of patients with this disorder — those with family history of mood disorders, for example. This notion is further supported by recent in vivo magnetic resonance spectroscopy findings of decreased levels of N-acetyl-aspartate, which is generally regarded as a measure of neuronal viability (Tsai and Coyle 1995), in the dorsolateral PFC of BD subjects compared to controls (Winsberg et al. 2000). Taken together with other recent evidence suggesting that lithium and valproate exert antiapoptotic and neuroprotective effects in a variety of experimental models (reviewed in Manji et al. 2000a) and preliminary in vivo evidence that lithium enhances neuronal viability in BD brain (Moore et al. 2000), these observations are building a compelling case to argue that some form of cellular injury evolves in the course of BD.

How do alterations in cAMP signaling fit with the subtle neuropathological abnormalities revealed in recent imaging and morphometric investigations of BD? Although disturbances in the cAMP cascade can affect a number of neuronal functions as discussed above, several intriguing lines of evidence suggest two possibilities that may be more significant in a neuropathological context: (i) these cAMP-signaling changes reflect molecular compensations to sustain neuronal integrity and function in the face of cellular stresses that could lead to injury and death; and (ii) the up-regulated $G\alpha_s$ functionality and/or cAMP formation may in itself predispose to cell death. On the one hand, increased cAMP levels following inhibition of phosphodiesterase type 4 (Hulley et al. 1995; Yamashita et al. 1997) and stimulation of PKA by cAMP analogues (De et al. 1994; Hulley et al. 1995; Nakao 1998) have been observed to exert neuroprotective effects in cultured neurons and animal models of neurogenerative diseases. In this regard, a variety of data support altered intracellular Ca^{2+} homeostasis in BD (Li et al.

2000), a change which may be trait-dependent (Emamghoreishi et al. 1997). Disruption of Ca^{2+} homeostasis plays an important role in the processes of cellular necrosis and apoptosis (Zhu et al. 2000). Thus, for example, the elevated PKA RIIβ and C subunit levels (Chang et al. 2000) and PKA activity (Fields et al. 1999) found in postmortem BD brain could reflect an endogenously activated neuroprotective response to processes affecting intracellular homeostasis and cellular resilience in subtypes of BD. CREB, which is phosphorylated by PKA, as well as Ca^{2+}/CaM kinases or MAPK, has been implicated as important molecular target central to such neuroprotective actions (Walton and Dragunow 2000). In a similar vein, recent findings indicate that antidepressants enhance cAMP-mediated signaling, CREB and brain-derived neurotropic factor expression (Duman et al. 1999, 2000), leading to increased neuroplasticity and neurogenesis that may be of importance in the therapeutic response to these agents. Thus, disturbances in cAMP signaling mechanisms in MDD may also impact on neuronal integrity and connectivity, although the nature of the disturbances in this disorder may differ qualitatively, quantitatively, and in the brain regions affected, in comparison with BD.

On the other hand, activation of receptors coupled to cAMP formation through G_s, and elevation of $G\alpha_s$ expression levels and stimulation of cAMP formation have been shown to activate neuronal cell death. In two molecular analyses in Caenorhabditis elegans (Korswagen et al. 1997; Berger et al. 1998), it was found that the induced increase in expression of gsa-1, the $G\alpha_s$ homologue expressed in this species, or expression of a constitutively active GTPase defective $G\alpha_s$ mutant (Q227L), resulted in rapid onset of neuronal degeneration. Particularly noteworthy was the observation that the effects of $G\alpha_s$ on neuronal degeneration and neural activity were neuronal cell type specific, reminiscent of the regionally specific neuronal and glial changes in selective cerebral cortical regions and laminae reported in recent morphometric analyses of postmortem BD brain described above, for example. The cytotoxic effect of $G\alpha_s$ is mediated by cAMP as it was dependent on the expression of the AC homologue, acy-1. Furthermore, it was unaffected by loss of the programmed cell death pathway, thus implicating a necrotic pattern of neurodegeneration (Korswagen et al. 1997; Berger et al. 1998). Although the exact mechanism of this $G\alpha_s$-induced cytotoxicity is still unknown, it may be related to the effect of $G\alpha_s$-mediated activation of membrane ion channels (Korswagen et al. 1997).

In other cell models, such as S49 mouse lymphoma cells, β-AR induced cell death signaling has been described, which is mediated through $G\alpha_s$ and cAMP dependent pathways (Gu et al. 2000; Yan et al. 2000). In contrast to the necrotic cell death induced by $G\alpha_s$ in C. elegans, in the S49 cell model, the mode of cell death is typically apoptotic and it can occur via either PKA-dependent (Yan et al. 2000) or PKA-independent (Gu et al. 2000) pathways. While the detailed mechanisms involved in the PKA-independent apoptosis remain to be elucidated, it has been shown to involve the c-Src family of tyrosine kinases. Of note, activation of other GPCRs in other cell models has been observed to induce apoptosis in a $G\alpha_s$-dependent but cAMP/PKA-independent manner (Gu et al. 2000). Such observations clearly highlight the possibility that sustained stimulation of receptors coupled to $G\alpha_s$ and/or processes that increase $G\alpha_s$-mediated signaling can lead to cell death and loss.

Conclusions

Since the change of focus from biochemical to molecular studies of mood disorders and from neurotransmitter function to postreceptor mechanisms, the dominant view emerging in this area of research is that disturbances in postreceptor signaling pathways play a major role in the pathophysiology of these disorders (Hudson et al. 1993; Warsh and Li 1996; Li et al. 2000). In this chapter, we have provided an overview of the evidence derived from postmortem brain studies supporting this notion from the perspective of cAMP signaling. In addition, we have highlighted the rapidly evolving ideas about the potential nature and impact of the cAMP signal transduction abnormalities on neuronal function and integrity in BD and MDD. There is compelling evidence supporting the notion of an enhanced state of cAMP signaling extending from the G protein level down to at least the primary protein target of cAMP, i.e, PKA, in BD brain. In contrast, most of the evidence gleaned from postmortem studies of MDD points towards a blunted cAMP signal transduction system. Because there is extensive cross-regulation among different intracellular signaling pathways, and an increasing body of evidence implicates abnormalities in other signal transduction systems (including the phosphoinositide system as reviewed by others in this volume) in mood disorders, one must ask if the disturbances in the cAMP signaling cascade result from dysregulation of one or more of the signal transduction pathways with which it interacts. Also to be answered are the questions of whether the observed disturbances represent adaptive or compensatory protective responses, or unmitigated pathophysiological aberrations that compromise neuronal function in those critical limbic brain regions central to mood regulation. Finally, a particularly intriguing and important question to be addressed is the extent to which the cAMP signaling abnormalities are related to and account for the morphological changes now being reported in postmortem brain of subjects with mood disorders. Although there has been considerable progress in establishing the importance of cAMP signaling disturbances in mood disorders, much remains to be done to gain a full understanding of the exact role that these disturbances play in their pathogenesis.

References

Amieux PS, Cummings DE, Motamed K, Brandon EP, Wailes LA, Le K, Idzerda RL, McKnight GS. Compensatory regulation of RIα protein levels in protein kinase A mutant mice. J Biol Chem 1997;272: 3993-3998.

Anand A, Charney DS. Norepinephrine dysfunction in depression. J Clin Psychiatry 2000;61: 16-24.

Andreopoulos S, Siu KP, Li PP, Warsh JJ. Reduced ADP-ribosylation of Gα_S in postmortem bipolar disorder temporal cortex. Biol Psychiatry 1997;41: 61S.

Andreopoulos S, Li PP, Siu KP, Warsh JJ. Characterization of α_s-immunoreactive ADP-ribosylated proteins in postmortem human brain. J Neurosci Res 1999;56: 632-643.

Berger AJ, Hart AC, Kaplan JM. Gα_S-induced neurodegeneration in Caenorhabditis elegans. J Neurosci 1998;18: 2871-2880.

Bornfeldt KE, Krebs EG. Crosstalk between protein kinase A and growth factor receptor signaling pathways in arterial smooth muscle. Cellular Signalling 1999;11: 465-477.

Callado LF, Meana JJ, Grijalba B, Pazos A, Sastre M, Sastre M, Garcia-Sevilla JA. Selective increase of α_{2A}-adrenoceptor agonist binding sites in brains of depressed suicide victims. J Neurochem 1998;70: 1114-1123.

Chang A, Li PP, Kish S, Warsh JJ. Altered postmortem temporal cortex cAMP-dependent protein kinase subunit levels in bipolar disorder. Abst Soc Neurosci 2000; 26: 2314.

Chang FH, Bourne HR. Cholera toxin induces cAMP-independent degradation of Gs. J Biol Chem 1989;264: 5352-5357.

Chen J, Rasenick MM. Chronic antidepressant treatment facilitates G protein activation of adenylyl cyclase without altering G protein content. J Pharmacol Exp Ther 1995;275: 509-517.

Chern Y. Regulation of adenylyl cyclase in the central nervous system. Cell Signal 2000;12: 195-204.

Chuang TT, Iacovelli L, Sallese M, De Blasi A. G protein-coupled receptors: heterologous regulation of homologous desensitization and its implications. Trends Pharmacol Sci 1996;17: 416-421.

Cotter D, Mackay D, Landau S, Kerwin R, Everall I. Reduced glial cell density and neuronal size in the anterior cingulate cortex in major depressive disorder. Arch Gen Psychiatry 2001;58: 545-53.

Cowburn RF, Marcusson JO, Eriksson A, Wiehager B, O'Neill C. Adenylyl cyclase activity and G-protein subunit levels in postmortem frontal cortex of suicide victims. Brain Res 1994;633: 297-304.

De A, Boyadjieva NI, Pastorcic M, Reddy BV, Sarkar DK. Cyclic AMP and ethanol interact to control apoptosis and differentiation in hypothalamic beta-endorphin neurons. J Biol Chem 1994;269: 26697-26705.

de Rooij J, Zwartkruis FJ, Verheijen MH, Cool RH, Nijman SM, Wittinghofer A, Bos JL. Epac is a Rap1 guanine-nucleotide-exchange factor directly activated by cyclic AMP. Nature 1998;396: 474-477.

Dowlatshahi D, MacQueen GM, Wang JF, Reiach JS, Young LT. G protein-coupled cyclic AMP signaling in postmortem brain of subjects with mood disorders: Effects of diagnosis, suicide, and treatment at the time of death. J Neurochem 1999;73: 1121-1126.

Drevets WC, Price JL, Simpson JR, Jr., Todd RD, Reich T, Vannier M, Raichle ME Subgenual prefrontal cortex abnormalities in mood disorders. Nature 1997;386: 824-827.

Drevets WC. Neuroimaging studies of mood disorders. Biol Psychiatry 2000;48: 813-829.

Duman RS, Malberg J, Thome J. Neural plasticity to stress and antidepressant treatment. Biol Psychiatry 1999;46: 1181-1191.

Duman RS, Malberg J, Nakagawa S, D'Sa C. Neuronal plasticity and survival in mood disorders. Biol Psychiatry 2000;48: 732-739.

Dwivedi Y, Pandey GN. Effects of subchronic administration of antidepressants and anxiolytics on levels of the alpha subunits of G proteins in the rat brain. J Neural Trans 1997;104: 747-760.

Dwivedi Y, Conley R, Roberts R, Tamminga C, Faludi G, et al. Reduced [^{3}H]cyclic AMP binding sites and PKA activity in the prefrontal cortex of suicide subjects. Abst Soc Neurosci 1999;25: 2097.

Dwivedi Y, Pandey GN. Adrenal glucocorticoids modulate [^{3}H] cyclic AMP binding to protein kinase A (PKA), cyclic AMP-dependent PKA activity, and protein levels of selective regulatory and catalytic subunit isoforms of PKA in rat brain. J Pharmacol Exp Ther 2000;294: 103-116.

Emamghoreishi M, Warsh JJ, Sibony D, Li PP. Lack of effect of chronic antidepressant treatment on Gs and Gi α-subunit protein and mRNA levels in the rat cerebral cortex. Neuropsychopharmacolgy 1996;15: 281-287.

Emamghoreishi M, Schlichter L, Li PP, Parikh S, Sen J, Kamble A, Warsh JJ. High intracellular calcium concentrations in transformed lymphoblasts from subjects with bipolar I disorder. Am J Psychiatry 1997;154: 976-982.

Extein I, Tallman J, Smith CC, Goodwin FK. Changes in lymphocyte beta-adrenergic receptors in depression and mania. Psychiatry Res 1979;1: 191-197.

Fields A, Li PP, Kish SJ, Warsh JJ. Increased cyclic AMP-dependent protein kinase activity in postmortem brain from patients with bipolar affective disorder. J Neurochem 1999;73: 1704-1710.

Friedman E, Wang HY. Receptor-mediated activation of G proteins is increased in postmortem brains of bipolar affective disorder subjects. J Neurochem 1996;67: 1145-1152.

Garcia-Sevilla JA, Escriba PV, Ozaita A, La Harpe R, Walzer C, Eytan A, Guimon J. Up-regulation of immunolabeled α_{2A}-adrenoceptors, G_i coupling proteins, and regulatory receptor kinases in the prefrontal cortex of depressed suicides. J Neurochem 1999;72: 282-291.

Gonzalez GA, Montminy MR. Cyclic AMP stimulates somatostatin gene transcription by phosphorylation of CREB at serine 133. Cell 1989;59: 675-680.

Gonzalez AM, Pascual J, Meana JJ, Barturen F, del Arco C, Pazos A, Garcia-Sevilla JA. Autoradiographic demonstration of increased α_2-adrenoceptor agonist binding sites in the hippocampus and frontal cortex of depressed suicide victims. J Neurochem 1994;63: 256-265.

Greengard P. Phosphorylated proteins as physiological effectors. Science 1978;199: 146-152.

Gu C, Ma YC, Benjamin J, Littman D, Chao MV, Huang XY. Apoptotic signaling through the β-adrenergic receptor. A new G_s effector pathway. J Biol Chem 2000;275: 20726-20733.

Holsboer F. Stress, hypercortisolism and corticosteroid receptors in depression: implications for therapy. J Affect Disord 2001;62: 77-91.

Hudson CJ, Young LT, Li PP, Warsh JJ. CNS signal transduction in the pathophysiology and pharmacotherapy of affective disorders and schizophrenia. Synapse 1993;13: 278-293.

Hulley P, Hartikka J, Lubbert H. Cyclic AMP promotes the survival of dopaminergic neurons in vitro and protects them from the toxic effects of MPP^+. J Neural Trans (Suppl) 1995;46: 217-228.

Jensen JB, Shimon H, Morka A. Abnormal phosphorylation in post-mortem brain tissue from bipolar patients. J Neural Transm 2000;107: 501-509.

Jewell-Motz EA, Liggett SB. G protein-coupled receptor kinase specificity for phosphorylation and desensitization of α_2-adrenergic receptor subtypes. J Biol Chem 1996;271: 18082-18087.

Jones DJ, Stavinoha WB. Microwave inactivation as a tool for studying the neuropharmacology of cyclic nucleotides. In: G. C. Palmer (ed). Neuropharmacology of Cyclic Nucleotides. Urban & Schwarzenberg, Baltimore: 1979; pp 253-281.

Jordan JD, Landau EM, Iyengar R. Signaling networks: the origins of cellular multitasking. Cell 2000;103: 193-200.

Korswagen HC, Park JH, Ohshima Y, Plasterk RH. An activating mutation in a Caenorhabditis elegans G_s protein induces neural degeneration. Genes & Development 1997;11: 1493-503.

Li PP, Young LT, Tam YK, Sibony D, Warsh JJ. Effects of chronic lithium and carbamazepine treatment on G-protein subunit expression in rat cerebral cortex. Biol Psychiatry 1993;34: 162-170.

Li PP, Andreopoulos A, Warsh JJ. Signal transduction abnormalities in bipolar affective disorder. In: M. E. A. Reith (ed). Cerebral Signal Transduction: From First to Fourth Messengers. Humana Press, Totowa: 2000; pp 283-309.

Li X, Greenwood AF, Powers R, Jope RS. Effects of postmortem interval, age, and Alzheimer's disease on G-proteins in human brain. Neurobiol. Aging 1996;17: 115-122.

Lowther S, Crompton MR, Katona CL, Horton RW. GTPγS and forskolin-stimulated adenylyl cyclase activity in post-mortem brain from depressed suicides and controls. Mol Psychiatry 1996;1: 470-477.

Lykouras E, Varsou E, Garelis E, Stefanis CN, Malliaras D. Plasma cyclic cAMP in manic-depressive illness. Acta Psychiat. Scand. 1978;57: 447-453.

Manji HK, Moore GJ, Chen G. Clinical and preclinical evidence for the neurotrophic effects of mood stabilizers: implications for the pathophysiology and treatment of manic-depressive illness. Biol Psychiatry 2000a;48: 740-754.

Manji HK, Moore GJ, Rajkowska G, Chen G. Neuroplasticity and cellular resilience in mood disorders. Mol Psychiatry 2000b;5: 578-593.

Mann JJ. Role of the serotonergic system in the pathogenesis of major depression and suicidal behavior. Neuropsychopharmacology 1999;2 Suppl: 99S - 105S.

Meana JJ, Barturen F, Garcia-Sevilla JA. α-Adrenoceptors in the brain of suicide victims: increased receptor density associated with major depression. Biol Psychiatry 1992;31: 471-490.

Milligan G, Unson CG, Wakelam JO. Cholera toxin treatment produces down-regulation of the α-subunit of the stimulatory guanine-nucleotide-binding protein (Gs). Biochem J 1989;262: 643-649.

Milligan G, Wakelam M. G proteins: Signal Transduction and Disease.; London, Academic Press, 1992.

Moore GJ, Bebchuk JM, Hasanat K, Chen G, Seraji-Bozorgzad N, Wilds IB, Faulk MW, Koch S, Glitz DA, Jolkovsky L, Manji HK. Lithium increases N-acetyl-aspartate in the human brain: in vivo evidence in support of bcl-2's neurotrophic effects? Biol Psychiatry 2000;48: 1-8.

Morris AJ, Malbon CC. Physiological regulation of G protein-linked signaling. Physiol Rev 1999;79: 1373-1430.

Nakao N. An increase in intracellular levels of cyclic AMP produces trophic effects on striatal neurons developing in culture. Neuroscience 1998;82: 1009-1020.

Ongur D, Drevets WC, Price JL. Glial reduction in the subgenual prefrontal cortex in mood disorders. Proc Natl Acad Sci USA 1998;95: 13290-13295.

Ordway GA, Widdowson PS, Smith KS, Halaris A. Agonist binding to α_2-adrenoceptors is elevated in the locus coeruleus from victims of suicide. J Neurochem 1994;63: 617-624.

Otten AD, McKnight GS. Overexpression of the type II regulatory subunit of the cAMP-dependent protein kinase eliminates the type I holoenzyme in mouse cells. J Biol Chem 1989;264: 20255-20260.

Ozawa H, Gsell W, Frolich L, Zochling R, Pantucek F, Beckmann H, Riederer P. Imbalance of the Gs and Gi/o function in post-mortem human brain of depressed patients. J Neural Trans 1993;94: 63-69.

Pacheco MA, Stockmeier C, Meltzer HY, Overholser JC, Dilley GE, Jope RS. Alterations in phosphoinositide signaling and G-protein levels in depressed suicide brain. Brain Res 1996;723: 37-45.

Pandey GN, Dysken MW, Garver DL, Davis JM. Beta-adrenergic receptor function in affective illness. Am J Psychiatry 1979;136: 675-678.

Rahman S, Li PP, Young LT, Kofman O, Kish SJ, Warsh JJ. Reduced [^{3}H]cyclic AMP binding in postmortem brain from subjects with bipolar affective disorder. J Neurochem 1997;68: 297-304.

Rajkowska G. Postmortem studies in mood disorders indicate altered numbers of neurons and glial cells. Biol Psychiatry 2000;48: 766-777.

Ram A, Guedj F, Cravchik A, Weinstein L, Cao Q, Badner JA, Goldin LR, Grisaru N, Manji HK, Belmaker RH, Gershon ES, Gejman PV. No abnormality in the gene for the G protein stimulatory α subunit in patients with bipolar disorder. Arch Gen Psychiatry 1997;54: 44-48.

Reiach JS, Li PP, Warsh JJ, Kish SJ, Young LT. Reduced adenylyl cyclase immunolabeling and activity in postmortem temporal cortex of depressed suicide victims. J Affective Disorders 1999;56: 141-151.

Seifert R, Wenzel-Seifert K, Lee TW, Gether U, Sanders-Bush E, Kobilka BK. Different effects of $G_s\alpha$ splice variants on β_2-adrenoreceptor-mediated signaling. The beta2-adrenoreceptor coupled to the long splice variant of $G_S\alpha$ has properties of a constitutively active receptor. J Biol Chem 1998;273: 5109-16.

Sheline YI. 3-D MRI studies of neuroanatomic changes in unipolar major depression: The role of stress and medical comorbidity. Biol Psychiatry 2000;48: 791-800.

Soares JC, Mann JJ. The anatomy of mood disorders - review of structural neuroimaging studies. Biol Psychiatry 1997;41: 86-106.

Spaulding SW. The ways in which hormones change cyclic adenosine 3',5'-monophosphate-dependent protein kinase subunits, and how such changes affect cell behavior. Endocr Rev 1993;14: 632-50.

Stoll AL, Renshaw PF, Yurgelun-Todd DA, Cohen BM. Neuroimaging in bipolar disorder: what have we learned? Biol Psychiatry 2000;48: 505-517.

Sutherland EW, Rall TW. Fractionation and characterization of a cyclic adenonsine ribonucleotide formed by tissue particles. J Biol Chem 1958;232: 1077-1091.

Toyoshige M, Okuya S, Rebois V. Choleragen catalyzes ADP-ribosylation of the stimulatory G protein heterotrimer but not its free α-subunit. Biochemistry 1994;33: 4865-4871.

Tsai G, Coyle JT. N-acetylaspartate in neuropsychiatric disorders. Prog in Neurobiol 1995;46: 531-540.

Walseth TF, Zhang HJ, Olson LK, Schroeder WA, Robertson RP. Increase in Gs and cyclic AMP generation in HIT cells. Evidence that the 45-kDa α-subunit of Gs has greater functional activity than the 52-kDa α-subunit. J Biol Chem 1989;264: 21106-21111.

Walton MR, Dragunow I. Is CREB a key to neuronal survival? Trends Neurosci 2000;23: 48-53.

Wang JF, Young LT, Li PP, Warsh JJ. Signal transduction abnormalities in bipolar disorder. In: L. T. Young and R. T. Joffe (eds). Bipolar Disorder: Biological Models and Their Clinical Application. Marcel-Dekker Inc., New York; 1997; pp 41-79.

Warsh JJ, Chiu AS, Li PP. Noradrenergic mechanisms in affective disorders: contributions on receptor research. In: A. K. Sen and T. Lee (eds). Receptors and Ligands in Psychiatry and Neurology. Cambridge University Press, Cambridge; 1988; pp 271-302.

Warsh JJ, Li PP. Second messenger systems and mood disorders. Curr Opin Psychiatry 1996;9: 23-29.

Warsh JJ, Li PP. Postmortem Brain Studies in Bipolar Disorder. In: J. C. Soares and S. Gershon (eds). Bipolar Disorders: Basic Mechanisms and Therapeutic Implications. Marcel Dekker, Inc., New York; 2000; pp 201-226.

Warsh JJ, Young LT, Li PP. Guanine nucleotide binding (G) protein disturbances in bipolar affective disorder. In: H. K. Manji, C. Bowden and R. H. Belmaker (eds). Mechanisms of Action of Antibipolar Drugs: Focus on lithium, carbamazepine and valproic acid. American Psychiatric Association Press, Inc., Washington; 2000; pp 299-319.

Winsberg ME, Sachs N, Tate DL, Adalsteinsson E, Spielman D, Ketter TA. Decreased dorsolateral prefrontal N-acetyl aspartate in bipolar disorder. Biol Psychiatry 2000;47: 475-481.

Yamamoto KK, Gonzalez GA, Biggs WH, 3rd, Montminy MR. Phosphorylation-induced binding and transcriptional efficacy of nuclear factor CREB. Nature 1988;334: 494-498.

Yamashita N, Hayashi A, Baba J, Sawa A. Rolipram, a phosphodiesterase-4-selective inhibitor, promotes the survival of cultured rat dopaminergic neurons. Jap J Pharmacol 1997;75: 155-159.

Yan L, Herrmann V, Hofer JK, Insel PA. Beta-adrenergic receptor/cAMP-mediated signaling and apoptosis of S49 lymphoma cells. Am J Physiol - Cell Physiol 2000;279: C1665-1674.

Young LT, Li PP, Kish SJ, Siu KP, Warsh JJ. Postmortem cerebral cortex Gs alpha-subunit levels are elevated in bipolar affective disorder. Brain Res 1991a;553: 323-326.

Young LT, Warsh JJ, Li PP, Siu KP, Becker L, Gilbert J, Hornykiewicz O, Kish SJ. Maturational and aging effects on guanine nucleotide binding protein immunoreactivity in human brain. Brain Res. Dev Brain Res 1991b;61: 243-248.

Young LT, Li PP, Kish SJ, Siu KP, Kamble A, Hornykiewicz O, Warsh JJ. Cerebral cortex $G_s\alpha$ protein levels and forskolin-stimulated cyclic AMP formation are increased in bipolar affective disorder. J Neurochem 1993;6: 890-898.

Young LT, Li PP, Kish SJ, Warsh JJ. Cerebral cortex β-adrenoceptor binding in bipolar affective disorder. J Affect Disord 1994;30: 89-92.

Young LT, Asghari V, Li PP, Kish SJ, Fahnestock M, Warsh JJ. Stimulatory G-protein α-subunit mRNA levels are not increased in autopsied cerebral cortex from patients with bipolar disorder. Brain Res Mol Brain Res 1996;42: 45-50.

Zhu LP, Yu XD, Ling S, Brown RA, Kuo TH. Mitochondrial Ca^{2+} homeostasis in the regulation of apoptotic and necrotic cell deaths. Cell Calcium 2000;28: 107-117.

19 MONOAMINE RECEPTORS IN POSTMORTEM BRAIN: Do Postmortem Brain Studies Cloud or Clarify our Understanding of the Affective Disorders?

Craig A. Stockmeier and George Jurjus

Abstract

Despite considerable research efforts, it has been difficult to reach definitive conclusions in postmortem brain tissue on monoamine receptors and transporters in suicide and affective disorders. Most studies cannot be directly compared for a number of methodological reasons. Critical issues to control in future postmortem brain studies include variables such as the cause of death (e.g. suicide), the specific psychiatric diagnoses of the subjects, long-term medication histories, psychoactive substance use by the subjects, smoking history, the hemisphere from which tissues were dissected, and the specific cytoarchitectonic region to be evaluated. Carefully controlled studies with larger numbers of subjects will ensure a greater likelihood of reaching a consensus on the influence of suicide and psychiatric history on monoaminergic markers in postmortem brain tissue.

Introduction and Methodological Considerations

Over the past two decades, a large number of studies have examined radioligand binding to monoamine receptors in postmortem brain tissue from subjects dying by suicide and/or having a depressive disorder. A major criticism often voiced at these receptor studies is that few observations in one laboratory have been confirmed by other laboratories. Prior to a detailed literature review, it will be useful to briefly consider several factors that may account for much of the purported variance between studies of postmortem brain tissue. It is premature to conclude that the use of postmortem human brain tissue is a flawed approach which inevitably yields conflicting results, prior to evaluating methodological issues across these studies and adopting controlled experimental approaches to dependably address the underlying pathophysiology of depressive disorders. Some of the critical issues to be considered when interpreting the studies of postmortem brain tissue include: the psychiatric status of the subject at the time of death and the underlying psychiatric disorder, whether "control" subjects were psychiatrically normal, the cause of death of the subjects (suicide or by other means), evolving criteria used to establish psychiatric diagnoses, the possible inclusion of subjects with concurrent psychoactive substance use disorders, the regional and hemispheric localization of the brain regions being studies, and the presence and duration of treatment with a psychotropic medication. These confounding variables are detailed in the chapter by Harrison and Everall in this book.

In many of the earlier studies of receptors in postmortem brain tissue, it was hypothesized that suicide *per se* or a depressive disorder would be associated primarily with increases in monoamine receptor densities, which might indicate decreases in functioning of norepinephrine or serotonin neurons consistent with the prevailing monoaminergic theory of depression or suicide. In support of these hypotheses, studies in experimental animals revealed that depletion or lesions of monoaminergic systems in the forebrain led to up-regulation or increases in number of certain monoaminergic receptors. There has been a gradual increase in methodological sophistication over the past twenty years in reports using postmortem brain tissue to study receptors in suicide and depression, yet consistent observations across laboratories have been elusive. This review will examine receptor and transporter studies of the serotonin, norepinephrine and dopamine systems in postmortem brain tissues. The review will follow the evolution of postmortem receptor studies that initially focused on suicide and gradually incorporated depressive disorders as psychiatric assessments were increasingly performed.

Serotonin-1 Receptors

Serotonin-1 receptors have been examined in a number of cerebral cortical and subcortical areas in suicide victims and subjects with a history of an affective disorder. Earlier studies used [^{3}H]serotonin to label all serotonin-1 receptors and many of the subjects had some evidence for depression or a depressive disorder. In suicide victims with or without evidence of depressive illness, no significant changes in serotonin-1 receptors were detected in the frontal convexity (Crow et al., 1984) or frontal cortex (Owen et al., 1986; Mann et al., 1986; McKeith et al., 1987; Cheetham et al., 1990).

The serotonin-1A receptor has been examined in several later studies using the more selective serotonin-1A receptor agonist [^{3}H]8-hydroxy-2-(di-n-propyl)aminotetralin (8-OH-DPAT) to label these receptors. The serotonin-1A receptor is either coupled to a GTP binding protein or free from such coupling. The serotonin-1A receptor agonist, 8-OH-DPAT, binds only to the coupled receptor, while the receptor antagonist WAY-100635 binds to both the coupled and free receptor (Gozlan et al., 1995; Palego et al., 1997).

Studies of serotonin-1A receptors in prefrontal cortex in suicide have yielded a mixture of observations. Matsubara et al. (1991) detected an increase in serotonin-1A receptors in prefrontal cortex (area 8,9) of suicide victims. Psychiatric information was not available on these subjects and the increase in serotonin-1A receptors was detected only in a relatively small number of male suicide victims dying from drug overdose or carbon monoxide (termed nonviolent) versus suicide victims dying from gunshot wounds, hanging, stabbing or falling from height (termed violent). In another study of suicide victims not characterized psychiatrically, Arango et al. (1995) reported increases in radioligand binding to serotonin-1A receptors confined to ventrolateral (areas 45 and 46) but not other areas (e.g. 8, 9, 11, 12, 24, 32) of prefrontal cortex of suicide victims. Four other studies report no significant changes in serotonin-1A receptors in prefrontal cortex, even when suicide victims were classified as either depressed or not depressed (Dillon et al., 1991, areas 8+9; Arranz et al., 1994, areas 9+10+11; Lowther et al., 1997, areas 10, 17/18; Stockmeier et al., 1997, area 10).

Serotonin-1A receptors have also been examined in the hippocampus of suicide victims. Joyce et al. (1993) reported an increase in these sites in the stratum pyramidale (CA1) of the hippocampus of suicide victims. In contrast, other studies, including those where subjects where psychiatrically characterized, did not detect significant changes in agonist binding to the serotonin-1A receptor in hippocampus (Dillon et al., 1991, Lowther et al., 1997; Stockmeier et al., 1997).

Animal studies reveal that serotonin-1A receptors located on serotonin cell bodies in the midbrain play an important role in the release of serotonin in the prefrontal cortex, a region implicated in the pathophysiology of major depression (MD, Rajkowska et al., 1999). Activation of serotonin-1A receptors in the midbrain inhibits the firing of serotonin neurons and diminishes the release of this neurotransmitter in prefrontal cortex (Aghajanian et al., 1987). Chronic treatment with antidepressant medications such as monoamine oxidase inhibitors and selective serotonin reuptake inhibitors (SSRI's) desensitizes serotonin-1A receptors in the midbrain (Blier and deMontigny, 1994). It has been hypothesized that serotonin-1A receptor inhibitory mechanisms in the midbrain are enhanced in MDD and may be targets for desensitization by antidepressant medications.

Diminished serotonergic mechanisms in the midbrain may indeed play a critical role in the pathophysiology of major depressive disorder (MDD) and suicide (Stockmeier et al., 1998). The binding of the agonist [^{3}H]8-OH-DPAT to serotonin-1A receptors is significantly increased in the dorsal and ventrolateral subnuclei of the dorsal raphe nucleus in the midbrain of suicide victims with MDD as compared to normal control subjects. Enhanced radioligand binding of an agonist to inhibitory serotonin-1A autoreceptors in the human dorsal raphe nucleus (DR) provides pharmacological evidence to support the hypothesis of diminished activity of serotonin neurons in suicide victims with MDD. It will be important to examine the contribution of suicide to this effect by studying serotonin-1A receptors in subjects with MDD not dying by suicide.

In addition to postmortem studies, serotonin-1A receptors have been evaluated in depression using positron emission tomography (PET). For a more complete review of PET studies of receptors and depression, readers are referred to Staley et al. (1998). A study by Sargent et al. (2000) examines antagonist binding to serotonin-1A receptors in subjects with MDD, excluding subjects with bipolar disorder. The binding of [^{11}C]4-(2'-methoxy-)-phenyl-1-[2'-(N-2"-pyridal)-p-fluorobenzamido] ethyl-piperazine ([^{11}C]WAY-100635) to serotonin-1A receptors is decreased in 15 unmedicated patients with MDD across several cortical regions, including medial temporal cortex, the temporal pole, orbitofrontal cortex, anterior cingulate cortex, insula and dorsolateral prefrontal cortex. These findings are bilaterally expressed except in the medial temporal cortex, ventral anterior cingulate cortex and the dorsolateral prefrontal cortex. Serotonin-1A receptor binding is similarly decreased in most of these regions in an additional cohort of depressed subjects medicated with paroxetine or sertraline. Interestingly, Sargent et al. (2000) do not show any change, or perhaps a decrease, in serotonin-1A receptors in the ventrolateral prefrontal cortex and raphe in unmedicated subjects, areas where Arango et al. (1995) and Stockmeier et al. (1998), respectively, saw significant increases in [^{3}H]8-OH-DPAT binding to serotonin-1A receptors.

Other evidence for altered serotonin-1A receptors in affective disorders comes from a PET study by Drevets et al. (1999). The binding of the serotonin-1A receptor antagonist WAY-100635 is diminished in the brainstem raphe and mesiotemporal cortex (hippocampus plus amygdala) of bipolar depressives and unipolar depressives with bipolar relatives. In contrast, there was no significant decrease in antagonist binding to serotonin-1A receptors in the raphe in non-medicated subjects with MDD (with or without a family history of bipolar disorder), as confirmed by Sargent et al. (2000).

The apparent discrepancies between the postmortem studies (Arango et al., 1995; Stockmeier et al., 1997; Stockmeier et al., 1998) and the PET studies in living patients could be resolved by one of several issues (see also the chapter by Pilowsky in this book) such as the psychiatric diagnoses of the subjects, the anatomic resolution of the imaging techniques, the choice of an agonist vs. an antagonist radioligand, and that subjects in the postmortem studies were suicide victims. The postmortem studies in cortex and raphe examined only suicide victims with MDD, whereas Drevets et al. (1999) noted significant changes in serotonin-1A receptors limited to subjects with bipolar disorder, and subjects with MDD who had a first degree relative with bipolar disorder. Sargent et al. (2000) excluded subjects with bipolar disorder.

The anatomic resolution of PET versus postmortem autoradiographic studies makes it difficult to directly compare results collected by these two methods. The anatomic location of the brainstem raphe in the PET images by Drevets et al. (1999) was guided by co-registered MRI images in the same subjects, although Sargent et al. (2000) did not appear to use MRI co-registration. The PET technique by Drevets et al. (1999) used a region of interest four times greater than the actual size of the midbrain raphe, and collected three slices of brain tissue at 2.4 mm per slice. The technique also involves blurring of the image over a total of six mm because of individual variation in position and size of the brainstem raphe. Consequently, the PET images of antagonist binding to serotonin-1A receptors very likely include both the dorsal and median raphe nuclei in the midbrain as well as the pons. In contrast, the postmortem study by Stockmeier et al. (1998) examined only the dorsal raphe nucleus within the midbrain as defined by tryptophan hydroxylase immunoreactivity in adjacent sections. In the postmortem study, the significant increase in agonist binding to serotonin-1A receptors was highly localized to only two of the five dorsal raphe subnuclei in the midbrain.

The choice of an agonist versus antagonist radioligand for the serotonin-1A receptors may affect the outcome of a receptor study. The two PET studies used the antagonist [^{11}C]WAY-100635, whereas postmortem studies use the agonist [^{3}H]8-OH-DPAT to measure serotonin-1A receptors. The serotonin-1A receptor exists either coupled to a GTP-binding protein (G-protein) or free from such coupling. The density of antagonist- labeled sites is 60-70% higher than agonist-labeled sites in human brain (Burnet et al., 1997). Radioligand agonists such as 8-OH-DPAT bind only to the coupled receptor, while antagonists such as WAY-100635 bind to both the coupled and free receptor. For example, in the dorsal raphe nucleus, the total number of serotonin-1A receptors (coupled plus free) may be unchanged in MDD, but there could be an increase in functional receptors coupled to G-proteins as measured with an agonist, accompanied by a corresponding decrease in free receptors. In evaluating serotonin-1A receptors in postmortem versus PET studies, it will be critical to use a serotonin-1A receptor antagonist as radioligand in cortical as well as subcortical areas.

Three studies have focused on other serotonin-1 receptor subtypes in affective disorders and suicide. Receptor gene knockout studies in mice suggest that the serotonin-1B receptor may contribute to aggressive/suicidal behavior, and serotonin-1D receptor functions as a terminal autoreceptor in human cortex (Maura et al., 1993; Ramboz et al., 1996). In a large study examining suicide, alcoholism, MDD and pathological aggression, however, there were no significant alterations in serotonin-1B receptor binding or genotype in the prefrontal cortex (area 9) observed in any of these psychiatric illnesses or behaviors (Huang et al., 1999). In prefrontal cortex (areas 9,10,11), a significant decrease in serotonin-1D binding affinity is reported in depressed suicide victims, and a significant increase in receptor number is reported in nondepressed suicide victims (Arranz et al., 1994). In contrast, Lowther et al. (1997) reports a significant increase in serotonin-1D receptors in globus pallidus, but not in the frontal cortex (area 10), and no changes in serotonin-1E/1F receptors. The functional significance of changes in serotonin-1D receptors in affective illness and/or suicide is yet to be determined.

Serotonin-2 Receptors

Serotonin-2A receptors have been examined in postmortem prefrontal cortex in affective disorders and suicide in 17 studies. Increases in serotonin-2A receptors are often cited as evidence to support the hypothesis of alterations in serotonergic neurotransmission in suicide and affective disorders. In six studies of suicide victims with no supporting psychiatric information, there were five reports of an increase in serotonin-2A receptors in prefrontal cortex as compared to age-matched non-suicide subjects (Stanley and Mann, 1983, areas 8+9; Mann et al., 1986, areas 8+9; Arora and Meltzer, 1989, areas 8+9; Arango et al., 1990, area 9; Turecki et al., 1999, areas 8+9). The study by Arango et al. (1990) included evidence from both receptor autoradiographic and homogenate binding studies for an increase in serotonin-2A receptors in area 9 but not area 38 (temporal pole). One report by Gross-Isseroff et al. (1990, areas 8,9) reported a decrease in the number of serotonin-2A receptors in suicide victims under the age of 50.

In contrast, in nine studies of suicide victims with some documentation of depressive or psychotic symptoms, two studies reported an increase in serotonin-2A receptors (Hrdina et al., 1993, area 9, amygdala; Laruelle et al., 1993, area 10), while seven studies detected no change in serotonin-2A receptors (Owen et al., 1983, Owen et al., 1986, areas 8+9; Crow et al., 1984, area 10; Cheetham et al., 1988, area 10; Lowther et al., 1994, area 10; Arranz et al., 1994, areas 9+10+11; Stockmeier et al., 1997, area 10). Lowther et al. (1994) is most notable among the studies reporting no significant change in serotonin-2A receptors in depressed suicide victims since six forebrain areas were examined in over 70 pairs of control subjects and suicide victims. These authors evaluated such variables as psychiatric diagnosis (endogenous depression or depressive syndromes vs. other diagnoses), recent history of antidepressant medication, and method of suicide ('violent' vs. 'nonviolent') and found no relationship between these variables and the number of serotonin-2A receptors.

Two other studies of postmortem tissue examined serotonin-2A receptors in prefrontal cortex of subjects with MDD, most of whom did not die by suicide (McKeith et al., 1987; Yates et al., 1990). McKeith et al. (1987, area 10) reported only a trend for an increase in serotonin-2A receptors in MDD, while Yates et al. (1990, area 9) noted a significant increase in serotonin-2A receptor number in a small number of subjects (n =

4) with MDD. A methodological complication of these two studies involves the use of non-radioactive ketanserin to measure the non-specific binding of [^{3}H]ketanserin. Ketanserin has high affinity for alpha-1 adrenergic receptors (Hoyer et al., 1987), in addition to serotonin-2A receptors, and both serotonin-2A and alpha-1 adrenergic receptors would be measured under these conditions.

Serotonin-2A receptors have also been examined in the hippocampus of suicide victims. An early study by Cheetham et al. (1988) reported a significant decrease in the number of serotonin-2A receptors in hippocampus of suicide victims with depression. However, a later study by this group (Lowther et al., 1994) and by others (Gross-Isseroff et al., 1990; Joyce et al., 1993) failed to replicate the observation of Cheetham et al. (1988). In the study by Stockmeier et al. (1997), there was a trend for an increase in [^{3}H]ketanserin binding in the hippocampus of suicide victims with MDD, however this difference did not reach statistical significance.

In conclusion, well-controlled postmortem studies in prefrontal cortex (area 10) and hippocampus reveal no significant wide-spread differences in serotonin-1A or -2A receptors in suicide victims with MDD as compared to normal control subjects. However, there may be evidence for serotonin-2A receptor changes in areas 8+9 - five studies show an increase in suicide, while three studies show either no change or a decrease in these sites.

There may be several reasons why no clear picture has emerged from these studies. There are several differences between the studies with regard to: 1) the precise depressive disorder experienced by the subjects, 2) the diagnostic criteria used to establish the depressive disorder, 3) the psychiatric status of the subject near the time of death, 4) the exclusion of a psychiatric diagnosis in "control" subjects, 5) the exclusion of subjects with current psychoactive substance use disorders, 6) the precise cytoarchitectonic brain region examined, 7) the hemisphere examined, 8) the presence and duration of treatment with antidepressant drugs of various classes, and 9) the pharmacological methodology to quantitate serotonin-1A and serotonin-2A receptors.

Imaging studies of serotonin-2A receptors have been performed in living subjects with MDD. An early single photon emission computed tomography (SPECT) study suggests a higher uptake bilaterally of radiolabeled ketanserin in the parietal cortex of subjects with MDD (D'haenen et al., 1992). Five positron emission tomography (PET) studies of serotonin-2A receptors have been published. Controlling for age of the subject groups, three PET studies see a significant decrease in binding potential or uptake of radiolabeled setoperone or altanserin in various cortical regions including cingulate, insular and inferior frontal cortex (Biver et al., 1997; Attar-Levy et al., 1999; Yatham et al., 2000). In contrast, two other PET studies using radiolabeled setoperone report no significant difference in serotonin-2A receptor binding potential in MDD (Meyer et al., 1999; Meyer et al., 2001). A complication of interpreting PET, as well as postmortem, studies of serotonin-2A receptors in MDD is the ability of antidepressant drugs such as desipramine, clomipramine or paroxetine to down-regulate serotonin-2A receptor binding (Attar-Levy et al., 1999; Yatham et al., 2000; Meyer et al., 2001).

The postmortem and PET studies of serotonin-2A receptors support at least two conclusions. Several studies suggest an increase in serotonin-2A receptors in prefrontal cortex highly localized to areas 8,9 in suicide victims. In PET studies of MDD, there appears to be a decrease in serotonin-2A receptor binding potential, albeit this site is highly susceptible to down-regulation by treatment with antidepressant drugs.

Other Serotonin Receptors

Serotonin-3 receptors have been described in the CNS and may play a role in anxiety and alcohol use disorders (Bloom and Morales, 1998). One study has examined serotonin-3 receptor binding in suicide victims. Mann et al. (1996) reported no significant change in the number or affinity of serotonin-3 receptors in temporal cortex of suicide victims, as compared to control subjects.

Serotonin Transporter

A role for serotonin in the pathophysiology of depression and suicide has additionally come from studies of the serotonin transporter. Several groups have investigated the serotonin transporter in projection regions of serotonergic cell bodies in postmortem tissue from suicide victims with a depressive disorder. Nearly all of these studies examined regions of cerebral cortex and the results are quite mixed. Early studies primarily focused on suicide victims and used [^{3}H]imipramine, a less than optimum radioligand for measuring the serotonin transporter. Those studies reported increases, decreases or no change in [^{3}H]imipramine binding to frontal cortex in suicide victims (Stanley et al., 1982; Crow et al., 1984; Owen et al., 1986; Arora and Meltzer, 1989; Gross-Isseroff et al., 1989; Lawrence et al., 1998). More recently, other radioligands including [^{3}H]paroxetine, [^{3}H]citalopram or [^{125}I]cyanoimipramine have been identified as superior ligands for measuring the serotonin transporter (Arranz and Marcusson, 1994; Gurevich and Joyce, 1996). Studies with these ligands, mostly in subjects with a depressive mood disorder, reveal either significant decreases (Leake et al., 1991; Laruelle et al., 1993; Joyce et al., 1993; Arango et al., 1995), or no change (Lawrence et al., 1990; Hrdina et al., 1993; Lawrence et al., 1998) in the serotonin transporter in cerebral cortex.

Serotonin transporter binding sites were recently evaluated in several prefrontal cortical regions in a large number of antidepressant medication-free suicide victims and subjects with a lifetime history of a major depressive episode (Mann et al. 2000). Subjects with a lifetime history of either unipolar or bipolar depression had significantly decreased radioligand binding to the serotonin transporter in ventral as well as dorsolateral prefrontal cortex. Confirming an earlier study (Arango et al., 1995), suicide victims had significantly lower binding to the serotonin transporter restricted to ventrolateral prefrontal cortex (Mann et al., 2000). The divergent observation in suicides versus depressives suggests that a more localized decrease in serotonergic function in ventral prefrontal cortex may be critical in the etiology of suicidal behavior, regardless of the complicating comorbid diagnoses.

The serotonin transporter is also expressed in the midbrain dorsal raphe nucleus. The pathophysiology of depression and suicide may involve alterations in the transporter at the serotonergic cell bodies of origin. In suicide victims with MDD and age-matched normal controls, there was no significant difference in the autoradiographic density of [^{3}H]paroxetine binding to the serotonin transporters in either the entire midbrain dorsal raphe nucleus or its subnuclei (Bligh-Glover et al., 2000).

Interestingly, functional studies suggest a desensitization of the serotonin transporter in the midbrain dorsal raphe nucleus after repeated administration of paroxetine (Pineyro et al., 1994). While four of the suicide victims examined by Bligh-Glover et al. (2000) had a history of therapy with antidepressant medications, only one had a prescription for an antidepressant medication in the last month of life.

A recent neuroimaging study describes potential alterations in the serotonin transporter in MDD. Malison et al. (1998) reported a decrease in the availability of the serotonin transporter in the brainstem in antidepressant-free subjects with MDD. The decrease in availability of the serotonin transporter was interpreted as further evidence for a serotonergic alteration in MDD. In the study by Malison et al. (1998), the image resolution does not permit localization of the transporter change to a specific region in the brainstem. One other postmortem study examined serotonin transporters in the midbrain of depressed suicide victims. Using the same radiolabeled cocaine derivative as Malison et al. (1998), Little et al. (1997) reported no significant change in depressed suicides in radioligand binding to the serotonin transporter in the midbrain substantia nigra or ventral tegmental area, and no significant change in mRNA for the serotonin transporter in the midbrain dorsal raphe nucleus or in the median raphe nucleus.

Hence, issues which may clarify the variance in the studies of the serotonin transporter in postmortem tissues from depressed and suicidal subjects include such variables as death by suicide, a homogeneous psychiatric diagnosis in subjects examined, the presence of a psychoactive substance use disorder, the length of time not taking an antidepressant medication, the radioligand selected to study the serotonin transporter, the hemisphere examined, and the specific cytoarchitectonic region of cerebral cortex or brainstem examined.

Noradrenergic Receptors

It has been hypothesized that diminished noradrenergic neuronal activity in human depression may be accompanied by up-regulated postsynaptic alpha- and beta-adrenergic receptors. Studies of postmortem tissue have provided evidence for altered presynaptic noradrenergic function in the locus coeruleus in MDD and suicide (Ordway et al., 1994; Arango et al., 1996; Klimek et al., 1997; Zhu et al., 1999). Additionally, in animal models, diminished function of noradrenergic neurons is associated with supersensitivity of noradrenergic receptors and associated second messengers (reviewed by Zhu et al., 1999).

Alpha-1 adrenergic receptors. A relatively small number of studies have examined [^{3}H]prazosin binding to alpha-1 adrenergic receptors in postmortem brain tissue in suicide and depression. In suicide victims, Gross-Isseroff et al. (1990) detected a significant decrease in alpha-1 adrenergic receptor binding across multiple laminae in several areas of prefrontal, temporal and parietal cortex, but not in the hippocampus or the parahippocampal gyrus. In contrast, Arango et al. (1993) reported that suicide victims had an increase in alpha-1 adrenergic binding localized just to layer IV-V of area 9, but not in the temporal pole (area 38). Psychiatric assessment information was not included for the suicide victims or the control subjects (Gross-Isseroff et al., 1990; Arango et al., 1993). In a larger group of suicide victims with depressive symptoms, De Paermentier et al. (1997) examined alpha-1 and alpha-1A + alpha-1D adrenergic receptors in homogenates of several brain regions. In subjects lacking antidepressant drug treatment for at least three months, there were no significant changes in the number of affinity of these receptors in alpha-1 adrenergic receptors in cerebral cortical (areas 11, 17/18, 21/22, 38), hippocampus or hypothalamus (De Paermentier et al., 1997). An older study reported a decrease in [^{3}H]WB4101 binding in the hippocampus of subjects with unipolar or another depressive illness, although WB4101 has high affinity for a variety of alpha-1 and alpha-2 adrenergic receptor subtypes (Crow et al., 1984).

Despite evidence for altered presynaptic noradrenergic function in the locus coeruleus in MDD and suicide (Ordway et al., 1994; Arango et al., 1996; Klimek et al., 1997; Zhu et al., 1999), there is little evidence that suicide victims, with or without documented evidence of depression, have consistent changes in alpha-1 adrenergic receptor binding in the prefrontal cortex or hippocampus.

Alpha-2 adrenergic receptors. The role of the alpha-2 adrenergic receptor in MDD has been strongly implicated by clinical reports that mirtazapine, a potent antagonist of alpha-2 adrenergic autoreceptors and heteroreceptors, as well as an antagonist of serotonin-2 and serotonin-3 receptors, is an effective antidepressant medication (Kent, 2000). Alpha-2 adrenergic receptors have been evaluated in suicide and depression in a considerable number of studies. At the cell bodies of origin in the locus coeruleus, there is evidence for increased binding of the agonist *p*-[^{125}I]iodoclonidine, but not the antagonist [^{3}H]yohimbine, to alpha-2A adrenergic presynaptic receptors (Ordway et al., 1994). The significance of the up-regulation in agonist binding is that these are the high-affinity state of the alpha -2 adrenergic receptor coupled to the G_i protein. The observation of increased alpha-2 binding in the locus coeruleus has been replicated in a second set of suicide victims with MDD (Ordway et al., 1999), and suggests an adaptive response of inhibitory autoreceptors to a noradrenergic deficit in depression. In the forebrain, there is mixed evidence for changes in alpha-2 adrenergic receptors in depression and suicide, depending on whether radiolabeled receptor antagonists or agonists are used. Comparing subjects with depressive disorders to control subjects, one study using a receptor antagonist reports that depressed subjects had a significant increase in alpha-2 adrenergic receptor binding in temporal cortex (De Paermentier et al., 1997), but no differences are detected using an antagonist in prefrontal or occipital cortex, hippocampus, hypothalamus or striatum (Crow et al., 1984; De Paermentier et al., 1997; Callado et al., 1998). One research group has publishes several reports that depressed subjects reveal an increase in agonist-labeled alpha-2 adrenergic receptors in prefrontal cortex and hippocampus (Meana and Garcia-Sevilla, 1987, area 9; Meana et al., 1992, area 9; Gonzales et al., 1994, areas 8+9, hippocampus; Callado et al., 1998, area 9). A recent elegant study by Garcia-Sevilla et al. (1999) examined levels of alpha-2A adrenergic receptor protein, $G\alpha_{i1/2}$ and G-protein coupled receptor kinase 2/3 in prefrontal cortex of antidepressant-free depressed suicide victims. In this study, significant increases in the immunolabeled receptor and associated proteins were detected in prefrontal cortex (area 9) of antidepressant-free but not antidepressant-treated subjects. It is not clear how the increase in antibody-labeled alpha-2A adrenergic receptors (presumably labeling both high and low affinity sites) in depressed suicides relates to the receptor binding studies by the same group where only agonist (high-affinity) but not antagonist radioligand binding is increased in depressed suicides. In contrast, two other groups have not detected significant changes in agonist labeling of alpha-2 adrenergic receptors in suicide victims (Arango et al., 1993, area 9) or suicide victims retrospectively determined to have MDD (Klimek et al., 1999, area 10, hippocampus).

Several issues may account for the discrepant observations by research groups of alpha-2 adrenergic postsynaptic receptors in cerebral cortex and hippocampus. The Garcia-Sevilla studies includes subjects with MDD, bipolar disorder and dysthymia, while subjects used by Klimek et al. (1999) were only diagnosed with MDD. For Klimek et al. (1999), retrospective psychiatric assessments for both the suicide victims and normal control subjects were made based on interviews of family members and only psychiatric subjects meeting criteria for MDD within the last two weeks of life were included. In the Garcia-Sevilla studies, diagnoses were assessed by chart reviews and it is unclear if depressive episodes were present just prior to death. Other differences between these studies include long-term medication or psychoactive substance histories of the subjects, varied postmortem intervals, and the specific cytoarchitectonic region evaluated. Finally, future postmortem studies of noradrenergic markers will need to match control and depressed subjects for smoking. A history of smoking near the time of death significantly suppresses levels of tyrosine hydroxylase by 50% and agonist binding to alpha-2 adrenergic receptor by 40% in the locus coeruleus (Klimek et al., 1999).

A recent neuroimaging study suggests that noradrenergic dysfunction is present in prefrontal cortex in MDD, despite the lack of consensus in postmortem studies examining alpha-2 adrenergic receptors in prefrontal cortex in depression. Fu et al. (2001) report that subjects with MDD have altered cerebral blood flow in the right superior prefrontal cortex (possibly areas 10/46) when performing a sustained attention task after an injection of clonidine. However, the effect of clonidine in the human prefrontal cortex could be mediated by either alpha-2 adrenergic or non-adrenergic imidazoline binding sites (Piletz et al., 2000a).

Binding sites have been identified in human brain for imidazoline compounds such as clonidine or idazoxan (reviewed by Piletz et al. 2000a,b). A subtype of the imidazoline binding site has been associated with monoamine oxidases A and B. Radioligand binding to the imidazoline-2 binding site and immunoreactivity for the 29-30 kDa imidazoline-2 binding site are decreased in the frontal cortex (area 9) in suicide victims (Garcia-Sevilla et al., 1996; Sastre and Garcia-Sevilla, 1997). Imidazoline receptor proteins are also decreased in the hippocampus of suicide victims with MDD, using the same antisera that Garcia Sevilla et al. (1996) used in cerebral cortex (Piletz et al., 2000b). The precise interactions between imidazoline binding sites and monoaminergic neurons in MDD and suicide remains to be determined.

Beta-adrenergic receptors. Several studies have measured beta-adrenergic receptor binding in a variety of forebrain areas in suicide and depression. Two research groups report increases in beta-adrenergic receptors in suicide. Mann et al. (1986) report a 73% increase in [^{3}H]dihydroalprenolol binding at one concentration in areas 8+9 in suicide victims dying by violent means (hanging, gunshot wound, hanging, fall from height, stabbing. In a follow up study by the same group using [^{125}I]pindolol, an autoradiographic binding study revealed that beta-adrenergic receptors were up-regulated in areas 9 and 38, with the increase in binding observed in area 9 representing an increase in the number (Bmax) of binding sites (Arango et al., 1990). Another group used [^{125} I]pindolol and reported an increase in beta-adrenergic receptor binding in the cingulate and superior frontal gyri in suicide victims, and the increase in binding was attributed to beta-1 adrenergic receptors (Biegon and Israeli, 1988).

In contrast to the two reports of increased beta-adrenergic receptors in suicide victims, four reports find no significant changes in beta-adrenergic receptors in suicide victims. Two early studies using a single concentration of [^{3}H]dihydroalprenolol report no significant effect of suicide on beta-adrenergic receptors in frontal cortex (Meyerson et al., 1982; Crow et al., 1984). Some of the subjects evaluated by Crow et al. (1984) included depressive suicide victims. Stockmeier and Meltzer (1991) evaluated beta-adrenergic receptors in prefrontal cortex (areas 8+9) of 22 pairs of control subjects and age-matched suicide victims. The number of high-affinity, low-capacity, beta-adrenergic receptor binding sites for [^{3}H]dihydroalprenolol was measured using a wide range of radioligand concentrations together with non-linear regression analysis. Under these conditions selective for beta-adrenergic receptors, there was no difference in the number or affinity of high-affinity beta-adrenergic (beta-1 plus beta-2) receptors between control subjects and suicide victims dying by either violent or non-violent means. Finally, a recent study by Klimek et al. (1999) did not detect a significant difference in either beta-1 or beta-2 adrenergic receptors in subfields of the hippocampus between psychiatrically normal control subjects or suicide victims with MDD.

Beta-adrenergic receptors were evaluated in postmortem brain tissue in one study of bipolar disorder. In ten pairs of subjects and controls, there was no significant effect of bipolar disorder on beta-adrenergic receptors in various cerebral cortical regions (Young et al., 1994).

There are three reports of a decrease in beta-adrenergic receptor binding in suicide victims or depression. An early report in subjects with a variety of depressive disorders reported a decrease in beta-adrenergic receptor binding in the hippocampus (Crow et al., 1984). Suicide victims free of antidepressant medications, with a depressive disorder and dying of violent means, had a significant decrease in beta-adrenergic receptors in prefrontal cortex (area 10) but not in hippocampus, as contrasted with age-matched comparison subjects (De Paermentier et al., 1990). Little et al. (1993) evaluated [^{125}I]pindolol binding in the right frontal pole (area 10) of suicide victims, and likewise reported a significant decrease in radioligand binding to beta-adrenergic receptors in suicide victims with mixed psychiatric disorders, as compared to comparison subjects.

In summary, there is little consistent evidence for significant changes in beta-adrenergic receptors in cortex or hippocampus in either suicide or in MDD. Earlier studies lacked homogeneous (or any) psychiatric grouping, and used less than optimal pharmacological definition of the beta-adrenergic receptor. While two groups examining suicide used adequate receptor autoradiographic techniques and verified autoradiographic data with homogenate studies (Biegon and Israeli, 1988; Arango et al., 1990), other groups examining psychiatrically heterogenous suicide victims have not replicated those results, and suicides with MDD may actually have a decrease in beta-adrenergic receptors in prefrontal cortex. Clearly, the number of studies on beta-adrenergic receptors in psychiatrically-characterized subjects has lagged significantly behind studies of alpha-2 adrenergic and serotonergic receptors.

Dopamine Receptors and Transporter

It has been hypothesized that dopamine function may be diminished in MDD. In several studies, dopamine function in MDD has been linked to psychomotor retardation (Van Praag and Korf, 1971; Willner, 1995). A recent well-controlled study by Martinot et al. (2001) demonstrated that the uptake of a dopamine precursor is diminished in the left caudate nucleus of depressed patients with psychomotor retardation, as opposed to depressed patients with high impulsivity or normal control subjects.

Three studies have examined dopamine receptors in suicide victims. There is no significant effect of MDD on ligand binding to dopamine-1 or dopamine-2/3 receptor, and no effect of suicide on mRNA for these receptors in the caudate, putamen or nucleus accumbens (Sumiyoshi et al., 1995; Bowden et al., 1997; Hurd et al., 1997). A neuroimaging study confirmed the observation of no significant difference in dopamine-2 receptors in the caudate of depressed patients, as compared to normal control subjects (Parsey et al., 2001). In addition, Parsey et al. (2001) report that amphetamine-induced decreases in receptor availability are not changed in depression.

The role of dopamine in depression has been examined in studies of the dopamine transporter in limbic and extrapyramidal motor regions of the CNS. Bowden et al. (1997) report no significant change in the dopamine transporter in the caudate, putamen or nucleus accumbens of non-medicated suicide victims with a diagnosis of a depressive disorder, as compared with gender-matched non-suicide control subjects. In contrast, radioligand binding to the dopamine transporter is significantly decreased in the central, accessory basal and basal nuclei of the amygdala in subjects with MDD, as compared to psychiatrically normal control subjects (Klimek et al., 2000). These data support the hypothesis that alterations in the mesolimbic dopamine system projecting to the amygdala, an important center in integrating emotions and stress, may play a role in the etiology of MDD.

Conclusions

The major question to be addressed in studies of postmortem brain tissue is determining the source of biological variability detected across studies of subject groups. The source of those variations may be due, either alone or in combination, to suicide, heterogeneity in grouping of depressive disorders, psychoactive substances of abuse, history of psychotropic medications, and experimental methodology in the laboratory. Studies on monoamine receptors in postmortem brain tissue in the 1980's and early 1990's typically involved tissues collected at a medical examiner's or coroner's office from subjects dying by suicide. Suicide victims were usually matched (by age and gender) with control subjects who had not died by suicide. It was generally hypothesized that suicide would be associated with increases in monoamine receptor densities. Consistent with the prevailing monoaminergic theory of depression, increases in receptor densities might reflect diminished functioning of norepinephrine or serotonin neurons. When available, the psychiatric and medication histories of the suicide victims were derived from medical records and police reports, including comments from the next-of-kin regarding the mental health status of the deceased. However, there remains little consensus between laboratories on the relationship between suicide or depressive disorders and a variety of monoaminergic markers. It would seem that the biological variability across studies is a natural effect of the great diversity of the groups of subjects in these studies. Variability in biological results has been experimentally compounded by methodological variations in sampling brain regions and assay parameters (e.g. different radioligands).

A variety of methods have been used in assessing the psychiatric history of suicide victims or subjects dying by natural causes. The presence of a psychiatric history has been deduced from on records at the coroner's office, hospital or doctors' records, comments from family members to police, and interviews of family members or others with close contact to the deceased. The presence or absence of psychiatric information in a coroner's report alone is not a particularly reliable means of assigning or discounting a psychiatric diagnosis. A suicide victim or an apparent control subject may not have sought appropriate treatment for a psychiatric illness. The absence of a documented diagnosis for depression or a psychoactive substance use disorder in coroner's files or medical records should not be over-interpreted as the absence of that disorder - the disorder may actually have existed and not been appropriately evaluated. A specific psychiatric diagnosis may be noted in medical records, but the reliability of that diagnosis may be low if the accompanying target symptoms are not included. Where psychiatric diagnoses are incorporated into postmortem brain studies, depressed subjects have usually been clustered into a depressive category (e.g. MDD, bipolar disorder, dysthymia, adjustment disorder). However, the presence of unique pharmacotherapies for and microscopic neuropathology of these disorders argues against such clustering in postmortem brain research.

A significant advance in postmortem brain studies of affective disorders has been the increasing use of informant-based retrospective psychiatric assessments in evaluating the mental health status of suicide victims and potentially normal control subjects. Family members or other informants who had frequent contact with the deceased are interviewed about the deceased using the same types of validated structured questionnaires used in the clinical setting (Stockmeier et al., 1997; Mann et al., 2000). An advantage of the informant method is having insight into the mental health status of the subject in the days immediately prior to their death. In a major contribution to postmortem brain studies, Kelly and Mann (1996) demonstrated excellent agreement between informant-based retrospective psychological assessments of deceased subjects and diagnoses by clinicians treating the same subjects before their deaths.

A major consideration in evaluating postmortem brain research in suicide and depressive disorders is the relative and unique contribution of suicide as opposed to depressive disorders. Suicide victims include a heterogeneous population of subjects suffering from a variety of single and comorbid psychiatric disorders including MDD, bipolar disorder, adjustment disorders, dysthymia, schizophrenia, psychoactive substance abuse or dependence, and personality disorders (Barraclough et al. 1974; Rich et al., 1986; Cheng, 1995). There is evidence to suggest that suicide and suicidal behavior, per se, involves specific neurochemical abnormalities that may only partially overlap with depressive disorders (Oquendo and Mann, 2000; Mann et al., 2000). Difficulties arise in interpreting postmortem studies of suicide and affective disorders because most studies are either among suicide victims (some with MDD) or among subjects with depression (some being suicides). The recent study by Mann et al. (2000) is the first to report that subjects with a depressive disorder appeared to have changes in the serotonin transporter which were regionally different from those detected among suicide victims. However, there are practical complicating issues in such studies since depression and suicide are often comorbid with psychoactive substance use disorders, which have their own unique effects on monoaminergic markers.

Several methodological issues must be considered in designing and interpreting postmortem studies of human brain tissue in suicide and depression. For example, several monoaminergic receptors and/or transporters can be affected by such issues as subject age, postmortem interval, antidepressant therapy, psychoactive substance use history, smoking history, and the hemisphere and precise regions sampled and examined. Recent studies have revealed the importance of regional and highly specific changes in neuron and glial density and size in MDD, bipolar disorder, schizophrenia and alcohol dependence (Ongur et al., 1998; Rajkowska et al., 1999; Rajkowska 2000; Rajkowska et al., 2001; Rajkowska, Miguel-Hidalgo and Stockmeier, unpublished observation).

Unique pathological changes in specifically defined regions argue against combining depressive disorders into a single group in studies of postmortem brain. In addition, it is likely that microanatomic inter-individual variability in cortical regions is a root cause of numerous examples of "failure to replicate" in postmortem brain research. Identifying cortical regions only by gross anatomic gyri and a simple Brodmann cytoarchitectonic map does not do justice to the significant anatomic variability in localization of comparable cortical regions across subjects (Rajkowska et al., 1995; Uylings et al., 2000).

Studies on the serotonin transporter and serotonin-2A receptor suggest that gene polymorphisms may or may not contribute to the expression levels of certain gene products and be relevant to suicide (Turecki et al., 1999; Bondy et al., 2000; Mann et al., 2000). However, the relative contribution of a particular gene polymorphism to a given receptor or transporter level may be small, in relation to other contributory factors such as environment, development, stress, or the exposure to psychotropic medications or drugs of abuse.

In summary, it is not possible to reach definitive conclusions on much of the literature on suicide and depression in postmortem brain tissue. Many studies cannot be directly compared since very different (or no) psychiatric assessment techniques were used. Different hemispheres were often examined, and gross anatomic rather than microscopic criteria were used to identify supposedly equivalent regions in subjects. In future studies, controlling for such variables as the cause of death (e.g. suicide), the specific psychiatric diagnoses of the subjects, long-term medication or psychoactive substance histories of the subjects, smoking history, postmortem interval and other markers of tissue integrity, the hemisphere from which tissues were dissected, and the specific cytoarchitectonic region to be evaluated will ensure a greater ability to evaluate the influence of suicide and psychiatric history on monoaminergic markers in postmortem brain tissue.

Acknowledgements

The authors would like to gratefully acknowledge the priceless contributions made by families consenting to donate brain tissue and be interviewed. In addition, the authors recognize the support of The National Institute of Mental Health (MH45488), The American Foundation for Suicide Prevention, Elizabeth K. Balraj, M.D., The Cuyahoga County Coroner, and the staff of The Cuyahoga County Coroner's Office, Cleveland, Ohio. The authors are also grateful to Drs. Gregory Ordway, Violetta Klimek and Grazyna Rajkowska for helpful discussions, and to Cynthia Cavett and M. Dolores Saenz for their editorial assistance.

References

Aghajanian GK, Sprouse JS, Rasmussen K. Physiology of the midbrain 5HT system. In: Meltzer HY (editor). Psychopharmacology: The Third Generation of Progress. Raven Press, New York, 1987; pp. 141-148.

Arango V, Ernsberger P, Marzuk PM, Chen JS, Tierney H, Stanley M, Reis DJ, Mann JJ. Autoradiographic demonstration of increased serotonin 5-HT2 and beta-adrenergic receptor binding sites in the brain of suicide victims. Arch Gen Psychiatry 1990;47: 1038-1047.

Arango V, Underwood MD, Mann JJ. Alterations in monoamine receptors in the brain of suicide victims. J Clin Psychopharmacol 1992;12: 8S-12S.

Arango V, Ernsberger P, Sved AF, Mann JJ. Quantitative autoradiography of alpha 1- and alpha 2-adrenergic receptors in the cerebral cortex of controls and suicide victims. Brain Res 1993;630: 271-282.

Arango V, Underwood MD, Gubbi AV, Mann JJ. Localized alterations in pre- and postsynaptic serotonin binding sites in the ventrolateral prefrontal cortex of suicide victims. Brain Res 1995; 688: 121-133.

Arango V, Underwood MD, Mann JJ. Fewer pigmented locus coeruleus neurons in suicide victims: preliminary results. Biol Psychiatry 1996; 39:112-120.

Arora RC, Meltzer HY. 3H-imipramine binding in the frontal cortex of suicides. Psychiatry Res 1989;30: 125-135.

Arora RC, Meltzer HY. Serotonergic measures in the brains of suicide victims. 5-HT2 binding sites in the frontal cortex of suicide victims and control subjects. Am J Psychiatry 1989;146: 730-736.

Arranz B, Marcusson J. [3H]paroxetine and [3H]citalopram as markers of the human brain 5-HT uptake site: a comparison study. J Neural Transm Gen Sect 1994;97:27-40.

Arranz B, Eriksson A, Mellerup E, Plenge P, Marcusson J. Brain 5-HT1A, 5-HT1D, and 5-HT2 receptors in suicide victims. Biol Psychiatry 1994;35: 457-463.

Attar-Levy D, Martinot JL, Blin J, Dao-Castellana MH, Crouzel C, Mazoyer B, Poirier MF, Bourdel MC, Aymard N, Syrota A, Feline A. The cortical serotonin2 receptors studied with positron-emission tomography and [18F]-setoperone during depressive illness and antidepressant treatment with clomipramine. Biol Psychiatry 1999;45: 180-186.

Barraclough B, Bunch J, Nelson B, Sainsbury P. A hundred cases of suicide: clinical aspects. Br J Psychiatry 1974;125:355-373.

Biegon A, Israeli M. Regionally selective increases in beta-adrenergic receptor density in the brains of suicide victims. Brain Res 1988;442: 199-203.

Biver F, Wikler D, Lotstra F, Damhaut P, Goldman S, Mendlewicz J. Serotonin 5-HT2 receptor imaging in major depression: focal changes in orbito-insular cortex. Br J Psychiatry 1997;171: 444-448.

Blier P, de Montigny C. Current advances and trends in the treatment of depression. Trends in Pharmacological Science 1994;15: 220-226.

Bligh-Glover W, Kolli T, Balraj E, Friedman L, Shapiro L, Dilley G, Stockmeier C. Subregional distribution of serotonin transporters in the midbrain of suicide victims with major depression. Biol Psychiatry 2000;47: 1015-1024.

Bondy B, Kuznik J, Baghai T, Schule C, Zwanzger P, Minov C, de Jonge S, Rupprecht R, Meyer H, Engel RR, Eisenmenger W, Ackenheil M. Lack of association of serotonin-2A receptor gene polymorphism (T102C) with suicidal ideation and suicide. Am J Med Genet 2000;96:831-835.

Bowden C, Cheetham SC, Lowther S, Katona CL, Crompton MR, Horton RW. Dopamine uptake sites, labelled with [3H]GBR12935, in brain samples from depressed suicides and controls. Eur Neuropsychopharmacol 1997;7: 247-252.

Bowden C, Theodorou AE, Cheetham SC, Lowther S, Katona CL, Crompton MR, Horton RW. Dopamine D1 and D2 receptor binding sites in brain samples from depressed suicides and controls. Brain Res 1997;752: 227-233.

Burnet PW, Eastwood SL, Harrison PJ. [3H]WAY-100635 for 5-HT1A receptor autoradiography in human brain: a comparison with [3H]8-OH-DPAT and demonstration of increased binding in the frontal cortex in schizophrenia. Neurochem Int 1997;30: 565-574.

Callado LF, Meana JJ, Grijalba B, Pazos A, Sastre M, Garcia-Sevilla JA. Selective increase of alpha2A-adrenoceptor agonist binding sites in brains of depressed suicide victims. J Neurochem 1998;70: 1114-1123.

Cheetham SC, Crompton MR, Katona CL, Horton RW. Brain 5-HT2 receptor binding sites in depressed suicide victims. Brain Res 1988;443: 272-280.

Cheetham SC, Crompton MR, Katona CL, Horton RW. Brain 5-HT1 binding sites in depressed suicides. Psychopharmacology (Berl) 1990;102: 544-548.

Cheng AT. Mental illness and suicide. A case-control study in east Taiwan. Arch Gen Psychiatry 1995;52:594-603.

Crow TJ, Cross AJ, Cooper SJ, Deakin JFW, Ferrier IM, Johnson JA, Joseph MH, Owen F, Poulter M, Lofthouse R, Corsellis JAN, Chambers DR, Blessed G, Perry EK, Perry RH, Tomlinson BE. Neurotransmitter receptors and monoamine metabolites in brain of patients with Alzheimer-type dementia and depression, and suicides. Neuropharmacology 1984;23: 1561-1569.

De Paermentier F, Mauger JM, Lowther S, Crompton MR, Katona CL, Horton RW. Brain alpha-adrenoceptors in depressed suicides. Brain Res 1997;757: 60-68.

De Paermentier F, Cheetham SC, Crompton MR, Katona CL, Horton RW. Brain beta-adrenoceptor binding sites in antidepressant-free depressed suicide victims. Brain Res 1990;525: 71-77.

D'haenen H, Bossuyt A, Mertens J, Bossuyt-Piron C, Gijsemans M, Kaufman L. SPECT imaging of serotonin2 receptors in depression. Psychiatry Res 1992;45: 227-237.

Dillon KA, Gross-Isseroff R, Israeli M, Biegon A. Autoradiographic analysis of serotonin 5-HT1A receptor binding in the human brain postmortem: effects of age and alcohol. Brain Res 1991;554: 56-64.

Drevets WC, Frank E, Price JC, Kupfer DJ, Holt D, Greer PJ, Huang Y, Gautier C, Mathis C. PET imaging of serotonin 1A receptor binding in depression. Biol Psychiatry 1999;46: 1375-1387.

Fu CH, Reed LJ, Meyer JH, Kennedy S, Houle S, Eisfeld BS, Brown GM. Noradrenergic dysfunction in the prefrontal cortex in depression. Biol Psychiatry 2001;49: 317-325.

Garcia-Sevilla JA, Escriba PV, Sastre M, Walzer C, Busquets X, Jaquet G, Reis DJ, Guimon J. Immunodetection and quantitation of imidazoline receptor proteins in platelets of patients with major depression and in brains of suicide victims. Arch Gen Psychiatry 1996;53: 803-810.

Garcia-Sevilla JA, Escriba PV, Ozaita A, La Harpe R, Walzer C, Eytan A, Guimon J. Up-regulation of immunolabeled alpha2A-adrenoceptors, Gi coupling proteins, and regulatory receptor kinases in the prefrontal cortex of depressed suicides. J Neurochem 1999;72: 282-291.

Gonzalez AM, Pascual J, Meana JJ, Barturen F, del Arco C, Pazos A, Garcia-Sevilla JA. Autoradiographic demonstration of increased alpha 2-adrenoceptor agonist binding sites in the hippocampus and frontal cortex of depressed suicide victims. J Neurochem 1994;63: 256-265.

Gozlan H, Thibault S, Laporte AM, Lima L, Hamon M. The selective 5-HT1A antagonist radioligand [3H]WAY 100635 labels both G-protein-coupled and free 5-HT1A receptors in rat brain membranes. Eur J Pharmacol 1995;288: 173-186.

Gross-Isseroff R, Israeli M, Biegon A. Autoradiographic analysis of tritiated imipramine binding in the human brain postmortem: effects of suicide. Arch Gen Psychiatry 1989;46: 237-241.

Gross-Isseroff R, Dillon KA, Fieldust SJ, Biegon A. Autoradiographic analysis of alpha 1-noradrenergic receptors in the human brain postmortem. Effect of suicide. Arch Gen Psychiatry 1990;47: 1049-1053.

Gross-Isseroff R, Salama D, Israeli M, Biegon A. Autoradiographic analysis of [3H]ketanserin binding in the human brain postmortem: effect of suicide. Brain Res 1990;507: 208-215.
Gurevich EV, Joyce JN. Comparison of [3H]paroxetine and [3H]cyanoimipramine for quantitative measurement of serotonin transporter sites in human brain. Neuropsychopharmacology 1996;14:309-323.
Hoyer D, Vos P, Closse A, Pazos A, Palacios JM, Davies H. [3H]ketanserin labels 5-HT2 receptors and alpha 1-adrenoceptors in human and pig brain membranes. Naunyn Schmiedebergs Arch Pharmacol 1987;335: 226-230.
Hrdina PD, Demeter E, Vu TB, Sùtùnyi P, Palkovits M. 5-HT uptake sites and 5-HT receptors in brain of antidepressant-free suicide victims/depressives: increase in 5-HT2 sites in cortex and amygdala. Brain Res 1993;614: 37-44.
Huang YY, Grailhe R, Arango V, Hen R, Mann JJ. Relationship of psychopathology to the human serotonin1B genotype and receptor binding kinetics in postmortem brain tissue. Neuropsychopharmacology 1999;21: 238-246.
Hurd YL, Herman MM, Hyde TM, Bigelow LB, Weinberger DR, Kleinman JE. Prodynorphin mRNA expression is increased in the patch vs matrix compartment of the caudate nucleus in suicide subjects. Mol Psychiatry 1997;2: 495-500.
Joyce JN, Shane A, Lexow N, Winokur A, Casanova MF, Kleinman JE. Serotonin uptake sites and serotonin receptors are altered in the limbic system of schizophrenics. Neuropsychopharmacology 1993;8: 315-336.
Kelly TM, Mann JJ. Validity of DSM-III-R diagnosis by psychological autopsy: a comparison with clinician ante-mortem diagnosis. Acta Psychiatr Scand 1996;94: 337-343.
Kent JM. SNaRIs, NaSSAs, and NaRIs: new agents for the treatment of depression. Lancet 2000;355(9207):911-918.
Klimek V, Stockmeier C, Overholser J, Meltzer HY, Kalka S, Dilley G, Ordway GA. Reduced levels of norepinephrine transporters in the locus coeruleus in major depression. J Neuroscience 1997;17: 8451-8458.
Klimek V, Rajkowska G, Luker SN, Dilley GE, Meltzer HY, Overholser JC, Stockmeier CA, Ordway GA. Brain noradrenergic receptors in major depression and schizophrenia. Neuropsychopharmacology 1999;21: 69-81.
Klimek V, Zhu M-Y, Dilley G, Overholser JC, Meltzer HY, Stockmeier CA, Ordway GA. Neurochemical evidence for a link between cigarette smoking and depression. Soc. Neuroscience Abstract 1999;25: 2139.
Klimek V, Schenck JE, Dilley G, Overholser JC, Meltzer HY, Stockmeier CA, Ordway GA. Low dopamine transporter in the amygdala in major depression. Soc. Neuroscience Abstract 2000;26: 1764.
Laruelle M, Abi-Dargham A, Casanova MF, Toti R, Weinberger DR, Kleinman JE. Selective abnormalities of prefrontal serotonergic receptors in schizophrenia. Arch Gen Psychiatry 1993;50: 810-818.
Laruelle M, Abi-Dargham A, Casanova MF, Toti R, Weinberger DR, Kleinman JE. Selective abnormalities of prefrontal serotonergic receptors in schizophrenia. A postmortem study. Arch Gen Psychiatry 1993;50: 810-818.
Lawrence KM, De Paermentier F, Cheetham SC, Crompton MR, Katona CL, Horton RW. Brain 5-HT uptake sites, labelled with [3H]paroxetine, in antidepressant-free depressed suicides. Brain Res 1990;526: 17-22.
Lawrence KM, Kanagasundaram M, Lowther S, Katona CL, Crompton MR, Horton RW. [3H] imipramine binding in brain samples from depressed suicides and controls: 5-HT uptake sites compared with sites defined by desmethylimipramine. J Affect Disord 1998;47: 105-112.
Leake A, Fairbairn AF, McKeith IG, Ferrier IN. Studies on the serotonin uptake binding site in major depressive disorder and control postmortem brain: neurochemical and clinical correlates. Psychiatry Res 1991;39: 155-165.
Little KY, Clark TB, Ranc J, Duncan GE. Beta-adrenergic receptor binding in frontal cortex from suicide victims. Biol Psychiatry 1993;39: 596-605.

Little KY, McLauglin DP, Ranc J, Gilmore J, Lopez JF, Watson SJ, Carroll FI, Butts JD. Serotonin transporter binding sites and mRNA levels in depressed persons committing suicide. Biol Psychiatry 1997;41: 1156-1164.

Lowther S, De Paermentier F, Cheetham SC, Crompton MR, Katona CL, Horton RW. 5-HT1A receptor binding sites in postmortem brain samples from depressed suicides and controls. J Affect Disord 1997;42: 199-207.

Lowther S, De Paermentier F, Crompton MR, Katona CL, Horton RW. Brain 5-HT2 receptors in suicide victims: violence of death, depression and effects of antidepressant treatment. Brain Res 1994;642: 281-289.

Lowther S, Katona CL, Crompton MR, Horton RW. 5-HT1D and 5-HT1E/1F binding sites in depressed suicides: increased 5-HT1D binding in globus pallidus but not cortex. Mol Psychiatry 1997;2: 314-321.

Malison RT, Price LH, Berman R, van Dyke CH, Pelton G, Carpenter L, Sanacora G, Owens MJ, Nemeroff CB, Rajeevan N, Baldwin RM, Seibyl JP, Innis RB, Charney DS. Reduced brain serotonin transporter availability in major depression as measured by [^{123}I]-2β-carbomethoxy-3β-(4-iodophenyl)tropane and Single Photon Emission Computed Tomography. Biol Psychiatry 1998;44: 1090-1098.

Mann JJ, Stanley M, McBride PA, McEwen BS. Increased serotonin2 and beta-adrenergic receptor binding in the frontal cortices of suicide victims. Arch Gen Psychiatry 1986;43: 954-959.

Mann JJ, Henteleff RA, Lagattuta TF, Perper JA, Li S, Arango V. Lower ^{3}H-paroxetine binding in cerebral cortex of suicide victims is partly due to fewer high affinity, non-transporter sites. J Neural Transm 1996;103: 1337-1350.

Mann JJ, Arango V, Henteleff RA, Lagattuta TF, Wong DT. Serotonin 5-HT3 receptor binding kinetics in the cortex of suicide victims are normal. J Neural Transm 1996;03: 165-171.

Mann JJ, Huang YY, Underwood MD, Kassir SA, Oppenheim S, Kelly TM, Dwork AJ, Arango V. A serotonin transporter gene promoter polymorphism (5-HTTLPR) and prefrontal cortical binding in major depression and suicide. Arch Gen Psychiatry 2000;57: 729-738.

Martinot M, Bragulat V, Artiges E, Dolle F, Hinnen F, Jouvent R, Martinot J. Decreased presynaptic dopamine function in the left caudate of depressed patients with affective flattening and psychomotor retardation. Am J Psychiatry 2001;158:314-316.

Matsubara S, Arora RC, Meltzer HY. Serotonergic measures in suicide brain: 5-HT-$_{1A}$ binding sites in frontal cortex of suicide victims. J Neural Transm 1991;85: 181-194

Maura G, Thellung S, Andrioli GC, Ruelle A, Raiteri M. Release-regulating serotonin 5-HT1D autoreceptors in human cerebral cortex. J Neurochem 1993;60: 1179-1182.

McKeith IG, Marshall EF, Ferrier IN, Armstrong MM, Kennedy WN, Perry RH, Perry EK, Eccleston D. 5-HT receptor binding in postmortem brain from patients with affective disorder. J Affect Disord 1987;13: 67-74.

Meana JJ, Barturen F, Garcia-Sevilla JA. Alpha 2-adrenoceptors in the brain of suicide victims: increased receptor density associated with major depression. Biol Psychiatry 1992;31: 471-490.

Meana JJ, Garcia-Sevilla JA. Increased alpha 2-adrenoceptor density in the frontal cortex of depressed suicide victims. J Neural Transm. 1987;70: 377-381.

Meyer JH, Kapur S, Eisfeld B, Brown GM, Houle S, DaSilva J, Wilson AA, Rafi-Tari S, Mayberg HS, Kennedy SH. The effect of paroxetine on 5-HT(2A) receptors in depression: An [18F]setoperone PET imaging study. Am J Psychiatry 2001;158: 78-85.

Meyer JH, Kapur S, Houle S, DaSilva J, Owczarek B, Brown GM, Wilson AA, Kennedy SH. Prefrontal cortex 5-HT2 receptors in depression: an [18F]setoperone PET imaging study. Am J Psychiatry 1999;156: 1029-1034.

Meyerson LR, Wennogle LP, Abel MS, Coupet J, Lippa AS, Rauh CE, Beer B. Human brain receptor alterations in suicide victims. Pharmacol Biochem Behav 1982;17: 159-163.

Bloom FE, Morales M. The central 5-HT3 receptor in CNS disorders. *Neurochem Res* 1998;23: 653-659.

Ongur D, Drevets WC, Price JL. Glial reduction in the subgenual prefrontal cortex in mood disorders. Proc Natl Acad Sci U S A 1998; 95:13290-13295.

Oquendo MA, Mann JJ. The biology of impulsivity and suicidality. Psychiatr Clin North Am 2000;2311-2325.

Ordway GA, Streator-Smith K, Haycock JW. Elevated tyrosine hydroxylase in the locus coeruleus of suicide victims. J Neurochem 1994;62: 680-685.

Ordway GA, Widdowson PS, Smith KS, Halaris A. Agonist binding to alpha 2-adrenoceptors is elevated in the locus coeruleus from victims of suicide. J Neurochem 1994;63: 617-624.

Ordway GA, Farley JT, Dilley GE, Overholser JC, Meltzer HY, Balraj EK, Stockmeier CA, Klimek V. Quantitative distribution of monoamine oxidase A in brainstem monoamine nuclei is normal in major depression. Brain Res 1999;847: 71-79.

Ordway GA, Schenck JE, Dilley GE, Overholser JC, Meltzer HY, Stockmeier CA, Halaris AE, Klimek V. Increased p-[125I]iodoclonidine binding to alpha-2 adrenoceptors in the locus coeruleus in major depression. Soc Neuroscience Abstract 1999;25: 2139.

Owen F, Chambers DR, Cooper SJ, Crow TJ, Johnson JA, Lofthouse R, Poulter M. Serotonergic mechanisms in brains of suicide victims. Brain Res 1986;362: 185-188.

Owen F, Cross AJ, Crow TJ, Deakin JFW, Ferrier IM, Lofthouse R, Poulter M. Brain 5-HT-2 receptors and suicide. Lancet 1983;2: 1256

Palego L, Marazziti D, Rotondo A, Batistini A, Lucacchini A, Naccarato AG, Bevilacqua G, Borsini F, Ladinsky H, Cassano GB. Further characterization of [3H]8-hydroxy-2-(di-n-propyl)aminotetralin binding sites in human brain postmortem. Neurochem Int 1997;30: 149-157.

Parsey RV, Oquendo MA, Zea-Ponce Y, Rodenhiser-Hill J, Kegeles LS, Pratap M, Cooper TB, Van Heertum RL, Mann JJ, Laruelle M. Dopamine D2 receptor availability and amphetamine-induced dopamine release in unipolar depression. Biol Psychiatry 2001; in press.

Piletz JE, Ordway GA, Zhu H, Duncan BJ, Halaris A. Autoradiographic Comparison of [^{3}H]-clonidine binding to non-adrenergic sites and alpha$_2$-adrenergic receptors in human brain. Neuropsychopharmacology 2000b;23: 697-708.

Piletz JE, Zhu H, Ordway G, Stockmeier C, Dilly G, Reis D, Halaris A. Imidazoline receptor proteins are decreased in the hippocampus of individuals with major depression. Biol Psychiatry 2000a;48: 910-919.

Pineyro G, Blier P, Dennis T, de Montigny C. Desensitization of the neuronal 5-HT carrier following its long-term blockade. J Neurosci 1994;14: 3036-3047.

Rajkowska G, Goldman-Rakic PS. Cytoarchitectonic definition of prefrontal areas in the normal human cortex: I. Remapping of areas 9 and 46 using quantitative criteria. Cereb Cortex 1995;5:307-322.

Rajkowska G, Miguel-Hidalgo JJ, Wei J, Dilley G, Pittman SD, Meltzer HY, Overholser JC, Roth BL, Stockmeier CA. Morphometric evidence for neuronal and glial prefrontal cell pathology in major depression. Biol Psychiatry 1999 45: 1085-1098.

Rajkowska G. Histopathology of the prefrontal cortex in major depression: what does it tell us about dysfunctional monoaminergic circuits? Prog Brain Res 2000;126: 397-412.

Rajkowska G, Halaris A, Selemon LD. Reductions in neuronal and glial density characterize the dorsolateral prefrontal cortex in bipolar disorder. Biol Psychiatry 2001; in press.

Ramboz S, Saudou F, Amara DA, Belzung C, Segu L, Misslin R, Buhot MC, Hen R. 5-HT1B receptor knock out--behavioral consequences. Behav Brain Res 1996; 73: 305-312.

Rich CL, Young D, Fowler RC. San Diego suicide study. I. Young vs old subjects. Arch Gen Psychiatry 1986;43:577-582.

Sargent PA, Kjaer KH, Bench CJ, Rabiner EA, Messa C, Meyer J, Gunn RN, Grasby PM, Cowen PJ. Brain serotonin1A receptor binding measured by positron emission tomography with [11C]WAY-100635: effects of depression and antidepressant treatment. Arch Gen Psychiatry 2000;57: 174-180.

Sastre M, Garcia-Sevilla JA. Densities of I2-imidazoline receptors, alpha 2-adrenoceptors and monoamine oxidase B in brains of suicide victims. Neurochem Int 1997;30: 63-72.

Staley JK, Malison RT, Innis RB. Imaging of the serotonergic system: interactions of neuroanatomical and functional abnormalities of depression. Biol Psychiatry 1998;44: 534-549.

Stanley M, Mann JJ. Increased serotonin-2 binding sites in frontal cortex of suicide victims. Lancet 1983;29: 214-216.

Stanley M, Mann JJ. Suicide and serotonin receptors. Lancet 1984;11;1: 349.

Stanley M, Virgilio J, Gershon S. Tritiated imipramine binding sites are decreased in the frontal cortex of suicides. Science 1982;216: 1337-1339.

Stockmeier CA, Meltzer HY. Beta-adrenergic receptor binding in frontal cortex of suicide victims. Biol Psychiatry 1991;29: 183-191.

Stockmeier CA, Dilley GE, Shapiro LA, Overholser JC, Thompson PA, Meltzer HY. Serotonin receptors in suicide victims with major depression. Neuropsychopharmacology 1997;16: 162-73.

Stockmeier CA, Shapiro LA, Dilley GE, Kolli TN, Friedman L, Rajkowska G. Increase in serotonin-1A autoreceptors in the midbrain of suicide victims with major depression-postmortem evidence for decreased serotonin activity. J Neurosci 1998;18: 7394-7401.

Sumiyoshi T, Stockmeier CA, Overholser JC, Thompson PA, Meltzer HY. Dopamine D4 receptors and effects of guanine nucleotides on [3H]raclopride binding in postmortem caudate nucleus of subjects with schizophrenia or major depression. Brain Res 1995;681: 109-116.

Turecki G, Briere R, Dewar K, Antonetti T, Lesage AD, Seguin M, Chawky N, Vanier C, Alda M, Joober R, Benkelfat C, Rouleau GA. Prediction of level of serotonin 2A receptor binding by serotonin receptor 2A genetic variation in postmortem brain samples from subjects who did or did not commit suicide. Am J Psychiatry 1999;156: 1456-1458.

Uylings HB, Sanz Arigita E, de Vos K, Smeets WJ, Pool CW, Amunts K, Rajkowska G, Zilles K. The importance of a human 3D database and atlas for studies of prefrontal and thalamic functions. Prog Brain Res 2000;126: 357-368.

Underwood MD, Khaibulina AA, Ellis SP, Moran A, Rice PM, Mann JJ, Arango V. Morphometry of the dorsal raphe nucleus serotonergic neurons in suicide victims. Biol Psychiatry 1999;46: 473-483.

van Praag HM, Korf J. Retarded depression and dopamine metabolism. Psychopharmacologia 1971;19: 199-203.

Willner P. Dopaminergic mechanisms in depression and mania. In: Bloom FE and Kupfer DJ (eds). Psychopharmacology: The Fourth Generation of Progress. Raven Press, Ltd., New York, NY,1995; pp 921-931.

Yates M, Leake A, Candy JM, Fairbairn AF, McKeith IG, Ferrier IN. 5HT2 receptor changes in major depression. Biol Psychiatry 1990;27: 489-496.

Yatham LN, Liddle PF, Shiah IS, Scarrow G, Lam RW, Adam MJ, Zis AP, Ruth TJ. Brain serotonin 2 receptors in major depression: a positron emission tomography study. Arch Gen Psychiatry 2000;57: 850-858.

Young LT, Li PP, Kish SJ, Warsh JJ. Cerebral cortex beta-adrenoceptor binding in bipolar affective disorder. J Affect Disord 1994;30: 89-92.

Zhu MY, Klimek V, Dilley GE, Haycock JW, Stockmeier C, Overholser JC, Meltzer HY, Ordway GA. Elevated levels of tyrosine hydroxylase in the locus coeruleus in major depression. Biol Psychiatry 1999;46: 1275-1286.

20 NON-MONOAMINERGIG TRANSMITTERS, GLIA CELL MARKERS, CELL ADHESION MOLECULES AND SYNAPTIC PROTEINS IN POSTMORTEM BRAIN TISSUE

Dan Rujescu and Peter Riederer

Abstract

Attempts to transfer findings gained in animal experiments to human depression have raised doubt that altered monoaminergic neurotransmission can fully explain the pathobiology of depression. This review gives a summary of postmortem findings in affective disorders with regard to non-monoaminergic neurotransmitter systems, glia cell markers, cell adhesion molecules and synaptic proteins. Only relatively few postmortem studies have examined these gene products in affective disorders and a systematic approach together with the application of modern techniques will hopefully reveal unexpected and exiting findings in the near future.

Introduction

Affective disorders form a heterogeneous disease group and are summarized due to syndromal similarity. Despite intensive research efforts there is little clarity on how to separate the different syndromes into subgroups. The clinically preferred division into bipolar versus unipolar depression is supported by family and genetic studies quite well, but there is disagreement whether the traditional subdivision of unipolar depression into reactive and endogenous depression is feasible or whether other classifications should be preferred (Parker, 2000). Additionally, the lack of (subgroup) specific biological markers complicates the investigation of the pathobiology. Despite this complex situation, the identification of the neuroanatomic structures involved in affective disorders evolved considerably also by introduction of structural and functional imaging during the last few years. Data from these research directions is comprehensive but not always clear. Hypotheses regarding a non-specific brain atrophy but also regarding specific structural abnormalities in limbic and associated structures have been tested. Meta analyses point towards a mild cerebellar and particularly cerebral atrophy. An increased ventricle/brain ratio and atrophy was found particularly in the frontal cortex, in the basal ganglia and in the cerebellar vermis (Videbech, 1997). Altogether, structural imaging data suggest that affective disorders may be accompanied by structural brain changes. This, as well as substantial methodological progress in the area of histology and molecular biology, has considerably increased the interest in postmortem studies during the last few years. This chapter gives a summary of postmortem findings in affective disorders with regard to non-monoaminergic transmitter systems, glia cell markers, cell adhesion molecules and synaptic proteins. Changes with a high probability to be secondary to antidepressive treatment are omitted in this chapter. Since no agreement regarding the exact neuroanatomical structures involved in the pathobiology of affective disorders has emerged yet, and most postmortem studies moreover concentrate on single brain areas, no strict division can be carried out with regard to neuroanatomical regions.

Non-Monoaminergic Transmitter Systems

The most common techniques used for the examination of pre- or postsynaptic proteins are based on radioligand binding either to tissue homogenates or to histological brain slices. Both methods have strengths and weaknesses. The use of tissue homogenates makes an extraction of the membrane fraction from a tissue block necessary. Since the morphology is destroyed, sum values from various neural and non-neural cell populations are examined. Many regions have a heterogeneous distribution of the proteins to be examined as demonstrated e.g. in the case of the serotonin transporter in the dorsal raphe nuclei (Bligh-Glover et al., 2000). The advantage of brain homogenates is that detailed pharmacokinetc investigations are possible and that the coupling of receptors to second messenger systems can be examined. Histological examinations of tissue slices which have the advantage of the exact anatomical topography, however, have the disadvantage that carrying out pharmacokinetic studies is very difficult.

In the neurobiological research of affective disorders there is a long tradition in the search for alterations particularly in the serotonergic and noradrenergic systems. Stimulated by the growing understanding of the effects of antidepressants, which interact with these monoaminergic systems, two hypotheses strongly influenced depression research: Schildkraut (1965) postulated alterations in noradrenergic neurotransmission as causative for depression, while Coppen (1965) formulated the serotonin hypothesis of depression. Attempts to transfer findings gained in animal experiments to human depression have raised doubt that only monoaminergic neurotransmisssion is involved in depression, since clear results, as in Parkinson disease, did not emerge (Bourne et al., 1968; Pare et al., 1969;

Shaw, 1967; Birkmayer and Riederer 1975). This prompted several investigations on further non-monoaminergic transmitter systems. Nevertheless, studies on the GABAergic and glutamatergic system and on different neuropeptide receptors and their ligands were carried out only occasionally and a systematic approach is needed.

GABAergic and Glutamatergic System

Only relatively few postmortem studies in affective disorders have examined the GABAergic and glutamatergic systems so far. An early postmortem work failed to find changes in activity of the GABA synthesizing enzyme glutamate decarboxylase while the $GABA_A$ binding sites were increased in the frontal but not in the temporal cortex (Cheetham et al., 1988). However, these findings could not be replicated subsequently (Harro et al., 1992). In a recent study, no changes in the GABA transporter binding sites were found in the frontal cortex (Sundman et al., 1997). There were also no differences in endozepines - a group of polypeptides, which can act as modulators at $GABA_A$ receptors (Rochet et al., 1998). Binding sites of the metabotropic $GABA_B$ receptors were examined in the frontal and temporal cortex as well as in the hippocampus. No change of the density of the binding sites in these regions was detected but the affinity in the temporal cortex was found to be increased (Cross et al., 1988). Altogether data on the GABAergic system still seem incomplete and a final statement on the involvement in depression would be premature. Particularly missing are studies on the differential expression of the various subunits and splice variants of the $GABA_A$ receptors, which undergo complex regulation.

The glutamatergic system is even less examined. Merely one study has examined the glycine binding site of NMDA receptors in the frontal cortex of a heterogeneous group of suicidal patients (Novak et al., 1995). No differences were found but whether the results can be generalized to affective disorders remains unclear. Again, data on the differential expression of subunits and splice variants of the various glutamate receptors is lacking.

Opiate System

Interesting, although incomplete, results are available for the opiate system. μ receptor binding was increased up to nine fold in the frontal and temporal cortex from young but not older patients with completed suicide (Gross-Isseroff et al., 1990). The up-regulation in the frontal cortex could be replicated. In addition receptor binding was elevated in the caudate nucleus but not in the thalamus (Gabilondo et al., 1995). Prodynorphin codes for peptides which bind at κ receptors. In patients with completed suicide prodynorphin mRNA was elevated in striosomes but not in the matrix of the caudate nucleus (Hurd et al., 1997). These results seem plausible since striosomes have connections to limbic regions while the matrix has connections to sensorimotor regions. It is still unclear whether these findings can be generalized from completed suicide to affective disorders, but there is indirect evidence since prodynorphin mRNA expression is not altered in schizophrenic patients (Hurd et al., 1997). In the same study, no differences were found regarding proenkephalin mRNA (Hurd et al., 1997).

HPA Axis

An increased activity of the HPA axis has been implicated in affective disorders by multiple studies and groups (Steckler et al., 1999). Postmortem results are congruent with this hypothesis and point towards increased corticotropin-releasing hormone (CRH) mRNA expression in the hypothalamic paraventricular nucleus (PVN) (Raadsheer et al., 1995). The number the CRH expressing neurons is increased fourfold and the number of CRH neurons which co-express arginin vasopressin (AVP) is three times higher (Raadsheer et al., 1994). The number of neurons showing AVP or oxytocin imunoreacivity in the PVN seems to be increased (Purba et al., 1996). AVP as well as oxytocin can potentiate the effects of CRH. Furthermore, a decreased density of CRH binding sites (corresponding to CRH receptors) was found in the frontal cortex (Nemeroff et al., 1988), but could not be replicated in a subsequent study (Hucks et al., 1997). Altogether, there is emerging evidence for changes of proteins involved in the HPA axis in the PVN but it is still unclear whether codirectional changes occur in other regions.

Neurokinins

Interest in the substance P receptor (neurokinin 1, NK1) followed the discovery of putative antidepressant proprieties of antagonists of this receptor (Rupniak and Kramer 1999). Overall binding at NK 1 receptors was unchanged in the cingulate gyrus but the relative binding in superficial compound to the low cortical layers was found to be decreased (Burnet and Harrison 2000). These results are hard to interpret. The authors speculate that these alterations could reflect the involvement of specific neural circuits expressing NK1 receptors in affective disorders.

Neuropeptide Y

Two newer studies report on unchanged neuropeptide Y (NPY) expression in the frontal cortex of patients with a past history of major depression (Ordway et al., 1995; Caberlotto and Hurd 1999), although a preliminary study found decreased NPY immunoreactivity in the frontal cortex and in the caudate nucleus but not in the temporal cortex and cerebellum (Widdowson et al., 1992). The NPY mRNA expression was decreased in the frontal cortex of patients with bipolar disorder (Caberlotto and Hurd 1999). Interestingly, NPY knock out mice show an increased ethanol intake and sensitivity (Thiele et al., 1998). Furthermore, a variation in the NPY gene was found to be associated with alcohol preference in humans (Kauhanen et al., 2000). It is tempting to speculate that the increased incidence of alcohol dependence in patients with affective disorders, and particularly with bipolar disorders, is associated with decreased NPY mRNA expression.

Cholecystokinin

Cholecystokinin (CKK) binding in the frontal cortex was examined in two studies, one of which also investigated the binding of vasoactive intestinal peptide (VIP) (Harro et al., 1992; Perry et al., 1981). While VIP binding did not differ, one of the studies (Harro et al., 1992) found increased binding sites in completed suicides, independent of a history of affective disorders. It is unclear whether these changes are of relevance for affective disorders or rather specific for suicidal behavior.

Glia Cell Markers

As reported in other chapters, glia cell reductions might be present in affective disorders (Ongur et al., 1998; Rajkowska et al., 1999). These results suggest examining markers of different glia cell subpopulations in greater detail.

A marker for astrocytes, GFAP (glial fibrillary acidic protein) immunoreactivity, was measured in the dorsolateral prefrontal cortex (Miguel-Hidalgo et al., 2000). Although no overall difference was found between controls and the entire patient group, GFAP immunoreactivity was lower in younger patients and higher in older patients regarding both, volume and density. However, this observation is difficult to interpret. Clearer results were found regarding an oligodendroglial marker (basic myelin protein) which was decreased in homogenates of the anterior frontal cortex (Honer et al., 1999).

Taking into account the promising cytoarchitectonic findings, further studies addressing markers of specific glia populations should be carried out.

Cell Adhesion Molecules

NCAM, L1, Thy 1 and ICAM 1 are members of the immune globulin (Ig) family of cell adhesion molecules (CAMs) and play a crucial role in cell-cell and cell-matrix adhesion and therefore in cell migration, synaptic plasticity, CNS development (NCAM, L1, Thy 1), and cerebral ischemia (ICAM 1). The interest in these molecules in unipolar affective disorder was stimulated by findings in schizophrenia.

The first observation of an increased hippocampal, but not cortical, NCAM expression in affective disorders was made already in 1985 (Jorgensen and Riederer 1985). NCAM occurs in different isoforms. The variable alternatively spliced exon (VASE) is 30 base pairs long and lies between exon 7 and 8. The secretory isoform, which expresses VASE and is associated with diminished neuroplasticity, was increased in the hippocampus of patients with bipolar disorder but not in completed suicides without a past history of psychosis (Vawter et al., 1998). The ratio of the secretory NCAM 115 kDa/105 kDa was also increased in the hippocampus of patients suffering from bipolar disorder (Vawter et al., 1999). The expression of L1 and Thy 1 in the frontal cortex in both, depression and bipolar disorder was unchanged (Webster et al., 1999). Interesting findings occurred in late life depression. Since in these patients increased numbers of vascular hyperintensities are observed in MRI scans, the expression of ICAM, a marker for ischema induced inflammation, was measured in the dorsolateral prefrontal and occipital cortex (Thomas et al., 2000). The expression was increased with the most pronounced differences in the frontal cortex (Thomas et al., 2000).

A simultaneous examination of all NCAM isoforms and of other adhesion molecules with modern molecular biology methods will contribute to further clarify if, and which, changes occur in affective disorders.

Synaptic Markers

While the expression of the synaptic marker SNAP25 (synaptosomal-associated protein, 25-kD) was increased in the hippocampus, no changes were found in the expression of SNAP25 or synaptophysin, another synaptic protein, in the frontal cortex (Jorgensen and Riederer 1985; Honer et al., 1999). GAP43 (growth-associated protein, 43-kD), a synaptic protein which is associated with neuroplasticity, was increased marginally in the frontal cortex of affective disorder patients (Honer et al., 1999).

Since neuroplastic changes in connectivity are plausible alterations in affective disorders, the systematic examination of further synaptic markers is needed.

A hypothesis free approach has been recently used to measure differential protein expression in unipolar affective disorder, bipolar affective disorder and schizophrenia (Johnston-Wilson et al., 2000). Still missing are simultaneous examinations of the complete mRNA expression by e.g. DNA arrays. These kinds of studies will hopefully reveal unexpected and exiting findings in the near future.

Acknowledgment

We are thankful to Christopher Murgatroyd for the review of the manuscript.

References

Birkmayer W, Riederer P. Biochemical postmortem findings in depressed patients. J Neural Transm 1975;37: 95-109.

Bligh-Glover W, Kolli TN, Shapiro-Kulnane L, Dilley GE, Friedman L, Balraj E, Rajkowska G, Stockmeier CA. The serotonin transporter in the midbrain of suicide victims with major depression. Biol Psychiatry 2000;47: 1015-1024.

Bourne HR, Bunney WE JR, Colburn RW, Davis JM, Davis JN, Shaw DM, Coppen AJ. Noradrenaline, 5-hydroxytryptamine, and 5-hydroxyindoleacetic acid in hindbrains of suicidal patients. Lancet 1968;2: 805-808.

Burnet PW, Harrison PJ. Substance P (NK1) receptors in the cingulate cortex in unipolar and bipolar mood disorder and schizophrenia. Biol Psychiatry 2000;47: 80-83.

Caberlotto L, Hurd YL. Reduced neuropeptide Y mRNA expression in the prefrontal cortex of subjects with bipolar disorder. Neuroreport 1999;10: 1747-1750.

Coppen A, Shaw DM, Malleson A, Eccleston E, Gundy G. Tryptamine metabolism in depression. Br J Psychiatry 1965;111: 993-998.

Cheetham SC, Crompton MR, Katona CL, Parker SJ, Horton RW. Brain GABAA/benzodiazepine binding sites and glutamic acid decarboxylase activity in depressed suicide victims. Brain Res 1988;460: 114-123.

Cross JA, Cheetham SC, Crompton MR, Katona CL, Horton RW. Brain GABAB binding sites in depressed suicide victims. Psychiatry Res 1988;26: 119-129.

Gabilondo AM, Meana JJ, Garcia-Sevilla JA. Increased density of mu-opioid receptors in the postmortem brain of suicide victims. Brain Res 1995;682: 245-250.

Gross-Isseroff R, Dillon KA, Israeli M, Biegon A. Regionally selective increases in mu opioid receptor density in the brains of suicide victims. Brain Res 1990;530: 312-316.

Harro J, Marcusson J, Oreland L. Alterations in brain cholecystokinin receptors in suicide victims. Eur Neuropsychopharmacol 1992;2: 57-63.

Honer WG, Falkai P, Chen C, Arango V, Mann JJ, Dwork AJ. Synaptic and plasticity-associated proteins in anterior frontal cortex in severe mental illness. Neuroscience 1999;91: 1247-1255.

Hucks D, Lowther S, Crompton MR, Katona CL, Horton RW. Corticotropin-releasing factor binding sites in cortex of depressed suicides. Psychopharmacology (Berl) 1997;134: 174-178.

Hurd YL, Herman MM, Hyde TM, Bigelow LB, Weinberger DR, Kleinman JE. Prodynorphin mRNA expression is increased in the patch vs matrix compartment of the caudate nucleus in suicide subjects. Mol Psychiatry 1997;2: 495-500.

Johnston-Wilson NL, Sims CD, Hofmann JP, Anderson L, Shore AD, Torrey EF, Yolken RH. Disease-specific alterations in frontal cortex brain proteins in schizophrenia, bipolar disorder, and major depressive disorder. The Stanley Neuropathology Consortium. Mol Psychiatry 2000;5: 142-149.

Jorgensen OS, Riederer P. Increased synaptic markers in hippocampus of depressed patients. J Neural Transm 1985;64: 55-66.

Kauhanen J, Karvonen MK, Pesonen U, Koulu M, Tuomainen TP, Uusitupa MI, Salonen JT. Neuropeptide Y polymorphism and alcohol consumption in middle-aged men. Am J Med Genet 2000;93: 117-121.

Miguel-Hidalgo JJ, Baucom C, Dilley G, Overholser JC, Meltzer HY, Stockmeier CA, Rajkowska G. Glial fibrillary acidic protein immunoreactivity in the prefrontal cortex distinguishes younger from older adults in major depressive disorder. Biol Psychiatry 2000;48: 861-873.

Nemeroff CB, Owens MJ, Bissette G, Andorn AC, Stanley M. Reduced corticotropin releasing factor binding sites in the frontal cortex of suicide victims. Arch Gen Psychiatry 1988;45: 577-579

Nowak G, Ordway GA, Paul IA. Alterations in the N-methyl-D-aspartate (NMDA) receptor complex in the frontal cortex of suicide victims. Brain Res 1995;675: 157-164.

Ongur D, Drevets WC, Price JL. Glial reduction in the subgenual prefrontal cortex in mood disorders. Proc Natl Acad Sci U S A 1998;95: 13290-13295.

Ordway GA, Stockmeier CA, Meltzer HY, Overholser JC, Jaconetta S, Widdowson PS. Neuropeptide Y in frontal cortex is not altered in major depression. J Neurochem 1995;65: 1646-1650.

Pare CM, Yeung DP, Price K, Stacey RS. 5-hydroxytryptamine, noradrenaline, and dopamine in brainstem, hypothalamus, and caudate nucleus of controls and of patients committing suicide by coal-gas poisoning. Lancet 1969;2: 133-135.

Parker G. Classifying depression: should paradigms lost be regained? Am J Psychiatry 2000;157: 1195-1203.

Perry RH, Dockray GJ, Dimaline R, Perry EK, Blessed G, Tomlinson BE. Neuropeptides in Alzheimer's disease, depression and schizophrenia. A post mortem analysis of vasoactive intestinal peptide and cholecystokinin in cerebral cortex. J Neurol Sci 1981;51: 465-472.

Raadsheer FC, Hoogendijk WJ, Stam FC, Tilders FJ, Swaab DF. Increased numbers of corticotropin-releasing hormone expressing neurons in the hypothalamic paraventricular nucleus of depressed patients. Neuroendocrinology 1994;60: 436-444.

Raadsheer FC, van Heerikhuize JJ, Lucassen PJ, Hoogendijk WJ, Tilders FJ, Swaab DF. Corticotropin-releasing hormone mRNA levels in the paraventricular nucleus of patients with Alzheimer's disease and depression. Am J Psychiatry 1995;152: 1372-1376.

Rajkowska G, Miguel-Hidalgo JJ, Wei J, Dilley G, Pittman SD, Meltzer HY, Overholser JC, Roth BL, Stockmeier CA. Morphometric evidence for neuronal and glial prefrontal cell pathology in major depression. Biol Psychiatry 1999;45: 1085-1098.

Rochet T, Tonon MC, Kopp N, Vaudry H, Miachon S. Evaluation of endozepine-like immunoreactivity in the frontal cortex of suicide victims. Neuroreport 1998;9: 53-56.

Rupniak NM, Kramer MS. Discovery of the antidepressant and anti-emetic efficacy of substance P receptor (NK1) antagonists. Trends Pharmacol Sci 1999;20: 485-490.

Shaw DM, Camps FE, Eccleston EG. 5-Hydroxytryptamine in the hind-brain of depressive suicides. Br J Psychiatry 1967;113: 1407-1411.

Schildkraut JJ. The catecholamine hypothesis of affective disorders: a review of supporting evidence. Am J Psychiatry 1965;122: 509-522.

Steckler T, Holsboer F, Reul JM. Glucocorticoids and depression. Baillieres Best Pract Res Clin Endocrinol Metab 1999;13: 597-614.

Sundman I, Allard P, Eriksson A, Marcusson J. GABA uptake sites in frontal cortex from suicide victims and in aging. Neuropsychobiology 1997;35: 11-15.

Thiele TE, Marsh DJ, Ste Marie L, Bernstein IL, Palmiter RD. Ethanol consumption and resistance are inversely related to neuropeptide Y levels. Nature 1998;396: 366-369.

Thomas AJ, Ferrier IN, Kalaria RN, Woodward SA, Ballard C, Oakley A, Perry RH, O'Brien JT. Elevation in late-life depression of intercellular adhesion molecule-1 expression in the dorsolateral prefrontal cortex. Am J Psychiatry 2000;157: 1682-1684.

Vawter MP, Hemperly JJ, Hyde TM, Bachus SE, Vanderputten DM, Howard AL, Cannon-Spoor HE, McCoy MT, Webster MJ, Kleinman JE, Freed WJ. VASE-containing N-CAM isoforms are increased in the hippocampus in bipolar disorder but not schizophrenia. Exp Neurol 1998;154: 1-11.

Vawter MP, Howard AL, Hyde TM, Kleinman JE, Freed WJ. Alterations of hippocampal secreted N-CAM in bipolar disorder and synaptophysin in schizophrenia. Mol Psychiatry 1999;4: 467-475.

Videbech P. MRI findings in patients with affective disorder: a meta-analysis. Acta Psychiatr Scand 1997;96: 157-168.

Webster MJ, Vawter MP, Freed WJ. Immunohistochemical localization of the cell adhesion molecules Thy-1 and L1 in the human prefrontal cortex patients with schizophrenia, bipolar disorder, and depression. Mol Psychiatry 1999;4: 46-52.

Widdowson PS, Ordway GA, Halaris AE. Reduced neuropeptide Y concentrations in suicide brain. J Neurochem 1992;59: 73-80.

CONCLUDING REMARKS

Daniel R. Weinberger

Scavenging through dead tissue in an effort to understand how a normal mental life becomes a mental illness has challenged researchers for over a century. It has proved a daunting challenge. Postmortem brain research has progressed along with other major developments that characterized psychiatry research in its last half century. These include revolutions in descriptive psychopathology and diagnosis, in genetics, molecular biology and neuroscience, and in the development of industrial technologies for analyzing gene and protein expression. In many respects, the research results of the second half of the century echo the words of Ferraro who summed up postmortem research in the first half of the century as follows: "...we have (not) succeeded.in establishing the specific pathology of schizophrenia. We have only succeeded in establishing concomitant ...pathology in the course of the schizophrenic syndrome (Ferraro, 1952)." There are many new findings, both anatomical and molecular, since the classic anatomical studies reviewed by Ferraro. There are still, however, no specific, diagnostic, or pathognomonic results. The traditional neuropathological definition of a brain disease, i.e. localization and characterization of "the lesion," has not been realized in primary psychiatric illness. It is reasonable to consider that the gross and cellular pathology of schizophrenia may never be characterizable in such classical neuropathological terms. The neuropathology of schizophrenia, rather than resulting from a specific histopathological process, may reflect combinatorial cellular responses to complex genetic programs and environmental stimuli across a lifetime.

One of the major developments in the past decade is the emergence of replicated results, in some cases by more than two independent groups of investigators. These results make it possible to assert that there are anatomical and molecular changes in the brains of patients with mental illness, particularly schizophrenia. The unresolved questions concern the mechanisms responsible for the findings and their relationship to etiology and pathophysiology.

Where Does the Field Stand?

Postmortem research in psychiatry has taken several major leaps forward over the past twenty years. These include: 1) the application of controlled methods of tissue preparation, 2) more rigorous anatomical sampling, e.g. unbiased quantitative cell counting and morphometry, 3) the application of techniques to study changes at the subcellular or molecular level, and 4) careful postmortem diagnosis. These advances have helped resolve many of the earlier problems that made replication difficult. Furthermore, it is now general practice to match samples carefully for age and sex, for postmortem interval, for pH, to use cytoarchitecturally characterized brain regions, and to include patient controls for other potential confounders, such as medication and chronic illness. While none of these refinements guarantees an artifact-free study, they do reduce the likelihood of a spurious result. It is important to note that controlling for the effects of chronic illness, including the impact of poor general health and nutrition, and for the effects of chronic psychotropic medications as well as decades of smoking is virtually impossible, as no control group experiences these factors in the same manner as do patients with chronic schizophrenia. It is also important to consider that implementing unbiased quantitative anatomical procedures does not ensure that negative results are valid. For example, it is very popular to use stereological procedures when counting absolute number of cells within a structure. This is a valid approach to comparing cell numbers in a well-delineated brain region. However, if the cellular changes involve only a subregion (e.g. a specific cortical cytoarchitectonic area) or specific connectivity zone of a structure, the boundaries of which may be difficult to define (e.g. dorsolateral caudate), counting all cellular elements in a structure may miss a subregional abnormality. Stereological counting procedures are not typically adapted to the distinct topography of the connectivity of a structure.

The various contributors to this volume summarize the past two decades of studies of postmortem tissue in psychiatric research, most of which has focused on schizophrenia. It is clear that there are many positive findings. Regardless of whether one measures the size of a structure (e.g. frontal cortical thickness), the number of cells it contains (e.g. medial dorsal thalamus), or the molecular constituents of those cells (e.g. GAD–67 mRNA), one can show differences in some of these measures between patients and controls. Moreover, some of the changes are replicable across different cohorts and laboratories. Noteworthy in cases of schizophrenia is evidence of reduced volume of neuronal processes and reduced abundance of connections in hippocampus and prefrontal cortex. This has been observed from a number of different perspectives, including somal size, neuropil volume, and various molecular markers of synapses and terminals (both pre and post synaptic).

While limited evidence also implicates other cortical and subcortical regions (which have generally been less widely studied), abnormalities in hippocampus and prefrontal cortex are supported by compelling data from several independent studies. Reductions in prefrontal and hippocampal neuropil volume also appear phenomenologically consistent with results of in vivo neuroimaging studies, which also report reduced cortical volume and reduced concentrations of N-acetyl aspartate, an intraneuronal chemical assayed with proton spectroscopy, which appears to be a sensitive marker of the abundance of neuronal processes (Bertolino and Weinberger, 1999).

There clearly is no single finding or molecular marker that stands out as unique to schizophrenia or likely to represent a primary neuropathological process. Rather, the data in hippocampus and in prefrontal cortex at least in schizophrenia implicate a pathology that results in overall less neuronal contacts, less terminal activity, and less intracellular signaling. It is curious that almost all the evidence points to less information processing and integration of signals at the cellular level, i.e. for most genes and proteins that have been studied, the evidence is of diminished transcriptional drive. This can be concluded from studies showing smaller neurons, fewer dendritic spines, shorter dendrites, reduced expression of synaptic terminal proteins (e.g. SNAP 25, synaptophysin, RAB3a, GAP-43, synapsin), reduced expression of trophic molecules (e.g. BDNF, LAMP, TRKs), and reduced expression or altered ratios of glutamate receptor subunits (e.g. GluR's, NMDARs). One of the remarkable results in a recent microarray gene expression study of schizophrenic brain was evidence of overall less abundant gene expression across the panoply of genes surveyed, particularly those encoding presynaptic terminal markers (Mirnics et al, 2000). What is most striking to me about these various changes is that they suggest that the neurons in the schizophrenic brain are not undergoing cellular stress or an assault on their integrity, processes which lead to increased transcription of many well-characterized genes. In most neurologic conditions associated with destruction or damage of neurons, in diseases or conditions associated with apoptosis, and in neurotoxin exposure (e.g. Alzheimer's disease, ageing, seizures, excitotoxity, alcohol neurotoxicity, etc.), neurons and glia that survive and even dying neurons increase expression of a number of genes and proteins aimed at molecular compensation and restoration. The induction of such "restoration genes" are characteristically not seen in schizophrenia (see Weinberger 1999, for further discussion). Much has been made of the lack of gliosis in schizophrenic brain failing to support hypotheses about neurodegeneration, but it is the lack of increased expression of genes involved in cellular responses to injury that most militates against hypotheses of neurotoxicity and neuronal destruction. As far as we can tell at this time, the cellular and molecular phenotype of schizophrenia is characterized by relative transcriptional hypoactivity of signal processing genes. To the extent that intraneuronal signaling - the means by which information is processed, stored, and utilized - translates into expression of genes and proteins and to remodeling of dendrites, terminals, and synaptic contacts, there is less of this going on in the schizophrenic brain.

The origin of decreased transcriptional drive in signal processing pathways is unclear. It very likely could reflect environmental factors reducing the neuronal information-processing load. These may include unstimulating environments, the impact of chronic illness, and the effects of chronic medication. It has become increasingly clear from studies in experimental animals that such environments have an impact on neuronal plasticity and generally are associated with regression of dendrites and spines and with reduced stimulus-linked transcriptional activity. While the effect of chronic medication is often assessed in animal models, even monkey models are inadequate to capture the life experience of a patient with chronic schizophrenia. Because most patients whose brains are studied die in the second half of their lives, it is important to study environmental effects in animals that also are in middle or late life. This has rarely been done. It is certainly conceivable that the older brain will make less dynamic molecular adaptations to the effects of the environment than the younger brain.

It is also possible that evidence of reduced transcriptional drive in signal processing pathways is a reflection of a molecular defect that is important to understanding the etiology of schizophrenia. Thus, it is conceivable that a genetic defect in critical cell signaling genes could bias the basic transcriptional machinery to misinterpret the molecular implications of certain cellular signals. The existence of such genetic defects in the germline would likely translate into abnormalities in brain development, which have also been implicated in schizophrenia. However, credible evidence of such a possibility is lacking and will have to await further investigation. It is also conceivable that the many nonspecific indicators of reduced synaptic activity and connectivity are the ramifications of a strategic molecular defect that initiates a cascade of secondary down regulations in gene and protein expression during development and in adulthood. Unfortunately, it is difficult to evaluate the primacy of any molecular finding in schizophrenic tissue unless there is a qualitative result (e.g. a missing protein, a novel molecular form), which seems very improbable at this time.

The evidence from postmortem studies of relative neuronal understimulation without neuronal loss or gliosis seems to fly in the face of a body of neuroimaging literature increasingly marshaled in support of a case for neurodegeneration. Clearly, the magnitude of the progressive changes in brain volume that have been reported with neuroimaging, on the order of 3-30% reductions in some structures over several years of illness, would be associated with marked reductions in neuronal numbers if neuronal loss were the basis for such MRI changes. In some of the imaging studies of patients with schizophrenia, the magnitude of the changes rivals that seen in patients with

Alzheimer's disease (Lasko et al, 2000). In patients with epilepsy, 25% volume loss in hippocampus on MRI is associated with approximately a 75% reduction in neuronal density (Lee et al, 1995). Clearly, neuronal loss even on a much smaller scale is not found in schizophrenia, though the reported reductions in hippocampal volume on MRI is in the 10-15% range. This makes it likely that the basis for the MRI changes in schizophrenia are either neuropil losses or something else. The magnitude of the neuropil reductions at postmortem examination also are much less than that reported on MRI, and it is unclear whether the magnitude of the neuropil losses observed in postmortem tissue (on the order of 5%) would be observable at the maroscopic level of MRI. Regardless, neuropil reductions without neuronal loss would suggest plastic, not toxic, effects, presumably involving regression of dendrites and terminals that might be reversible. The possibility that the MRI changes reflect primarily physiological changes (e.g. changes in tissue perfusion, hydration, or lipid content), related to medication, nicotine, and nutritional effects, also cannot be ruled out. Clearly, alcohol use, weight changes, and hormonal variations all have been shown to reversibly affect MRI measurements. Evidence that brain structures may actually get bigger and smaller and then bigger again within the same patient over short periods of time add further support to this possibility (Garver et al, 2000). In my view, while there is much enthusiasm for invoking a neurodegenerative explanation for illness progression in schizophrenia, the scientific evidence in support of this mechanism is very weak.

Where is the Field Going?

Postmortem research in psychiatry is charging ahead at the subcellular level. While there are still gross structures to be measured and cells to be counted, it is debatable whether such morphometric results will further advance our understanding of the etiology of mental illness, and it is even less likely that new treatments will emerge from such observations. The continued effort to measure and count follows a long and questionable tradition of looking for "a lesion." The molecular approach, though also in search of a smoking gun, focuses on the expression at the tissue and subcellular level of genes and proteins. The goal is to characterize a cellular phenotype that would implicate specific molecular mechanisms.

Postmortem research will be critical for understanding the molecular biology of mental illness in two respects: it will help identify candidate molecules that may be linked to the pathology, and it will test the cellular implications of candidate genes. Thus, postmortem research will inform the clinical search for genetic mechanisms and will test the cellular implications of findings from clinical genetics. At the level of brain, molecular changes that account for risk versus those associated with manifest illness cannot be distinguished, because they are observed in ill people. However, the genes implicated can be evaluated in at risk family members and their relationship to susceptibility so determined. Thus, postmortem molecular findings may represent a priority list of candidate genes for clinical genetics. There are caveats, however. For example, it is important to remember that gene expression in tissue reflects the dynamic regulation of gene function, while nucleotide sequence in genomic DNA represents the structural architecture of inheritance and the blueprint for cell mechanics. Thus, differences in gene or protein expression found in tissue may not directly reflect a mutation in genomic DNA. It certainly is conceivable that molecular abnormalities in postmortem tissue reflect cellular adaptations to genetic defects that are several regulatory steps removed from the molecules found to best differentiate patients from controls.

Postmortem studies also offer a unique window into the cellular implications of genes discovered in the clinic. It is very likely that within the next few years several genes will be confirmed as susceptibility alleles for schizophrenia. The challenge will then be to understand how these alleles change the biology of critical cells and lead to increased risk. Postmortem analyses of the spatial and temporal expression of these genes, regulation of their protein products, identification of post-transcriptional modifications, and characterization of their effects on other genes and proteins will help clarify the cell biology and identify potential new diagnostic and therapeutic targets.

In addition to assays based on candidate genes and proteins, the capability to profile the expression of thousands of genes and proteins in the same sample, the so-called array approach, is gathering enthusiastic converts. Since candidate molecules based on current research findings in ill patients represent less than one percent of the genome, it is reasonable to assume that surveying large numbers of genes and proteins at once will lead to new findings that could not have been predicted from the limited biological information available about mental illness. Preliminary results using the array approach have been encouraging (Mirnics et al, 2000). It will be important to develop sensitive and reliable arrays, which are still in the future, and to validate methods for array analyses in other diseases with established molecular neuropathology.

In my view, the most promising application of array methodology will not be in the comparison of ill to well cases. Such comparisons are likely to involve too much individual variation, both in patients and in controls. Expression array case-control comparisons also will likely miss many differences related to genomic sequence, as most arrays measure abundance of a transcript or product, not point variations in nucleotide sequence. Thus, a gene with a coding polymorphim in a single base pair (a SNP) that translates into a functional difference in a protein without changing regulation of its expression might be missed by an array analysis. If a subcellular defect in schizophrenia were an unusual splice variant or other postransciptional modification, this also would be missed by a generic array. An alternative and more promising approach is to compare cases based on a genotype distinction. This has been illustrated in a recent study of breast cancer comparing gene expression profiles of tissue with BRCA-1 and BRCA-2 mutations (Hedenfalk et al, 2001). This approach also could be applied to tissue from patients with mental illness, or to specific cell populations, categorized, for instance, by a susceptibility genotype. A promising example of categorizing patients by genotype is illustrated with the gene for catechol-O-methyl transferase (COMT). COMT is the first gene associated with susceptibility for a mental illness where it has been possible to understand the biological mechanism by which risk is enhanced (Egan et al, 2001). Thus, it has been shown that COMT genotype affects the efficiency and efficacy of dopamine mediated information processing in dorsolateral prefrontal cortex of human beings, and this effect interacts with other risk factors related to prefrontal information processing in schizophrenia to add additional risk (Egan et al, 2001). Using COMT genotype as a grouping variable to compare genes and proteins differentially expressed in prefrontal cortex may reveal a pathway of molecular adaptations that are affected by or mediate the effects of the polymorphism and that could be targets for remediation.

Finally, new methods in tissue preparation and molecular extraction will allow implementation of these approaches not just at the level of a block of tissue, or a slice, but at the level of individual cells. This will make it possible to map the molecular anatomy of putatively critical circuits (e.g. the "trisynaptic" pathway of the hippocampus, or local circuits in prefrontal cortex) so that gene and protein expression can be viewed in the context of a complex functional system. I imagine that these approaches will bring us much closer towards a cell biology of mental illness, where expression of genes and proteins will be parsed along several dimensions, including genetic background and variations across specific cell populations and circuits.

References

Bertolino A, Weinberger DR. Proton magnetic resonance spectroscopy in schizophrenia. Eur J Radiol 1999;30:132-141.

Egan MF, Goldberg TE, Kolachana BS, Callicott JH, Mazzanti CM, Straub RE, Goldman D, Weinberger DR. Effect of COMT Val108/158Met genotype on frontal lobe function and risk for schizophrenia. Proc Nat Acad of Sci (USA) 2001;in press.

Ferraro A. Discussion. In: Rosenberg and Sellier (eds). Proceedings of the First International Congress of Neuropathology, Vol 1. Turin, Rome, 1952; pp 630-636.

Garver DL, Nair TR, Christensen JD, Holcomb JA, Kingsbury SJ. Brain and ventricle instability during psychotic episodes of schizophrenia. Schiz. Res 2000;44:11-23.

Hedenfalk I, Duggan D, Chen Y, Radmacher M, Bittner M, Simon R, Meltzer P, Gusterson B, Esteller M, Kallioniemi OP, Wilford B, Borg A, Trent J. Gene-expression profiles in hereditary breast cancer. New Eng J Med 2001;344; 539-548.

Laakso MP, Frisoni GB, Kononen M, Mikkonen M, Beltramello A, Geroldi C, Bianchetti A, Trabucchi M, Soininen H, Aronen HJ. Hippocampus and entorhinal cortex in frontotemporal dementia and Alzheimer's disease: A morphometric MRI study. Biol Psychiatry 2000;47:1056-1063.

Lee N, Tien RD, Lewis DV, Friedman AH, Felsberg GJ, Crain B, Hulette C, Osumi AK, Smith JS, VanLandingham KE et al. Fast spin-echo, magnetic resonance imaging-measured hippocampal volume: Correlation with neuronal density in anterior temporal lobectomy patients. Epilepsia 1995;36:899-904

Mirnics K, Middleton FA, Marquez A, Lewis DA, Levitt P. Molecular characterization of schizophrenia viewed by microarray analysis of gene expression in prefrontal cortex. Neuron 2000;28:53-67.

Weinberger DR. Cell biology of the hippocampal formation in schizophrenia. Biol Psychiatry 1999;45:395-402.

INDEX

N

P

R

S

T

W